Commercial Observation Satellites

AT THE LEADING EDGE OF
GLOBAL TRANSPARENCY

Edited by
John C. Baker
Kevin M. O'Connell
Ray A. Williamson

RAND

The research described in this report was internally sponsored by RAND's National Security Research Division.

Founded in 1934, the American Society for Photogrammetry and Remote Sensing (ASPRS) is a scientific association serving over 7,000 professional members around the world. Our mission is to advance knowledge and improve understanding of mapping sciences to promote the responsible applications of photogrammetry, remote sensing, geographic information systems (GIS), and supporting technologies.

RAND is a nonprofit institution that helps improve policy and decisionmaking through research and analysis. RAND® is a registered trademark. RAND's publications do not necessarily reflect the opinions of its research sponsors.

Library of Congress Cataloging-in-Publication Data
Commercial observation satellites: at the leading edge of global transparency / edited by John C. Baker, Kevin M. O'Connell, and Ray A. Williamson.
    p. cm.
MR-1229
Includes bibliographical references.
ISBN 0-8330-2872-3 (hardcover)
ISBN 0-8330-2951-7 (paperback)
    1. Scientific satellites. 2. Outer space—Civilian use—Government policy.
3. Artificial satellites in remote sensing—Government policy. 4. Space industrialization—Government policy. 5. National security—Government policy.
    I. Baker, John C., 1949– II. O'Connell, Kevin M, 1959– III. Williamson, Ray A., 1938–

TL798.S3 C587 2000
621.3678—dc21

                                                                                                                 00-042512

© **Copyright 2001 RAND**

All rights reserved. No part of this book may be reproduced in any form by any electronic or mechanical means (including photocopying, recording, or information storage and retrieval) without permission in writing from RAND.

Cover Credits
Design—Eileen Delson La Russo, RAND
Front cover—IKONOS spacecraft graphic (courtesy Space Imaging, Inc.)
Inset photo, IKONOS 4-m multispectral image of Bejing's Forbidden City and Tiananmen Square, 10/22/99
   (courtesy Space Imaging, Inc.)
Inset photo, SPOT 10-m multispectral image of forest fires surrounding Los Alamos, NM, 5/13/00
   (© CNES2000/courtesy SPOT Image Corporation)
Back cover—Landsat image of Amsterdam, Holland, 10/18/99, (courtesy of EROS Data Center, USGS)
RADARSAT data © Canadian Space Agency/Agence spatiale canadienne (CSA) 1996, 1997. Received by the Canada Centre for Remote Sensing. Processed and distributed by RADARSAT International. Sentry photograph provided courtesy of IOSAT Inc./Sentry.

Published 2001 by RAND and ASPRS

RAND Offices: 1700 Main Street, P.O. Box 2138, Santa Monica, CA 90407-2138
1200 South Hayes Street, Arlington, VA 22202-5050
RAND URL: http://www.rand.org
To order through RAND or to obtain additional information,
contact Distribution Services: Tel: 310-451-7002; Fax: 310-451-6915; Email: order@rand.org

ASPRS Offices: 5410 Grosvenor Lane, #210, Bethesda, MD 20814-2160
ASPRS URL: http://www.asprs.org
To order through ASPRS or to obtain additional information,
contact ASPRS Distribution Center, Tel: 301-617-7812; Fax: 301-206-9789;
Email: ASPRSPub@PMDS.com

Foreword

Marta Macias Brown

I am sure that my late husband, George E. Brown, Jr., would have enjoyed reading and discussing this book. Those who knew George knew that he had a deep interest in the potential uses of space-based remote sensing, as well as a strong conviction that remote sensing could play a constructive role in preserving peace in the world and improving the quality of life of the world's citizens. It was that conviction that drove his efforts in Congress to promote the development of both public- and private-sector remote sensing systems, and in particular the Landsat system.

During his 35 years in the House of Representatives, George probably did more than any single Member of Congress to nurture and sustain the Federal government's investment in civilian land remote sensing technology and applications. He recognized the unique benefits that the Landsat system had delivered in terms of increased understanding of the Earth's surface, creation of a plethora of new applications to benefit both the private and public sectors, development of a user community conversant in the uses of remote sensing data, and promotion of international collaboration and data sharing.

Given all of those benefits, George believed it to be self-evident that the Landsat system should be put on a stable institutional and financial footing, and the turbulence that buffeted the program during the 1980s and 1990s was a source of some anguish for him. Never one to sit idly by in such a situation, he introduced legislation that eventually was enacted as the Land Remote Sensing Policy Act of 1992. It was a comprehensive approach that addressed all of the major issues confronting the Landsat program: management, funding, spacecraft options, and data policy. I know that George would be pleased to see his efforts bearing fruit in the spectacular success of the Landsat 7 spacecraft, which has been producing high-quality Earth observation images since its launch in April 1999.

The Land Remote Sensing Policy Act of 1992 was notable in another respect—it clearly recognized the potential benefits that could flow from the emergence of a commercially viable remote sensing industry, and it codified in law a set of procedures for the licensing and operation of such commercial systems. While the commercial remote sensing industry has suffered from a variety of "growing pains" over the past decade—including the challenges we still face in getting payloads successfully deployed into orbit—it appears that private-sector systems are here to stay. George recognized that the development of commercial remote sensing systems inevitably would lead to a whole host of policy issues that would have to be addressed, including the implications of growing global transparency. I know that he would have looked forward to reading and debating the analyses of those issues that are contained in this book.

Of course, his colleagues knew that George brought his own core values to such debates. In the area of remote sensing, one of those core values was his belief in an open society. He strongly believed that such openness contributed to peace and freedom both at home and around the world, and he was a steadfast opponent of what he considered to be arbitrary over-classification and secrecy dictates. It was view that attempts to block the dissemination of civil and commercial remote sensing imagery were ultimately doomed to failure as the capabilities for producing such imagery spread around the globe, and that rather than threatening the security of the nation, openness would ultimately enhance it.

In closing, I would like to express my appreciation to the authors of this book for their decision to dedicate it to the memory of my husband, George E. Brown, Jr. He loved a good book and a thoughtful analysis, and I know that he would have found both in this volume.

Preface

Overhead observation of human activities and natural developments on the Earth's surface long ago captured human imagination. These early hopes became reality in the mid-nineteenth century as the first aerial photographs were taken from a captive balloon. The rapid growth of aviation in the early twentieth century spurred the development of aerial photography and photo-interpretation for military, scientific, and commercial purposes. The space age subsequently ushered in another new era in remote sensing as satellites and manned spacecraft provided a unique vantage point for Earth observation for both military and civilian purposes. We believe that the advent of commercial observation satellites at the start of the new millennium is the next major step in a rapidly evolving ability to acquire timely information on what is occurring at almost any location on the Earth's surface—what we view as growing global transparency.

This book focuses on the nexus of technology and politics in seeking to understand the broader political, security, and market implications of commercial observation satellites. It offers a policy-oriented focus that is relevant to those in government, industry, nongovernmental organizations, and academia who are concerned with the potential benefits and possible risks of this emerging information technology. The book explores the policy tensions both in the United States and abroad that arise from the dual-purpose nature of high-resolution satellite imagery data, which can be applied to both civilian and military uses. Several chapters highlight the distinctive approaches that many countries take to the commercial use of remote sensing satellites, as well as their disparate views on the security implications of growing global transparency. Finally, special attention is given to assessing the implications of the unprecedented access to the high-quality satellite imagery increasingly available to governments, businesses, news media, international and nongovernmental organizations, and even individuals.

Many chapters in this book build on and extend the work of other scholars and analysts who have laid much of the intellectual foundation in this area. We would be remiss not to recognize the seminal historical contribution of our RAND colleagues and others in developing satellite photoreconnaissance for security purposes.[1] And two more recent efforts also deserve special recognition. One is the groundbreaking work of the Carnegie Endowment's Commercial Observation Satellite Project in the late 1980s, which anticipated many of the fundamental policy issues arising today.[2] The other work is the public revelation of the formerly classified CORONA imaging satellite program, which is ably recounted in the volume published by the American Society for Photogrammetry and Remote Sensing.[3] These works provide a strong conceptual basis and historical context for many of the chapters in this book.

The book's themes and observations also benefit from a series of workshops and informal discussions that have provided intellectual stimulus for the co-editors and book authors in thinking through the complex issues posed by the advent of high-resolution commercial and civilian observation satellites. The first workshop, in May 1998, was cosponsored by the George Washington University's Space Policy Institute and the National Air and Space Museum. The Institute's study of Dual-Purpose Satellite Technologies was critical in supplying important insights regarding these issues.

A larger conference addressing similar issues was hosted by Carnegie Endowment's Project on Transparency in May 1999. Finally, RAND held a workshop that brought together most of the book's authors, along with several experts from industry and government, in October 1999 to define key issues and to consider alternative perspectives. Throughout this period, our thinking benefited greatly from the informal discussions and luncheon debates of the "Eye Street Group" over the implications of imaging satellites for global transparency.

John C. Baker, Kevin M. O'Connell, and Ray A. Williamson
January 2001

1. For a historical review of RAND's role in early U.S. planning for space operations, including Earth observation, see Merton E. Davies and William R. Harris, *RAND's Role in the Evolution of Balloon and Satellite Observation Systems and Related U.S. Space Technology* (Santa Monica, Calif.: RAND, R-3692-RC, 1988).
2. Michael Krepon, Peter D. Zimmerman, Leonard S. Spector, and Mary Umberger, eds., *Commercial Observation Satellites and International Security* (New York: St. Martin's Press, 1990).
3. Robert A. McDonald, ed., *CORONA Between the Sun and the Earth: The First NRO Reconnaissance Eye in Space* (Bethesda, Md.: American Society for Photogrammetry and Remote Sensing, 1997).

Acknowledgments

Books often acquire a life of their own and this one is no exception. It began as an idea several years ago that caught the imagination of the three editors who wondered whether the imminent advent of commercial observation satellites would be more than a seemingly inevitable technological milestone. On greater reflection, we came to believe that these new satellites have much to say about how the world is changing at the start of the new millennium with the rapid growth of global transparency based on unprecedented access to information. In this case, high-resolution satellite imagery is not only potentially available to all governments but also accessible by most nonstate actors, including the news media, nongovernmental and international organizations, and even small groups and individuals. Understanding the broader political, military, and economic consequences of this new technological development is the main focus of this book.

Our book rests on the combined work of a multinational group of experts and scholars who have contributed thoughtful chapters on the diverse issues raised by commercial observation satellites. We offer special thanks to the many authors for their diligence and patience in seeing this project through. Each author played an important role in writing (and rewriting) their chapters to fit within the broad framework of the book envisioned by the editors without sacrificing their unique perspectives. We encourage our colleagues to continue their good work on this subject.

The editors are particularly grateful to RAND for providing the necessary financial support needed to move this book from the idea stage to its realization as a published work. RAND colleagues who played an important role supporting this book at various points in the lengthy process include Jeff Isaacson, Stuart Johnson, Robert Nurick, and Rachel Swanger. An important step on the way was the workshop RAND held in October 1999 in Washington, D.C., which brought together many of the book authors with outside experts and provided an invaluable opportunity to review authors' early drafts and to identify convergent and divergent perspectives. The editors greatly appreciated the presentations and comments provided by Mark Brender, Richard Buenneke, Jeff Harris, Robert McDonald, Kurt Schwoppe, David Thibault, and Charles Wooldridge during the workshop.

RAND and the American Society for Photogrammetry and Remote Sensing (ASPRS) have taken the unusual step of choosing to copublish this large volume. This approach reflects their mutual recognition of the substantial added value that came from combining their efforts on this important subject. We give special recognition and thanks to James R. Plasker, the executive director at ASPRS, for his steadfast support of its publication, as well as gratitude to the ASPRS board of directors for committing their organization to publishing this volume.

Although this large book has been an editing and production challenge for both RAND and ASPRS, the publication experts in each organization approached their tasks with a relentless dedication to producing a high-quality product. They also exhibited an impressive sense of cooperation. The editors especially thank Phyllis Gilmore and Daniel Sheehan for their great diligence in editing the manuscript, as well as Carolyn A. Staab for her excellent work as the layout designer and coordinator for this large and complex book. Kimberly A. Tilley, ASPRS's communication director, was a source of constant encouragement and good advice throughout this long project. Many individuals in RAND's publication department played important roles in helping to produce and publish this book, including Jane Ryan, Paul Murphy, Judy Lewis, and John Warren, as well as Richard Wright, Joanne Selby, and Shirley Hall. Eileen Delson La Russo did an excellent job designing the book cover.

A special note of gratitude is owed to Marta Macias Brown, who provided the foreword, which gives well-deserved recognition to the leading role that her husband, the late Rep. George E. Brown, Jr. (D-Calif.), played in shaping U.S. public policy on civilian and commercial Earth observation satellite systems. We also greatly appreciate the assistance provided by Richard Obermann.

Acknowledgements

The editors are particularly indebted to the several manuscript reviewers who graciously made available their thoughts and time to enhance the substantive contents of this book. David Frelinger and Dennis Gormley reviewed the larger manuscript while particular chapters benefited from more specific reviews made by David Adamson, John Hussey, John Jensen, Thomas Lillesand, and Frank Pabian. The thankless task of proofreading and other administrative tasks was made much easier by contributions from Yahya Dehqanzada, whose dedication was unflagging, as well as Holly Carter and Kim Gavin. This project also benefited regularly from administrative support at RAND provided at various times by Mickeeta Brooks, Karen Echeverri, Shivaulie Feliciano, and Viki Halabuk.

Our work on this book was greatly enhanced by contacts and cooperation from a broad range of organizations that are at the heart of the rapidly evolving commercial and civilian remote sensing satellite sector. These include individuals at the following government organizations: the Commerce Department's National Oceanic and Atmospheric Administration (NOAA), the National Aeronautics and Space Administration (NASA), the National Imagery and Mapping Agency (NIMA), the National Reconnaissance Office (NRO), the Office of the Secretary of Defense (C3I), the State Department, and the U.S. Geological Survey (USGS). In addition, many commercial firms and civilian satellite imagery organizations provided important insights and assistance to this book project. Thus, we are very grateful to the senior managers and staff experts at EarthWatch, Orbital Imaging Corporation, Radarsat International, Space Imaging, Inc., SPIN-2, and SPOT Image, as well as at value-added firms, particularly Cambridge Research Associates, EarthData International, Earth Satellite Corporation, Eastman Kodak, ERDAS, ESRI, and Mitretek.

Table of Contents

Commercial Observation Satellites: At the Leading Edge of Global Transparency

Foreword v
Preface vii
Acknowledgements xi
List of Figures & Tables xix

1 Introduction 1
   John C. Baker, Ray A. Williamson and Kevin M. O'Connell

Section I: The Policymaking Context

2 The Origins and Evolution of Openness in Overhead Global Observations 17
   Richard S. Leghorn and Gregg Herken

3 Remote Sensing Policy and the Development of Commercial Remote Sensing 37
   Ray A. Williamson

4 From Space Imagery to Information: Commercial Remote Sensing Market Factors and Trends 53
   Kevin O'Connell and Beth E. Lachman

5 Emerging Technologies: Emerging Issues for Space Remote Sensing ... 79
   Bob Preston

6 Security Implications of Commercial Satellite Imagery 101
John C. Baker and Dana J. Johnson

Section II: National Remote Sensing Programs & Policies

7 U. S. Remote Sensing Programs and Policies 139
Kevin M. O'Connell and Gregory Hilgenberg

8 Russian Remote Sensing Programs and Policies 165
George J. Tahu

9 The French Pioneering Approach to Global Transparency 187
Isabelle Sourbès-Verger and Xavier Pasco

10 Japanese Remote Sensing Policy at a Crossroads 205
Kazuto Suzuki

11 Commercial Observation Satellites in the Middle East and the
Persian Gulf
Gerald M. Steinberg 225

12 The Indian Space Program 247
Deborah Foster

13 Canada's Remote Sensing Program and Policies 263
Michel Bourbonniere and Louis Haeck

Section III: Remote Sensing Applications to International Problems

14 Supporting the Dayton Peace Talks 297
Richard G. Johnson

15 Imagery and Mapping Support to the Ecuador-Peru Peace Process ... 311
John Gates and John Weikel

16 Keeping and Eye on the Islands: Cooperative Remote
Monitoring in the South China Sea 327
Vipin Gupta and Adam Bernstein

17 The Role of Commercial Satellite Imagery in Locating South
Asian Nuclear Test Sites 361
David Albright and Corey Gay Hinderstein

18 Nongovernmental Use of Commercial Satellite Imagery for Achieving Nuclear Nonproliferation Goals: Perspectives and Case Studies 381
David Albright and Corey Gay Hinderstein

19 Supporting Humanitarian Relief Operations 403
Einar Bjorgo

Section IV: Emerging International Policy Issues

20 The Global Politics of Commercial Observation Satellites 433
Ann M. Florini and Yahya A. Dehqanzada

21 How Open Will the Skies Really Be? 449
Mark David Gabriele

22 The Globalization of Transparency: The Use of Commercial Satellite Imagery by Nongovernmental Organizations 463
Karen T. Litfin

23 Remote Sensing Technology and the News Media 485
Steven Livingston

24 Space Remote Sensing Regulatory Landscape 501
Robert Preston

25 New Users and Established Experts: Bridging the Knowledge Gap in Interpreting Commercial Satellite Imagery 533
John C. Baker

26 Conclusions 559
John C. Baker, Kevin M. O'Connell, and Ray A. Williamson

Appendices

A List of Abbreviations 576
B The Past, Present and Future of the Medium- and High-Resolution Satellite World 579
William E. Stoney
C Color Plates 589
D About the Authors 617
E Selected Bibliography 623

Index 631

List of Figures & Tables

Figures

- 4.1 Digital Orthoimage with Road Network Overlay, Austin, Texas
- 4.2 Early 1990s Market Projections for 2000 Remote Sensing Market
- 4.3 Projection of the Worldwide Revenues for Commercial Satellite Imaging
- 5.1 Synthetic Aperture Radar Image: Pipeline Over the Rio Grande, Near Albuquerque, New Mexico
- 5.2 False Color SAR Image: Barstow, California, and Vicinity, April 12, 1994
- 5.3 SAR Interferogram: A Portion of the Rutford Ice Stream in Antarctica
- 5.4 Notional Hyperspectral Data Cube
- 5.5 Appearance of Actual Data
- 5.6 Spectra of Four Different Materials
- 9.1 ERS Ground Receiving Stations
- 9.2 SPOT Family Sensors
- 9.3 SPOT Ground Receiving Stations
- 9.4 Distribution of Spot Image Revenues by Geographical Region
- 9.5 Distribution of Spot Image Revenues by Applications in 1998
- 10.1 Administrative Structure for Remote Sensing
- 13.1 Oil Spill off Japan's Coast, January 11, 1997
- 13.2 San Diego Harbor Area, March 20, 1996

13.3 Red River Valley Flood Area, May 6, 1997

13.4 Sentry Mobile Ground Receiving Station

13.5 Airfield at Goma, Zaire, November 12, 1996

13.6 Signs of Refugee Activity at Goma, Zaire, November 22, 1996

14.1 Process Flow for Mapping and Imagery Support to Negotiations

14.2 Perspective View Screen Shot on PowerScene™

14.3 Printed Map with Boundaries and Buffer Zones

14.4 Factors in Choosing Technology for Mapping and Imagery Tasks

14.5 Negotiated Lines and Buffer Zones, Gorazde Corridor, Bosnia-Herzegovina

15.1 Cordillera del Condor Map Used by Brazilian Arbiter Capt. Braz Dias in 1945

15.2 Small-Scale NIMA Aeronautical Chart

15.3 Cenepa River Valley

15.4 Georectified Mosaic of RADARSAT Images in the Cordillera del Condor

15.5 PowerScene™ Screen Capture of the Northern End of the Cordillera del Condor

15.6 Members of Ecuador-Peru Border Commission and a Border Monument

15.7 Areas of New 1:25,000-Scale Mapping Done in 1999

16.1 Islands, Reefs, and Shoals in the Spratly Island Region

16.2 February 1, 1995, Philippine Aerial Photographs Showing a Ship Reportedly Located at Mischief Reef

16.3 Spring 1997 Philippine Aerial Photographs Showing Two Ships Reportedly Photographed in the South China Sea and Identified as Chinese

16.4 Aerial Photograph, Taken a Few Months After China Took Control of Mischief Reef, Showing One of Four Building Clusters China Constructed and Occupied

16.5 Aerial Photograph Showing Another Chinese Building Cluster on Mischief Reef

16.6 Aerial Photograph of Construction at a Chinese-Occupied Area on Mischief Reef

16.7 Aerial Photograph Showing New Construction Activity at a Second Chinese-Occupied Area on Mischief Reef

16.8 November 7, 1998, Philippine Aerial Photograph Showing Two Ships Reportedly Anchored at Mischief Reef

16.9 December 10, 1998, Philippine Aerial Photograph Showing Three Ships Reportedly Anchored at Mischief Reef

16.10 September 4, 1997, IRS-1C Satellite Image of Mischief Reef

16.11 November 10, 1997, IRS-1C Satellite Image of Mischief Reef

16.12 November 26, 1999, IKONOS 1-m Resolution Satellite Image of Mischief Reef

16.13 Details of 16.12: Southern and Northern Sites

List of Figures & Tables

16.14 Details of 16.12: Western and Eastern Sites
16.15 Close-up of Ships in the Interior Waters at Mischief Reef
16.16 September 4, 1997, RADARSAT-1 Image of Thitu Island
16.17 February 1998 Aerial Photograph of Thitu Island
16.18 September 4, 1997, RADARSAT-1 Image of Chinese-Occupied Subi Reef Aerial Photograph of Subi Reef
16.19 Aerial Photographs of the Fortified Chinese Structures on Subi Reef
16.20 September 4, 1997, IRS-1C Satellite Image of Commodore Reef
16.21 March 15, 1998, IRS-1C Satellite Image of Alicoa Annie
16.22 March 10, 1998, IRS-1C Satellite Image of Yuan Anha
17.1   Geographic Coverage of the Archived March 25, 1995, SPOT Image and May 24, 1992, KVR-1000 Image
17.2   Annotated Helicopter Photo of the Crater
17.3   KVR-1000 Perspective View of the Large Circular Depression
17.4   Before, During, and After the Tests on May 28, 1998
17.5   Geographic Location of Pakistan's May 28 Nuclear Test Site
17.6   KVR-1000, 2-m Resolution Satellite Image taken February 25, 1998
17.7   Closer View from KVR-1000 Image of Pakistan's First Nuclear Test Site
17.8   Pakistan's First Nuclear Test Site
17.9   Instrumentation Bunker, Forward Recording Site
17.10 Concealed Tunnel Portal at Pakistan's First Nuclear Test Site
17.11 IRS-1C 5-m Resolution Image
18.1   A Small Research Reactor at North Korea's Main Nuclear Site, January 1989
18.2   A 1-m IKONOS Image Showing the Dome of the Khushab Reactor, Near Sargodha, Pakistan
18.3   A Heavy Water Plant Just South of the Khushab Reactor
18.4   Aerial Photograph of the Rocky Flats Site, October 1964
18.5   The Yongbyon Nuclear Center, 1989
18.6   The Yongbyon Nuclear Center, 2000
18.7   Plutonium Separation Building at Yongbyon, 1989
18.8   Plutonium Separation Building at Yongbyon, 2000
18.9   Rashdiya and Vicinity, December 1991
18.10 Rashdiya Layout
18.11 Tuwaitha Site, December 1990

18.12 Schematic of Tuwaitha Site
19.1   Types of Imagery Products Needed at Different Phases of Relief Operations
19.2   KVR-1000 Satellite Image of Refugee Camp, Thailand
19.3   Detail View of Refugee Camp Shown in Figure 19.2
19.4   High-Resolution IKONOS Image of Beldangi Refugee Camps, Nepal, May 15, 2000
19.5   Detail View of Refugee Camps Shown in Figure 19.4
19.6   Space Map of Refugee Hosting Area in Chiapas, Mexico, January 13, 1992
19.7   Deforested Regions in Refugee Hosting Areas in Chiapas, Mexico
19.8   Deforestation over Time (1993–1997) in Chiapas, Mexico, as Recorded in Satellite Imagery
19.9   Space Map of Qala en Nahal Refugee Settlement Plan, Sudan, January 20, 1992
19.10 Um Burush Refugee Camp (image ©1996 Sovinformsputnik)
19.11 IKONOS Image of the Ifo Refugee Camp, Dadaab Settlements, Kenya, May 25, 2000
19.12 Detail View of Refugee Camp Shown in Figure 19.11
19.13 Satellite Image of Kosovar Refugee Hosting Area, Albania and Macedonia
19.14 Satellite Image of Neprosteno Refugee Camp
19.15 Three-dimensional Perspective of Kukës Refugee Camps, Albania
23.1   Russian Satellite Image Purporting to Show the May 1998 Indian Nuclear Test Site That Appeared in Newsweek
23.2   Mass Burial Site near Izbica, Kosovo
A.1   The History of 1- to 30-m Land Imaging Satellites
A.2   The Next Five Years
A.3   Spectral Band Coverage

Tables

4.1   Observation Capabilities of Manned Aircraft
4.2   Observation Capabilities of Imaging Satellites
4.3   Remote Sensing Market Segments
4.4   Commercial Observation Business Models Are Changing
6.1   Civilian Satellite Imagery Use in the Gulf War
6.2   Security Applications of Commercial Satellite Imagery
7.1   U.S. Civil Observation Satellites
7.2   Landsat Satellite Series

List of Figures & Tables

- 7.3 Sensor Characteristics of the Landsat Series
- 7.4 U.S. Commercial Remote Sensing Licensees
- 8.1 Commercially Available Russian Earth-Observation Instruments
- 9.1 Shareholders of Spot Image, 1982
- 9.2 Shareholders of Spot Image, 1985
- 9.3 Turnover (in million FF)
- 9.4 Shareholders of Spot Image, 1999
- 10.1 Data Distribution Schemes
- 10.2 Major Remote Sensing Satellites
- 10.3 Budget for Remote Sensing Related Activities
- 11.1 Israeli Military Satellite Imaging Systems
- 11.2 Israeli Commercial Satellite Imaging Systems
- 12.1 Indian Remote Sensing Satellites (IRS Series)
- 16.1 Selection of Islands in the South China Sea for Detailed Image Analysis
- 16.2 The Strengths and Weaknesses of Aerial and Satellite Imaging for Monitoring the South China Sea
- 20.1 Approximate Ground Resolution for Target Detection, Identification, Description and Analysis
- 20.2 Estimated Costs of Entering the Market for High-Resolution Satellite Imagery
- 20.3 Overview of the Land Observation Microsatellite Industry
- 21.1 The Effects of Image Delivery Delays on the Strategic Balance Between Two Nations
- 23.1 Camcorders: U.S. Sales to Dealers
- 23.2 Cellular Telephone Expansion: 1985–1998
- 23.3 Measures to Distort Information for an Enemy
- 25.1 Comparison of Traditional and New Users of Overhead Imagery Data
- 25.2 Imagery Data–Interpretation Techniques, Technologies, and Processes
- 25.3 Imagery Data and Deliberate Errors
- 25.4 Imagery Data and Inadvertent Errors
- 25.5 Methods for Enhancing Commercial Satellite Imagery Credibility
- A.1 Data Characteristics of Current and Planned Land Imaging Satellites

Please note: The figures reproduced in this book are from many sources and not always of the highest quality. However, to help illustrate points made in the text, we have used the best images available to us at the time of publication.

1
Introduction

John C. Baker, Ray A. Williamson, and Kevin M. O'Connell

Commercial observation satellites are emblematic of information age technologies at the start of the new millennium; they promise to bolster global transparency by offering unprecedented access to accurate and timely information on important developments. Uninhibited by political borders, these earth-orbiting satellites collect images of human activities and the natural environment, images that will be available globally on a commercial basis. A broad range of government agencies and nonstate actors, including international organizations, the news media, businesses, and nongovernmental organizations (NGOs), will gain access to satellite images of sharpness and quality previously available only from U.S., Russian, and French military reconnaissance satellites. Even individuals and small groups will be able to purchase images off the Internet. For the most part, the imagery data produced by earth-observation satellites will provide information to meet wide-ranging civil and commercial remote sensing needs, such as surveying natural resources, monitoring the health of crops and vegetation, and supporting humanitarian and disaster relief operations.

However, commercial earth-observation satellites are also a dual-use technology with important national-security applications because they feature major improvements over the earlier civilian satellites in image sharpness (feature resolution) and timely delivery of imagery data. Higher-resolution images can help warn countries of military buildups by threatening neighbors and can support negotiations to settle their territorial disputes. Hence, diplomats, military planners, and intelligence analysts around the world are likely to welcome commercial observation satellites as an important new source of information.

Unfortunately, not all earth-observation data purchasers have benign intentions. Some governments will undoubtedly use satellite images for aggressive purposes against other countries or even against groups within their own countries. And worrisome nonstate actors, such as terrorists and narco-criminal groups, could learn to use satellite images for their own harmful purposes. Thus, the dual-use nature of commercial earth-observation satellites poses major policy issues for the United States and other national governments that support and regulate these new imaging satellites.

Governments will also play an important role in determining the economic prospects for commercial observation satellites. To succeed, the new commercial earth-observation enterprises must make the transition from being primarily providers of satellite imagery data to offering geospatial information products and services tailored to a broad range of traditional and new customers. A successful transition would secure them an integral role in the knowledge-based economy. National governments, as both regulators of commercial observation satellites and major customers for imagery products, will exert a crucial influence on how the emerging global market for earth-observation data and information products develops.

The U.S. experience to date highlights the challenges of designing and implementing policies that regulate commercial observation satellites in ways that both promote this fledgling industry and safeguard national-security interests, while also addressing the concerns of other countries. U.S. and foreign policymakers now face a series of difficult choices: how best to deal with the dual-use nature of high-resolution imagery, how to balance public needs and private interests in producing and consuming earth-observation data, how to redefine the relationship between governments and various nonstate actors in using satellite imagery, and how to reconcile the need for international cooperation with the inevitable pressures arising from commercial competition.

This book assembles a multidisciplinary group of policy analysts and experts in different imagery applications to offer the reader a broad perspective on the political, security, and market implications of commercial earth-observation satellites. Each chapter is written to reach nonspecialists while still

offering insights on emerging issues of concern to policymakers and commercial practitioners. The chapters are grouped into the following main parts:
- **Section I: The Policymaking Context** provides essential historical background and an overview of the major policy issues and technical trends associated with commercial and civilian observation satellites. Chapters in this section explore the broader issues that growing access to satellite imagery, which significantly contributes to global transparency, poses for policymakers.
- **Section II: National Remote Sensing Programs and Policies** relies on both U.S. and foreign experts to provide readers with a detailed assessment of U.S. and selected non-U.S. commercial and civilian remote sensing programs. This section includes chapters devoted to countries that possess mature earth-observation satellite programs (i.e., the United States, Russia, France, and Japan), as well as other countries with rapidly emerging programs (i.e., India, Canada, and Israel).
- **Section III: Case Studies in Remote Sensing Applications** offers in-depth assessments by remote sensing experts that demonstrate the potential benefits and limitations of the new commercial and civil government satellite imagery for helping policymakers and experts. The case studies address a range of international security problems, including resolving territorial disputes, monitoring nuclear proliferation, supporting humanitarian relief activities, and formulating a regional cooperative monitoring regime.
- **Section IV: Emerging International Policy Issues** focuses on the salient policy issues created by rapidly expanding global and public access to higher-resolution satellite imagery. This section highlights the distinctive interests of and roles played by national governments, NGOs, and the news media. It also addresses the broader international politics surrounding commercial observation satellites.

Advent of Commercial Observation Satellites

In the first years of the new millennium, the users of satellite imagery can look forward to having access to the data from more than two or three dozen civil and commercial observation satellites, offering a range of image types and resolutions (see Appendix B). These new satellites continue a trend that began in 1972 when the United States launched the world's first civilian remote sensing satellite, Landsat.[1] This pioneering satellite collected digital multispectral (i.e., color) images at about 80-m resolution.[2] Images at this resolution are sufficient for routinely monitoring natural features and broad human influences on the earth's surface, such as urbanization, but much too coarse for identifying details, such as vehicles or even individual buildings. In comparison, the new generation of commercial and civilian earth-observation satellites is capable of collecting images at much higher resolutions. For example, the world's first truly commercial earth-observation satellite, Space Imaging's IKONOS, successfully launched in September 1999, collects panchromatic (black and white) images with better than 1-m resolution, which, under the right conditions of lighting and contrast, is sufficient to detect objects as small as the average card table. The IKONOS satellite can also collect multispectral images at about 4-m resolution.

A clear distinction originally existed between civilian remote sensing satellites, such as Landsat, and the imaging satellites the United States and Soviet Union relied on for military reconnaissance and intelligence-gathering purposes, such as the U.S. CORONA military reconnaissance satellite system. The civilian satellites mainly focused on acquiring lower resolution, multispectral imagery that was better suited to scientific and civil needs; the military satellites mostly collected high-resolution panchromatic images to monitor worldwide military developments and strategic events. Military users attached high priority to the timely acquisition and processing of satellite imagery data to satisfy their need for adequate warning of any enemy attack or for supporting policymakers involved in foreign policy crises. In comparison, civilian users were more willing to wait weeks and even months to receive high-quality imagery quality at low cost, even if timeliness was sacrificed. It is important to note that governments, throughout the Cold War, treated military images as highly classified national-security information and never released them publicly. In contrast, civilian satellite imagery data, which are largely governed by a scientific ethos, have been openly published and globally shared with relatively few restrictions.[3]

Commercial earth-observation satellites now blur such long-standing differences between civil and military imaging satellites. These new satellites are being financed, built, and operated by private firms seeking to create profitable businesses by selling satellite imagery and services. They also see these data as invaluable inputs for a broad range of geospatial information products that various commercial and government customers desire.

Similarly, the distinction between civilian (government-owned) and commercial (government licensed but privately owned) observation satellites is becoming less clear cut. Civilian earth-observation enterprises in Europe, Canada, and Russia are increasingly focused on selling data in the commercial satellite imagery market to help cover their operating costs. In comparison, U.S. commercial remote sensing firms are looking to U.S. government agencies as some of their more important customers for imagery products and services. Thus, much of this book's discussion on commercial earth-observation satellites will also apply to their civilian counterparts.

A confluence of trends—political, technological, and economic—has encouraged firms to enter the nascent earth-observation data marketplace some 40 years after the first Sputnik satellite achieved orbit. The primary inhibiting factor was not technology barriers but policy constraints. The end of the U.S.-Soviet confrontation had the effect of relaxing earlier Cold War restrictions on satellite imaging technologies and expanding public access to higher-resolution images. By the early 1990s, Moscow was allowing the sale of declassified imagery (with resolutions as sharp as 2 m) from Russian military satellites in the international marketplace. Washington soon responded to avoid placing U.S. industry at a disadvantage in the emerging global market. Building on earlier legislation, the 1992 Land Remote Sensing Policy Act set forth provisions for licensing commercial earth-observation satellites. The first such license (for WorldView's 3-m resolution satellite[4]) was issued in January 1993, in the last days of the Bush administration. Because of their national-security implications, license applications for satellites of much higher resolution challenged the U.S. policy process. However, with strong encouragement from industry and Congress, the Clinton administration issued Presidential Decision Directive (PDD) 23 in March 1994, which set specific regulatory guidelines for U.S. commercial remote sensing enterprises. These licenses permit U.S. firms to sell high-resolution satellite imagery and to seek government approval for selling remote sensing technologies or even turnkey imaging satellite systems to foreign buyers. Less than a year later, the U.S. government also announced its decision to declassify over 800,000 CORONA spy satellite images that had been acquired between 1960 and 1972.

Technology push also spurred greater industry interest in developing commercial observation satellites and marketing imagery products. Advances in satellite and optical sensor technologies allow the development of imaging satellites that are substantially smaller, cheaper, and more agile than the relatively large and expensive Landsat and other civilian observation satellites. Both types of imaging satellites could produce higher-resolution panchromatic (i.e., black and white) and multispectral (i.e., color) images. And some firms hope to develop commercial customers for other types of imaging technologies, such as radar and hyperspectral sensors.

Equally important has been a rapid improvement in enabling technologies that reduce the technical and cost barriers for a potentially broader range of customers in using satellite imagery. These technologies include more-affordable computing power, user-friendly software for processing and displaying satellite images, and less-expensive data-storage systems. Furthermore, commercial remote sensing firms have been quick to take advantage of the Internet for permitting potential customers to search and order satellite images from their large and growing imagery databases.

The political and technological trends for commercial earth-observation satellites are generally encouraging. However, the contribution of commercial observation satellites to global transparency also depends on their economic prospects, and the commercial viability of the remote sensing firms is far from certain at this early stage. Convinced that a potentially strong domestic and international marketplace exists for satellite imagery data and information products, several U.S. firms invested heavily in the first generation of commercial observation satellites. They lead the way in creating the necessary ground infrastructure and sophisticated imagery archives and in creating the multinational partnerships needed to break into the global marketplace. Their actions provided added impetus for non-U.S. remote sensing enterprises to explore their own options for becoming more engaged in the international market for earth-observation data. A key question, however, is whether firms that have substantial experience in building and operating the earth-observation satellite systems can reinvent themselves to be highly competitive in the emerging information economy, with its growing emphasis on e-commerce as a model for information products and services.

Global Transparency and Commercial Observation Satellites

One of the new millennium's defining features is rapidly growing global transparency. This trend is driven by a combination of factors, including more-open political and economic institutions, rising expectations about public access to information, and an explosion of information technologies. Key technologies include global telecommunication networks, the Internet and World Wide Web, commercial and civil observation satellites, and other enabling technologies that encourage worldwide connectivity and awareness. Commercial observation satellites are playing a leading role in expanding transparency on a global basis because they offer a broad range of actors (e.g., governments, corporations, the news media, NGOs, and even individuals) an unprecedented ability to acquire relevant information on natural and human developments occurring nearly anywhere in the world.

Earth-observation satellites uniquely offer global coverage as they operate beyond the limits of national sovereignty.[5] Most other information sources are constrained by borders or are scarcely found in remote regions and war zones. Space-based observation systems can also penetrate denied areas to obtain invaluable information where authoritarian regimes, or internal conflicts, substantially limit the amount of information available to the outside world. For example, these data can be used to reveal environmental disasters, detect mass graves, and identify questionable activities occurring at nuclear facilities and military sites—despite the efforts of host governments or warring factions to restrict external access to the territory they control. Even if some facilities are placed underground or concealed by other measures, indications of their existence and activities are likely to appear on satellite imagery over time.

In the past, Landsat and SPOT earth-observation satellites captured public attention when they returned images in 1986 that revealed the extent of the Chernobyl nuclear reactor disaster, despite the stonewalling of Soviet officials at the time. Such satellites have also provided convincing evidence on the disturbing rate of deforestation in the Amazon and other tropical forests and the effects of oil pollution along coastal areas. In comparison, commercial observation satellites will offer higher-resolution, stereo capabilities, and the relatively rapid delivery of imagery data. Furthermore, as a result of commercial marketing, imagery data will be substantially more accessible to a broad range of countries and nonstate actors.

Thus, these new imaging satellites could bolster global transparency by collecting timely images that focus international attention on war preparations among regional adversaries or provide timely data to assess the nature of humanitarian emergencies occurring around the world. The contribution to global transparency does not depend on a single satellite but arises from the cumulative impact of a growing constellation of commercial and civilian observation satellites that are slated to become operational over the next few years (see Appendix B). International access to a large number of observation satellites using various types of imaging sensors (e.g., electro-optical, radar, thermal) will substantially enhance global transparency by increasing the chances that major events occurring around the world will be captured on overhead imagery in a timely fashion.

The new commercial earth-observation satellites also raise important questions on the policymaking relationship between governments and non-state actors. Private and nongovernmental actors with a strong interest in using satellite imagery for public policy purposes include the news media and a broad range of NGOs concerned with public policy issues, including arms control and nonproliferation experts, environmental groups, and international NGOs concerned with human rights and humanitarian relief activities, all of whom could use satellite imagery to advance their agendas. These new "imagery activists" will use information derived from satellite images for independent assessments that can either confirm or challenge the accuracy of government assessments. Debates over imagery credibility could arise in specific cases as these new users, who have relatively limited experience in image interpretation, build their proficiency in analyzing satellite imagery data.

The expert contributors to this book generally agree that global transparency is inevitable, but they pointedly disagree on whether this trend will increase or diminish international security. Many of the chapter authors share the view that—on balance—global transparency, and commercial observation satellites in particular, will enhance international security. This view holds that a security benefit will arise from having expanded access to earth-observation satellites that can help anticipate and document when countries are engaging in aggressive behavior or when governments are undertaking large-scale internal atrocities. However, an alternative perspective contends that greater transparency may actually erode the national security of countries located in regions where the risk of war is high by diminishing the uncertainties that help dissuade military conflicts. To some degree, these alternative perspectives on the broader security implications of global transparency help to frame the debate over the dual-use nature of commercial satellite imagery.

Dual-Use Aspects of Commercial Observation Satellites

As noted earlier, commercial earth-observation satellites are intrinsically a dual-use technology. Their military potential will only grow as more commercial observation satellites become operational. The 1990–1991 Gulf War highlighted the growing importance of civilian observation satellites to military activities. The United States and its Coalition allies benefited from Landsat and SPOT imagery data that were used to support various military missions and to provide a way of sharing imagery data among all partners. At the same time, steps were taken to deny Iraqi forces access to the same imagery sources during the Desert Shield and Desert Storm operations, as well as imagery from U.S. and European weather satellites. The Gulf War experience encouraged several countries to seek to acquire their own observation satellite systems or to gain access to commercial observation satellites. Although a few countries, such as the United States, Russia, and France, have their own military reconnaissance satellites, most countries do not. Thus, the military and intelligence establishments of many countries will probably take advantage of commercial observation satellites as a new source of overhead information.

Along with their potential military applications, commercial and civilian observation satellites offer a new instrument for supporting peace negotiations and encouraging regional conflict resolution. Civilian satellite imagery and three-dimensional visualization technologies helped the U.S. negotiators at the 1995 Dayton peace talks resolve complicated territorial issues that were obstructing the final peace settlement for Bosnia. Similarly, the United States provided civilian imagery and mapping support to facilitate the successful negotiations that ended the long-standing border dispute between Ecuador and Peru. These cases suggest that opportunities could exist for creating cooperative monitoring regimes using commercial observation satellites to reduce the potential risk of regional conflicts arising over geographical flashpoints elsewhere in the world.

Thus, policymakers face a new challenge in taking advantage of the benefits commercial observation satellites promise while limiting the potentially harmful uses that can be made of their data. Any attempt to restrict the use of commercial satellite imagery for military and intelligence purposes will be complicated by the difficulty of separating potentially aggressive uses of imagery data from legitimate defensive needs for timely information on activities occurring beyond a state's boundaries. More specifically, any interest the U.S. government may have in using administrative measures to prevent U.S. commercial earth-observation satellite companies from collecting or selling imagery because of military or foreign policy raises contentious constitutional issues involving first amendment rights.

Uncertain Markets and Competing Government Policies

Although the political and technological trends are generally encouraging for commercial observation satellites, greater uncertainty surrounds the economic prospects. The perception that most countries are shifting to knowledge-based economies only encourages the notion of a rapidly growing international market for satellite imagery and information products over the next few years. Nonetheless, commercial imaging satellite firms confront stiff competition in selling geospatial information products and services at home and abroad. Aerial photography firms have long dominated the remote sensing market and are pursuing their own innovative paths in the data marketplace. Land-based surveys using Global Positioning System (GPS) data and geographic information systems (GIS) both complement and compete with satellite imagery in providing affordable geospatial information.

At the same time, international competition in providing geospatial information products and services is likely to be fierce, particularly because many civilian observation satellites are also focused on establishing a niche in the emerging global market for satellite imagery, even though they are usually state-owned or rely on substantial government subsidies. Furthermore, international interest in acquiring imaging satellites is steadily growing as countries such as Australia, Brazil, Israel, South Korea, and Taiwan make progress in developing or purchasing their own civilian and commercial observation satellites.

National governments will be critical to the long-term success of the commercial observation satellites. Governments can play multiple and concurrent roles because they serve as regulators, patrons, customers, and even competitors in some cases. On the positive side, government policies and programs can promote the beneficial contributions of commercial observation satellites while limiting the chances that their imagery data will be used for harmful or aggressive purposes. Yet governments can also inhibit the development of the fledgling commercial remote sensing industry. Government policies can unintentionally undercut the commercial viability of startup firms through excessive regulations. Struggling private firms can also become too dependent on government sales rather than taking the hard and risky efforts needed to develop the commercial market for satellite imagery products and services. Thus, national governments must reconcile competing policy objectives in dealing with commercial remote sensing enterprises.

This book offers an in-depth assessment of the policy implications of a new information technology—commercial observation satellites—that is coming of age at the start of the new millennium. The various chapters suggest that these imaging satellites can enhance global transparency, encourage economic development, and contribute to international security. However, the book contributors also recognize that commercial satellite imagery can be put to more harmful uses. Thus, the primary challenge is to find ways of hedging against the potential risks that unprecedented global access to overhead imagery poses without diminishing the many positive contributions that commercial observation satellites can offer for civil, commercial, and international uses.

Notes

1. This book assesses the broader implications of commercial observation satellites, which are owned and operated by private firms seeking to generate revenues by selling satellite imagery data and services. These firms are playing a leading role in encouraging global transparency by making satellite imagery, including high-resolution images, broadly available to public and private consumers of imagery data. However, this book does not underestimate the continuing importance of civilian observation satellites, such as Landsat and the Satellite Pour l'Observation de la Terre (SPOT), which are owned or heavily subsidized by governments. Imagery data from most civilian observation satellites are available in the marketplace as government organizations seek to recover at least some of their investments and operating costs. Although major differences exist between commercial and civilian observation satellite enterprises, these distinctions are less important to growing global transparency than the fact that both groups produce unclassified satellite images that can be acquired by a broad range of domestic and international users.
2. *Resolution* generally refers to the size of the smallest object that can be distinguished in an image from its surroundings. Higher-resolution images (e.g., 1 m) enable imagery analysts to detect and identify smaller objects, such as vehicles and small buildings, whereas lower-resolution images (e.g., 20 m or larger) are mainly limited to distinguishing such objects as airports, population centers, and large natural features.
3. With the exception of imagery data gathered by U.S. government satellites, most of these data are protected by copyright and use-related licensing provisions.
4. Unfortunately, that satellite failed in orbit a few days after its launch in December 1997.
5. National sovereignty extends only to the airspace above a state. The 1967 Outer Space Treaty guarantees the right of all states to use outer space for peaceful purposes. However, cloud cover and dark of night significantly limit the use of optical sensors. Certain regions at certain times of year are particularly difficult to image from space because of darkness or cloud cover. Radar, on the other hand, can pierce both cloud cover and darkness. Furthermore, given their orbital dynamics, imaging satellites are only occasionally in the right overhead position.

Section I: The Policymaking Context

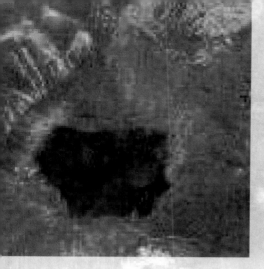

A confluence of trends—political, technological, and economic—has encouraged the development of a new generation of imaging satellites that will be owned and operated by commercial firms rather than by governments. More than four decades after the Soviet Union first orbited the Sputnik satellite in 1957, a nascent marketplace for satellite imagery data is emerging. Civilian agencies, businesses, scientific centers, and defense establishments all anticipate gaining another valuable source of earth-observation data. Commercial observation satellites also promise to bolster global transparency by substantially expanding public and international access to high-resolution satellite images. This trend is most notable in the growing interest among news media and nongovernmental organizations in applying commercial satellite imagery to their information needs.

However, policymakers in the United States and elsewhere are ambivalent about encouraging commercial observation satellites and the greater global transparency that expanded public access to commercial satellite imagery provides. Commercial observation satellites raise several major policy issues, including:

- *Government policy effects*: How do civilian remote sensing programs and government policies encourage or inhibit the development of commercial observation satellites?

- *Market prospects:* Is there a viable commercial market for satellite imagery, particularly given alternative technologies and data sources?
- *Dual-use utility:* Will rapidly growing global transparency, which includes unprecedented access to high-resolution commercial satellite imagery, enhance or diminish the security of countries, particularly in unstable regions?

The chapters in Section I offer background on the policymaking context for commercial observation satellites. They provide a useful historical review of U.S. decisionmaking on whether and how to commercialize satellite imaging. In addition, these chapters explore the market, technology, and national-security implications of expanding global access to this new type of information technology.

The belated appearance of commercial observation satellites results more from a basic policy change than from technological impediments. As discussed in Chapter 2, the unique qualities of satellite imagery were recognized early on as a means of increasing global security. In this chapter, Richard Leghorn, a leading participant in the U.S. policy debates on this issue in the 1950s and 1960s, and Gregg Herken, a historian specializing in the Cold War era, recount how President Eisenhower and other senior officials seriously considered the value of global transparency in the form of the "Open Skies" proposal and other measures to manage U.S.-Soviet relations. However, the perceived national-security imperatives at the time favored the proponents of greater secrecy. With the end of the Cold War, the earlier restrictions on public access to imagery data have been substantially relaxed. Thus, the authors see a new opportunity to reconsider the "road not taken" as Washington and Moscow substantially loosen earlier restrictions on broad access to imaging satellite technologies and high-resolution images.

In Chapter 3, Ray Williamson reviews the evolution of satellite remote sensing with particular attention to assessing the conditions that have encouraged the commercialization of satellite imagery. His chapter highlights the development and modernization of the U.S. Landsat satellite as the world's longest-running civilian remote sensing satellite program, as well as the subsequent emergence of non-U.S. imaging satellites, including France's Satellite Pour l'Observation de la Terre (SPOT) series. He also analyzes how advances in information technologies and remote sensing applications have created more-favorable market conditions for commercial remote sensing than existed during the failed U.S. government attempt to privatize the Landsat program in the 1980s.

The remaining chapters in Section I assess key policymaking issues regarding the market prospects, technology trends, and national-security implications of commercial observation satellites. Kevin O'Connell and Beth Lachman assess the market trends for commercial satellite imagery and related spatial technologies in Chapter 4. Although a major international market already exists for remote sensing, satellite imagery data providers currently account for only a small portion of the total remote sensing revenues. Most of the revenues are generated by airborne providers of overhead photography and digital imagery and by the value-added firms that process and analyze both airborne and satellite data to produce information products. The authors also note that national governments will exert a major influence on the business focus and fortunes of commercial imaging satellite firms because governments continue to play a critical part in regulating these firms and are the primary consumers of their imagery data and services.

The policy issues surrounding commercial observation satellites are rapidly evolving because technological advances are creating the possibility of commercial observation satellites with improved sensors based on different types of imaging technologies. In Chapter 5, Bob Preston analyzes two types of emerging technologies: synthetic aperture radar and multi- and hyperspectral imaging. These sensors are expanding the utility of remote sensing applications beyond the traditional emphasis on high-resolution panchromatic (e.g., black and white) images. His chapter describes the unique characteristics of these imaging technologies and discusses the particular licensing and operational control issues that they pose for policymakers.

Finally, in Chapter 6, John Baker and Dana Johnson analyze commercial observation satellites as a dual-purpose technology, paying particular attention to the U.S. national-security implications of this new information technology. This chapter reviews the significant contribution that civilian remote sensing satellites made to the military operations of the Coalition forces in the Gulf War. With their higher resolution and more timely imagery, commercial observation satellites could become an important source of geospatial information for U.S. and allied combat or peace operations. However, the same commercial imagery could also be exploited by potential adversaries, including rogue states and terrorists, provided that they can translate access to imagery data into the information crucial for their operations. Either way, the authors contend that commercial satellite imagery poses some important organizational and doctrinal challenges for U.S. military planners.

The Origins and Evolution of Openness in Overhead Global Observations

Richard S. Leghorn and Gregg Herken

The subject of this book—commercial observation satellites and global transparency—evokes a sense of déjà vu. While the current impetus for open observation of the earth comes from commercial and, most recently, journalistic considerations, it is useful to recall that the original impetus for global transparency arose from security concerns.

In the mid-1950s, a new national security paradigm emerged in the United States. Instead of pursuing military superiority vis-à-vis the then–Soviet Union ad infinitum, the United States began to consider the possibility that its security could be based on openness in matters of armaments and on a stable system of nuclear deterrents. This approach called for mutual disclosure of military information and verification by overflights—the very opposite of secrecy. This alternative approach to security came to naught in the face of the overarching Soviet reliance on secrecy. Nonetheless, U.S. initiatives toward global transparency during the 1950s and early 1960s provide insights into national and international security issues posed by today's commercial observation satellites.

The Postwar World and the Drive for Secrecy

Safeguarding against surprise attack became a critical concern for the major powers in the aftermath of World War II. Both the Americans and the Soviets had devastating experiences from surprise attacks. For a whole generation of

Americans, the Japanese attack on Pearl Harbor was a life-shaping event. The lesson learned from the events of December 7, 1941, was that sneak attack was how modern wars began, as a bolt from the blue.[1] North Korea's massive invasion of South Korea in 1950, followed by the surprise Chinese invasion, reinforced these anxieties. The Russians, and later the Soviets, also suffered from surprise attack, most notably in the Russo-Japanese war and with the surprise German invasion in June 1941.

In the immediate postwar era, these experiences of surprise attack, coupled with deep fears about armament buildups and new military technologies, especially nuclear weapons, led both sides to seek security in significant measure through secrecy. The Soviets retreated behind their Iron Curtain and substantially blocked the flow of information and people between East and West. For the Western powers, the "ringing down" of the Iron Curtain, as Winston Churchill characterized it in a 1946 speech at Fulton, Missouri, turned the Soviet Union and its allies into a vast "denied" territory that could be used as a staging area for a surprise attack on the West (see Herken, 1981, pp. 143–144).

In turn, the United States turned to clandestine means to penetrate Soviet secrecy and thus ensure its own security. Aerial reconnaissance offered the best means of gaining information about threatening Soviet military capabilities. However, overflights of Soviet territories would risk interception and, in peacetime, violated international law.[2] The Soviets might also have construed overflights as a prelude to an allied offensive or used them as a *casus belli*. The United States, therefore, turned its efforts to overflights that would be hard to detect or intercept. These methods included high-altitude, unarmed aircraft; camera-carrying balloons; and, later, reconnaissance satellites.[3]

In a secret report completed in May 1946, just months after the end of World War II, analysts at Project RAND, a "think tank" Douglas Aircraft and the U.S. Army Air Forces had recently created, explored the feasibility of an "experimental world-circling spaceship" (Douglas Aircraft Company, 1946).[4] One anticipated application of such a craft would be photographing the earth below. However, at the time of the RAND report and a similar Navy study, the notion of satellite reconnaissance was speculative. There was neither a rocket booster of sufficient power nor the government support necessary for such a program.

Nonetheless, the value of overhead photography to gauge the strength of the enemy had been convincingly proven in the war just ended (U.S. Strategic Bombing Survey, 1945).[5] LTC Richard S. Leghorn, a combat reconnaissance pilot and group commander in World War II, was an early advocate of peacetime aerial reconnaissance. Leghorn had flown reconnaissance missions before and after the Allied invasion of France. He had also participated in the organization and operation of the aerial collection of scientific and photographic information during the atomic tests on Bikini in mid-1946.

In December 1946, at the dedication of the Air Force–funded Optical Research Laboratory at Boston University, Leghorn advocated "pre–D-day photography" taken from aircraft flying at an extremely high altitude (Leghorn, 1946). He argued that such reconnaissance might be decisive in preparing to win or, better still, in preventing a future atomic war. Leghorn observed that the nature of atomic warfare made it "essential that we have prior knowledge of the possibility of an attack, for defensive action against it must be taken before it is launched." (Leghorn, 1946, p. 55.) Unfortunately, he noted, that conventional espionage was not held to be an act of war, but overflights of sovereign national territory were still so considered under international law. Accordingly, unauthorized aerial overflights were illegal under international law in peacetime—at least until such time as "thinking on this subject is changed." (Leghorn, 1946, p. 55.)

Leghorn therefore proposed development of a special aircraft for long-range prehostility reconnaissance that would be hard to detect and intercept. Such an aircraft could operate in peacetime under compelling circumstances with less risk of provoking hostility. He also argued that international tensions might actually be eased, not increased, by mutual peacetime overflights, since they would ameliorate fears of surprise attack.

In December 1948, during a highly classified Air Force conference on aerial surveillance held at a Topeka, Kansas, air base and hosted by the Strategic Air Command, Leghorn reemphasized the ideas he had originally outlined in Boston. Now a civilian, Leghorn had just returned from Czechoslovakia, where he had had first-hand experience of Soviet enforcement of Iron Curtain secrecy. Whereas at Boston he had noted that overhead reconnaissance of Russia was feasible and that the capability should be developed, Leghorn urged at the Topeka conference "that actual missions be carried out." (Leghorn, 1948, p. 10.) In addition to aerial photography, radar could be used to image intelligence targets during nighttime missions or bad weather, he postulated (Leghorn, 1948).

While the emphasis at the Topeka conference was on the feasibility and necessity of secret overflights of Soviet territory, participants also discussed a hitherto little-recognized purpose for overhead reconnaissance: In the then-unlikely event that the Cold War antagonists would someday agree to reduce their stockpiles of weapons, aerial surveillance could be used to police and verify arms limitation agreements.

The Soviets' acquisition of a nuclear capability in the late summer of 1949, followed less than a year later by North Korea's surprise attack against its southern neighbor,[6] reconfirmed U.S. fears of surprise attack and accelerated the U.S. search for a reliable and effective method of obtaining warning of any impending attack. In December 1950, the U.S. Joint Chiefs of Staff asked President Harry Truman to reconsider the prohibition against overflights of Russia (Hall, 1996, p. 113).[7] Truman approved two overflights. In October 1952, after various delays, a camera-carrying RB-47B flew over and photographed a Soviet air base in the Siberian Arctic that was believed to be a likely staging area for a surprise atomic attack on the U.S. mainland. The suspected massed formation of Soviet bombers was not found. Perhaps it is more important that this established a precedent for subsequent overflights, when deemed justified as a matter of overriding national interest.

In late 1951, then—Col Bernard A. Schriever, later a four-star general in charge of Air Force missile and space programs, headed the Development Planning Activities on the Air Staff. Schriever arranged for Leghorn, who had been recalled to active duty at Wright Field during the Korean War, to be transferred to the Air Staff at the Pentagon to head planning for the development of future reconnaissance and intelligence capabilities.[8] Leghorn, with the help of a small task force drawn from throughout the Air Force, reviewed not only his work at Wright Field but also all relevant studies, including RAND reports and the so-called Beacon Hill Report,[9] and solicited imaginative suggestions from colleagues.[10]

The result of this effort was a draft plan for the Air Staff suggesting that development efforts focus, first, on special-purpose, single-engine, high-altitude aircraft for reconnaissance of specific objectives; second, on high-altitude "weather" photographic balloons to provide area coverage; and third, on observation satellites, beginning with coverage of large areas and, in subsequent designs, coverage of specific targets in more detail.

Satellites overcame problems inherent in aerial reconnaissance: They provided much greater coverage;[11] satellite observation did not necessarily violate existing international law and was therefore less politically vulnerable; and satellites, unlike aircraft and balloons, were impossible to intercept with then-current technology.

Amidst all this planning for secret aerial reconnaissance, seen as critical to U.S. security, there was also recognition of the broader implications of overflights. In a January 1955 article for *U.S. News and World Report*, Leghorn repeated his earlier arguments that clandestine overflights of Russia were both urgently necessary and immediately feasible (Leghorn, 1955).[12] At the same time, he envisioned a future when satellites might regularly scan the earth's surface to report on armaments and potential aggressors. Secret aerial surveillance would contribute to U.S. security in the short term, but open surveillance could contribute to long-term world security.

1955 Open Skies Proposal

President Dwight Eisenhower presented the Open Skies proposal on July 21, 1955, at the Geneva summit of Soviet, British, French, and U.S. heads of state. This was the first significant governmental recognition that openness could contribute to world security. In his speech at the summit, Eisenhower especially addressed the delegates of the Soviet Union, suggesting that the two countries exchange complete blueprints of their military establishment. Eisenhower then proposed that the United States and the Soviet Union agree to aerial inspections:

Next, to provide within our countries facilities for aerial photography to the other country—we to provide you the facilities within our country, ample facilities for aerial reconnaissance, where you can make all the pictures you choose and take them to your own country to study; you to provide exactly the same facilities for us and we to make these examinations, and by this step to convince the world that we are providing as between ourselves against the possibilities of great surprise attack, thus lessening danger and relaxing tensions. (Eisenhower, 1955.)[13]

The concepts that Leghorn and colleagues had been developing and advocating within the Air Force and the government since the end of World War II were now in the public arena.[14]

Eisenhower emphasized that his Open Skies proposal would reassure each side against surprise attack, but he also had a broader political objective. Mutual overflights would contribute to confidence-building and would help establish better relations between the United States and the Soviet Union. This, in turn, would contribute to progress in arms control and opening Soviet society.[15] However, the Soviets did not accept Eisenhower's proposal. In contrast to Eisenhower's approach of using overflights to create a workable inspection system, and then working on agreements for mutual arms reductions, the Soviets had proposed an exchange of ground observation posts on both sides of the Iron Curtain and had insisted that arms limitation agreements should come first. They rejected the Open Skies proposal out of hand as a brazen attempt to obtain information about both possible targets and the weaknesses of the Soviet military. The Soviets also feared, justifiably or not, that such knowledge might lead to a preemptive attack by the West before they could mount a sufficient second-strike deterrent.[16] Ironically, the Open Skies proposal, intended to reassure against surprise attack, rekindled Soviet fears in this regard.

The acceptance of openness required then, and still requires, a fundamentally different approach to security issues, one based on disclosure and verification of armaments plus a balance of deterrents, rather than pursuit of military superiority. Whether the advent of global transparency will lead to a revision of international security concepts is an issue that the successful launch of high-resolution commercial satellites is bringing to the fore.[17]

Continuing Efforts to Penetrate Soviet Secrecy: The U-2 and the Beginnings of the Corona Program

With the failure of Open Skies, the Soviets continued their policy of extraordinary secrecy. Eisenhower recalled in his memoirs that "Khrushchev's own purpose was evident—*at all costs to keep the USSR a closed society*." (Eisenhower, 1963, pp. 521–522.) Americans accelerated their own secret efforts to acquire information on Soviet military capabilities. On July 4, 1956, a year after the Soviets had rejected the Open Skies proposal at Geneva, the United States conducted the first U-2 overflight of the Soviet Union. Contrary to U.S. hopes and expectations, Soviet radar detected the U-2 overflight, but Russian MiGs were unable to intercept the high-flying reconnaissance plane, and Soviet surface-to-air missiles were not operational until 1960. The Soviets tracked U-2 flights from the beginning but, to avoid acknowledging domestically and internationally their military inferiority, chose not to oppose U-2 overflights publicly until they had the means to oppose the flights militarily.

On their part, the Soviets continued developing more advanced weapon systems, successfully testing an intercontinental ballistic missile in August 1957. Less than two months later, on October 4, 1957, Russia launched Sputnik, the first earth-orbiting satellite.

The Soviets' technological coup led to Western dismay and to official U.S. scrutiny of its existing military and civilian satellite plans and programs. Although the initial Air Force development plan for a reconnaissance satellite had been approved in 1956, the program had not received sufficient funding, and little had been accomplished. Budgetary constraints were in part to blame, but the lack of support for the development of a military reconnaissance satellite has also been attributed to the Eisenhower administration's commitment to the "peaceful" uses of space and avoidance of military uses. Arguably, space reconnaissance would not, in itself, be "unpeaceful," because of its potential support of arms control and confidence-building measures. And satellite observation could be considered a defensive measure consistent with the United Nations (UN) charter. On the other hand, a reconnaissance satellite would not have a strictly civilian or scientific purpose. Thus, economic and policy constraints had slowed progress in the Air Force's underfunded satellite reconnaissance program (Oder, Fitzpatrick, and Worthman, 1988, pp. 4–7).[18]

The successful Sputnik launch revived the U.S. reconnaissance satellite program. In February 1958, President Eisenhower approved an accelerated and reorganized program for developing a covert photographic satellite, building on the infrastructure of the Air Force WS-117L system. The Air Force program continued as the Discoverer scientific program, as a cover for the covert reconnaissance effort and also for the development of future electronic "readout" techniques. The camera and film recovery subsystems for the covert reconnaissance program were placed under Central Intelligence Agency (CIA) control, as were the management of operations and security arrangements. Richard Bissell of the CIA, who was in charge of the U-2 program, was appointed director of what became the Corona Project (Hall, 1997).

1958 Surprise Attack Conference

The Eisenhower administration backed the development of the U-2 and then the reconnaissance satellite on a priority basis as an essential means of strengthening national security. There appeared to be few alternatives. Following the failed Open Skies proposal at the 1955 Geneva conference, both the United States and the Soviet Union accelerated development and production of armaments out of fear the other would win the arms race. While technological developments certainly contributed to the continuing arms race, they also provided potential solutions, both to national and international security issues. Satellite observation in particular might provide an inspection system to verify arms control agreements.

At the November–December 1958 Surprise Attack Conference in Geneva, Merton E. Davies of RAND, a member of the U.S. technical delegation, introduced a written proposal to use satellite observations to enforce arms control agreements. As at the Geneva summit, the advocates of openness argued that it was a means of enhancing world security. Open, mutually available information on military capabilities might lead to stable mutual deterrence. But, just as at Geneva, the initiative on openness went nowhere. The Soviets wanted arms limitation to precede any discussion on overhead observation, and the United States refused to accept this approach.[19]

The Corona Program and U.S. Policy Debates on Secrecy versus Openness

President Eisenhower had approved the Corona satellite program as a highly secret endeavor in February 1958, following the launch of Sputnik the previous October. The CIA-directed crash program developed a new rocket, a new camera, new film, and advanced techniques for recovering objects from space. Eisenhower never gave up on the program, despite 12 failed attempts to launch the reconnaissance satellite before its first successful flight in August 1960.

It proved more straightforward to resolve the unprecedented technical challenges of satellite reconnaissance than those of policy. While there was consensus within the government that reconnaissance satellites were a critical component of U.S. national security, given the secretive and closed Soviet society, differing views emerged on how open the United States should be about its observation satellites. During the Eisenhower administration, the government developed and strongly supported a "space for peace" policy, but there was uncertainty over whether open acknowledgment of reconnaissance satellites would advance or hinder this policy and national security.

The approach Eisenhower's advisors supported was formulated in early 1955, even before his Open Skies proposal later that year. The emphasis was on establishing the legality and peaceful nature of space observations. Cargill Hall describes this decision:

> [In] the spring of 1955 the president's closest advisors determined, if at all possible, to keep outer space a region open to all, where the spacecraft of any state might overfly all states, a region free of military posturing. By adopting a policy that favored a legal regime for outer space analogous to that of the high seas, the United States might make possible the precedent of "freedom of space" with all that that implied for overflight. (Hall, 1995, p. 221.)

The Eisenhower administration took care not to undermine this goal. The administration emphasized the peaceful, scientific purposes of its space initiatives and provided considerable information about these programs, especially after Sputnik, both to encourage public support and to demonstrate its commitment to the peaceful uses of space and to the concept of "freedom of space" (Stares, 1985, p. 62). Indeed, although Sputnik caused public distress in the United States about falling behind the Soviet Union in space exploration, then–Secretary of Defense Donald Quarles, in private at least, welcomed the Soviet initiative because of the precedent it established. Sputnik's overflight of many nations, with no opposition, could be seen as helping to establish "freedom of space," a concept the Eisenhower administration considered critical to the legal justification for reconnaissance satellites (Hall, 1995, p. 228).

The administration did not publicly acknowledge the CIA's reconnaissance satellite program, authorized in February 1958 and commencing operations in August 1960. The prevailing view was that, until the legitimacy of space flight was firmly established in international law, disclosure might invite political and even military attack and threaten the Corona program that was so vital to national security.

There were, though, other views on this issue. Richard Leghorn, and later the Department of State, advocated a more proactive role for the government in establishing the legitimacy of all informational overflights and especially in defining and championing a peaceful role for reconnaissance satellites for arms control verification and for confidence building. In 1955 and again in 1956, Leghorn had written and circulated memoranda urging the government to organize a political action program to preempt Soviet political countermeasures. His ideas were well-known, at least within the Air Force (Leghorn, 1956; Oder, Fitzpatrick, and Worthman, 1988, pp. 8–9).

Again in January 1960, while serving as the Technical Deputy to the Joint Disarmament Study directed by Charles A. Coolidge for then–Secretary of State Christian A. Herter, Leghorn urged the government to develop a "space political action program." Having in mind Eisenhower's Open Skies purposes to begin to open Soviet society and provide a basis for arms control agreements, the purpose of this program would be to bring about worldwide acceptance of the concept that information satellites can be of great utilitarian value and can contribute significantly to securing the peace. They will be an up-to-date approach to "open skies." (Leghorn, 1960a).

To forestall anticipated Soviet political opposition and buttress U.S. determination regarding openness, Leghorn's political action plan included measures to educate the public about the many uses of information-gathering satellites, a proposal to ban weapons of mass destruction in space, diplomatic initiatives to build international support, and confidential overtures to the Soviet Union to underscore both the U.S. capabilities and its commitment to "open space." Leghorn argued that

> The general advancement of space law must be subordinated, however, to gaining acceptance of the unimpaired use of space for information purposes to help the enforcement of peace. However, this issue and space law have an important interaction, namely that copious and free use of space for many purposes—including arms information gathering and transmission—can establish space law through the exercise of a "common law" approach. This would achieve much of our general policy goal without the frustrations and complications from attempting prematurely to write Space Statutes. (Leghorn, 1960a.)

In Leghorn's view, complete and confident openness on the existence, peaceful purpose, and potentials of reconnaissance satellites was a necessary first step toward a new world security regime, such as that originally envisioned in Eisenhower's 1955 Open Skies proposal.

The U-2 Incident and the Continuing Debate on Openness versus Secrecy

The Soviet downing of a U-2 in May 1960 occurred a few months after Leghorn's memorandum and just two weeks before the four-power conference. At the conference, Khrushchev's tirade against the U.S. violation of its airspace and international law highlighted U.S. vulnerability to Soviet political countermeasures. The Eisenhower administration initially denied that the U-2 was engaged in photographic reconnaissance within the Soviet Union, but the Soviets soon presented irrefutable evidence, having captured Francis Gary Powers and possessing not just parts of the downed plane but also the film. The Eisenhower administration's clumsy denials dramatically accented that secrecy and attempted cover-ups had their own risks.[20]

Immediately after the U-2 incident, Leghorn had continued to point out that opportunities still existed to seize the political initiative and work for acceptance of openness. On May 12, he sent to the president and other ranking members of the government a proposal for the imminent summit conference concerning aircraft *and* satellite photography. He suggested that the president acknowledge that the United States had "most reluctantly" continued with aerial reconnaissance after the Soviet Union rejected the 1955 Open Skies proposal and that he emphasize the contribution of such reconnaissance to disarmament and world security. Leghorn further suggested that the United States should provide the Soviet Union with a sample of U-2 photography and should offer the Soviet Union the use of our air bases and the opportunity to make unarmed overflights itself of the United States. Finally, Leghorn urged the president to announce that the United States was about to launch an observation satellite and that the information gained from this satellite would be openly available. Leghorn concluded his proposal with the assertion that, "Timely arms information, openly available to all, is the key to world military stability and the subsequent disarmament of nations." (Leghorn, 1960b.)

President Eisenhower's acknowledgment of Leghorn's proposals is interesting. Nikita Khrushchev had stormed out of the four-power conference, ending any hopes the Eisenhower administration had of achieving more openness in security matters and a framework for arms control. On May 23, Eisenhower wrote to Leghorn that he regretted "that the intransigence of one individual prevented the holding of a Summit Conference for discussion of suggestions such as the one made by you." (Eisenhower, 1960.) The same day, Eisenhower gave way to Soviet pressure, telling Secretary of State Herter "that he had no thought whatsoever of permitting more [U-2 flights over the Soviet Union]." (Ambrose, 1990, p. 515.)

The U-2 incident demonstrated the political risks of secrecy but did not alter Eisenhower's approach to satellite reconnaissance. The Eisenhower administration did not make the reconnaissance satellite program public. It pressed on with Corona as a highly classified project and was unwilling to risk Soviet political reactions that might end the program. Vigorous Soviet political reaction had ended aerial overflights once U-2 flights could not be denied. Presumably, another factor in this decision was continuance of the U-2's highly secret management structure, which was devoid of committee reviews and the like and had proven most effective for a deadline-driven program incorporating so many technological innovations.

At first the Soviets appeared willing to accept the legitimacy of satellite reconnaissance. At a meeting with Eisenhower and General Charles de Gaulle at the Paris summit in May 1960, Khrushchev asserted that "any nation in the world who wanted to photograph the Soviet areas by satellite was completely free to do so."[21] However, the Soviet regime soon changed its mind, claiming that the illegality of aircraft photography extended to satellite photography. In an October 1960 article, after the successful August Corona mission, G. Zhukov, the Soviet marshal and former defense minister, defined the Soviet position:

> It should be noted that American plans of space espionage directed against the security of the U.S.S.R. and the other Socialist countries are incompatible with the generally recognized principles and rules of international law, designed to protect the security of states against encroachments from outside, including outer space. In the past, considerations of state security have been of decisive importance in determining the air space regime. Today the same considerations must underlie the regime of outer space. (Zhukov, 1960, p. 55.)

The United States had lost any opportunity for a successful political initiative to advance "openness" of satellite observations as a contributor to world stability and security.

The Decision to Maintain Secrecy

In the early Kennedy years, the Department of State continued to work toward legitimizing reconnaissance satellite activities within the framework of international law. It briefed close U.S. allies on the purposes of the satellite reconnaissance program, building support for its position in the UN. The debate whether the U.S. policy should go beyond a cautious, legalistic approach continued within the Kennedy administration. In the history of the Corona program written under the auspices of the National Reconnaissance Office, the authors described the contending points of view in 1962 as ranging from "full revelation to complete secrecy." (Oder, Fitzpatrick, and Worthman, 1988, p. 67.) The dilemma confronting government officials was articulated by the Department of State's legal advisor, Abram Chayes:

> The question was this: How was one to protect our satellite reconnaissance operations politically? Everybody had the U-2 incident in mind.... And in the State Department on the whole we thought you gained something politically by being somewhat more open about our operations and *developing a climate of legitimacy* about them instead of trying to keep them completely secret. The opposite view was let's keep them very, very secret. First of all, nobody knows about them.... And secondly, to the extent that this gets out in public, it forces the

Russians to make a challenge of some kind. . . . And that we could never defend against a political challenge because there was so much power in this idea of peaceful uses of outer space, and this might not be regarded as a peaceful use and so on. (Cited in Stares, 1985, p. 63.)

By mid-1962, the Kennedy administration came down decisively on the side of strictest secrecy. It had already limited the information available on space launches and discouraged press coverage of space activities. In March 1962, then–Deputy Secretary of Defense Roswell Gilpatric issued what came to be known as the "Gilpatric Directive," prohibiting advance announcements and press coverage of all military space launchings. The National Security Council accepted this directive "as sound practice" in July 1962 (Oder, Fitzpatrick, and Worthman, 1988, p. 67). At the same time, though, National Security Council Action 2454 stated that the U.S. policy on space was to make no distinction between civil and military observations from space and to continue to work toward the acceptance of space observations as legitimate.[22]

The opportunity to establish the legitimacy of space observations in international law seemingly improved when the Soviet Union's opposition evaporated after the successful launch of its own reconnaissance satellite later in 1962.[23] The U.S. policy on secrecy remained unchanged, however. From its start, the structure and security of the Corona program had followed those of the U-2, and the concept of openness lacked sufficient commitment to alter an approach that had proved effective for at least the short term.[24]

The Era of "Tacit Open Skies"

By 1963, the United States and the Soviet Union each had reconnaissance satellites, and each side was well aware of the other's information-gathering capabilities.[25] Yet openness was no longer on either the United States or the international agenda. There were no public debates on the issue or serious public consideration of the proposition that openness could enhance world security, something Eisenhower had argued at the Geneva four-power conference in 1955. Instead, the United States and the Soviet Union entered an era of "tacit open skies." Mutual arms observations from space became a reality but remained highly classified. In the U.S. government, there was worry about openly challenging the vulnerability of the Soviet secret society. "Let sleeping bears lie" seems to have been the operative policy.

Even though tacit, this open skies regime nonetheless enabled both sides to realize the benefits of arms information in achieving stable mutual nuclear deterrents. In 1967, President Lyndon Johnson, in an off-the-record meeting, told a group of local government officials that

> we've spent thirty-five or forty billion dollars on the space program. And if nothing else had come out of it except the knowledge we've gained from space photography, it would be worth ten times what the whole program has cost. Because tonight we know how many missiles the enemy has and, it turned out, our guesses were way off. We were doing things we didn't need to do. We were building things we didn't need to build. We were harboring fears we didn't need to harbor.[26]

In the 1972 Strategic Arms Limitation Treaty, the United States and the Soviet Union agreed that each party could monitor compliance with the treaty's arms limitation provisions through "national technical means of verification," a euphemism for reconnaissance satellites. It was only in 1978, during the Carter administration, that the United States officially acknowledged even the fact that reconnaissance satellites existed (McDonald, 1997b).

A de facto "open space" duopoly continued into the 1990s. Ongoing satellite reconnaissance continued secretly, but world events increasingly favored greater openness. Indeed, at the beginning of the decade, President George Bush spearheaded a U.S. effort to revive the aerial observation feature of Eisenhower's Open Skies proposal. In an age of satellites, a new open skies proposal for aerial overflights appeared irrelevant to some critics. In certain situations, though, aircraft can provide more timely information on specific locations. Certainly, there was renewed interest in aerial overflights among many nations. Negotiations for this new open-skies treaty continued over the next two years. Initially disinterested, the Soviets came to accept the open skies proposal in principle. The treaty permits mutual aerial overflights by aircraft equipped with specific photographic and other sensors to collect imagery over the territories of the other signatories (Gabriele, 1997). In 1992, all members of NATO and all former Warsaw Pact nations signed the treaty. Although Russia, Ukraine, and Belarus have not yet ratified it, the United States, Russia, Ukraine, and several other countries are participating in demonstration flights under the treaty's provisions. If the concept of openness has not been accepted fully, the 1992 Open Skies treaty indicates that there has been some progress toward its realization.

Other events have also favored openness. The growing use of Landsat imagery from the mid-1970s, and later SPOT imagery, to gauge and predict crop yields, led in turn to the dawning realization that higher-resolution commercial observation satellites might contribute to the economic well-being of individual states.[27]

The fall of the Berlin Wall in November 1989 signaled the breakup of the Soviet empire and the possibility of a more-open Soviet society. In 1992, the Soviets began an aggressive international marketing program for high-resolution satellite photos (Gorin, 1997, p. 98). In 1994, President Bill Clinton issued the Presidential Decision Directive (PDD) on Foreign Access to Remote Sensing Space Capabilities (PDD-23), which authorized commercial high-resolution satellite photography with "shutter control," as the Soviets had done two years previously. The 1995 U.S. decision to declassify photographs and details of the Corona program, which had continued until 1972, was another step on the path toward more openness (McDonald, 1997b). The world began moving decisively toward a new era—one in which openness is gaining dramatically on secrecy.

The Present

U.S. and Russian operation of high-resolution commercial satellites has changed the balance but not resolved the tensions between openness and secrecy.[28] The question of when military secrecy should predominate is still on the table, as demonstrated by the issue of shutter control (see Chapter 24 in this volume) and the question of image resolution.[29] Also, concerns about the protection of commercial and industrial secrets will certainly increase with the launch of the new satellites. Further, issues of journalistic freedom and the First Amendment will inevitably arise. Another concern is how rogue states and nonstate actors, such as terrorists and drug cartels, might exploit the information made available from commercial satellites. The arguments supporting secrecy, however, are not as compelling as they were during the early years of the Cold War, when the opposition of the Soviet Union and the possibility of antisatellite weapons threatened space reconnaissance.

The United States and the Soviet Union originally developed secret means for satellite observation to meet national security concerns. In the context of the Cold War, secrecy was a critical component of security. For security officials, however, open skies became a reality after 1963 because each side reaped benefits from its "secret" information gathering, and each side could estimate what the other could be learning from satellite observations.

While the use of high-resolution commercial satellites is raising many complex issues, an overriding concern for military secrecy need not dominate. It may even be possible to revive Eisenhower's concept that open observations of the earth can contribute to world security by facilitating stable military deterrents through either tacit or explicit arms controls and reductions. Transparency of arms information is also being enhanced by other information technologies, such as phone, fax, camcorders, and the Internet, particularly when relayed by satellites.[30] Ironically, in the early days of the Cold War, the limitations of technology made true openness impossible, whatever the dictates of policy. Now, however, technology is the driving force behind openness, and it is policy—and perhaps imagination—that lags behind. As military secrecy becomes less and less feasible, a new international security paradigm becomes possible, and the world can address issues of information and markets increasingly free from excessive demands for military secrecy.

Acknowledgments

The authors would like to thank Cargill Hall for his guidance on historical issues in this essay and numerous helpful suggestions, especially on the policies of the Eisenhower administration, and Merton Davies for his insightful comments. We also thank Kathleen Sommers Luchs for editorial assistance.

Notes

1. During World War II, the unanticipated German offensive known as the Battle of the Bulge reinforced fears of surprise attacks among Americans and their commander, Gen Dwight Eisenhower.
2. Under the UN Charter and international law, overflights could be justified in certain situations of self-defense or in support of UN peacekeeping operations, and it was presumably this reasoning that led President Truman to authorize flights over the Soviet Union during the Korean conflict. This issue is discussed in Hall (1996), p. 113.
3. A recent historical review of U.S. overflight activities leading up to the Corona satellite program can be found in Peebles (1997). Paul Worthman's comments about balloon reconnaissance appear in Rostow (1982), pp. 189–194. RAND began studying satellite and balloon reconnaissance in 1946, see Davies and Harris (1988).
4. For an account of this and subsequent reports, see Davies and Harris (1988), pp. 8ff.
5. As reprinted in U.S. Strategic Bombing Survey (1987). This report is discussed in Hall (1996), p. 109.
6. Peebles (1997), pp. 16–17, emphasized the effects of the surprise attack on Korea: On June 25, 1950 the North Korean army launched a surprise attack on South Korea. Then, on November 26, Communist Chinese forces unexpectedly intervened in the war. Their onslaught drove U.S. ground troops south in a headlong retreat. Many in the U.S. government and military believed the Korean War was the prelude to a Soviet invasion of Western Europe and an air attack against the U.S. main-

7 land. This was considered no more improbable than the Japanese attack on Pearl Harbor had been.

7 As noted above, overflights during the Korean War could be justified under international law, although, as Hall observes, "Whether recognized as lawful or not, however, if the aircraft were shot down, it promised an international incident of the first magnitude." (Hall, 1996, p. 113.)

8 Schriever had been impressed by Leghorn's earlier studies to modify the English Canberra bomber, later to become the RB-57 combat reconnaissance plane, for prehostility reconnaissance at altitudes above 63,000 feet and penetration over 700 miles. This penetration was estimated to cover over 80 percent of Soviet and allied intelligence targets.

9 For a brief account of the Beacon Hill Report, see Hall (1998), Ch. 4.

10 For example, David Griggs of the California Institute of Technology, then serving as chief scientist for the Air Force, suggested a polonium-powered reconnaissance aircraft. Although this was interesting because of the aircraft's small size and long range, the idea was rejected after Atomic Energy Commission studies concluded that it would take approximately one third the capacity of the Savannah Nuclear Facility (then under construction) to build the weapon-grade nuclear fuel and also because, if one crashed, its radioactive debris would create an area uninhabitable for generations.

11 For example, the initial photographic satellite mission in 1960 covered as much area as all the U-2 flights combined.

12 Leghorn had returned to civilian life in January 1953 and did not know that President Eisenhower had approved development of the highly secret U-2 in November 1954.

13 For an account of the conference, see Rostow (1982).

14 The origins of the Open Skies concept are discussed in Oder, Fitzpatrick, and Worthman (1988), pp. 5–10; McDonald (1997a), p. 150; Hall (1997), pp. 25–58; and Peebles (1997). These histories ascribe the origins of the concept to Leghorn.

15 For a discussion of Eisenhower's larger purpose in proposing mutual aerial overflights, see Appleby (1987), pp. 132–133.

16 Sergei Khrushchev, the son of the Soviet leader Nikita Khrushchev, confirmed these Soviet fears in an interview with R. Cargill Hall and Richard S. Leghorn (Khrushchev, 1995).

17 IKONOS, the first of several planned commercial high-resolution satellites, was successfully launched on September 24, 1999.

18 The funding problems of the Air Force reconnaissance satellite program are also discussed in Davies and Harris (1988), pp. 73–74, 94–95.

19 For an account of the conference, see "Aerial and space inspection at the Surprise Attack Conference, 1958," in Davies and Murray (1971), App. D.

20 See Ambrose (1990), pp. 508–511, for an account of the disastrous handling of the U-2 incident. In considering Eisenhower's decisions, Ambrose writes on p. 497 that Eisenhower "made a series of mistakes . . . brought on by his fetish for secrecy."

21 Quoted in Rostow (1982), p. 11.

22 Hall (1999) clarified the Kennedy administration's space policy.
23 For an account of the development of the first Soviet satellite and the about-turn in Soviet propaganda after its own satellite became operational, see Gorin (1997).
24 This chapter provides historical background for today's issues relating to open, as opposed to secret, satellite observations. The reader may wish to speculate about "the road not taken;" that is, if open overhead observation had become a reality in the late 1950s or mid-1960s, what might the effects have been on the arms race and such crises as the Bay of Pigs, Berlin, Afghanistan, and others?
25 Even before launching their own reconnaissance satellites in mid-1962, the Soviets were certainly aware of U.S. capabilities. They had captured balloon systems in 1958 and U-2 photographic systems in 1960 and thus had detailed information on the cameras and film U.S. aircraft were using. The Soviets were also able to track the trajectory of the satellite and what areas it passed over. With this information, its analysts could easily assess the capabilities of the Corona camera and estimate just what information the United States gained through satellite observations.
26 Quoted in Burrows (1986), p. vii.
27 See, for example, Chapter 3 in this volume and Liverman et al. (1998). In 1986, the news media used civilian satellite imagery to observe the Chernobyl nuclear accident and to get around the Soviet government's efforts to limit information. See Florini (1998), p. 55, and Wriston (1992), pp. 13, 662.
28 IKONOS was the first of several planned commercial high-resolution satellites by U.S. firms.
29 Several observers have reported the actual resolution of military satellites in operation to be substantially better than the 1-m resolution consistent with the PDD-23 guidance on commercial satellites. For example, back in 1963, when the Atlas Agena D orbited the first of a new generation of satellites, the camera had a reported resolution of 18 in. See Peebles (1997), pp. 136–137.
30 For a discussion of the effects of these technologies on global transparency, see Chapter 20 in this volume.

References

Ambrose, Stephen E., *Eisenhower: Soldier and President*, New York: Simon & Schuster, 1990.

Appleby, Charles A., *Eisenhower and Arms Control, 1953–1961: A Balance of Risks*, Vol. I, Ph.D. dissertation, Johns Hopkins University, 1987.

Burrows, William E., *Deep Black: Space Espionage and National Security*, Random House, New York, 1986.

Davies, Merton E., and Bruce C. Murray, *The View from Space: Photographic Exploration of the Planets*, New York: Columbia University Press, 1971.

Davies, Merton E., and William R. Harris, *RAND's Role in the Evolution of Balloon and Satellite Observation Systems and Related U.S. Space Technology*, Santa Monica, Calif.: RAND, 1988.

Douglas Aircraft Company, "Preliminary Design of an Experimental World-Circling Spaceship, Santa Monica, Calif., SM-11827, May 2, 1946.

Eisenhower, Dwight D., "Statement by President Eisenhower at the Geneva Conference of Heads of Government: Aerial Inspection and Exchange of Military Blueprints, July 21, 1955," in U.S. Department of State, *Documents on Disarmament, 1945–1959*, Vol. I, Washington, D.C.: U.S. Government Printing Office, August 1960, pp. 487–488.

Eisenhower, Dwight D., *Mandate for Change 1953–1956*, Garden City, New York: Doubleday, 1963.

Eisenhower, Dwight D., letter to Richard S. Leghorn, May 23, 1960. Personal papers of Richard S. Leghorn.

Florini, Ann, "The End of Secrecy," *Foreign* Policy, No. 111, Summer 1998, p. 55.

Gabriele, Mark David, *The Treaty on Open Skies and Its Practical Applications and Implications for the United States*, Ph.D. dissertation, Santa Monica, Calif.: The RAND Graduate School, RGSD-143, 1997.

Gorin, Peter A., "ZENIT: Corona's Soviet Counterpart," in Robert A. McDonald, ed., *Corona: Between the Sun and the Earth*, Bethesda, Md.: The American Society for Photogrammetry and Remote Sensing, 1997, pp. 85–107.

Hall, R. Cargill, "Origins of U.S. Space Policy: Eisenhower, Open Skies, and Freedom of Space," in John M. Logsdon et al, eds., *Exploring the Unknown: Selected Documents in the History of the U.S. Civil Space Program*, Vol. I: *Organizing for Exploration*, Washington, D.C.: National Aeronautics and Space Administration, NASA History Office, 1995, pp. 213–229.

---, "From Concept to National Policy: Strategic Reconnaissance in the Cold War," *Prologue,* Summer 1996.

---, "Post War Strategic Reconnaissance and the Genesis of Project Corona," in Robert A. McDonald, ed., *Corona: Between the Sun and the Earth*, Bethesda, Md.: The American Society for Photogrammetry and Remote Sensing, 1997, pp. 25–58.

---, "Post-War Strategic Reconnaissance and the Genesis of Project CORONA," Ch. 4 in Dwayne A. Day, John M. Logsdon, and Brian Latell, eds., *Eye in the Sky: The Story of the Corona Spy Satellites*, Washington D.C.: Smithsonian Institution Press, 1998.

---, "Interview with Robert S. McNamara," Washington, D.C.: National Reconnaissance Office Oral History Program, March 25, 1999.

Herken, Gregg, *The Winning Weapon: The Atomic Bomb in the Cold War, 1945–1950*, New York: Knopf, 1981.

Katz, Amrom H., ed., *Selected Readings in Aerial Reconnaissance: A Reissue of a Collection of Papers from 1946 and 1948*, Santa Monica, Calif.: RAND, P-2762, 1963.

Khrushchev, Sergei, interview by R. Cargill Hall and Richard S. Leghorn, Providence, R.I.: Brown University, July 5, 1995.

Leghorn, Richard S., "Objectives for Research and Development in Military Aerial Reconnaissance," paper delivered at a symposium at the Boston University Optical Research Laboratory, Boston, December 13, 1946, in Katz (1963), pp. 39–55.

---, "Aerial Reconnaissance," paper presented to the Reconnaissance Symposium, Topeka Air Force Base, December 3, 1948, in Katz (1963), pp. 6–12.

---, "No Need to Bomb Cities to Win War," *U.S. News & World Report*, January 28, 1955, pp. 79–95.

---, "Political Action and Unauthorized Overflight of the USSR," July 26, 1956. Personal papers of Richard S. Leghorn.

---, "U.S. Political Action for Reconnaissance Satellites," January 8, 1960a, forwarded to Christian A. Herter, Secretary of State, by Charles A. Coolidge, Director, Joint Disarmament Study. Personal papers of Richard S. Leghorn.

---, "A Proposal for Achieving Open Aerospace Photography the World Around," May 12, 1960b. Personal papers of Richard S. Leghorn.

Liverman, Diana, ed., *People and Pixels: Linking Remote Sensing and Social Science*, Washington, D.C.: National Academy Press, 1998.

McDonald, Robert A., "Corona's Pioneers," in Robert A., McDonald, ed., *Corona: Between the Sun and the Earth*, Bethesda, Md.: American Society for Photogrammetry and Remote Sensing, 1997a, p. 150.

---, "The Declassification Decision: Opening the Cold War Sky to the Public," in Robert A. McDonald, ed., *Corona: Between the Sun and the Earth*, Bethesda, Md.: American Society for Photogrammetry and Remote Sensing, 1997b, pp. 169–176.

Oder, Frederic C., James C. Fitzpatrick, and Paul E. Worthman in *The Corona Story*, National Reconnaissance Office, 1988 [Declassified 1997].

Peebles, Curtis, *The Corona Project*, Annapolis, Maryland: Naval Institute Press, 1997.

Rostow, W. W., *Open Skies: Eisenhower's Proposal of July 21, 1955*, Austin: University of Texas Press, 1982.

Stares, Paul B., *The Militarization of Space*, Ithaca, N.Y.: Cornell University Press, 1985.

U.S. Strategic Bombing Survey, Summary Report (European War), September 30, 1945, as reprinted in U.S. Strategic Bombing Surveys (European War and Pacific War), October 1987.

Worthman, Paul, in W. W. Rostow, *Open Skies: Eisenhower's Proposal of July 21, 1955*, Austin: University of Texas Press, 1982, pp. 189–194.

Wriston, Walter B., *The Twilight of Sovereignty*, New York: Charles Scribner's Sons, 1992.

Zhukov, G., "Space Espionage Plans and International Law," *International Affairs*, October 1960.

Remote Sensing Policy and the Development of Commercial Remote Sensing

Ray A. Williamson

First conceived in the late 1960s and 1970s as an experimental device for studying and managing earth's resources, the Landsat system became the focus of an ambitious experiment in the 1980s to transfer government-developed technology to private ownership and operation. The experiment ultimately failed, in large part because the supportive information infrastructure was not yet in place and because policymakers ignored the lack of a sufficient market for data. In the 1990s, the Landsat program was returned to the government to aid scientific efforts to understand the global environment and to support national security needs. Scientists supported by the National Aeronautics and Space Administration (NASA) and other entities have used Landsat data to support their studies of land processes since 1972. More recently, the Landsat system has been integrated into NASA's Mission to Planet Earth (NASA, 1996) and is being used to collect as much medium-resolution data about earth's surface as possible.

Although the privatization[1] experiment failed, nearly taking the Landsat program with it, the long experience working with Landsat data demonstrated the utility of land remote sensing and ultimately led to a new, more sustainable thrust toward a marketplace of remote sensing data and information. Commercial development of land remote sensing continues independently of the Landsat program in the efforts of several companies to serve the information needs of a wide diversity of existing and potential future data customers. Although much of the policy focus has been directed toward high-resolution systems in the 1990s, the data marketplace, if it continues to expand, will also continue to include low- and moderate-resolution data in a variety of spectral bands selected to serve specific information needs. Efforts to sell data from new moderate- and low-resolution sensors are much more likely to be successful than the experiment with Landsat was because they are based on the pull of the marketplace rather than the push of federal government policy. A key component of the evolution from programs centered on supporting government needs to private-sector initiatives is the growing understanding that land remote sensing could have a significant role in the rapidly expanding global information marketplace.

Early Landsat Policy

The Landsat system, taken for granted today, had a rocky beginning. In the early 1960s, experience with the Television Infrared Observing Satellite (TIROS) series of meteorological satellites and the classified Corona reconnaissance program (Day et al., 1997; McDonald, 1995), as well as many years of practical use of aerial photography, indicated both the value and the feasibility of routine, consistent satellite observations of the earth's surface. If sufficient spatial resolution could be achieved and if the right spectral bands were chosen, data acquired under similar lighting conditions would make powerful tools for understanding earth's biophysical systems and for managing and exploiting its resources.

Despite the allure of such data for a wide variety of useful applications, policy backing was slow to come for several reasons (Congressional Office of Technology Assessment [OTA], 1982, pp. 95–96):

1. Concern over the reaction of other nations. The national security community was concerned about competition with its classified efforts and the possible international reaction if a civilian system were to collect high-resolution data.

2. Lack of a lead agency with operational responsibility for land remote sensing. In the 1960s, no agency had a clear lead responsibility for mounting a land remote sensing program.

3. Doubts over demonstration of economic benefits. Proponents were unable to demonstrate a clear economic benefit compared to other methods of gathering and using land surface data. In the mid-1960s,

when a civilian land remote sensing program was first being debated, attitudes toward new space ventures had already begun to shift from the expansionist vision of Apollo to a concern for cost-effectiveness.

4. Competition from alternative systems. Some observers promoted high-altitude aircraft survey instead, considering it more cost-effective and flexible. Further, they believed that relying on aircraft platforms rather than satellites would ease foreign concerns about satellite overflight (Katz, 1976).

5. Lack of an organized community of remote sensing data users. In contrast with the earlier case of satellite telecommunications, in which a strong commercial market for long-distance communications already existed, there was only a tiny market for remotely sensed data.

In one form or another, these issues, first raised at the early stages of the Landsat program, have been part of the continuing debate over the Landsat system ever since. Only in the early 1990s did consensus over the future of multispectral remote sensing finally emerge.

In the mid-1960s, NASA began to plan for a land remote sensing system. In 1967, the Department of the Interior (DOI) attempted to become the lead agency for a system by announcing the Earth Resources Observation Satellite (EROS) program, focusing primarily on mapping and geology. However, DOI had not obtained the concurrence of the White House and its agencies, including the Bureau of the Budget, the Office of Science and Technology, and the National Security Council (Mack, 1990, Ch. 5). DOI's attempt failed, leaving NASA, which has a clear role in space research and development thanks to the National Aeronautics and Space Act of 1958 (24 U.S.C. 2451, 102c-1), in charge of research and development for land remote sensing. NASA continued to refine its effort, called the Earth Resources Technology Satellite (ERTS) program.

ERTS-1 was launched on July 23, 1972. It carried a return-beam vidicon camera for mapping and a multispectral scanner (MSS). The latter instrument, which collected data of 80-m resolution in four spectral bands along a 185-km-wide swath, was found to be useful for geological exploration and resource monitoring and became the basis for the current thematic mapper (TM) instrument aboard Landsats 4 and 5.

Landsats 2 and 3 followed ERTS-1 in 1975 and 1978, respectively. Both carried the MSS and an updated version of the return-beam vidicon. All three satellites also carried tape recorders, enabling NASA to download data directly to U.S. receiving stations when the satellite passed within range. During the 1970s, NASA's Landsat program included not only scientific investigations into the detailed relationship between reflectance and the character and condition of earth's surface but also applications of the data for resource management. These programs developed much of the know-how, including software algorithms, that supported later uses of Landsat data in other disciplines. Landsat data were made available directly from NASA (generally for free) or through the EROS Data Center (EDC) for $200 per scene, the EDC-set cost of reproduction and distribution.

These application programs also assisted in developing the nascent market for Landsat data and in establishing a new industry devoted to converting raw digital data into useful information. This "value-added" industry became a principal source of innovation for developing methods to "tease" useful information from Landsat data.

One of the key policy thrusts of the Landsat program was to make the data widely available to nearly all potential users, regardless of political affiliation. During the Cold War, Landsat data played an important role in demonstrating the open interchange of information and ideas to the world community. Hence, during the 1970s, despite the existence of onboard tape recorders that allowed NASA to download most data to its receivers, the United States also helped to establish Landsat receiving stations in 10 other countries. This enabled NASA to collect Landsat data beyond the borders of the United States if the tape recorders failed, as they all eventually did. More important, however, these receiving stations also spread the experience with remotely sensed data, building small local markets for data. As a result of these efforts, the Landsat program began to build a cadre of data users who were experienced in analyzing and applying remotely sensed multispectral data.

In the mid-1970s, NASA began developing the more-capable TM, which would fly on Landsats 4 and 5. The TM collects 30-m-resolution data in six visible and near-infrared spectral bands along the same 185-km swath as the MSS and carries an additional thermal band of 120-m resolution. In the late 1970s, the Carter administration decided that the system was ready for operational status and decided to transfer operational control to the National Oceanic and Atmospheric Administration (NOAA) (White House, 1979). Administration officials considered that NOAA, which had a successful history of managing the geostationary and polar-orbiting environmental satellites, was much better suited to managing operational systems. Some believed that, under

NOAA's management, the user base for data would eventually mature to the point that private firms could fund, develop, and operate their own remote sensing systems for government and private markets. In their view, additional experience with 30-m-resolution data from Landsats 4 and 5 would help pave the way.

The Office of Management and Budget directed that system operating costs would be recovered by data sales, which meant that data prices would have to increase. In October 1981, prices for MSS data increased more than 300 percent, to $650 per scene, resulting in a significant drop in the number of scenes purchased (OTA, 1984, p. 60).

With the Reagan administration, the policy focus shifted to more rapid privatization of the system, and officials called for "transferring the responsibility [for Landsat] to the private sector as soon as possible" (Wright, 1981). NASA launched Landsat 4 in July 1982, ushering in the use of data from the TM instrument. In 1983, NOAA took over full responsibility for Landsat operations. To cover its costs and prepare customers for commercial prices, NOAA raised the price of data a second time, leading to much lower data sales (Draeger et al., 1997).

Receiving no strong opposition within Congress to private-sector transfer of the Landsat system, the Reagan administration pressed ahead. However, neither the small market for Landsat data sales nor three studies the Department of Commerce commissioned on the matter supported privatizing the Landsat system (House, 1983). Nevertheless, the likelihood that the Reagan administration would cancel operation of the system altogether with the demise of Landsat 5, or even sooner, caused Landsat system supporters within Congress and elsewhere to assist in crafting the best possible policy toward transfer to a private operator. In 1983 and 1984, Congress held a series of hearings on the issue, concluding that Landsat was ready for a phased transfer to private-sector development and operation (House, 1983). Among other things, supporters worried that continued uncertainty over the future of the Landsat system would hinder the development of operational uses of remotely sensed land data. Potential users were reluctant to invest in the hardware and software to process Landsat data if operation of the system would be discontinued in the near future.

Continuity of a source of similar data became an important factor in the debate over the Landsat system. Using remotely sensed data to detect changes in surface conditions requires that the data retain similar technical characteristics over time. Privatization proponents argued that involving the private sector in system operation would ensure continuity and in time would bring down the high costs of data. Scientists and educators nevertheless worried that continuing the trend toward high data prices in the near term would make it difficult or even impossible for them to continue pursuing the research and educational activities for which remotely sensed data were crucial (OTA, 1984, pp. 60–61).

On January 3, 1984, the Department of Commerce issued a request for proposals for a private firm to assume operational control of the system, market the data, and prepare to assume full responsibility for the Landsat system in a few years. Congress crafted a bill that laid out the legal terms for such a transfer; it was passed and signed into law on July 17, 1984. The Earth Observing Satellite Corporation (EOSAT), Inc., a new company formed by RCA and Hughes Aircraft Company, won the competitive bidding process in August 1984 and soon took over operation of the system.

According to the plan, NOAA would work with EOSAT to develop Landsat 6 and 7, which EOSAT would operate. EOSAT would put some of its capital at risk by partially funding both satellites, each of which would be designed to last five years. In 1985, officials expected Landsat 6 to be ready for launch in 1990 or 1991, followed five years later by Landsat 7. Yet, the terms of financing still needed to be worked out. During the late 1980s, Congress, the administration, and EOSAT made several abortive attempts to find a funding plan acceptable to all parties.

Because the Landsat system had been built by the government, the 1984 act required EOSAT to "make unenhanced data available to all potential users on a nondiscriminatory basis" (15 U.S.C. 4242, 402 b(2)). To permit a level playing field for other value-added services, the law also required the licensee to "notify the Secretary [of Commerce] of any 'value added' activities . . . and provide the Secretary with a plan for compliance with the provisions of this Act concerning nondiscriminatory access" (15 U.S.C. 4242, 402 b (9)). The two provisions inhibited EOSAT's ability to establish commercial sales policies that might favor one client over another and protected the value-added community from the possibility that EOSAT could compete with it for sales of enhanced data products. These policies also inhibited EOSAT's control over sales terms and conditions (Gabrynowicz, 1993, p. 321). Sales growth of unenhanced Landsat data was steady but slow.

Although the 1984 act supported the idea of subsidizing the program enough to ensure its commercial success, Landsat's operation was nearly terminated several times for lack of a few million dollars of operating funds (House, 1992, pp. 20–23). Further, disagreements among EOSAT, the administration, and Congress concerning continued Landsat funding delayed a decision to fund the Landsat system until the spring of 1987 (Radzanowski, 1991, pp. 6–7).

Ultimately, EOSAT, the administration, and Congress resolved the faltering transfer effort by agreeing to develop only Landsat 6 under the subsidy terms of the 1984 act. President George Bush "directed the National Space Council and the Office of Management and Budget to review options with the intention of continuing Landsat-type data collections after Landsat 6" (White House, 1989). NOAA and EOSAT planned to launch Landsat 6 in 1992. The federal government provided most of the funding for building and launching Landsat 6. Assuming that Landsat 6 successfully reached orbit and operated as designed, this plan still left the United States with the prospect of entering the late 1990s with no capability to collect Landsat data.

Much of the dispute over the future of the Landsat system involved differing views of its nature. In collecting moderate-resolution data, the Landsat system serves both public and private interests. On the one hand, it provides U.S. government agencies with data for carrying out mandated responsibilities in their pursuit of scientific research, managing federal lands, and maintaining public safety (OTA, 1994, pp. 45–47). On the other, data from the system also have direct economic value in the search for oil, gas, and minerals and for managing private lands. In part, the argument over the future of Landsat concerned which use had primacy. If the use of the public good was more important, the system should remain in government hands. If the system predominantly served private interests, or could in the near future, it was appropriate to take immediate steps to transfer the system to private hands.

Complicating the question was the fact that proponents of Landsat technology faced the same problem in the 1980s that they had experienced in the 1960s. Even though the federal government as a whole was, and remains, the largest customer for Landsat data, no single agency was willing to commit sufficient operating funds to continue system operations. System operating costs were estimated to be $25 to 40 million, of which only about $10 million could be recovered through data sales. Unlike the situation with the weather satellites, for which NOAA had a clear mandate to provide satellite data for the U.S. National Weather Service, the Department of Commerce had no internal requirement to collect remotely sensed land data. NOAA was selected because of its experience in operating satellite systems. However, because NOAA itself had no operational demand for such data, it had no internal constituency for building follow-on systems. Furthermore, the authorization and appropriations committees of Congress overseeing NOAA's operations provided relatively little support for long-term operation of Landsat. This lack of commitment to a continuously operated remote sensing system undercut what little confidence data customers had in the Landsat system. Especially, customers needing repetitive data were unwilling to develop the necessary infrastructure, train personnel, and make other investments that depended on the delivery of Landsat data.

Non-U.S. Entrants

As the United States continued to debate the future of the Landsat system, other countries, which recognized the economic and social values of acquiring land data, were building their own systems. The French space agency, le Centre National d'Études Spatiales, had begun planning for the Satellite Pour l'Observation de la Terre (SPOT) remote sensing satellite system in the late 1970s. From the first, French planners envisioned the SPOT system as a government-developed, commercially operated system, and set up SPOT Image, S.A., to operate it and to develop a marketing strategy (see Chapter 9 in this volume). This commercial stance led SPOT planners to design their system around operational rather than scientific needs, which resulted in data of higher resolution but fewer spectral bands.

As detailed in Section II of this volume, other countries also developed remote sensing systems and began to market data in the late 1980s and early 1990s, as a way of recovering their investment costs. These efforts signaled to U.S. policymakers that other countries saw remote sensing as an important component of their stance toward technology development. Many observers worried that the United States had lost an important lead in remote sensing technology.

The Land Remote Sensing Policy Act of 1992

By the early 1990s, considerable pressure had built to return the Landsat system to government operation, aided by four circumstances. First, data from Landsat and SPOT proved extremely important in planning U.S. maneuvers in the 1990–1991 Gulf War. Among other things, the data provided the basis for creating up-to-date maps of the Persian Gulf (Gordon, 1991). The maps had the distinct advantage that, in working with the Gulf coalition, the United States could share them openly with allies. Second, proponents of maintaining the U.S. stake in remote sensing worried that failing to develop Landsat 7 would leave SPOT Image in control of the international market for multispectral satellite data. Third, global change researchers began to realize how important Landsat data are for following environmental change. By 1992, the Landsat archives had accumulated 20 years of multispectral data that could be mined for land-cover change information throughout the world. The European Space Agency's successful operation of the European Remote Sensing–1 (ERS-1) radar satellite and construction of ERS-2, as well as France's development of the Helios military reconnaissance satellite, demonstrated that Europe intended to continue investing in remote sensing technology. Fourth, the attempt to commercialize the Landsat system had faltered badly, and policymakers began to feel that no private company was likely to be able to provide equivalent data on the scale federal agencies needed (Knauss, 1990).

As a result of these and other pressures to continue collecting Landsat data, the administration, with the strong support of Congress, moved in 1992 to transfer operational control of the Landsat system from NOAA and EOSAT to the Department of Defense (DoD) and NASA (White House, 1992). Under the Landsat management plan negotiated between DoD and NASA, DoD would have funded development of the spacecraft and its instruments, and NASA was to fund construction of the ground-data processing and operating systems, operate the satellite, and provide for distribution of Landsat data. The Land Remote Sensing Policy Act of 1992 (PL 102-555; 106 Stat. 4163-4180), signed into law in October, codified the management plan and authorized approximately equal funding from each agency for the operational life of Landsat 7.

The act reaffirmed congressional interest in the "continuous collection and utilization of land remote sensing data from space" in the belief that such data are of major benefit in studying and understanding human impacts on the global environment, in managing the Earth's natural resources, in carrying out national security functions, and in planning and conducting many other activities of scientific, economic, and social importance." (15 U.S.C. 5601, Sec. 2.)

Thus, continuity of data collection became of paramount importance, especially because, by 1992, Landsat 4 was capable of transmitting very few data to ground stations, and Landsat 5, though still healthy, could have failed at any time.

EOSAT expected to continue supplying Landsat data through Landsat 6. Landsat 6, which was under construction, was to carry an Enhanced Thematic Mapper (ETM) that had better radiometric calibration and included an additional "sharpening" panchromatic band of 15-m resolution, allowing the instrument to deliver data with sharpness nearly equivalent to that of SPOT data. This additional capability had been studied in the mid-1970s but dropped as a result of national security restrictions.

Initial NASA and DoD plans called for Landsat 7 to carry an ETM Plus, an improved version of the ETM that was aboard Landsat 6. Later, the two agencies also began to consider including a new sensor, the High Resolution Multispectral Stereo Imager, which would collect 5-m-resolution data of particular interest to DoD. However, because of the high costs of developing and operating this instrument, it was eventually dropped from the satellite, and DoD dropped out of the partnership.

In September 1993, Landsat 6 was launched but failed to reach orbit. This failure added to concerns that high system costs for Landsat 7 would delay satellite development and might cause appropriators to cancel the project. Hence, with White House help in resolving the funding impasse with DoD, NASA agreed to fund Landsat development, carrying only the planned 30-m-resolution ETM Plus. DoD transferred $90 million to NASA to assist in developing the satellite and sensor.

In early 1994, the question of whether NASA or some other agency would operate Landsat 7 had not been resolved. NASA planned to use Landsat data to support its research into land use and land change as part of the U.S. Global Change Research Program. However, Landsat data also supported many government operational programs and the data needs of state and local governments, the U.S. private sector, and foreign entities. In May 1994, the Clinton administration resolved the outstanding issue of procurement and operational control of the Landsat system by assigning it to NASA, NOAA, and DOI. Under this plan, NASA would procure the satellite, NOAA would manage and operate the spacecraft and ground system, and DOI would archive and distribute the data at the marginal cost of reproduction (White House, 1994). This institutional arrangement, because it divided responsibilities for the system, was vulnerable to the congressional appropriations process and to changes of leadership within the agencies and the White House.

By 1998, NOAA's role in Landsat had disappeared, and the U.S. Geological Survey (USGS) was given the entire operational role. Landsat 7 was successfully launched on April 15, 1999. Data are now available from the USGS's EDC in Sioux Falls, South Dakota. NASA and USGS are collecting about 250 scenes a day, a rate much greater than achieved with Landsats 4 and 5. At that rate, Landsat 7 will be capable of imaging most of the world's land and coastal areas at least four times a year. With a panchromatic band of 15-m resolution and improved radiometric calibration, Landsat 7 are data twice as sharp as earlier TM imagery and are better calibrated.

EDC makes Landsat 7 data available at the cost of reproduction. This amounts to $600 a scene for minimally corrected imagery data. Officials expect Landsat 7 data to contribute to better-managed federal resources, more-accurate and more-detailed analysis of environmental change, and a better understanding of earth's environmental processes. Value-added firms will be able to process Landsat 7 data to their customers' specifications, making a profit on the transaction and contributing to the national economy. The rapidly increasing archive of Landsat data will contribute to a much better understanding of broad-scale environmental change throughout the world.

Building the Infrastructure for a Data Market

Building the market for remotely sensed data has required time; patience; and considerable, mostly uncoordinated, effort from many different quarters. One of the impediments to developing a data market was the absence of a supportive information infrastructure. During the 1970s and early 1980s, data users had to rely on expensive mainframe computers to process and analyze Landsat digital data. Hence, considerable analysis was carried out using traditional visual techniques on hard-copy images.

That condition began to change in the early 1980s with the development of software tailored to process Landsat scenes that would run on workstations, rather than mainframe computers. By the late 1980s, inexpensive personal computers became powerful enough to process Landsat data efficiently. In addition, the software industry was producing more capable, more user-friendly software.

One of the biggest drivers toward usability of remotely sensed data was the advent of geographic information system (GIS) software that allows users of spatial data to create layers of information in a two-dimensional format and to manipulate them with relative ease (see Chapter 4 in this volume). For the most part, this development occurred independently of remote sensing analysis, but the latter has greatly benefited from it (OTA, 1994, pp. 53–59).

Until commercial satellite systems are delivering data on a routine basis, it will be impossible to assess the precise growth potential for remotely sensed data. Nevertheless, the creation of commercial image-processing software and the development and widespread marketing of compact disk-read only memory (CD-ROM) disks and readers for multimedia presentations have made remotely sensed data more accessible to a wider base of customers. CD-ROM disks are capable of storing massive amounts of digital data, making it possible to distribute Landsat images and other data with considerable ease. CD-ROM readers are relatively inexpensive and are now standard on personal computers. Equipment capable of creating CD-ROMs is also becoming common in the marketplace. The ease of creating CD-ROM disks has encouraged companies to tailor remote sensing digital information products for convenient, cost-effective distribution.

Development of the Internet and database software have had a marked effect on market potential by supporting the search for and transfer of large data files quickly and efficiently by electronic means. Internet access to databases and specific data content also makes it attractive for the data seller to create sample browser images for each scene, so customers can determine for themselves whether or not the scene in question will serve the need in mind. The Internet also promotes electronic queries of databases and electronic ordering. Such net-based solutions, whether through the Internet or through dedicated, private networks, have improved the ability of data suppliers to deliver data quickly and efficiently, vastly improving the timeliness of data delivery.

Another aspect of developing the market for remotely sensed data is the creation of a wide variety of data applications, data products, and the software necessary to analyze them more effectively. Firms are developing software for a wide variety of map-making and remote sensing applications.

Throughout the history of the Landsat system, the value-added industry has been the primary interface between the data producer and the ultimate user of information generated from remotely sensed data. Most have been relatively small firms serving the needs of industry; state and local governments; and, of course, federal agencies. They, too, have helped build the data market by providing innovative solutions to data users' information needs.

As noted above, EDC offers data from Landsat 7 for the cost of reproduction and delivery. This policy is governed by the Paperwork Reduction Act of

1995 (44 U.S.C. 35) and the Office of Management and Budget Circular A-130, which states that information collected for government needs should be provided to the public at no more than the cost of dissemination, without restrictions on use or redistribution. This policy is based on the rationale that the public has already paid for data collection to meet a public need. U.S. policy for Landsat 7 is in sharp contrast with policies established by the European Space Agency, Canada, and India, which sell data from government and hybrid government-private multispectral and synthetic aperture radar systems to recover part of the system costs (International Space University, 1997). Making Landsat data broadly available at relatively low cost should enhance the overall market for commercial remotely sensed data of much higher resolution. Because Landsat data cover a relatively wide 185-km swath in seven spectral bands, data customers are able to use inexpensive Landsat data to establish a firm analytical basis for their higher-resolution data needs. The open availability of other supportive satellite data through NASA's Earth Observation System Data and Information System should further enhance the commercial market for remotely sensed data and data products of all kinds (NASA, 1997).

Conclusions

It is still too early to tell which new commercial satellite remote sensing ventures will succeed. Nevertheless, the Landsat program provides an instructive case for transferring technology that was originally developed within the government for government needs to the private sector. It also illustrates the myriad forces at work developing a new market for goods and services. Government policy enabled the initial development of Landsat technology, promoted it internationally, and made possible the creation of advanced sensors for the new private systems. Ultimately, however, external factors not directly related to remote sensing have played a larger part in developing conditions that make possible a private market for remotely sensed data. The end of the Cold War, which has led to greater openness within the national security community, played a major role in removing technical barriers to building commercial systems. The largest influence, however, has been in the development and heavy marketing of information technologies for business and entertainment—GIS, personal computers, and information storage and distribution technologies. The Internet, the result of a set of government policies different from those influencing the development of space technology, has also supported these developments.

Although the success of such U.S. firms as Space Imaging, Orbimage, and EarthWatch remains to be seen, the inflow of private capital to support remote sensing ventures reflects an increasing vitality in this space sector. Especially at the end of the century, it is important to remember that these companies, as well as many other government-developed systems throughout the world, have built on the Landsat experience in producing consistent, reproducible images of the earth's surface.

The failed attempt to commercialize Landsat technology illustrates one of the great lessons of technology transfer to the marketplace: The infrastructure has to be in place before new technologies can result in successful commercial ventures. Despite the policy failures of the Landsat program, it also shows that government technology efforts can be effective in developing the basic technologies, testing them extensively, and building the knowledge base. However, a new market cannot be legislated. Private industry must find its own way into the marketplace.

Acknowledgments

This paper is based substantially on a paper originally published as "The Landsat Legacy: Remote Sensing Policy and the Development of Commercial Remote Sensing" (Williamson, 1997). This adaptation was carried out as part of a Space Policy Institute project, Dual-Purpose Space Technologies: Opportunities and Challenges for U.S. Policy Making.

Notes

1. Often also termed *commercialization*. Strictly speaking, privatization generally involves some measure of transfer of operational responsibility or even ownership of government property and systems to private hands.

References

Congressional Office of Technology Assessment, *Civilian Space Policy and Applications*, Washington, D.C.: U.S. Government Printing Office, OTA-STI-177, 1982.

---, *Remote Sensing and the Private Sector*, Washington, D.C.: U.S. Government Printing Office, OTA-TM-ISC-20, 1984.

---, *Remotely Sensed Data: Technology, Management, and Markets*, Washington, D.C.: U.S. Government Printing Office, OTA-ISS-604, 1994.

Day, Dwayne A., John M. Logsdon, and Brian Latell, eds., *Eye in the Sky: The Story of the Corona Spy Satellites*, Washington, D.C.: Smithsonian Institution Press, 1997.

Draeger, William C., Thomas M. Holm, Donald T. Lauer, and R. J. Thompson, "The

Availability of Landsat Data: Past, Present, and Future," *Photogrammetric Engineering & Remote Sensing*, July 1997.

Gabrynowicz, J., "The Promise and Problems of the Land Remote Sensing Policy Act of 1992," *Space Policy*, Vol. 9, No. 4, 1993, pp. 319–328.

Gordon, B., "Statement Before the U.S. House of Representatives, Joint Hearing of the Committee on Science, Space, and Technology and the House Permanent Select Committee on Intelligence, Scientific, Military, and Commercial Applications of the Landsat Program," Hearing Report 102-61, June 26, 1991.

House—*see* U.S. House of Representatives.

International Space University, *Toward an Integrated International Data Policy Framework for Earth Observations: A Workshop Report*, Illkirch, France: International Space University, 1997.

Katz, A. H., "A Retrospective on Earth-Resource Surveys: Arguments about Technology, Analysis, Politics, and Bureaucracy," *Photogrammetric Engineering & Remote Sensing*, No. 42, 1976, pp. 189–199.

Knauss, J., Under Secretary for Oceans and Atmosphere, Department of Commerce, "Statement before the Senate Committee on Commerce, Science, and Transportation," June 12, 1990.

Mack, P., *Viewing the Earth: The Social Construction of the Landsat Satellite System*, Cambridge, Mass.: The MIT Press, 1990.

McDonald, R. A., "CORONA: Success for Space Reconnaissance, a Look into the Cold War, and a Revolution for Intelligence," *Photogrammetric Engineering & Remote Sensing*, Vol. 41, 1995, pp. 189–199.

NASA—*see* National Aeronautics and Space Administration.

National Aeronautics and Space Administration, *Mission to Planet Earth Space Science Research Plan*, Washington, D.C., 1996.

---, *MTPE/EOS Data Products Handbook*, Vol. 1, Greenbelt, Md.: NASA Goddard Space Flight Center, 1997.

OTA—*see* Congressional Office of Technology Assessment.

Radzanowski, David P., *The Future of Land Remote Sensing System (Landsat)*, Washington, D.C.: Library of Congress, Congressional Research Service, 91-685-SPR, 1991.

U.S. House of Representatives, Committee on Science and Technology, *The Commercialization of Meteorological and Land Remote-Sensing Satellites*, Hearings, 98th Congress, Hearing Report 98-53, 1983.

---, *National Landsat Policy Act of 1992*, Report 102-539:20-23, 1992.

The White House, *Presidential Directive NSC 54*, Washington, D.C., November 16, 1979.

---, Statement on Landsat, Washington, D.C., June 1, 1989.

---, *National Space Policy Directive 5*, Washington, D.C., February 13, 1992.

---, *Presidential Decision Directive NSTC-3*, Washington, D.C., May 5, 1994.

Williamson, Ray A., "The Landsat Legacy: Remote Sensing Policy and the Development of Commercial Remote Sensing," *Photogrammetric Engineering & Remote Sensing*, Vol. 63, No. 7, July 1997, pp. 877–885.

Wright, J., Under Secretary for Oceans and Atmosphere, Department of Commerce, Statement to the Subcommittee on Space Science and Applications of the House Committee on Science and Technology, and the Subcommittee on Science, Technology, and Space of the Senate Committee on Commerce, Science, and Transportation, July 22-23, 1981.

4
From Space Imagery to Information: Commercial Remote Sensing Market Factors and Trends

Kevin O'Connell and Beth E. Lachman

Commercial remote sensing markets and how they contribute to transparency are complex subjects. Remote sensing markets involve everything from the sale of raw collected imagery to value-added imagery products and even information products derived from the use of imagery, for which the end user could discern no apparent connection with a space-based image. The contribution that remote sensing makes to transparency can extend from a single image of a missile site or a humanitarian emergency to a more sophisticated analysis—as seen through multiple images—of a military research and development program. Therefore, clarity and taxonomy are essential to understanding exactly what is going on. Most of this book is about the commercialization of space-based remote sensing imagery and of the ground technical and human resources necessary to exploit it.[1] Yet, imaging satellites are only part of a broader remote sensing market that includes data from space, airborne, terrestrial, and other sensing resources. This remote sensing market in turn is part of the broader spatial technologies market,[2] further complicating our understanding of this marketplace.

Although some observers focus on imagery sales as the product of commercial remote sensing, firms talk about the larger slate of value-added products as the largest component of remote sensing markets. Failure to capture both raw imagery sales and secondary information products results in underestimating or incompletely understanding the market. The emphasis on the newly available high-resolution commercial satellite imagery is understandable given the novelty of its availability in the public and commercial sectors; however, lower-resolution satellite data are likely to remain major market components for some time, particularly as inputs for geographic information systems (GISs). Other components of remote sensing include development and sales of ground stations, infrastructure, and software. Analysts and observers need to pay careful attention to these diverse features of the remote sensing industry as they assess market prospects for the new commercial remote sensing satellite firms. All of this is complicated by the emerging nature of remote sensing as a commercial business.

This chapter focuses on several key questions in assessing the changing nature of the commercial remote sensing marketplace, and the role that satellite imagery is likely to play. It begins with a concise discussion of the initial impetus for commercialization of remote sensing satellite systems and then focuses on identifying the external and internal factors that shape the commercial remote sensing market. The chapter highlights the complex connection between remote sensing and the broader spatial technology marketplace by analyzing the role that GISs play as an enabling technology and a potential example of spatial market development. The chapter also reviews the changing perceptions and trends associated with the market for satellite and aerial imagery data products, and concludes with some observations on the frequently overlooked implications of the commercial remote sensing market for global transparency—a larger focus of this book.

The Comparative Value of Space Imaging

Since much of this chapter focuses on the competition between space-based sources of remote sensing and aerial sources of remote sensing, it is important to look first at the relative advantages and disadvantages that space imaging offers compared with manned aircraft. Certainly, if space-based remote sensing is going to compete effectively in a broader marketplace, it must take advantage of the factors that make it unique in that marketplace. Tables 4.1 and 4.2 outline the relative advantages and disadvantages of space-based remote sensing compared to manned aircraft.

Although other sources of spatial data exist, commercial observation satellites and manned aircraft often directly compete as providers of remote sensing data, information products, and services. However, as Tables 4.1 and 4.2

Table 4.1—Observation Capabilities of Manned Aircraft

Advantages	Disadvantages
Flexibility in data collection (fly when and where desired)	Limited-area coverage (need for multiple flight lines)
Flexible revisit	Global coverage is impractical
Very high resolution achievable	More difficult change detection (with variable sun angle)
Linear coverage possible (can fly along pipelines and highways)	Difficult to access all areas (especially remote areas)
Faster hardware acquisition	Airspace restrictions (especially denied areas)
Lower instrument cost	Varying resolution (with altitude)
Fly advanced instruments earlier (testbed for space instruments)	Data delivery delays (especially if film-based system)
Ease of servicing and repair	Pilot requirements (limited flight time)

Source: Adapted from David L. Glackin and Gerard R. Peltzer, 1999, p. xviii.

Table 4.2—Observation Capabilities of Imaging Satellites

Advantages	Disadvantages
Global access (with polar orbit)	Relatively fixed flight path
Global coverage in 12 hours (with wide-swath day/night instrument)	Limited revisit frequency
Large-area (synoptic or simultaneous) coverage	Fixed orbit parameters (inflexible in time and space)
Ease of change detection (with sun-synchronous orbit)	Very high resolution infeasible (higher costs and limits on optics, etc.)
Imagery usually digital (i.e., more immediate utility)	Repair not possible; launch failure a possibility
Potentially short delivery time to user (with appropriate communications system)	Development time for instruments is usually lengthy (must be space qualified)
Fixed resolution (with fixed orbit altitude)	Higher instrument cost
No overflight restrictions	Longer hardware development time
Unmanned system avoids the risk of the loss of human life	Communications network required

Source: Adapted from David L. Glackin and Gerard R. Peltzer, 1999, p. xviii.

illustrate, each type of imaging technology possesses unique features for collecting overhead imagery that are complementary and could help the remote sensing industry expand its presence within the larger spatial technologies marketplace.

Initial Impetus for Commercialization

Why commercialize space-based remote sensing, an activity traditionally held under the exclusive purview of governments? Given that the market has not evolved in the direction anticipated in the early 1990s, it is important to revisit the initial rationale underlying U.S. decisions and other choices made at that time.

The market for commercial remote sensing has evolved in directions not fully appreciated during the early 1990s when the U.S. government made several important decisions on the commercialization of remote sensing satellite systems. To understand this evolution more thoroughly, it is useful to review the rationale underlying earlier U.S. policy decisions encouraging commercialization.

While many observers point to 1994 as the critical decision point for the commercialization of imaging satellites, this is only one important date in the trend toward privatization and commercialization, not only in the United States —as evidenced in the 1984 Land Remote Sensing Commercialization Act and the 1992 Land Remote Sensing Policy Act (see Chapter 3 in this volume)— but in France, Russia, and elsewhere.

Many factors contributed to the U.S. decision to allow the commercialization of satellite remote sensing data and technology.[3] Government concerns about the health of the U.S. aerospace industry emerged during a period of decreased funding for civil and military space programs in the early 1990s. Ironically, Cold War investments in space had lowered costs and improved reliability to the point that other countries had begun to expand their use of space technologies, thereby creating the conditions for a viable market for space products and services.

The significant support that space-based capabilities provided for the coalition victory against Iraq in the Gulf War also stimulated the interests of foreign governments in remote sensing technologies. Decreasing costs for space technologies and increasing launch availability encouraged foreign governments to develop imaging satellite systems for their own national purposes, whether resource management, international prestige, or security.

Finally, the increasing sales of satellite remote sensing data from such systems as the Satellite Pour l'Observation de la Terre (SPOT) and the emergence of Russia as a data provider prompted the belief that the U.S. industry could extend remote sensing into the emerging set of commercial information tools and could take advantage of U.S. technological strength and market depth for both public and private benefit.

Presidential Decision Directive 23 (PDD-23), the enabling document, outlined a vision and the goals for U.S. commercial remote sensing from space (see Chapter 7 in this volume). Certain aspects of the PDD are well known, such as the objective of balancing U.S. commercial and national security interests and the tiered structure that it created for the protection of U.S. satellite technology by requiring the case-by-case review of proposed exports of turnkey systems and sensitive technologies compared with the relatively few restrictions imposed on the sale of remote sensing data.

PDD-23 was partly based on the expectation that private remote sensing satellite systems were likely to play a growing role in a rapidly expanding international market for remote sensing. Given the underlying concerns discussed above, and rather than concede the market to foreign enterprises, U.S. policymakers decided to encourage the U.S. aerospace industry to seek a dominant share of the projected worldwide market for the sale of remote sensing systems and data products. In essence, PDD-23 attempted to encourage the broad availability of satellite imagery data and information products within an environment conducive to the free flow of information yet to retain government control during periods of crisis or other foreign policy and security concerns. The result is that the burden of proof for the need to restrict access to U.S. commercial satellite imagery data and information, therefore, is now placed on the government.

Commercial Remote Sensing Market Factors

Growth in the market for commercial satellite imagery will depend, like many other goods, on both external and internal market factors that affect supply and demand. Examining these factors is important to gain a better appreciation of the opportunities and challenges that exist for the new commercial remote sensing enterprises attempting to secure a niche in a highly competitive marketplace.

EXTERNAL FACTORS

Government Regulations. Because of their dual-use nature, commercial remote sensing satellites present governments with a difficult set of trade-offs and choices. The unprecedented availability of high-resolution commercial satellite imagery data demonstrates that many governments favor unobstructed sale of high-resolution imagery, in part to encourage the growth of indigenous high-technology industries and in part to recoup some of the enormous costs of developing and operating national remote sensing satellites. On the other hand, governments also fear that unfettered access to high-resolution satellite imagery may undermine either their security or that of their allies (see Chapter 20 in this volume).

To balance these competing interests, governments in different countries have chosen one of two paths. Some spacefaring nations, including France, India, and Russia, have opted to maintain government control over the operation of satellites and the release of imagery. In contrast, nations such as Canada, Israel, and the United States have allowed commercial ownership and operation of satellites with the stipulation that, when necessary, state authorities may temporarily suspend collection and dissemination of satellite imagery.[4]

Given the security and military applications of high-resolution imagery, there is little argument against some form of government oversight. Given the potential security and military applications of high-resolution imagery, there is strong interest in the U.S. government for closely scrutinizing decisions on commercial remote sensing satellites. Satellite operators fear, however, that arbitrary or excessive government interference may undermine their economic competitiveness in a market already saturated with alternative sources of data and information. With proposals for commercialization of even higher-resolution imagery, radar, and hyperspectral data, U.S. industry is pushing the policy, market, and technology envelopes, albeit with uncertain results. While government regulations are necessary, they need to be unambiguous and, more importantly, to be used responsibly, to limit potential harm to the commercial satellite industry. Ironically, while the loudest complaints center on the U.S. "shutter control" regulations, these are the most clearly and publicly elaborated set of rules on remote sensing controls in the world. Most countries have not yet detailed their specific controls on remote sensing systems, at least not publicly.

U.S. Government. The implementation of policy and regulation by U.S. agencies during the past five years has been complex and not always predictable (see Chapter 7 in this volume). To be fair, while PDD-23 defined a vision and U.S. national objectives for remote sensing commercialization, it did not offer an implementation plan. Policy is being crafted on an ad hoc basis, rather than systematically and according to a U.S. strategy. As with many other areas of technology policy, government policy development is moving more slowly than the technology. While "traditional" remote sensing policy issues continue to challenge government agencies, cutting-edge policy issues are surfacing in such areas as foreign market access, competing international data policies, and intellectual property rights.

Data Access Restrictions. National leaders and defense planners around the world are concerned about the dual-use nature of satellite imagery. Fearful that high-resolution imagery may expose state secrets, which could undermine national security or simply embarrass national authorities, governments, especially those involved in civil conflicts or with large numbers of discontented citizens, may draft legislation to restrict access to satellite remote sensing data within the territories under their control. Governments continue to have numerous policy and regulatory instruments at their disposal to control the dissemination of satellite remote sensing information, or at least believe they do, so the likelihood of data access restrictions in this area is high. Such policy development will, by definition, stifle the emergence of a commercial market.

MARKET FACTORS

Broader Spatial Marketplace. One of the most significant impacts for the remote sensing industry is its place in the broader spatial technology marketplace and how the trends in this broader marketplace influence the prospects for remote sensing. Other spatial technologies, such as global positioning systems (GPS) and GIS, are entering the mainstream of our society, as current trends in the evolution of the technologies and expansion of the user base demonstrate. Such technologies also have had significant market growth.[5] Three current trends are helping to shape the future for spatial technologies and the spatial marketplace. These trends are also starting to shape remote sensing markets. First, GIS and spatial technologies are merging with one another and into the broader information technology (IT) infrastructure. This is helping to open the doors for additional remote sensing applications, such as remote sensing data being used with GIS and 3-D imagery for tourist applications through the World Wide Web. Such technologies, including remote sensing, are also being integrated into new and diverse spatial applications, which also will help increase demand for remote sensing products. There also is an increase in the development of national, state, and local government spatial data infrastructures, as well as cross organizational data sharing,[6] such as the development of framework data (see "Texas Strategic Mapping Program" sidebar). The resulting knowledge, sharing, and use of spatial technologies will also likely increase the interest and demand for remote sensing products. All these trends will continue. However, it is unclear how much the commercial remote sensing satellite part of this market will benefit

> **Texas Strategic Mapping Program (StratMap)**
>
> Many national, state, and local governments are developing framework data so their agencies and other organizations can benefit from the shared creation, maintenance, and use of spatial data.[7]
>
> Framework data are a widely available source of basic geographic data. The framework consists of the basic spatial data that government agencies and other participating organizations most need, create, and use. Such framework efforts, which often include extensive remote sensing data, will help increase user familiarity, technology developments, and market developments for remote sensing data.
>
> To see how significant such developments are, consider the framework effort in Texas. The state's $40-million framework development effort, the Texas Strategic Mapping Program (StratMap), is a cost-sharing program to develop digital geographic data layers in partnership with public- and private-sector entities. This framework has seven data layers: digital orthoimages, digital elevation models, contours (hypsography), soil surveys, water features (hydrology), transportation, and political boundaries. StratMap includes more than 17,000 1-m–resolution digital orthoimages that cover the entire state, created from aerial photography data rather than satellite data. All these data are available free to anyone and have been used extensively by commercial entities and the public (see Texas Geographic Information Council, 1999).
>
> The public availability of such data in Texas will likely increase user knowledge about, use of, and demand for remote sensing data. Similarly, international framework and data-sharing efforts will increase user knowledge and demand for remote sensing data.

from such trends, given market factors such as the widespread availability of low cost substitutes for space-based remote sensing data.

Competition. Beyond the need to introduce a new set of information-age products, commercial remote sensing satellite providers confront intense competition from many directions, especially from the well-established providers of aerial imagery. Market pressures in the already competitive spatial market are already intense, and the general information market, in which satellite imagery providers must compete if they are to be successful, is even more intense. They must also still compete with the considerable government investment in space observation, which tends to skew the market dynamics. Some dimensions of these competitions are discussed below.

Price. The current price of high-resolution satellite images, $1,000 to over $4,000, is well beyond the means of most data buyers, especially if they have less expensive alternatives (Dehqanzada and Florini, 2000, p. 19). Some optimism remains that prices will decrease as the number of operational satellites increases. But until substantial reductions in imagery prices occur, the market for space-based data will likely remain small and underdeveloped. One way that firms are trying to mitigate this problem is to market derivative information products and services that do not involve the specific sale of an image.

Availability of Substitutes. While some space observation data are novelties, remote sensing overall—especially airborne photography—is not. For years, data users have relied on alternative sources of data to carry out their day-to-day activities. Aerial imagery and ground-based data collection can, in many cases, be a very attractive substitute for satellite imagery. Airborne imagery collectors are taking advantage of improving technologies, including digital cameras and sophisticated navigational systems, to improve their operations and reduce costs. Airborne firms have moved toward digitization, consolidating data sets, and developing large-scale databases of their image data, for example. In some cases, aerial imagery providers can offer products unavailable from commercial remote sensing satellite firms, such as submeter imagery with resolutions as high as 1 foot or better. Similarly, ground-based data collectors can rely on vehicles and people on foot, usually equipped with advanced data-recording and position-location technologies, to collect affordable data. Thus, despite the claims of some satellite operators, it is rather unlikely that satellite imagery will easily replace existing sources of imagery and data, especially because providers of the latter can often collect more-detailed information at reduced costs for many jobs.

Timeliness. It usually takes anywhere from two days to well over four weeks before customers receive imagery from vendors. This time lapse clearly is a major problem for users with rigid time constraints. Data providers are well aware of this and are trying to meet the challenge. Most of the new commercial satellites have a revisit cycle (the time it takes to pass twice over any site on earth) of one to five days. However, as present companies deploy more and more satellites and new companies enter the market, the time it takes to image any location on earth is likely to decrease considerably. Moreover, satellite operators claim that once the imagery is collected, it can be downloaded, processed, and distributed to customers within one to four hours. The implication is that, in the near future, a customer could receive imagery as little as 24 to 48 hours, on average, after placing an order (Dehqanzada and Florini, 2000, p. 20), once the satellite has collected the desired image. Of course, cloud cover will continue to be the major factor in limiting opportunities for collecting timely satellite images, particularly for such regions as northern Europe and northeast Asia where cloud cover can persist for protracted periods. However, the availability of commercial imaging radar satellites could greatly improve the chances of acquiring timely imagery of desired locations regardless of local weather conditions.

Marketing to New and Existing Users. An important impediment to the growth of satellite imagery data and information products within the larger remote sensing market is the challenge of convincing existing users of remote sensing data that commercial satellite imagery data offer better value than existing data sources. These sophisticated users need to be convinced that information products derived from satellite data are readily available, are affordable, and can be specifically applied to their particular information needs. If convinced that commercial imaging satellites offer a cost-effective solution to meeting their data needs, these experts will become the early adopters of commercial satellite imagery data and will lead the way in promoting innovative applications to new market segments. They are likely to be found in those regions where the civil and commercial applications for using satellite and aerial remote sensing data and information are the most mature, such as North America and Europe.

A parallel challenge is educating new government, business, and nongovernmental users about how satellite remote sensing data is relevant to their particular information needs. Part of this process requires developing proven remote sensing applications and successful marketing to potential new users. The future growth of the satellite imagery portion of the market with the larger remote sensing market depends, therefore, on the ability of satellite operators to make the value and utility of space-based data known to new consumers and to convince established users to reconsider satellite imagery as a cost-effective alternative to their current data sources.

Tensions Between Data and Technology Sales. At the nexus of the two kinds of competitions—market-based and government—lies the tension between data and technology. We have already mentioned that governments are likely to alter the market demand—and an important component of demand at that—by virtue of their own satellite acquisition programs.[8] Yet the interest of spacefaring governments to further their own interests by selling space technologies—or even turnkey satellite systems—to others with such interests also affects that demand. Aggressive sales of satellite remote sensing technologies could place downward pressure on the price of commercial remote sensing products.

Government Satellite Remote Sensing Programs and Parastatal Providers. This category includes not only traditional government satellite remote sensing programs but also firms that, while described as "commercial," are heavily subsidized by their governments. The government-industry model within which other countries will commercialize remote sensing is clearly different from the model now emerging in the United States (see Chapters 9, 10, 12, and 13 in this volume). We refer to these firms as parastatal providers because these enterprises are unlikely to behave like truly commercial firms that must respond to market trends and competition or else suffer the consequences of business failure.

While satellite remote sensing is becoming more commercial, national motives for developing remote sensing capabilities often transcend economics. Other governments are pursuing independent remote sensing capabilities, with a variety of motives, including such traditional uses of remote sensing—resource management, security, intelligence, industrial-base development—and others. Foreign governments, some frustrated with U.S. cooperation in this area, desire the political leverage historically gained from U.S. sharing of data and technology. Some simply wish to depend less on other countries for this kind of information.

At present, about a dozen government-owned or government-subsidized sources of satellite remote sensing data exist. By 2010, that number is forecast to grow to over 40 satellites. This new development has two implications: First, spending scarce national resources either on developing and operating national systems or on subsidizing unprofitable parastatal entities may proportionately reduce a state's funds to purchase commercial imagery. This in turn may inhibit the growth of the remote sensing market. Second, given that these parastatal imagery programs are generally driven by factors other than pure economics (e.g., security, resource management, research and development [R&D], national pride), they are less likely to be very sensitive to the surges and tides of the imagery market. As a result, parastatal organizations have only a small incentive to invest time and resources in further developing the commercial imagery market. Moreover, because national organizations can price their products below average cost, they may be able to drive out true commercial entities, which are more committed to developing the imagery market and more sensitive to the needs of imagery consumers other than their primary national clients.

A Complex Connection Between Remote Sensing and GIS

The discussion thus far has attempted to convey the idea that a substantial commercial market for space-based remote sensing is unlikely to develop unless the output—imagery data—expands into other information products and services. We explore this notion in greater detail by focusing on the complex connection that is developing among satellite imagery data and GIS within the broader trends for spatial technologies.[9] Assessing the connection between remote sensing satellite imagery and GIS is important for two reasons.

> **How GIS Users Use Remote Sensing Data**
>
> Satellite and digital aerial remote sensing technologies provide an important way to collect image data automatically for transfer into a GIS. GIS users apply remote sensing data in many ways, e.g., to create, verify, and update a GIS database. For example, a state or local government may use remote sensing data to create a database of roads in its jurisdiction. Figure 4.1 shows a road network overlaid on a digital ortho-image in Austin, Texas, which could be used to examine and correct the GIS database of roads for the city. Traditionally, creating such street network databases required digitizing them using existing maps. However, maps are often out of date, especially in fast-growing metropolitan areas, so imagery data provide a useful resource for updating GIS databases. For example, Loudoun County, Virginia, pays for a yearly aerial survey and uses the remote sensing data to update its street network, land-use, and other GIS databases.
>
> Many people still use GIS without directly using remote sensing data; however, the underlying databases, such as urban land-use types, may have been created or updated using remote sensing data. It is important to recognize that end users, such as decisionmakers, do not care about the data itself. They want information.
>
> Since nonimage GIS data have traditionally been easier to use than image data, many GIS users still use GIS data sets that are mostly or completely nonimage. The remote sensing data themselves are not used because they still require an extensive amount of work to use and to create smart GIS data sets. Such issues have inhibited widespread use of image data.
>
> Users still need significant technical training and knowledge to use image data within GIS applications. Industry is trying to address this by simplifying image processing for GIS applications. However, processing images more quickly and easily still presents significant technical challenges. For example, image analysts are needed to interpret and extract features from image data. Automating this process is technically very difficult.

First, GIS is a key enabling technology for expanding the use of current and future remote sensing satellite data. Second, the history of the development of the GIS marketplace, which has steadily become integrated into the broader information economy, is a likely parallel for how the remote sensing marketplace could develop under the right circumstances.

GIS AS AN ENABLING TECHNOLOGY

Other technology trends affect the potential commercial use of remote sensing, often in favorable ways. As information and computer technologies have advanced, they have helped facilitate the increasing demand for and use of remote sensing data. For example, both the scope and capabilities of the data processing technologies needed for handling remote sensing data have improved exponentially as computer hardware processing power has grown (Glackin and Peltzer, 1999, p. 55).

The development of GIS and the entire spatial technology industry has significantly influenced and will continue to influence the demand for remote sensing data. GIS is an enabling technology for current and future uses of remote sensing data; it is a platform for integrating remote sensing data with other data to create useful information for decisionmakers and other customers who really do not care about the data itself (for example, see "How GIS Users Use Remote Sensing Data" sidebar).

The technical advances that have made GIS more accessible, cheaper, and easier to use have helped the GIS marketplace grow and the diversity of users increase. Many desktop systems incorporate GIS that less-technical users can work with. In addition, the benefits of using GISs have been recognized. More and more users throughout

Figure 4.1—Digital Orthoimage with Road Network Overlay, Austin, Texas (courtesy StratMap, Texas Natural Resource Information Systems) (see color plate)

industry, governments, and nongovernmental organizations (NGOs) have found GIS and spatial technologies to be cost effective and useful. When implemented and used properly, GIS technologies save money and increase operational efficiency by reducing costs, allowing more work with fewer resources, providing better services, improving decisionmaking, etc.[10] Governments, NGOs, and industry strive for improvements in these areas. GIS technologies, along with other spatial technologies, are also becoming more integrated with the evolving IT infrastructure. Spatial technologies are likely to follow broader IT growth rates.

REMOTE SENSING MARKET DEVELOPMENT MAY PARALLEL GIS EXPERIENCE

The development of computer GIS systems began over 30 years ago. In many cases, governments and others initially developed and used GISs for environmental, land use, natural resource, and transportation applications. GISs have become a standard tool in a wide range of other applications, including mapping, national security, utilities, facility siting and management, urban planning and analysis, telecommunications, and commercial marketing. The systems are used extensively in government, industry, and academia. Many industry applications were developed because GIS is a useful tool for planning, analysis, and management activities.

Historically, most civilian government applications did not use image data in GIS applications because the data required much more technical sophistication, computing power, and data storage than did vector data. Military applications did tend to use image data because the demand was greater and more computing power and technical expertise were available to handle the image data.

By the end of the 1990s, however, based on rapid advances in computer technology and the increasing availability of user-friendly software, GIS has essentially become a desktop system for most users. As GIS technologies have gotten easier to use, less-technical people in less-technical application areas have begun using them.[11] For example, managers of chain restaurants have found that GIS tools customized for fast-food marketing are fairly easy to use and implement without specialized training (Rubinstein, 1998). Because of such advances, the GIS marketplace expanded during the late 1990s from limited and specialized niche markets into a very broad marketplace with broad diversity in the types of users, number of market sectors, and depth of sector penetration. GISs have become standard operating tools in many areas, and they are now common tools in many environmental, natural resource, utility and transportation management, planning, and analysis efforts.

However, in this development process, it is important to distinguish between traditional, nonimage GIS applications and image ones. Even in the late 1990s, most image applications (the ones that use remote sensing data) still requires a high degree of technical sophistication and computing power for processing the data, which still limits the use of remote sensing data.

Many different data and technology developments help spur this market development for GIS. One such example, the Topologically Integrated Geographic Encoding and Referencing System (TIGER) data, is briefly presented here because it represents a useful analogy for the way the remote sensing markets might develop in the future. The U.S. Census Bureau developed the TIGER database in the 1980s. TIGER data include the location of every street, river, and railroad for the entire United States at a very low cost. The extensive availability of such low-cost data helped to create more interest and demand in GIS and spur the development and use of GIS in the late 1980s and early 1990s, especially within the transportation and local and state government market segments.

POTENTIAL IMPLICATIONS FOR THE REMOTE SENSING MARKET

As GIS and other spatial technologies merge into the broader IT infrastructure, remote sensing is also becoming more integrated into this broader marketplace. Remote sensing use should increase because of the enabling GIS technologies.

GIS is a key platform for integrating remote sensing data with other data to create useful information for decisionmakers and other customers. More users are becoming comfortable with spatial applications, such as using desktop GIS systems and World Wide Web-based spatial applications. Such trends increase the numbers of users of spatial technology and of different applications in which it is being used. Similarly, the many spatial data-sharing partnerships and clearinghouse efforts and the wide availability of public GIS data that includes remote sensing data, such as the Texas digital orthoimages, increases the use and familiarity of remote sensing data. Namely, the public availability of such data in Texas will likely increase user knowledge, use, and demand for remote sensing data, just as the TIGER data helped spur demand for spatial street network databases and related GIS applications. As more people become familiar and comfortable with the technology and skilled in using spatial data, demand for such data, including the remote sensing data, will increase.

However, as indicated above, some factors still limit the ability of the remote sensing market to grow as quickly and as large as the GIS and broader spatial market. One key challenge, discussed earlier, is the difficulty of creating smart GIS data sets out of remote sensing data. In addition, some applications, such as automobile navigation systems, do not necessarily need image data. Despite such difficulties, the remote sensing market is likely to parallel that of the GIS market but at a slower rate. Although remote sensing data are currently used mostly in limited ways in niche markets, these data are likely to penetrate the market more pervasively as the technology develops and becomes more commonplace and easier to use, as happened with GIS. Remote sensing data will eventually be used by people with average computer skills and by others with desktop systems and through the World Wide Web, as GIS capabilities are beginning to be used now.

A significant amount of R&D has been invested in automated feature extraction techniques, but computers still cannot do the task very well without a large amount of human help. Similarly, interpreting and analyzing multispectral data for different application sectors still need a large amount of R&D. In agriculture applications, for example, it will take years, even decades, of R&D to figure out which spectral band produces data that can help identify which pest is on which crop. Such R&D is needed to develop specific application information for the agricultural market. Similar issues will need to be addressed for other markets to help increase the penetration of remote sensing data into the broader spatial marketplace.

Looking at Past, Present, and Future Trends in Remote Sensing Markets

THE EARLY DAYS: AN UNFULFILLED OPTIMISM

In the early 1990s, estimates of the demand for satellite remote sensing data and derivative products were highly inflated. As Figure 4.2 indicates, projections of the annual size of the satellite imagery market by 2000 ranged from a relatively conservative $2 billion to an astonishing $20 billion (Gabbard et al., 1996, p. 18; White House, 1994, p. 1). Several factors contributed to these overly optimistic forecasts of the satellite imagery market. First, the end of the Cold War removed many of the security and other barriers that had previously hampered the growth of a remote sensing market. Nations in both the West and the East were concerned about the dual-use nature of satellite data and so had severely restricted access to high-resolution imagery and the associated technologies to a few national security agencies. The collapse of the Soviet Union in 1991 and the end of the 50-year-old East-West conflict led to a loosening of these restrictions.

Second, the size of the overall remote sensing market at the time (approximately $1.7 billion worldwide in 1995) led many to believe that a robust market for space-based remote sensing data already existed. Early investors in high-resolution satellites were convinced that their systems could capture a significant portion of a growing remote sensing market. Most investors were confident that their satellites would better meet many of the needs of imagery

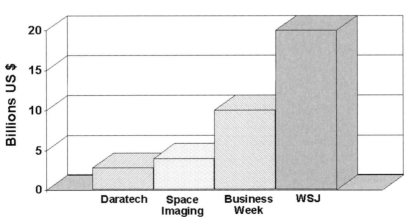

Figure 4.2—Early 1990s Market Projections for 2000 Remote Sensing Market

consumers, particularly given their expectation that planned high-resolution platforms possessed intrinsic advantages over aerial systems, such as better spatial and spectral resolutions, larger swath widths, greater timeliness, access to remote regions of the planet, and nonintrusiveness. Moreover, a great deal of optimism arose that other potential consumers of remote sensing data, whose needs could not be satisfied by the existing sources, would begin buying remote sensing data once the new generation of land-observation satellites became operational.

A third reason for inflated estimates in the early 1990s was the early success of the French company, SPOT Image, which began selling imagery in 1986 after the launch of the first SPOT satellite. Between 1986 and 1991, SPOT Image reported average annual revenue growth rates of 42 percent, which propelled the newcomer rapidly past Landsat as a data provider. Industry observers falsely believed that SPOT Image represented the first in a potential wave of imagery vendors that could reap large revenues in a rapidly growing commercial market. Unfortunately, SPOT Image's example was not representative of a new boom market; rather, it was simply a cost-effective alternative to Landsat that had come along at the right time to meet an unmet, but quite limited, demand. It would take several more years for the demand to increase enough so that more competitors could expect to operate profitably.

Fourth, technological advances in acquiring, storing, and processing remotely sensed data and in its quick and efficient electronic transfer further improved the outlook for the growth of the remote sensing market. For many years, the absence of a supportive information infrastructure prevented the broader public from using imagery. During the 1970s and 1980s, data users had to rely on large, expensive mainframe computers to process and analyze digital satellite images (Williamson, 1997, p. 883). The development of powerful desktop computers and easy-to-use GIS software in the early 1990s gave imagery consumers the ability to manipulate and customize large data files to address their specific needs. In addition, the advent of the Internet and CD-ROMs greatly enhanced the ability of imagery providers to market and distribute their products.

CURRENT TRENDS: UNCERTAIN MARKET OPPORTUNITIES

The evolution of the commercial remote sensing industry has taken a very different path from the one envisioned in the early 1990s. Plagued by nearly 20 launch delays, one launch failure, and one satellite malfunction, the imaging satellite industry has not fully lived up to early expectations (Stoney 1999,

p. 1). Of the nine imaging satellites with a 3-m–resolution or better planned for deployment by the year 2000, only one, Space Imaging's IKONOS, is operational.

As a result, estimated worldwide revenues for the commercial satellite imaging market (see Figure 4.3) will total only about $173 million in 2000, less than 1 percent of the most conservative estimates of the early 1990s. Including both aerial and satellite imaging in the commercial remote sensing market increases the estimate for 2000 substantially, to over $2.8 billion, with aerial data providers accounting for nearly 94 percent of the market.

To be fair, assessing the true size of remote sensing markets can be complicated. Note the earlier caveat about clarity of terms and taxonomy. Thus, the revenue projects presented in Figure 4.3 probably underestimate the value-added products and services generated by the satellite remote sensing market, as well as the broader use of satellite remote sensing data as an input to GIS databases.

Observers of this market should note the wonderfully diverse set of business models that have proliferated, each hoping to capture niches of economic value while minimizing business risk. The emerging companies are continuing to adapt to the market, including forging new links to complementary industries,

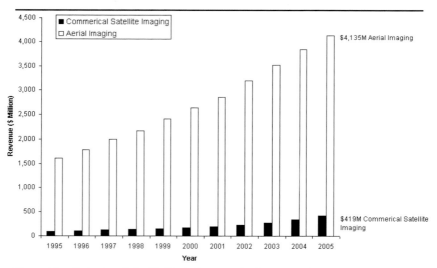

Figure 4.3—Projection of the Worldwide Revenues for Commercial Satellite Imaging (Source: Figure derived from Frost & Sullivan's estimates of worldwide revenues of about $173 million for the satellite imaging market and about $2.64 billion for the aerial imaging market. Frost & Sullivan, 1999, pp. 2-3 and 3-2.)

such as airborne remote sensing and GIS, with the latter growing dramatically and sure to be a source of demand for remote sensing data.[12] On a bright note, the optimism of the early 1990s has been replaced by a more-informed analytic sense of when space-based remote sensing can support commercial applications and when it cannot.

While devoting most of their attention to commercial applications, private remote sensing satellite companies have been helped by a stronger—but not yet complete—government commitment to purchase their services and products, especially imagery data. In the United States, for example, federal government agencies are working to understand their role as customers in this market and are explicitly considering how their data-policy decisions compete directly with private actors (see Chapter 7 in this volume). Commercial remote sensing has emerged as an important component of NASA's Earth Science Enterprise (formerly known as the Mission to Planet Earth) (Pace, Sponberg, and Macauley, 1999), as well as within the U.S. National Reconnaissance Office's plans for its Future Imagery Architecture and for the related tasking, processing, exploiting, and disseminating of commercial imagery by the National Imagery and Mapping Agency. While U.S. and other governments continue to pursue remote sensing capabilities for their own reasons, nonspacefaring nations are beginning to assess how best to take advantage of commercial remote sensing for their own purposes.

Where are we today? Remote sensing markets remain complicated, based on continued government ambivalence about the commercialization of satellite remote sensing and the complexities of competition described earlier. Because satellite remote sensing has a legacy as a tool for military and intelligence organizations and continues to be useful to these government agencies, governments continue to have a larger-than-expected role in a market that is becoming commercial by design and by attraction, as related geospatial markets strongly pull on new and diverse information sources. Satellite remote sensing, it might be said, is being commercialized in a dynamic but volatile environment: Dynamic because of the competitive nature of the business and the dramatic growth in downstream geospatial and information markets, but volatile because of government controls, government competition, and the possibility of technical or market failure.

POSSIBILITIES FOR THE FUTURE

Where is the commercial remote sensing market headed? There is little doubt that the market for space-based imagery and its associated products and services will continue to grow, although the pace and scope of that growth remain uncertain. Table 4.3 identifies some of the areas in which the market might grow and the subjects on which satellite remote sensing firms are focusing. A number of factors will ultimately determine whether or not satellite imaging will find a prominent place among the many other types of IT that have so fundamentally altered the social, economic, political, and security fabrics of the world.

Yet remote sensing commercialization remains in transition. Businesses will have to navigate important policy and regulatory issues as they try to grow a unique value-laden business. While they will continue to take advantage of the broader trends in IT, the new commercial remote sensing satellite enterprises will also have to adapt increasingly to emerging information-business models rather than hold on to the aerospace models that had previously characterized this market. Table 4.4 captures the kinds of transitions that space imagery companies are undertaking and must continue to undertake.

Commercial observation providers, then, must really evolve rapidly toward being information providers and must adapt to the rules that govern success in information markets. While some consolidation of space and aerial firms is occurring, the early notion that space providers would capture the more-established aerial market has not yet taken place and should not be expected to. Firms must continue to push for global partnerships to ensure market access and navigation of the unique rules that govern high-technology sales and data policy within different countries. The strain between the old business models—largely driven by aerospace rules and norms—and the new ones—which include rapid exchange of information between companies and nationalities—is substantial yet essential to the transition.

Table 4.3—Remote Sensing Market Segments

Mapping	News, media, and entertainment
National security	Insurance (e.g., risk and damage assessments)
Extraction industries (e.g., energy, mining)	Utilities and communications
Agriculture	Real estate and property finance
Forestry	Transportation
Natural resource management and land use planning	Travel and tourism
Environmental monitoring and other environmental applications	Outdoor recreation and sports
Disaster warning, assessment, and response	

Table 4.4—Commercial Observation Business Models Are Changing

Traditional Image Data Provider	Information Age Image Company
Produces data	Produces information
Satellite data only	Multiple data sources
Isolated	Partnerships
Limited pricing and access policies	Flexible pricing and access
Knows data only	Focused on customer needs
Only collects, processes, and disseminates data	Integrated into broader spatial technology/IT marketplace
	Vertical market expertise

Remote Sensing Markets: Concluding Observations

Whether commercial observation satellites will succeed or fail at becoming a competitive source of spatial data for governments, businesses, and even consumers is an open question at this point. Substantial advances in imaging technologies, along with powerful information technologies, such as GIS and three-dimensional visualization capabilities, have given rise to high expectations about the ability of commercial satellite imagery to become an integral component of the emerging information-based world economy. The loosening of government restrictions that once limited high-resolution satellite imagery and sophisticated image processing and display technologies to military and intelligence community users has been equally important.

COMPETING FOR A NICHE IN THE SPATIAL MARKETPLACE

Space-based remote sensing will surely occupy some niche within this spatial environment, although the competition for access will be stiff. However, aside from the persistent complexities of government regulations, space-based remote sensing will compete for "market access" with government remote sensing systems, aerial photography, terrestrial, and other sources of information. These systems were historically a tool of Cold War intelligence systems and the industry that built them. The question now is whether the satellite remote sensing industry can make the difficult transition to the demands and rules of the information economy. The competition between space and alternative information sources (e.g., airborne or ground-based data providers) extends well beyond the technical issue of spatial resolution, into the equally complex areas of service, cost, and application development.

Internationally, the competition extends into the even-more-complicated areas of competing with government systems and parastatal enterprises, nondiscriminatory market access, and foreign regulation of information. The outcome of these key competitions is far from certain. One way or the other, however, success will depend on a combination of predictable government policies that encourage the market growth of the new commercial remote sensing firms and the determination of the firms' top managers to make the difficult transition from seeing themselves as satellite imagery producers to seeing themselves as providers of spatial information products and services tailored to the specific needs of a broad range of customers.

REMOTE SENSING MARKETS—IMPLICATIONS FOR TRANSPARENCY

And what about transparency? Whatever role satellite data providers secure in the larger remote sensing markets, it is clear that space-based commercial remote sensing represents a new, and potentially very important, tool that provides independent access to the statesman, intelligence officer, military operator, NGO official, and citizen to aid their assessment of security and governance issues around the world. The role of information in statecraft and warfare, by all accounts, is growing (see Arquilla and Ronfeldt, 1999). Information technologies—including satellite remote sensing—are creating greater transparency in political and military affairs; what is uncertain is whether remote sensing, by virtue of its visual characteristics, represents a revolutionary, as opposed to an evolutionary, tool.

How much transparency? We can dissect this question a number of ways. For one, the early effects of providers of high-resolution imagery, such as Space Imaging's IKONOS satellite or Terraserver's Internet-based imagery distribution system, have been of sufficient interest to ensure that the world will always have at least one system to capture the events of the day. The decreasing costs of space capabilities, increased global interest in microsats (i.e., smaller and cheaper imaging satellites, see Chapter 20 in this volume), and the capturing effect of high-resolution satellite imagery for some problems have lead us to believe that there will always be at least some commercial capability that allows independent access from above.[13]

Beyond one or two satellites, the emergence of a true commercial market for space-based remote sensing has the strong potential to create transparency in many different ways. First, and simplest, the emergence of such a market will create the incentives for a diverse set of actors to launch satellites, thereby increasing capability, perhaps dramatically. The sheer volume of imagery will increase, likely generating an entirely new slate of downstream applications for imagery analysis, including those related to security. Moreover, while these imaging satellite systems may be diverse in terms of their specific capabilities, they will also likely be diverse in their ownership, in terms of where they are located, and therefore how they are controlled.

Although multiple high-resolution satellite systems would provide interesting conditions, a commercial market would likely create the conditions for individuals or institutions to gather and analyze a set of diverse capabilities (e.g., different data sets, different resolutions, different phenomena) that would allow even the casual (if not reasonably wealthy) user to have a substantial database on specific activities on the Earth, including those of interest to security analysts and organizations. Photographic imagery, radar, and hyperspectral data, along with a robust set of commercial processing tools, will provide a complex understanding of the earth to those who make the investment and create the processes for use of the information.

While the emergence of parastatal enterprises in this remote sensing market is expected to be strong—both for reasons of government control and government protection of industry—and has ramifications for the nature of true commercial competition, their presence will also affect the degree of transparency that is bound to emerge. As captives of their government funding, parastatals are less likely to pursue a broad range of imagery customers —including NGOs and the media—than would a true commercial entity. They are also less likely to sell or provide advanced capabilities in the market because of national policy constraints, or because better capabilities are simply less essential to their declared missions.

Unsaid in all of this, of course, is the extent to which transparency can ultimately be controlled. While a world with fewer satellites means the possibility that information can be controlled—witness the discussion in the U.S. context of "shutter control," which will likely meet a U.S. Constitutional challenge—a world with greater numbers of and more geographically diverse imaging satellites creates the likelihood that no one nation will be able to control all the data. Rather than focusing on satellites as the locus of control, countries will have to become much more accustomed to living in a transparent world.

Notes

1   In this broader context, remote sensing refers to the use of such overhead sources as aerial platforms, imaging satellites, and manned orbiting platforms,

such as the space shuttle, to collect data on the earth. More specifically, satellite remote sensing concerns the use of satellite systems to collect the imagery data. The commercialization of remote sensing also has different meanings for different countries, as discussed in the various chapters in Section II of this volume. In this chapter, the term refers to the private firms, rather than government agencies, having responsibility for acquiring and operating earth observation satellite systems, as well as selling the resulting imagery data and derived information products and services in the geospatial marketplace.

2   Spatial technologies include any technology that facilitates the collection and use of digital spatial data and information. Such technologies include GIS, Global Positioning System (GPS), Computer-Aided Design (CAD), remote sensing, and 3-D Visualization, etc.

3   For a general discussion of the market for commercial remote sensing, see Gabbard, O'Connell, and Park (1996).

4   The Canadian government enacted laws regulating its commercial remote sensing industry in 1999 that are somewhat similar to those of the United States. They will be in place by the time the first privately owned Canadian satellite, RADARSAT-2, begins operation in 2002. (See Chapter 13 in this volume).

5   For example, according to Frost & Sullivan (1999, p. 4-3), total worldwide revenues for GIS software and services market were $976 million in 1998 and growing rapidly. The National Academy of Public Administration reported in 1998 that total sales of GIS products and services in all of North America were under $1 billion in 1991; however, by 1998, the U.S. budget for GIS alone was estimated to be $4.2 to 4.5 billion.

6   For example, see the U.S. National Spatial Data Infrastructure (NSDI).

7   For an example of extensive framework efforts within U.S. state and local governments, see Sommers, 1999.

8   Our assumption here is that spending government funds on acquisition programs likely means forgone sales in the commercial data markets, which is only partially true, especially for countries with strong needs for remote sensing data.

9   The term GIS has become so commonplace that different people often use it in many different ways. However, the original specific technical definition of GIS was a computer tool for handling geographically referenced information, consisting of both spatial and descriptive data. A GIS is designed to handle large volumes of data from a variety of sources and to provide for the efficient storage, retrieval, editing, manipulation, analysis, and display of the data according to user specifications. A GIS facilitates analysis of spatially referenced data by providing a variety of operations, such as overlay, juxtaposition, generalization, and interpolation.

10 Benefit-cost ratios have been calculated in a number of different application areas and have been found to range from 1:1 to as high as 10:1 or more. (Bernhardsen, 1999, p. 23).

11 It is important to note that commercial-off-the-shelf systems for GIS analysis still require some technical knowledge. As with any data-analysis tool, there are po-

tential problems with and caveats about data quality, model development, analysis, and interpretation. Unfortunately, the visual display of GIS and image data may lead some to assume that the data are always accurate. However, significant data quality issues arise, especially for image data that require more interpretation and processing. For more information about these issues, see Minnesota Planning (1999) and Congalton and Green (1998).
12 For an excellent discussion of the economic, technological, and societal trends, and how they relate to geographic information, see NAPA (1998).
13 Obviously, not all security issues lend themselves to visibility through a high-resolution image. However, overhead imagery is often an important element in many security issues.

References

Arquilla, John, and David Ronfeldt, *The Emergence of Noopolitik: Toward an American Information Strategy*, Santa Monica, Calif.: RAND, MR-1033-OSD, 1999.

Bernhardsen, Tor, *Geographic Information Systems: An Introduction*, 2nd ed., New York: John Wiley & Sons, Inc., 1999.

Congalton, Russell G., and Kass Green, *Assessing the Accuracy of Remotely Sensed Data: Principles and Practices*, CRC Press—Lewis Publishers, October 1998.

Dehqanzada, Yahya A., and Ann M. Florini, *Secrets for Sale: How Commercial Satellite Imagery Will Change the World*, Washington, D.C.: Carnegie Endowment for International Peace, March 2000.

Divis, Dee Ann, "Access Locally, Digitize Globally," *Geospatial Solutions*, July 2000.

Frost & Sullivan, "World Remote-Sensing Data and GIS Software Markets," Mountain View, Calif., 1999.

Gabbard, C. Bryan, Kevin O'Connell, George Shin-Kyu Park, and Peter J. E. Stan, *Emerging Markets of the Information Age: A Case Study in Remote Sensing Data and Technology*, Santa Monica, Calif.: RAND, DB-176-CIRA, March 1996.

Glackin, David L., and Gerald R. Peltzer, *Civil, Commercial, and International Remote Sensing Systems and Geoprocessing*, El Segundo, Calif.: The Aerospace Press, 1999.

Hada, H. et al, "Bringing Mobile Sensors and Global Correction Services On Line," *GPS WORLD*, May 2000.

Hodges, Mark, "See Data In-depth," *Computer Graphics World*, May 2000.

Kline, Karen D., John E. Estes, and Timothy W. Foresman, "Synergy: The Importance of Relationships," University of California, Santa Barbara, Calif., 4th GSDI Conference, Cape Town, South Africa, March 13–15, 2000.

Lachman, Beth, E., "Public-Private Partnerships for Data Sharing: A Dynamic Environment," Santa Monica, Calif.: RAND, DRU-2259-NASA/OSTP, April 2000.

Lee, Young K., "Korea—Geographic Information Systems—Market Research Reports: Industry Sector Analyses," International Trade Administration, U.S. Department of Commerce, 1999.

Minnesota Planning, *Positional Accuracy Handbook*, St. Paul, Minn., October 1999.

NAPA—See National Academy of Public Administration.

National Academy of Public Administration, *Geographic Information for the 21st Century: Building A Strategy for the Nation*, Washington, D.C., January 1998.

NSDI—See U.S. National Spatial Data Infrastructure

Pace, Scott, Brant Sponberg, and Molly Macauley, *Data Policy Issues and Barriers to Using Commercial Resources for Mission to Planet Earth*, Santa Monica, Calif.: RAND, DB-247-NASA/OSTP, 1999.

Rubenstein, Ed, "Chains chart their courses of actions with geographic information systems," *Nation's Restaurant News*, February 9, 1998.

Sommers, Rebecca, "Framework Data Survey: Preliminary Report," *GeoInfo Systems*, supplement, September 1999.

Stoney, William E., "Summary of Land Imaging Satellites Planned to be Operational by 2003," See Appendix B in this volume.

Texas Geographic Information Council, "Geographic Information Framework for Texas: Resolutions for Action," Austin, Texas, January 1999.

U.S. National Spatial Data Infrastructure. Web site, 2000. Available at http://fgdc.er.usgs.gov/nsdi/nsdi.html (last accessed November 30, 2000).

Warford, Robert L., and Mary Elaine Lora, "Managing Petroleum Data," *ArcUser*, April–June, 1999.

The White House, Office of the Press Secretary, Statement, Washington, D.C., March 10, 1994, p. 1.

Williamson, Ray A., "The Landsat Legacy: Remote Sensing Policy and the Development of Commercial Remote Sensing," *Photogrammetric Engineering & Remote Sensing*, Vol. 63, No. 7, July 1997.

Wilson, J. D., "Mobile Technology Takes GIS to the Field," *GEOWorld*, June 2000.

5
Emerging Technologies: Emerging Issues for Space Remote Sensing

Bob Preston

This chapter reviews two emerging technologies in commercial remote sensing: radar imagery and imaging spectroscopy. While both have substantial histories in civil and military applications, they are recent additions to commercial space remote sensing. They also present new regulatory issues for commercial licenses. The emerging issues are a consequence of the change in nature of the products from the initial precedent of commercial imagery satellites. This chapter will briefly describe the technical attributes and applications of both technologies. In the process, it will place them roughly in historical and international context and comment on markets for their products. The chapter will then identify policy issues and their constituent interests and describe the range of strategies available to implement policy. Finally, for strategies that include a role for commercial remote sensing, the chapter will discuss licensing and operational control issues for these technologies.

Description

Before describing emerging technologies for earth remote sensing, it might help to describe existing remote sensing technology in similar terms. The description throughout this section will characterize these technologies by the nature of the sense provided at a distance. The conventional notion of remote sensing is of photography from a distance. In this respect, it is like a one-eyed sense of sight at a distance. In contrast, stereo image pairs provide the equivalent of two-eyed (binocular) vision. For panchromatic remote sensing, where the product is a map of brightness over a scene, this sense of sight is like black-and-white photography. The information content is in the shape, size, and shade of objects in the scene. For multispectral remote sensing, this sense of sight is like color photography, except that the colors may be a false-color translation of wavelength bands not visible to the eye. The added information content is the thematic information contained in the color bands selected. Most multispectral instruments intended for land remote sensing have wavelength bands selected for sensitivity to broad categories of land-use information, such as water content or vegetation. These technologies are easily understood in the familiar terms of vision. The emerging technologies for remote sensing differ fundamentally from conventional remote sensing in the kind of sense at a distance they provide. The differences raise policy and regulatory issues that are not tractable if the technologies are thought of only in familiar vision or imagery terms.

RADAR

A radar image is a map of radar brightness over some area observed by a radar. The radar brightness of an element of a scene is an expression of how well that element behaves as a radar antenna directing energy back at the radar. This will depend on size, shape, materials (particularly their conductivity and dielectric constant), and orientation of objects in the pixel and on the wavelength, polarization, and angle of illumination from the radar. Objects that act like a corner reflector (e.g., buildings, mountains) create bright spots in a radar image.

The sharpness of a radar image depends on the range and angle resolution of the radar. Its range resolution will depend on the bandwidth of spectrum that is available and is used for modulating the radar signal. Wider bandwidth modulation allows finer range resolution. Angle resolution depends on the size of the radar antenna in the same way that the instantaneous field of view of a lens depends on the size of the lens aperture. Large antennas can resolve small angles. At long microwave wavelengths, antennas large enough for fine angle resolution become physically unrealizable. The alternative to a large physical antenna is a large synthetic antenna made up of multiple smaller antennas or a single smaller antenna moved over the physical extent needed. The last approach is known as synthetic aperture radar (SAR). The limit to angle resolution for a synthetic aperture is set by the limit on coherence of the different signals to be combined over the physical transit. That limit may be determined by the coherence of the radar signal source, by the coherence of the elements in the scene over the time of illumination (motion blurs and aspect angles change), or by the duration of illumination of the scene by the radar's antenna (sharper images may require the radar to stare at or spotlight the scene rather than scan past it). While sharpness of the radar image will determine the size of features recognizable by shape or configuration, it is not the sole metric of utility for a radar image.[1]

A radar image will generally look very different from an optical image. Figure 5.1 is an example[2] from the Sandia National Laboratory Ku-band (15 GHz) SAR carried by the Sandia Twin Otter aircraft. The image is of a pipeline over the Rio Grande near Albuquerque, New Mexico, taken at a spatial resolution of 1 m.

While the differences in a radar image may make the image seem less familiar or appealing, the information contained in the radar data may be exploited to reveal aspects of the scene not visible in the image. Some characterizations of radar remote sensing miss this. An otherwise excellent primer on the military uses of civil remote sensing satellites described radar imagery

Figure 5.1—Synthetic Aperture Radar Image: Pipeline Over the Rio Grande, Near Albuquerque, New Mexico (courtesy Sandia National Laboratory)

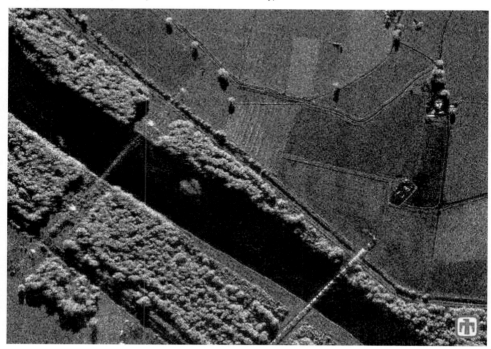

in the following terms: "Their final products have the appearance of black-and-white photographs, but they can be colorized, for example to display soil characteristics of particular interest." (OTA, 1993, App. C, p. 151.) Figure 5.2 shows a sample colorized radar image, which is described fully in the Color Plates section of this volume.

In the ability to infer moisture, conductivity, and texture information from some radar images, radar imagery resembles a long-distance sense of touch. For example, an oil slick in the ocean will appear darker than the surrounding water in a radar image, "feeling" the relative smoothness of oil over water. Beyond its texture-sensing capability, radar provides significant technical and operational advantages over optical imagery.

The technical advantages accrue from the active ranging inherent in radar measurement. Radars make range measurements with precision independent of distance, depending only on the bandwidth and signal strength available. This can provide accurate location of elements in the scene. This is akin to having a long-distance tape measure in addition to the sense of touch. SAR data also include an element of information in the phase, or relative timing,

Figure 5.2—False Color SAR Image: Barstow, California, and Vicinity, April 12, 1994 (courtesy of JPL) (see color plate)

of the signal from one element relative to others. Phase information is not ordinarily visible in a finished image (although it can leave picture artifacts, such as speckle, that are often processed out). Instead of discarding or obscuring it, the phase history data can be used to detect minute changes in a scene or to make extremely precise interferometric measurements of the terrain profile in a scene (with multiple radars or multiple passes of a single radar).[3]

For a sample of interferometric SAR data, see Figure 5.3. It is a radar interferogram of a portion of the Rutford ice stream in Antarctica, based on two ERS-1 images taken six days apart. The fringe pattern (color cycle) is essentially a map of ice flow velocity, with one fringe representing 28 mm of range change along the radar line of sight.

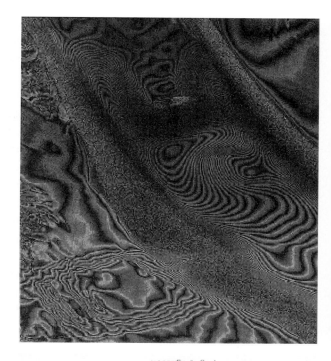

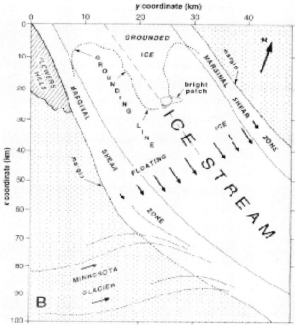

Figure 5.3—SAR Interferogram: A Portion of the Rutford Ice Stream in Antarctica (courtesy of JPL) (see color plate)

Although the portrayal of movement in the interferogram may be unfamiliar, a view of the scene consisting only of the images that were the source of the ice movement information would obscure the valuable information available from this radar sensing technique.

For a historical instance in which this kind of long-distance tape measure could have provided a profound military advantage over high-resolution panchromatic imagery, consider the World War II Allied attempt to break out of the beachhead established at Anzio, Italy, in January, 1944:

> [T]he 3rd Infantry Division spearheading the assault bogged down. Features that had been interpreted as hedgerows on aerial photos were actually drainage ditches overgrown with brush and proved impassable for armor—a very costly mistake. Meanwhile, the U.S. 1st Armored Division encountered marshy ground that gave way beneath the heavy tanks, and steep-walled ditches impossible to cross by combat vehicles. The division, therefore, could not follow and exploit sizable advances by the British 1st Division in this area. The attack lost its spearheading armor force not to the enemy but to the terrain. (Winters et al., 1998, p. 206.)

Had the Allies had the technical advantages of high-resolution (especially interferometric) imaging radar instead of aerial photography, they could have provided engineering equipment to breach the obstacles. Alternatively, they could have selected alternative routes, an alternative location for the amphibious landing in the first place, or an alternative strategy entirely.

The most glaring operational advantage of radar imagery is its freedom from external illumination. This is also the source of a cost disadvantage. The radar requires greater power from its platform to provide its own active illumination. In this respect, it is like flash photography, but with the significant additional advantage of penetrating haze and moisture, depending on the wavelength of the radar and the quantity of moisture. For example, foliage contains enough moisture to attenuate radar signals severely at the shorter wavelengths normally used for high-resolution radar imagery. The military advantages of visibility at night and in bad weather are obvious. For civil scientific inquiry, there are tropical regions where only radar can reliably penetrate weather.

To see the full breadth of civil utility of radar remote sensing of the earth, combine the possibilities of day or night, all-weather flash photography with a built-in tape measure and a long-distance sense of touch and texture. In civil applications and scientific research, radar remote sensing can be useful in many of the land-use and cartographic applications familiar from passive optical remote sensing, including the soil moisture and vegetation indications possible with multispectral optical data.[4] In addition, radar remote sensing can provide broad-area geodetic measurements precise enough for use in volcanic, seismic, and other land movement, both for research and potentially for public safety. It can measure snow pack and ice volume and water content; glacier volume, topography, and movement; floodplain inundation; and ocean currents and eddies for fisheries. It can track oil slicks, icebergs, and ships (and their wakes) at sea.

Radar was born in military use and in war, and imaging radar developed first for military reconnaissance, but the civil applications described above rest on several decades of development in the international civilian science community. The technology basis for building and using radar remote sensing is well distributed around the world. Notable centers of capability for radar remote sensing from space include Canada, Europe, Russia, Japan, and the United States. Including countries with airborne imaging radar capabilities that could be reasonably sent into space would add a few more to the list. Any U.S. policy for radar remote sensing from space should recognize a significant international dimension.

While the civil, scientific, and potential public-safety uses of radar remote sensing from space seem clear enough, commercial markets big enough for viable business cases without substantial subsidy seem limited now to national-security customers. This could change over time if conventional, electro-optical remote sensing satellites develop a big enough market for remote sensing products to support the peculiar advantages of radar sensing. For now, the only U.S. commercial licensee for space radar remote sensing based its business case on national-security uses and customers, listing explicitly surveillance of denied territory, borders, territorial waters, and smuggling in the "global security market." (Tatoian, 1999.) International efforts in space radar remote sensing for civil uses continue to be essentially government funded, with limited cost recovery from commercial data sales.

IMAGING SPECTROMETRY

This "emerging" technology for space remote sensing has a far older and more dominantly civil history than does radar. Measurement of spectra is a venerable scientific technique for determining the physical makeup of unknown or distant materials. Space spectroscopy has been used for centuries;

with roots in the Renaissance, it has been seriously used since the early 19th century. Its modern, civil use in space occurs now in planetary science and operational meteorology, both for other planets and for earth. For example, passive sounding of the atmosphere by weather satellites at microwave, millimeter wave, and infrared wavelengths for constituent profiles relies on imaging radiometers sampling the sharp edges of the constituents' spectral lines. The technology has acquired a new gloss recently with the emergence of the descriptive labels hyperspectral imaging (HSI) and ultraspectral imaging (USI) as apparent advances beyond the familiar multispectral imaging (MSI) of Landsat and others.[5]

Despite the impression that a move from multispectral to hyper- or ultraspectral sensing is a change in degree, the difference is more one of kind. Multispectral images can improve the contrast and visibility of features in a panchromatic image by color coding pixels with a hue determined by the presence of energy in broad bands of wavelengths. The bands selected are typically thematically associated with broad classes of material or activity, such as the presence of moisture or vegetation. The improvement in contrast can make objects or areas too small to resolve spatially stand out starkly against a background of a thematically different area.[6] The Landsat multispectral instrument, descriptively called the thematic mapper, produces an image or map. The information contained in the image is derived from interpreting the configuration of multiple picture elements. Where the handful of coarse bands in multispectral imagery provide something like color photography, these high-spectral-resolution technologies provide something more like a long-distance sense of smell or taste.

Although the product of hyperspectral imagery is often shown as an image or map identifying thematic information like that of a multispectral product, there is substantial difference in the underlying data sensed. A hyperspectral data set is often portrayed as a cube (see Figure 5.4) as a strictly notional representation. Here, the layer on top looks like a color picture of the scene. The spectral information content is portrayed by the rainbow of colors underneath, which are intended to indicate different spectra at each picture element. At any point in the scene, the actual data look more like Figure 5.5, a sampling of the spectrum of the materials sensed in a single picture element.

In contrast to multispectral imagery, emerging remote sensing technology for imaging spectrometry offers the possibility of sniffing out the specific

5 Emerging Technologies: Emerging Issues for Space Remote Sensing

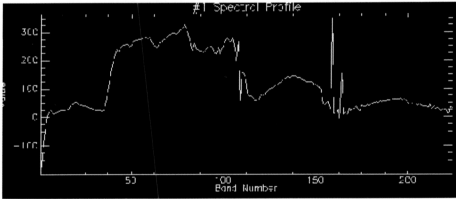

Figure 5.4, top—Notional Hyperspectral Data Cube (see color plate)
Figure 5.5, bottom—Appearance of Actual Data

material contents of individual picture elements, not just broad thematic indications. With this technology, spatial resolution has most to do with the precision of locating the material in the scene rather than with providing a sharper, less jagged edge for the outlines of a shape to be interpreted.[7] Constituent materials can be detected well below the spatial resolution of the image. Practical experience to date in detecting constituent materials present in only a fraction of a picture element has produced workable detection of the constituent with fill factors on the order of 10 to 25 percent, if received signal levels are adequate (McGregor et al., 1998; Keller et al., 1998). Attempting to understand spectral utility from the perspective of imagery interpretability scales and the well-established habits of panchromatic photointerpretation—in which the imagery interpreter attaches great importance to the sharpness and clarity of the image—is misleading. The question is not how sharp the picture is and how clear an object or activity of military interest is, but rather, what can you "smell"? This fundamental difference also has important implications for devising suitable regulatory safeguards.

This spectral "sense of smell" at a distance receives a combination of emitted and reflected energy from constituent materials in a picture element, both as the sensor receives it directly and as diffusely reflected through multiple paths (including from nearby picture elements). The energy received also includes diffuse reflections from materials in the atmospheric path between the sensor and the scene. Each diffuse reflection flavors the spectrum of reflected energy. The combined received spectra—direct, diffuse, and path—are all filtered through the spectral response of the atmosphere as well. The effect of the atmosphere's window and path scattering can be compensated for to a degree if some spectral measurements are devoted to sensing the intervening atmosphere. The mixture of direct and diffuse reflection from a picture element may help to disclose the presence of elements in a scene that might otherwise be obscured by tree canopies. In this respect, spectral sensing is quite like the sense of smell's precision in distinguishing odors and ambiguity in locating their precise source. Even the processing used to detect constituent components of a spectrum is similar to the way the sense of smell's chemical receptors are matched to particular chemical sources of odor.

The spectra of different constituent materials combined in a picture element may be distinguished from one another if the spectra are different enough from each other and if the bands selected for sensing adequately capture the differences. The definition of adequacy is not a straightforward combination of spectral bandwidth and numbers, although the number of bands must generally be on the order of or greater than the number of constituent materials. Adequacy is sensitive to the number and kinds of particular materials in a scene of interest. These naturally depend on the user's problem as much as on the scene.

To visualize the process of detecting constituent materials in the composite spectrum from a picture element, consider the spectra of four different materials shown in Figure 5.6.[8] The spectrum in the upper right corner is of desert varnish, a mineral mixture made up of Birnessite, hematite, and clays on sandstone. The others are spectra of different kinds of vegetation, each recognizably distinctive from the others. It is easy to see that a multispectral sampling of a few broad bands could distinguish the minerals from the various kinds of vegetation. However, distinguishing the different kinds of vegetation present in a spectrum sample requires high enough fidelity to recognize the more subtle differences between their broadly similar spectra. The civil and commercial value of this class of spectral imaging is precisely the ability to distinguish among broadly similar materials like these. The process of detecting a material in the composite spectrum is the mathematical correlation of the composite with the spectrum of the material of interest. When the fidelity

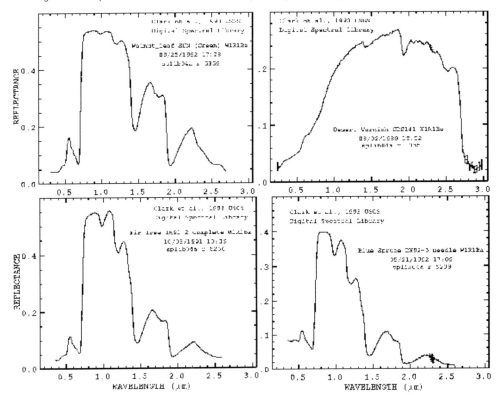

Figure 5.6—Spectra of Four Different Materials (courtesy U.S. Geological Survey)

of the spectrum sample is lowered far enough (by reducing the number or coverage of bands), correlation cannot distinguish one material from another.

The key point to be taken from this dependence on scene and problem is that regulation of spectral remote sensing on the basis of either spectral bandwidths or numbers is likely to be clumsy and ineffective (just as it was for spatial resolution in the earlier paragraph). Protection from the unpleasant consequences of undesired observation will depend more effectively on obscuration, interference with sensing, and deception based on detailed understanding of the opponent's sensing capabilities to observe and deduce.

Applications of imaging spectrometry include any problem in which the presence or absence of a material needs to be determined economically at a distance (for example, agriculture, forestry, fishing, environmental regulation and remediation, land-use management, and prospecting). These are some of the same market applications that conventional remote sensing addresses but that remain to be proven profitable for satellite-based sensing. Despite many interesting applications, markets for hyperspectral data as a commercial commodity are also necessarily undeveloped. However, spectral sensing could easily become a valuable means of product differentiation in the development of a competitive market for space remote sensing of any kind.

In the military case, the applications of spectral sensing include reconnaissance, identifying friend and foe, assessing of battle damage, etc. Spectral sensing offers competing sides in military conflict substantial opportunities for penetrating and perpetrating deception, depending on their relative skill in sensing and managing the opponent's perception. Spectral sensing does not offer a magic lens to penetrate camouflage, concealment, and deception if your opponent understands his observables and your skill. If he does, spectral sensing is just another means available to him to manage your perception.

Although spectral sensing as a precise scientific tool is old, it is quite new and immature as a general technology for large-scale use in surveillance and intelligence, both civil and military. The apparent novelty of spectral sensing seems largely a consequence of recent advances in general-purpose computing and communication technology to handle the volumes of data it can produce in large-scale use compared with narrow scientific or specialized instrumentation. Hyperspectral imaging sensors are widely available for experimentation and commercial use on aircraft platforms. The technology is readily adaptable to spacecraft platforms. The necessary processing capability is widespread. Algorithms are well-developed in open literature, and suitable software is commercially available for desktop computers. International centers of activity and capability include the United States, Canada, Australia, Europe, Japan, and China.

Policy Issues

Other chapters in this book describe the content of evolving U.S. policy for commercial remote sensing policy (Chapter 3), as well as analyze the constitutional, legislative, and policy bases that the U.S. government looks to for regulating commercial remote sensing operations (Chapter 24). In comparison, this chapter reviews the diverse range of national interests that any government must weigh in defining the goals of a national remote sensing policy. It also discusses the different strategies that governments might pursue to realize those interests with special attention to the U.S. case.

Reasonable goals of a national remote sensing policy could include:
- protecting military forces and operations
- protecting intelligence sources and methods
- providing for public good in environmental protection, public safety, and other uses of remote sensing when markets fail to capture externalities
- advancing science
- encouraging economic growth and productivity
- supporting a valued industrial base through diversification
- encouraging transparency and accountability (domestically and internationally).

The challenge in crafting policy might be seen as balancing among these goals. Ideally policy could seek congruence among them, advancing them all. To identify opportunities for balance or congruence, one should first examine the distinct perspectives of security (military and intelligence), economic, and civil interests in the policy and then consider strategies in the context of the emerging technologies.

NATIONAL SECURITY

The military interest in remote sensing policy is information advantage in military operations. Operational advantage from information generally accrues from avoiding being surprised and from surprising the opponent. Surprise allows relative concentration of forces in space and time. If one side has relatively greater force, it is correspondingly less sensitive to surprise. (U.S. military practice has generally emphasized overwhelming advantage over surprise.)

Operational advantage may also accrue from an advantage in operational tempo. Advantage in operational tempo may occur from the ability to generate and apply force more quickly than the opponent can react or from the ability to apply it without pause for night, weather, logistics, or diplomacy, which would allow the opponent to rest, resupply, and reconstitute forces. Information advantage may accrue from more timely access to better sources of information and better use of available information. Better sensors are not necessarily an advantage if the opponent understands how to counter or deceive them.

The intelligence community's interest in remote sensing policy is information advantage in intelligence collection and covert operations. In intelligence collection, the advantage accrues from protecting sources and methods of collection. In covert operation, it accrues from avoiding detection. Both of these resemble military interests in creating and avoiding surprise, differing at times in time scale and period of activity, the intelligence community generally has a longer time scale and more continuous need. The activities may also differ in reach and intensity, with the military's need for reach often shorter and the intensity of effort and counters greater in actual conflict. In time of conflict, the two communities' interests coincide. The intelligence community's resources are a major element of the military community's operational means. Beyond shared resources, the two communities also share the industrial base that supplies their needs for sensing, interpreting, and using the results of sensing.

Emerging technologies for spectral and radar remote sensing impinge on both communities' opportunities for relative advantage over adversaries in surprise and tempo. Advantage in spectral sensing will complicate operational security and deception for the opponent. Undisclosed advantage may penetrate deception. This type of advantage may accrue from knowledge of observables, skill in data interpretation, operational employment, or sensor capability. Advantage in radar sensing will restrict the opponent's flexibility and enable advantage in tempo. Undisclosed advantage in radar sensing may also help penetrate deception. This type of advantage may accrue from similar sources, except that the waveform used will necessarily disclose some sensor capability and its use will disclose operational employment to an opponent with his own signals intelligence capability.

CIVIL

The civil interests in remote sensing policy include science, public safety, environmental protection, other nonmarket or premarket public goods, cost recovery or private sector transition, civil rights (privacy, transparency or accountability of government, and press freedoms), regulation of commerce, and encouragement of economic development. Issues and measures for regulation and civil rights are discussed in Chapter 24 in this volume. The earlier sections of this chapter described applications for radar and spectral sensing that support science, public safety, and environmental interests. What may not be immediately obvious in those interests and applications is that there are no clean divisions in sensor capabilities between civil and national-security applications. There may be enough preponderance of security interest over civil utility in the thermal, emissive (as opposed to reflective) spectral bands to justify some restrictions there until greater civil utility becomes clear. Otherwise, there is no dividing line for spatial resolution, spectral resolution, spectral sampling, observation angle, time of day, polarization, timeliness of data, and geolocation accuracy that separates legitimate and valuable civil needs from national-security concerns. If national-security concerns are to be protected from these sensing technologies without undue harm to civil uses or high likelihood of a successful constitutional challenge to the means of protection, civil sensing should be regulated via operational controls on content, not via design controls on sensor capabilities.

COMMERCIAL

Commercial interests in remote sensing policy include providers of sensing from space and air platforms and suppliers of sensing equipment and processing capabilities, particularly of value-added data services and products. Providers are concerned about the effects of policy on market development (enabling or barring new products and applications), on their competitive advantage through product differentiation and efficiency, on financial community and customer confidence resulting from regulatory stability and transparency, and on subsidy and other trade barriers. These emerging technologies for remote sensing have unknown and undeveloped market potential, but they clearly represent opportunity for market and product development and product differentiation. In considering commercial interests, policymakers should also note that this industry, while small, may have a substantial multiplicative effect on the development of the general information-based economy. Finally, policymakers might have a general preference for lightly regulated markets over direct involvement (where markets can support business cases) as more rapid and efficient developers and sustainers of capability compared with government-directed activity, civil or military.

Strategies

The strategies that governments might adopt to serve the varying interests listed above can be described in a number of ways. One natural description is in terms of the means used. Although the division of means here uses the same terms as the division of interests, it is important to note that the means are not aligned with the interests with the same name. In this way, a security means strategy might employ coercion, destruction, disruption, deception, or delay. A civil means strategy might use licensing conditions, operational or "shutter" controls, export controls, or subsidy, including the extreme case of preemption by supplying a free good as barrier to market entry, as occurred with the Global Positioning System navigation and timing utility. A market means strategy would use and regulate commercial competitive activity. All three means could support security interests. Civil and commercial means could advance civil or commercial interests. Although policy discussion usually focuses primarily on civil means, an effective strategy is likely to be mixed, not relying solely on any single category of means. In considering policy for these emerging remote sensing technologies, this section examines contributing means within these three different categories as elements of a combined strategy seeking to advance all the interests.

SECURITY MEANS

The previous discussion of military interests noted that the availability of overwhelming force can mitigate advantage in information. Although overwhelming force is any commander's preference and will certainly be available in many contingencies, it is probably not fiscally or logistically realistic in all contingencies. Rather, the first element and final resort of military response to remote sensing by others ought to be the operational skill in maneuver, speed, and deception taught by the enduring principles of war. U.S. forces should teach and train for operations under observation by smart, learning opponents, with particular attention to understanding all the opponent's means of observation and managing both the opponent's perceptions and his means of observation.

Among security means available for managing an opponent's and any supporting third parties' means of observation are camouflage, concealment, decoys, deceptive activity, false-target generation, blinding (jamming), and physical attack. None of them should be omitted from consideration. The last of these, when applied to satellites, invokes considerable emotional response in policy discussion and some legal limits when the means of attack could be construed as an element of a missile defense. The remainder all depend critically on understanding the limits and abilities of the various means of observation, particularly in combination. None of the advanced remote sensing technologies discussed in this chapter is immune to any of them. However, greater multiplicity of sources and phenomenologies will complicate the task for security means. And competitive innovation (both market-driven innovation and adaptation by opponents) will make that task evolve continuously.

While listing security means, there is one that might go unnoticed as an element of a commercial means strategy: to use commercial suppliers as intelligence sources. Awareness of others' interests in particular areas at particular times with particular means and qualities of information sought is a valuable source of intelligence in its own right. Intelligence derives both from what is sought and what is not sought. For foreign purchases, routine intelligence monitoring of this kind of information seems within the scope of U.S. remote sensing law and easily within constitutional authorities of the Executive Branch. For domestic purchases, police surveillance would likely first require judicial determination of probable cause. In other countries the laws may differ, but any purchaser of a remote sensing product should expect that government awareness of the request could fall within normal police and security powers of the state.

CIVIL MEANS

Current U.S. (and most international) policy includes an element of market preemption in government subsidy of land and atmospheric remote sensing. Preemption by subsidy was predominant in the past, partly out of necessity because governments had to assist in the development of the technology and because markets large enough to support the high cost of space have been slow to develop. Preemption continues in the data niche served by the Landsat program. The United States has chosen not to preempt or even to lead after pioneering efforts in civil space radar remote sensing, although this is a technology where U.S. industry and civil researchers have considerable competitive advantage at the moment. It has been leading in the development of spectral sensing, but has not committed to preemption.

The United States has a long history of using export controls on almost all space technology, considering it to be fundamentally dual purpose.[9] Recent U.S. political activities, such as the Cox Commission report (Cox, 1999) and congressional transfer of export control authority for satellites from the Department of Commerce to the Department of State (Taylor, 1999), have treated space items more like nuclear weapon secrets than global commercial commodities. In the short term, this political demonization of space products has acted to encourage the movement offshore of advanced remote sensing capability. RADARSAT-2 is a prime example.[10] In the longer term, more thoughtful use of export controls will still have to operate within severe limitations of a globalized space industry. Unilateral action will generally be ineffective or counterproductive.

COMMERCIAL MEANS

The U.S. policy in Presidential Decision Directive 23 represents a first, partial commitment to use market means rather than preemption for remote sensing. Aside from benefit to U.S. industry, the market means approach could serve security interests to the degree that entities under U.S. jurisdiction with safe or controllable capabilities dominate the marketplace. The licensing of 1-m resolution panchromatic systems enabled the U.S. industry to enter the market competitively. However, initial U.S. licensing attempts for commercial hyperspectral and radar sensing have been ineffective in specifying safe systems (from a security perspective) and counterproductive for the ability of U.S. firms to create or dominate the market.

Radar license conditions on resolution, grazing angles, phase history data, polarization, and geolocation accuracy have effectively ruled out the only potential market apparently viable: international national-security customers (Tatoian, 1999, slide 18). Hyperspectral license conditions seem likely to remove even civil customers by including limits on spatial resolution, spectral resolution, and coregistration with higher-resolution panchromatic instruments (National Oceanic and Atmospheric Administration [NOAA], not dated).

NOAA's proposed restraints on spatial resolution miss the point for hyperspectral licenses. The restraints may provide no protection for military and intelligence activities unless they also diminish the spectral signal below the level needed to detect constituent materials within a pixel. Otherwise, location of the material of interest to the precision of the degraded pixel size is still likely to be more than useful for military exploitation. If the uniformly degraded spatial resolution and signal level prevent subpixel detection of military materials, they will do the same for civil objects, degrading the opportunity to develop a commercial market. Fundamentally, as highlighted in the earlier discussion of spectroscopy, imagery-based, spatial-resolution criteria are not appropriate for regulating this phenomenology.

The earlier section also discussed the problem with spectral resolution limits. The ability to detect a material in a scene depends on the spectral signatures of the materials in the scene. The difference between the signatures is not defined by spectral resolution. Any a priori restriction on spectral resolution coarse enough to prevent the detection of materials of military interest will also prevent the discrimination of materials of civil interest. Spectral resolution limits cannot discriminate legitimate civil use from dangerous military capabilities. This kind of license limit will also degrade commercial product and market development.

In the hyperspectral case, viable commercial markets remain to be identified and developed. A purely commercial means strategy is probably premature. However, the net effect of NOAA's current hyperspectral restrictions will hamper U.S. companies in developing markets for space remote sensing, leaving them to airborne platforms and international sources. If spectral information becomes a competitive advantage for international sources, this first attempt at hyperspectral regulation will have been counterproductive to the policy goal of encouraging development of U.S. space remote sensing industry.

If a commercial means strategy is to work, it needs to be able to encourage U.S. commercial success without endangering national-security. License conditions on product performance necessarily limit the ability of U.S. firms to compete. Unilateral performance limits aimed at reducing military utility remove a market sector without protecting security from international sources of similar capabilities. Even if military customers could be written off in a commercial strategy and if protection could be provided by other means against international sources, performance limits would still have to leave a viable civil customer base. Unfortunately, with radar and spectral sensing, there are no distinctively military performance parameters to limit. In the end, the only effective regulatory controls (to protect security interests) on commercial radar and spectral sensing must be operational controls, i.e. shutter control.

Ways Forward

For any kind of remote sensing from space, policy should use a mixed strategy of security, civil, and commercial means. The strategy should be able to advance security, civil, and commercial interests without pitting them against each other.

In any mixed strategy, security means must be available. Many of them should be available within the scope of general-purpose military forces. Some may require new capabilities to jam, dazzle, or deceive sensors. Fundamental to any of these means is a heightened awareness and attention to opposing capabilities to sense. The ease or difficulty of the security means can be influenced by effective coordination with the civil and commercial means. The purpose of that coordination is less to limit U.S. commercial capability than to influence international capability and increase understanding of, control over, and ability to cope with opponents' capabilities. To do so, coordination should enhance rather than limit U.S. commercial capability.

With the emerging technologies of radar and spectral sensing, it is early yet to gauge a preference for or reliance on civil preemptive measures or commercial market devices. Until market potentials are better developed and understood, civil means should help develop the technology and new products. With either civil or commercial systems, effective operational controls will be needed. With such controls, an effective way ahead can be found for radar and spectral sensing. Without them, current attempts at regulation are destructive for civil and commercial interests and ineffective or dangerous for security interests.

Notes

1. For an introduction to SAR imagery from space, see Elachi (1982). For extensive examples of radar imagery and applications, see the Jet Propulsion Laboratory's (JPL's) Radar Web site (http://southport.jpl.nasa.gov/) and Sandia National Laboratory's Web site (http://www.sandia.gov/radar/imagery.html). For a comprehensive treatment of radar imagery in remote sensing, see Henderson and Lewis (1998).
2. This image is also available on Sandia's Web site, at http://www.sandia.gov/radar/imageryku.html (last accessed July 10, 2000).
3. For an introduction to interferometric SAR, see Hansen and Johnson (1995) or Massonnet (1997).
4. For a recent survey of radar science and civil applications see JPL (1998). For a slightly older survey of civil operational uses with illustrative products, see Montgomery (1996).
5. These two terms of art describe increasingly finer spectral resolution and increasing numbers of spectral samples for elements sensed in a scene. MSI generally refers to spectral resolution on the order of one-fifth of wavelength and a half dozen or so bands. HSI generally refers to instruments with spectral resolution at least on the order of one-fiftieth of the wavelength and on the order of one or a few hundred spectral samples. USI generally refers to spectral resolution on the order of a thousandth or less of the wavelength and thousands of spectral samples. For an introduction to hyperspectral data sets and their uses, see (Barr, 1995).
6. For example, the cover of Preston (1994) shows a Landsat multispectral image of the 1991 Persian Gulf War that reveals patterns of military vehicle activity and infantry fighting positions smaller than the image's spatial resolution, made visible by the contrast in spectral response between soil disturbed by the human activity and background desert.
7. Spatial resolution also interacts importantly with the collection of adequate signal

strength in each picture element to detect interesting spectra against the combination of background noise in the sensor and competing spectra in the picture element. Also, coarse spatial resolution spectral data sets can be sharpened with sharper panchromatic data of the same scene. Preliminary experimentation suggests that hyperspectral imagery sharpened with panchromatic data can achieve results with spectral errors in the vicinity of 5 percent off truth (Patterson et al., 1998).

8 This image is also available on the U.S. Geological Survey Digital Spectral Library's Web site, at http://speclab.cr.usgs.gov/spectral.lib04/spectral-lib.desc+plots.html (last accessed June 15, 2000).

9 For the history of past efforts at U.S. export controls on space remote sensing, see Preston (1994), Ch. 2.

10 When faced with extraterritorial application of U.S. export controls on data transfer that would require Canadian firms to violate Canadian human-rights law, the Canadian government directed its industry not to use U.S. suppliers for RADARSAT-2 (Scoffield, 1999).

References

Barr, Samuel, "Special Applications of Spectral Sensing," *Proceedings of the 1995 International Symposium on Spectral Sensing Research*, 1995. Available at http://ltpwww.gsfc.nasa.gov/ISSSR-95/speciala.htm (last accessed June 15, 2000).

Cox, Christopher, "U.S. National Security and Military/Commercial Concerns with the People's Republic of China," U.S. House of Representatives, Select Committee on U.S. National Security and Military/Commercial Concerns with the People's Republic of China, 1999. Available at http://www.house.gov/coxreport/ (last accessed June 15, 2000).

Dixon, Timothy H., ed., "SAR Interferometry and Surface Change Detection: Report of a Workshop held in Boulder, Colorado : February 3–4, 1994," July 1995. Available at http://southport.jpl.nasa.gov/scienceapps/dixon/index.html (last accessed June 15, 2000).

Elachi, Charles, "Radar Images of the Earth from Space," *Scientific American*, Vol. 247, No. 6, December 1982, pp. 54–61.

Evans, Diane L., and Mahta Moghaddam, eds., LightSAR Science Requirements and Mission Enhancements: Report of the LightSAR Science Working Group (LSWG), Pasadena, Calif.: Jet Propulsion Laboratory, D-13945, March 1998.

Gerth, Jeff, and James Risen, "Reports Show Scientist Gave U.S. Radar Secrets to China," *New York Times*, May 10, 1999.

Goldstein, R. M., H. A. Zebker, and C. Werner, "Satellite Radar Interferometry: Two-Dimensional Phase Unwrapping," *Radio Science*, 23, 1988, pp. 713–720.

Goldstein, R. M., H. Engelhardt, B. Kamb, and R. M. Frolich, "Satellite Radar Interferometry for Monitoring Ice Sheet Motion: Application to an Antarctic Ice Stream," *Science*, No. 262, 1993, pp. 1,525–1,530.

Hansen, John V. E., and Peter B. Johnson, "Interferometric Synthetic Aperture Radar: Applications to Crisis Support," *Proceedings of the 1995 International Symposium* on *Spectral Sensing Research*, 1995. Available at http://ltpwww.gsfc.nasa.gov/ISSSR-95/interfer.htm (last accessed June 15, 2000).

Henderson, Floyd M., and Anthony J. Lewis, eds., *Manual of Remote Sensing: Principles and Applications of Imaging Radar*, New York: John Wiley & Sons, 1998.

Keller, Robert A., et al., "Sensor Design Considerations for HSI Remote Sensing" *Imaging Spectrometry*, IV, 1998, pp. 2–12.

Massonnet, Didier, "Satellite Radar Interferometry," *Scientific American*, February 1997. Available at http://www.sciam.com/0297issue/0297massonnet.html (last accessed June 15, 2000).

McGregor, R. Daniel, et al., "Target Detection Using HSI systems: An Update of TRW Results," *Imaging Spectrometry*, IV, 1998, pp. 52–65

Montgomery, Donald, ed., *Operational Use of Civil Space-based Synthetic Aperture Radar*, Jet Propulsion Laboratory, Interagency Ad Hoc Working Group on SAR, JPL Publication 96-16, August 21, 1996.

National Oceanic and Atmospheric Administration, "General Conditions for Private Remote Sensing Space System Licenses," no date. Available at http://www.licensing.noaa.gov/eolicense.htm (last accessed June 15, 2000).

Patterson, Tim J., et al., "Extension of Spatial Sharpening Techniques to Hyperspectral Data," *Algorithms for Multispectral and Hyperspectral Imagery*, IV, 1998.

Preston, Bob, *Plowshares and Power*, Washington, D.C.: National Defense University Press, 1994.

Scoffield, Heather, "Ottawa to Cut U.S. Out of Satellite Project: Manley Takes Radarsat Business to Europe," *The Globe and Mail*, August 11, 1999.

Tatoian, James, "Building Your Own Commercial Remote Sensing System: A Small Company Perspective," slides presented at 3rd National Space Forum: Space Based Remote Sensing—Seeing and Shaping the Planet, Washington, D.C.: National Academy of Sciences, June 2–3, 1999. Available at http://www.astronautical.org/conferences/spaceforum99/ (last accessed June 15, 2000).

Taylor, Chuck, "Secrets in the sky: Government faces off with global space industry," *Seattle Times*, May 2, 1999.

U.S. Congress, Office of Technology Assessment, *The Future of Remote Sensing from Space: Civilian Satellite Systems and Applications*, OTA-ISC-558, July 1993, p. 151.

U.S. Geological Survey Digital Spectral Library: Sample Descriptions and Plots, January 8, 2000. Available at http://speclab.cr.usgs.gov/spectral.lib04/spectral-lib.desc+plots.html (last accessed June 15, 2000.

Winters, Harold A., Gerald E. Galloway, Jr., William J. Reynolds, and David W. Rhyne, *Battling the Elements: Weather and Terrain in the Conduct of War*, Baltimore, Md.: The Johns Hopkins University Press, 1998.

6
Security Implications of Commercial Satellite Imagery

John C. Baker and Dana J. Johnson

Commercial satellite imagery has important security implications for the United States and other countries. Although the new observation satellites were mainly designed with commercial and civilian applications in mind, these imaging satellites offer a dual-purpose technology that is increasingly well suited for supporting various military applications. The advent of higher-resolution satellite imagery has created new opportunities for countries to enhance their national security and to resolve conflicts using satellite images to help settle territorial disputes. Yet, there is also a risk that aggressive governments and terrorists could exploit this emerging information technology for more harmful purposes. However, assessing the potential security risks that could arise from growing global access to high-resolution satellite imagery is complicated. Acquiring an overhead image is only the first step in a complex decisionmaking process that begins with accurately identifying the potential data within a given image and then translating its information value into a significant military or intelligence advantage.

This chapter analyzes the military security issues that commercial satellite imagery poses, paying special attention to the implications for the United States. It begins by focusing on the modest role that Landsat and Satellite Pour l'Observation de la Terre (SPOT) images played during the 1990–1991 Gulf War. Against this background, the chapter broadly assesses the new strengths and continuing shortcomings of commercial and civilian observation satellites for supporting military missions and other security-related uses. It pays special attention to some of the doctrinal and organizational challenges that commercial satellite imagery poses for the U.S. military and intelligence communities. The chapter then assesses the threats that could arise if aggressive states or nonstate actors, such as terrorist groups, were to attempt to exploit their unprecedented access to high-resolution satellite imagery. It distinguishes between the near-term and long-term likelihood of threats arising for U.S. military forces from foreign military adversaries enjoying access to improving commercial and civilian observation satellites. Finally, the chapter reviews U.S. options for preventing or limiting the access of potential adversaries to timely satellite imagery during a crisis or armed conflict.

Gulf War Experience with Civilian Satellite Imagery

The military utility of civilian and commercial satellite imagery was clearly demonstrated in the 1990–1991 Gulf War (see Table 6.1). The U.S. Department of Defense (DoD) spent up to $6 million to purchase Landsat and SPOT images, even though this was a relatively small portion of all the satellite imagery that U.S. forces used during the conflict. Although the images civilian observation satellites provided was of lower resolution than that available from military satellites, the civilian satellites offered broad-area coverage and multispectral imagery data that were of use to military planners (Congress, 1991, p. 28; Anson and Cummings, 1992, p. 131).

SUPPORTING MILITARY OPERATIONS

The most prominent use of civilian imagery for supporting combat operations in the Gulf War was updating military maps. Military involvement in the specific area of operations had been unanticipated. As a result, the maps the U.S. forces took with them to Saudi Arabia in 1990 were 10 to 30 years old and did not even cover the area completely (McDonnell, 1996, p. 175). This unsatisfactory situation prompted the Defense Mapping Agency to take advantage of Landsat's broad-area coverage to produce 122 image maps of Saudi Arabia, Iraq, and Kuwait (Winnefeld, Niblack, and Johnson, 1994, p. 201; Congress, 1991, p. 18). These image maps were a vital interim product for supporting the Desert Shield deployments while standard topographic line maps were being updated and produced with more-detailed features.

Table 6.1—Civilian Satellite Imagery Use in the Gulf War

Imagery Application	Type	Coalition Force User(s)
Image maps for initial Desert Shield force deployments	Landsat multispectral imagery	U.S. Defense Mapping Agency produced 122 image maps for Coalition Force users
Coastal features analysis for planning amphibious operations	Landsat and SPOT multispectral imagery	U.S. Navy and Marine Corps planners
Terrain analysis to determine constraints on ground forces maneuverability	Landsat and SPOT multispectral imagery	U.S. Army planners and higher-echelon military planners
Mission planning for air attacks	SPOT panchromatic and multispectral imagery	U.S. Air Force and, to a lesser degree, U.S. Navy and Marine Corps aviators
Environmental monitoring of Kuwaiti oil-well fires and oil spills into the Persian Gulf	Landsat multispectral, SPOT panchromatic and multispectral imagery; Advanced Very High Resolution Radiometer (AVHRR) imagery; Space Shuttle photographs	U.S. and foreign governments; environmental nongovernmental organizations

U.S. military planners took advantage of the multispectral characteristics of Landsat and SPOT imagery data to analyze terrain and coastline features. Multispectral imagery aided area delimitation analyses, such as evaluating terrain suitability for ground vehicles to identify impassable areas that Coalition Forces should avoid or that would present barriers to Iraqi forces attempting to retreat (Kutyna, 1998, p. 112). And U.S. Marine Corps planners took advantage the ability of multispectral imagery to penetrate shallow water to generate bathymetric and hydrographic information on Kuwait City's coastline to support amphibious operations (Winnefeld, Niblack, and Johnson, 1994, p. 201).

Landsat and SPOT imagery data were important mission planning tools for U.S. air operations in the Gulf War. Integrating overhead color images with digital terrain elevation data provided U.S. pilots three-dimensional views of planned missions. The result was a more realistic perception of terrain features and colors that pilots would encounter in flight, which greatly increased the chances of successful target strikes.

Finally, Landsat and SPOT imagery were also used during the Gulf War to monitor the environmental damage caused when Saddam Hussein's forces set fire to more than 500 Kuwaiti oil wells and caused large-scale oil spills that polluted the Persian Gulf waters. Initial evidence of oil spills emanating from oil storage facilities and terminals along the Kuwaiti coastline came from

AVHRR images from lower-resolution U.S. environmental monitoring satellites. Medium-resolution Landsat images were used to track the spread of the growing oil slicks.[1] The panoramic views that Landsat and AVHRR wide-area coverage provided captured world attention both by showing the broad extent of the fires and by providing dramatic evidence that the heavy smoke plumes, which rose to high altitudes, spread the effects of pollution to distant regions.

SHORTCOMINGS OF CIVILIAN IMAGERY IN THE GULF WAR

Although civilian satellite imagery proved its worth in the Gulf War, particularly as a source of unclassified geospatial information that could be readily distributed to combat forces and shared with allies, it also had certain limitations from the perspective of military users. The most obvious shortcoming was low resolution: There was a limit to what could be detected and identified in these images. Another limitation was that the imagery data was not very timely. At that time, civilian observation satellites focused on acquiring high-quality multispectral images, rather than on images with high spatial resolution or on rapid delivery of imagery data. Further, the handful of civilian satellites covered targets of interest relatively infrequently; in particular, Landsat even lacked the imaging flexibility of the SPOT sensor. However, the time lags during the Gulf War also reflected a larger problem with imagery data dissemination that constrained the Coalition Forces from timely receipt and exploitation of the data-intensive satellite images. Rather than being received directly by ground stations in the field, civilian imagery data were received and processed at sites outside of the Persian Gulf and then flown to military users in Saudi Arabia (Congress, 1991, p. 28).

COMMERCIAL IMAGERY AND THE RISKS TO U.S. MILITARY OPERATIONS

The Gulf War raised concerns that other countries could take advantage of commercial satellite imagery to support their own military activities. Iraq reportedly purchased SPOT imagery of Kuwait prior to its invasion (Gordon and Trainor, 1995, p. 16). Following that invasion, the United States and France took steps to deny Iraq access to additional imagery data from either Landsat or SPOT. Although the United States did not formally impose "shutter controls," DoD officials took steps to delay the public release of Landsat imagery taken over the Persian Gulf to ensure that the overhead imagery data would lose much of its military value.[2] Similarly, France took steps to deny Iraq access to SPOT imagery that could potentially be used to threaten the Coalition Forces, which included French combat units.

But did Iraq's lack of access to commercial satellite imagery substantially affect the course of the Gulf War? Some analysts believe that Iraq's inability to obtain commercial satellite imagery, combined with the air superiority of the Coalition Forces, which prevented Iraqi air reconnaissance, ensured that Baghdad was blinded to the Coalition Force's buildup of some 270,000 troops in preparation for the westward flanking maneuver across the desert into Iraq. Studies have revealed that SPOT and Landsat imagery acquired at the time contained sufficient information to have provided Iraqi military planners with at least general indications that a large-scale redeployment of the Coalition Forces was under way in 1991 (Preston, 1994, pp. 35–42; Gupta, 1998). Although Landsat's 30-m resolution, and even SPOT's 10-m resolution, was too coarse to detect such specific military equipment as tanks and artillery, experienced imagery analysts could have exploited this lower-resolution imagery to identify the standardized field deployments of combat units by observing the ground scars and vehicle activity patterns imprinted on the desert terrain by large-scale military operations (Preston, 1994, pp. 35–42; Gupta, 1998; Congress, 1991, p. 56). Hence, civilian satellite imagery at the time of Desert Storm had the capability to detect and identify the important "Hail Mary" military maneuver that GEN Norman Schwarzkopf was counting on for a decisive attack around entrenched Iraqi ground forces.

To some observers, this episode strongly suggests the potential risks of exposure that higher-resolution commercial satellite imagery is likely to pose for modern military operations. However, taking any such lesson from the Gulf War experience is questionable for at least two reasons (Grundhauser, 1998). First, there is the question of whether Saddam Hussein would have accepted information based the commercial satellite imagery as credible even if Iraq had access to such information. Would he have believed that this ambiguous data was convincing evidence that his overall battlefield strategy was flawed? Or would the Iraqi leader have been more likely to dismiss the imagery's significance, suspecting that the data the U.S. and French satellites produced had been manipulated to deceive him? Second, even if Saddam Hussein had been convinced that preparations for a flanking maneuver were under way, it is doubtful whether Iraqi forces could have undertaken effective military counterattacks, given the battlefield dominance of the Coalition Forces. In particular, the complete air superiority that the Coalition Forces enjoyed at the time made an Iraqi counterattack very vulnerable to being devastated, regardless of whatever information the Iraqis might have acquired about the buildup of Coalition Forces.

GULF WAR LEGACY

The Gulf War experience therefore highlights the potential contributions and shortcomings of using civilian and commercial satellite imagery to support military operations. U.S. military planners gained greater operational experience in making use of nonmilitary satellite imagery to address combat requirements within the context of coalition warfare. However, possible Iraqi access to civilian satellite imagery also raised concerns over the potential risks for U.S. and allied military operations arising from foreign adversaries having access to existing and projected civilian and commercial observation satellites.

Furthermore, the less-than-satisfactory Desert Storm experience in obtaining civilian satellite imagery bolstered DoD interest in acquiring EAGLE VISION, a mobile station for receiving and processing unclassified satellite imagery (e.g., SPOT and Landsat) that was being built by the French firm Matra under a U.S. Air Force foreign technology acquisition program (Ackerman, 1995). However, DoD experts were also quick to note that, despite the contributions Landsat and SPOT imagery made in the Gulf War, future nonmilitary imagery needed to offer higher resolution, greater timeliness, and improved geolocation capabilities (i.e., precision in determining where the satellite image falls on the earth's surface) to be more useful for supporting military operations (Congress, 1991, p. 18).

Improving Commercial Observation Satellites

Although the resolution of civilian satellite imagery has significantly improved in recent years, relatively low spatial resolution and somewhat uncertain timeliness have limited the usefulness of this imagery for detecting and identifying objects of national security significance. However, a new generation of commercial and improved civilian observation satellites is becoming operational that at least partially addresses these challenges.

Key Advances in the New Observation Satellites

The new generation of nonmilitary observation satellites offers major advances in many performance characteristics over the earlier Landsat and SPOT civilian observation satellites:

- Higher resolution—Most of the new systems will collect data of much higher spatial resolution (less than 5 m) than the 10- to 30-m resolutions that were the best earlier civilian observation satellites offered.[3]

Over the next few years, several commercial observation satellites offering 1-m resolution are expected to become operational (see Appendix B). Plans also exist for commercial radar imaging satellites that will offer high-resolution imagery data.

- Broad range of spectral capabilities—In addition to the traditional panchromatic (black and white) imagery, the new generation of commercial and civilian observation satellites will also be able to produce multispectral images. These color images can be merged with the panchromatic images to produce even higher quality "pan-sharpened" imagery. In addition, some planned satellites will carry hyperspectral sensors that could collect data from over 100 to 200 different spectral bands. One important military application for such sensors is distinguishing between camouflage concealing military equipment and natural vegetation.
- Growing numbers of satellites—Current projections suggest that two or three times the current number of commercial and civilian observation satellites could go into orbit during the first decade of the 21st century (see Appendix B), depending on the rates of commercial and technical progress (Caceres, 2000). This substantial growth is due in part to the existence of numerous national government and commercial remote sensing programs and to plans for them to operate more than one remote sensing satellite in an effort to improve global coverage and reduce the time between revisits.

SOME LIMITATIONS ON MILITARY UTILITY

Given the rapid advances in the spatial and spectral resolutions of commercial and civilian imaging satellites, timeliness is fast becoming the main drawback for military and security applications. Several factors influence the timeliness of the collection, processing, and delivery to military and security users:

- Revisit time—Remote sensing satellites in low earth orbits are not positioned to image the desired surface location or target at all times. In some cases, such as Landsat, it takes about 16 days before the satellite passes over a particular location again because the fixed-imaging sensor system does not permit off-nadir imaging of desired locations. However, most civilian and commercial imaging satellites, such as SPOT and IKONOS, have adopted camera-pointing technologies that permit off-nadir viewing. These more-sophisticated systems present substantially more opportunities to acquire images of any given location, so

the satellite can acquire an image of a site roughly every 2 or 3 days, depending on the latitude of the target.
- Environmental constraints—Weather continues to be the predominant constraint on timeliness. Many commercial and civilian observation satellites use electro-optical sensors, which require unobstructed, daytime views to acquire useful images. Clouds can obscure the entire target location or, at least potentially, important portions of the scene, thus degrading the utility of electro-optical sensors. Some civilian imaging satellites, however, now have radar sensors that can acquire images at night and during most adverse weather conditions.
- Imagery data delivery time—The time it takes to order, task, collect, receive, process, and deliver the satellite imagery data is usually longer than is desirable for many military and intelligence applications. Under the best of circumstances (the satellite is in the correct orbital location to acquire the desired image and that there are no weather problems), several days can elapse from receiving the collection order to data delivery. Such delays may be acceptable for most commercial and civilian imagery uses but substantially diminish the potential value of the information for many military applications.

Constraints on the timeliness of satellite imagery data, therefore, are one of the key characteristics distinguishing civilian and military imaging satellites. However, as Table 6.2 indicates, not every military mission requires rapid delivery of satellite imagery data. Many military activities, such as cartography or planning missions for cruise missiles and aircraft against fixed enemy targets, can be done methodically, during peacetime. During a crisis or in combat, however, the demand for timely satellite imagery data substantially increases. Timely imagery data are usually needed to support battlefield missions, such as targeting the enemy's mobile units or assessing target damage after an attack. Governments having military reconnaissance satellites are usually willing to incur the added costs of using sophisticated technologies and dedicating additional human resources to accelerate the processes associated with collecting, receiving, processing, and analyzing imagery data.

Commercial firms have begun assigning a higher priority to prompt data acquisition but lack a compelling business case for investing in regular delivery of high-resolution satellite imagery data within 24 hours to most private-sector users. Still, in the long run, firms are likely to incorporate rapid, highly automated information processing and communication technologies into their

Table 6.2—Security Applications of Commercial Satellite Imagery

Mission Application	Requirements	Coverage
Military mapping, charting, and geodesy (MC&G)	Large scenes, high-resolution panchromatic imagery, stereo imagery for creating digital elevation models, high geolocation accuracy	Relatively infrequent
Broad-area search	Medium-resolution panchromatic or radar imagery covering very large areas	Periodic coverage to detect new facilities or activities
Targeting and bomb damage assessment	High-resolution panchromatic or radar imagery	Greater timeliness to support military combat operations
Mission planning for air and cruise-missile operations	High- to medium-resolution panchromatic and multispectral imagery and stereo imagery for digital elevation models	As required to support peacetime military planning or greater timeliness to support combat operations
Arms control and non-proliferation monitoring	High-resolution panchromatic, multispectral, or radar imagery	Periodic coverage under normal circumstances; intense cover in exceptional cases (e.g., treaty violations)
Supporting diplomatic negotiations on territorial disputes	High-resolution panchromatic, multispectral, or radar imagery with accurate geolocation	As required to support the diplomatic negotiations

Sources: Adapted from OTA (1993), pp. 81–84, and 145–165.

systems, which will improve their efficiency and responsiveness for a broad range of civil, commercial, and national security users.

Security-Related Uses for Commercial Satellite Imagery

The new commercial and civilian observation satellites do feature improvements that support a range of security-related uses. Table 6.2 highlights the potential relevance of nonmilitary imaging satellites to a broad range of U.S. national security missions, including military missions, for which overhead imagery data is an important source of information.

SUPPORTING MILITARY MISSIONS

New-generation satellites have much more potential for supporting military missions than the earlier Landsat and SPOT satellites had during the Gulf War. Although the usefulness of commercial satellite imagery varies according to specific mission requirements for imagery data, some broad conclusions are possible.

First, this imagery has enough spatial and spectral resolution to satisfy many of the imaging requirements for a broad range of military and intelligence missions. Spatial resolutions on the new and planned satellites approach 1-m (or better), which is good enough for detecting and identifying many military objects.[4] In some cases, such as mission planning, commercial and civilian observation satellites can produce the large multispectral image scenes that are very useful in mapping out flight profiles for attack aircraft and long-range cruise missiles. Three-dimensional visualization models generated from a combination of satellite imagery and digital elevation models allow pilots to rehearse combat missions in advance, thus increasing the chances of success.

Second, timeliness continues to inhibit greater military reliance on commercial and civilian satellite imagery data. This is less important for such mission areas as updating military maps and nautical charts in peacetime. But for such other missions as targeting mobile targets or assessing bomb damage, successful combat operations require very timely imagery. Hence, existing commercial and civilian observation satellites generally do not provide imagery quickly enough to support tactical operations.

Third, commercial and civilian satellite imagery also has some special advantages for supporting modern military operations. One advantage is that there is much less reservation about sharing images from nonmilitary imaging satellites with allies and other organizations.[5] In addition, unclassified imagery reduces the impediments to sharing important information between military organizations and the diverse civilian organizations that are often involved in vital activities associated with complex humanitarian emergencies, such as Kosovo.

Thus, commercial and civilian satellite imagery is a useful source of information for military planners. In some cases, it simply supplements classified sources of intelligence information; in others, it can usefully substitute for classified imagery and can be widely shared with coalition partners or even with nonmilitary organizations.

OTHER SECURITY-RELATED APPLICATIONS

Commercial satellite imagery has broad utility for security-related applications. One area is intelligence-gathering missions. For at least the near term, commercial observation satellites are unlikely to offer much help in fulfilling the needs of high-priority intelligence monitoring missions (e.g., indications and warning, or science and technology assessments), particularly for countries that already possess sophisticated military and intelligence reconnaissance capabilities. However, commercial satellite imagery is well suited to supporting intelligence missions that are less demanding or less urgent. These missions can include broad-area searches to detect significant new activities or facility construction in foreign countries of concern or, more routinely, monitoring military forces to identify the deployments of a foreign country.

Commercial imagery could also play an important role in arms control and nonproliferation monitoring. It could be used for monitoring foreign weapon testing and industrial facilities that are suspected sites for developing weapons of mass destruction or long-range missile delivery systems. The growing availability of commercial satellite imagery with higher spatial resolutions (or a wider range of spectral bands) is involving new actors in arms control treaty verification and nonproliferation monitoring. Commercial satellite imagery offers a means for "compliance bystanders"—countries that previously have played a less-active role in treaty implementation and compliance—to become more proactive in international deliberations (Dunn and Robertson, 1997, pp. 11–15). Nongovernmental organizations are also taking advantage of commercial and civilian observation satellites to study facilities associated with the production and testing of nuclear weapons (see Chapters 17 and 18 in this volume).

Similarly, the satellites offer a new tool to help diplomats resolve the territorial disputes that can fuel regional conflicts. Imagery data from such satellites have the major advantage of being unclassified and highly accessible to all parties to a negotiation, including the regional rivals. Chapters 14 and 15 in this volume highlight how the United States has used commercial satellite imagery and sophisticated image-processing technologies to help diplomats settle territorial disputes that were complicating the Dayton peace negotiations in 1995, which cleared the way for the Bosnia peace settlement. The United States later used similar means to help negotiators reach an end to the long-standing border dispute between Ecuador and Peru. In the future, commercial observation satellites could mitigate the risks of regional conflicts over territorial disagreements, such as the disputed Spratly Islands in the South China Sea, by offering a better means for countries to engage in cooperative monitoring (see Chapter 16 in this volume).

Challenges for the U.S. Military

Advances in commercial and civilian observation satellites pose a series of challenges for U.S. military planners, who must confront both the opportunities and risks that come with the rapid expansion in global access to overhead imagery data. One challenge involves overcoming the conceptual and organizational impediments that prevent the U.S. military from fully benefiting from advances in commercial and civilian imaging satellites to support peacetime and wartime military missions while helping to reduce costs. An equally important challenge involves assessing and countering potential threats to the United States, its military forces, or allies from foreign adversaries that now have growing access to satellite imagery with higher resolution and other improvements. Both sets of challenges have important implications for U.S. military doctrine, organizational approaches, and even global partnerships.

CHALLENGES TO U.S. MILITARY DOCTRINE

It is safe to say that the exploitation of commercial imagery is not addressed in existing U.S. military doctrine documents, although its absence does not necessarily indicate lack of attention to the issue.[6] For example, counterspace, surveillance, reconnaissance, and weather services are considered to be basic air and space power functions (USAF, 1997, Ch. 3), and commercial remote sensing would contribute to their implementation. Furthermore, the success of the Air Force's EAGLE VISION effort should bode well for similar efforts in the future. All the military services consider the intelligence, surveillance, and reconnaissance functions to be critical for providing precise, timely information to warfighters. Commercially available imagery can help identify and locate adversary military forces and facilities and give insights into the adversary's capabilities and intentions. It can also enable commanders to conduct battlefield assessments and develop concepts of operation and can provide insights into the terrain onto which forces may be deployed (USAF, 1999, pp. 26, 31). Space-based imagery offers the theater commander in chief and the joint force commander the opportunity to undertake surveillance of areas of interest from a great distance, particularly when hostile situations may not allow conventional forces to gather information.

Although not directly addressed, the proliferation of imagery and imagery products has specific doctrinal implications, as expressed in an article written in 1990, prior to the start of the Gulf War:

> However, as space surveillance improves ... we and our foes will start to feel a cramp in our terrestrial (land, sea, and air) combat styles. We will have to start routinely adapting our terrestrial operations to the reality of observation... [A]chieving strategic (global, long-term) surprise will become increasingly difficult unless we use methods which do not require massing of military forces. Catching a foe off guard will demand imaginative strokes that cannot be detected from space ...to dislocate the enemy's posture and plans. Small military forces whose presence or intentions can be concealed will usually be more effective

than much larger forces that cannot be hidden .... Tactical (local, short-term) surprise can be achieved another way—as long as space observation data must be relayed to earth for analysis before it can be acted on. The detection-analysis-decision cycle creates a significant time lag, which can be exploited for tactical action. Instead of fooling the satellites, one need only act so fast that the enemy cannot react in time. However, this "window" will narrow when terrestrial commanders get direct, real-time access to space intelligence . . . . (Noyes, 1990, p. 3.)

In this sense, the emergence of commercial imaging satellites offering imagery with much improved spatial and spectral resolution has not surprised U.S. military planners. Rather, it has been generally assumed that future U.S. military activities and combat operations would be undertaken in an era of growing global transparency whereby substantially greater information, including overhead imagery, will be broadly available to friends and foes alike.

Actively incorporating the possibility of global access to rapidly improving commercial imagery into joint doctrine, concepts of operation, and other elements of campaign and mission planning will be critical to the successful conduct of future U.S. military operations. Such an effort will also influence professional military education, training, and operational exercises and may encourage closer working relationships with commercial providers. An added doctrinal benefit will be increasing commanders' understanding of the benefits and risks of commercial imagery, thereby enhancing their ability to create favorable operational and tactical situations in their theaters.

ORGANIZATIONAL AND CULTURAL CHANGE

Commercial imagery poses opportunities for initiating organizational and cultural change within the U.S. national security community as well as enhancing the U.S. security advantage over potential military adversaries. Existing intelligence organizations, such as the National Imagery and Mapping Agency (NIMA)[7] and the National Reconnaissance Office (NRO), are coming to grips with the effects of commercial availability of imagery of a level of quality and resolution previously not available on the open market (see Chapter 7 in this volume).

NIMA has devoted substantial effort to developing a commercial imagery strategy and has allocated some of its limited resources to modernizing its imagery analysis systems to be able to take advantage of the emerging commercial satellite imagery. However, given serious funding constraints, NIMA will likely face an uphill battle to stay current with improvements in commercial imagery systems, products, and applications.

In comparison, NRO is responsible for "the unique and innovative technology, large-scale systems engineering, development and acquisition, and operation of space reconnaissance systems and related intelligence activities" for the nation (NRO, 2000). The reconnaissance satellites NRO designs, builds, and operates have been critical to U.S. national security. As with NIMA, NRO has found that the advent of commercially available imagery at or near the same level of quality that its own systems provide is posing difficult institutional and cultural issues. The new generation of commercial and civilian imaging satellites therefore has potentially important implications for formulating the future NRO satellite architecture.

Operationally and in concert with NIMA, NRO, and other agencies, U.S. and coalition military planners need to consider the security implications of having or not having access to commercial imagery for theater campaign planning and operations. The planners then need to develop an appropriate strategy for securing and maintaining that access and denying it to military adversaries. Such an effort is likely to be a complicated task involving potential organizational and cultural change, given the multinational and competitive nature of commercial imagery sources, the governmental (and nongovernmental) institutions involved, and the need to reach political consensus regarding coalition objectives. The complexity and sensitivity of this task require that the national security community consider approaches to "shaping the battlespace," before a conflict begins, through legal, regulatory, diplomatic, or market-driven mechanisms. Furthermore, operational exercises are needed to address the procedural aspects of including commercial imagery providers and applications in theater campaign planning. This will enable the national security community to gain maximum leverage from its organic resources and those of its allies and the industry.

GLOBAL PARTNERSHIPS

One aspect of the potential organizational and cultural change that may occur from an expanded relationship between the national security community and the commercial imagery providers is found in recent initiatives by the U.S. Space Command (USSPACECOM), the military organization responsible for the warfighting aspects of military space activities and systems.[8] USSPACECOM (1997, 1998) advocates "Global Partnerships," which is key to implementing the vision. Global Partnerships is defined as follows:

> a concept for leveraging domestic and international resources from the military, civil, commercial, intelligence, and national communities to

strengthen the DoD's space capabilities. Additionally, these efforts will enhance confidence in coalition warfare through closer cooperation with our allies in space. (USSPACECOM, 1998, p. 99)

Such partnerships are intended to decrease budgetary pressure on military infrastructure by adding to the DoD's resources and satisfying its requirements with both organic and inorganic solutions. They will enable the U.S. to "gain increased battlespace awareness and information connectivity in a cost-effective manner" (USSPACECOM, 1997, p. 12).

Part of the Global Partnership initiative lies in the recognition that DoD can no longer afford to rely solely on unique military space capabilities. Remote sensing partnerships with civil agencies, companies, and international organizations may help share costs and risks for military advantage. However, these partnerships may require finding common ground among very different organizational cultures and perspectives that may have little or no interest in military concerns (Johnson, Pace, and Gabbard, 1998, pp. 71–72). USSPACECOM and the military services should not assume that access to commercial imagery is guaranteed simply because it can be purchased on the international marketplace. Rather, while the ability of commercially available imagery to enhance military operations and forces appears to be clear, the partnership relationship will require the military community to adapt to give the commercial industry a strong incentive to provide adequate support when required, rather than to give priority to competing commercial interests.

The Potential Risks of Commercial Satellite Imagery

Although new commercial and civilian observation satellites can support a broad range of U.S. security-related missions, potential adversaries could also exploit higher-resolution satellite imagery in an attempt to threaten the United States or others. Thus, U.S. planners should consider four types of security risks associated with growing global transparency featuring worldwide access to satellite imagery: (1) risks from imagery use by aggressive states seeking to threaten the United States, their military forces, other states, or even internal groups; (2) risks from imagery use by dangerous nonstate actors (e.g., terrorists); (3) risks of exposing time-sensitive military operations; and (4) the unintended risks of exposing intelligence sources and methods.

RISKS FROM AGGRESSIVE STATES

U.S. military forces abroad could be at risk from foreign military adversaries seeking to deny or deter the United States from projecting its military power into a regional conflict. Under such circumstances, a military adversary might take advantage of commercial satellite imagery as another important way to monitor U.S. and allied military forces. The U.S. Air Force's Aggressor Space Applications Project has concluded that commercial satellite imagery is readily accessible to foreign adversaries and that they can analyze higher-resolution imagery data to obtain detailed information for supporting military operations and paramilitary operations against U.S. force deployments (Space Warfare Center, no date).

In theory, the range of political-military contingencies in which adversaries might use commercial or civilian satellite imagery to threaten U.S. or allied forces includes the following:

- U.S. military forces are engaged in combat or peace operations or are at risk of becoming engaged in armed conflict, for example, Central Europe (Bosnia, Kosovo), or Persian Gulf (Iraq) or special operations (hostage rescue, counterterrorism).
- High-risk regional crises are under way that could directly or indirectly involve U.S. military forces, for example, East Asia (Korean Peninsula and the Taiwan Straits), Middle East (Israel under external attack), or counter-drug operations (Columbia).
- Regional crises are under way that do not directly involve U.S. military forces, for example, South Asia (India-Pakistan military conflict), but that could upset regional stability.
- Other contingencies exist that do not feature U.S. military involvement, for example, hostage rescue operations by foreign countries or humanitarian emergencies resulting from internal conflicts.

Potential adversaries might take advantage of higher-resolution panchromatic or multispectral imagery to identify possible routes for planning ground or air attacks on U.S. or allied military forces. Adversaries could also use nonmilitary satellite imagery to identify high-value targets, particularly during the early phase of a contingency (U.S. military forces tend to be somewhat more vulnerable during the force insertion and buildup stage).

Analysis of timely, high-resolution satellite imagery could provide an adversary's military planners with valuable information on the particular order of battle for U.S. and allied troop deployments in forward areas, the nature and configuration of logistical centers, and the location of major command and control centers. A military adversary could potentially exploit such information to identify possible high-value targets and to detect changes in the disposition of U.S. troops that might give away preparations for combat operations, thus diminishing the chances that U.S. forces will achieve operational or tactical surprise in pursuing their military objectives.

Aggressive states could also use commercial and civilian satellite imagery in planning or executing military attacks against neighboring countries. Satellite imagery analysis can contribute to the broad range of military missions outlined in Table 6.2 in supporting cross-border operations, including identifying avenues of attack for invading forces and defining targets for air and missile strikes. As Chapter 21 in this volume discusses, regional instability could be exacerbated if an information asymmetry among the regional powers enables an aggressive state to exploit an advantage in access to satellite imagery for supporting its military planning and operations against its neighbors.[9] In these situations, U.S. policymakers will probably be concerned that U.S. commercial observation satellites could become militarily significant to the conflict and possibly even targets of forceful responses from one of the warring parties.

Finally, aggressive regimes might also take advantage of information derived from satellite imagery to move against insurgent groups or to undertake ethnic cleansing activities. Their intelligence analysts could find higher-resolution panchromatic and multispectral imagery useful for locating rebel bases. Furthermore, large groups of refugees hiding in forests and mountains to escape government-backed forces engaged in ethnic cleansing or genocide could be at risk of being detected on satellite imagery.

RISKS FROM NONSTATE ACTORS

Another potential risk associated with high-resolution commercial satellite imagery is that some nonstate actors, such as terrorist groups, will use it for malicious purposes. In theory, commercial satellite imagery is another information technology that terrorist groups could exploit to support illegal and threatening activities. By taking to use front organizations or individuals to order and receive the desired imagery data, terrorists could probably acquire commercial satellite imagery of most locations throughout the world with little likelihood of being detected by legal authorities.

However, the ability to obtain commercial satellite imagery with little risk does not necessarily mean that terrorists have a strong need for or interest in acquiring overhead imagery. For them, the value of an overhead view of a planned target is likely to be marginal at best; most terrorist targets appear to be relatively vulnerable to attack and approachable for gathering preattack intelligence using traditional means, including inside sources. The additional value to terrorists of overhead imagery of their intended targets appears to be questionable (Oberg, 1999). For example, it seems unlikely that the availability of commercial satellite imagery would have substantially contributed to the most devastating terrorist attacks that have occurred on U.S. targets in recent years: the bombings of the federal building in Oklahoma City, 1995; the Khobar Towers in Dhahran, Saudi Arabia, 1996; the U.S. embassies in Kenya and Tanzania, 1998; the World Trade Center in New York, 1993; and, most recently, the USS *Cole* in Yemen, 2000.

This limitation on the utility of the overhead imagery for terrorist groups is not that they lack technical prowess. RAND analysts and others have noted the growing technological sophistication of the "new terrorists," who are adept at using the Internet and other information technologies to support their widespread activities. Some terrorist groups have highlighted their growing professionalism by developing innovative methods, including technologies, to improve the effectiveness of their weapons against the technologies and countermeasures of legal authorities (Lesser et al., 1999, pp. 28–37; Hoffman, 1997). Nonetheless, by their very nature, terrorists seem to have much less need for satellite imagery because they usually enjoy direct access to their planned targets. Furthermore, the information terrorists seem to require on their targets is much more detailed and timely than what existing and planned commercial observation satellites can provide. Thus, it is not surprising to note that, even though articles on commercial satellite imagery often contain a passing reference to the risk of high-resolution satellite imagery being acquired by terrorists, the experts who have assessed the growing technological proficiency of terrorist groups do not tend to include commercial satellite imagery among the likely technologies of interest (Lesser et al., 1999, pp. 28–37; Hoffman, 1997; Bowers and Keys, 1998).[10]

Similarly, other criminal nonstate actors, such as drug traffickers and hostage takers, are unlikely to have much use for satellite imagery to support their operations. They also will probably rely on such tried-and-true methods as bribery and infiltration to obtain information on their intended targets and to outmaneuver government forces.

RISK OF EXPOSING TIME-SENSITIVE MILITARY OPERATIONS

A potentially worrisome contingency is that commercial or civilian satellite imagery could divulge U.S. military preparations for undertaking operations that depend on tactical or operational surprise to succeed, such as hostage rescue or counterterrorism missions. Such time-sensitive operations might be deliberately exposed by the actions of the adversary or another government or group allied with the foreign adversary. In addition, such noncombatants as the news media could be using commercial satellite imagery to help monitor the military conflict and could inadvertently expose such operations. These potential risks are one reason the U.S. government has insisted on its authority to impose shutter controls on the collection and dissemination of high-resolution U.S. commercial satellite imagery in times of emergency (see Chapter 24 in this volume).

RISK OF EXPOSING INTELLIGENCE SOURCES AND METHODS

A somewhat different national security concern is that the growing availability of commercial and civilian satellite imagery could diminish the utility of "national technical means"—secret imaging satellite systems—for monitoring rogue states or other governments that could be engaged in illegal, suspicious, or otherwise threatening activities. Some intelligence experts worry that sensitive sources and methods could be exposed if other governments are able to gain insights from higher-resolution commercial satellite imagery that are suggestive of the operations and performance capabilities of classified imaging satellites. These experts worry that such countries could use these insights to adopt more effective methods of concealing their illegal activities, which would degrade the utility of national technical means for providing vital intelligence on foreign threats (Dunn and Robertson, 1997, pp. 17–18; McDonald, 1999).

This risk could be a particular concern for governments that use classified monitoring capabilities to detect and identify illegal activities related to arms control and disarmament agreements or nonproliferation regimes. Some U.S. arms control experts are concerned that the benefits of having a broad range of users of commercial satellite imagery are questionable if they give countries violating arms control or nonproliferation norms greater insights into the specific types of imagery observables that the U.S. intelligence community uses to monitor suspicious or illegal activities. There is, however, substantial uncertainty about what practical insights a state seeking to conceal illegal weapon activities could gain from having access to commercial satellite imagery that its intelligence agencies would not already have obtained through other sources.

Assessing the Likelihood of Military Risks Occurring

Any realistic assessment of the potential military risks for the United States and its allies that could arise from growing global access to improved commercial and civilian observation satellites must distinguish the possibility of a threat from the likelihood that it will occur. Today, few foreign militaries possess anything like the capabilities, expertise, and experience necessary to translate satellite imagery data into information that directly contributes to a significant military advantage on the battlefield, such as long-range precision strikes. Thus, it is important to distinguish between the near-term impediments that potential foreign military adversaries confront in trying to extract military advantage from satellite imagery and the long-term trends that could help reduce these impediments enough to allow foreign adversaries to use commercial satellite imagery effectively.

IMPEDIMENTS TO TRANSLATING IMAGERY ACCESS INTO MILITARY ADVANTAGE

Assessing the ability of aggressive states to translate their new access to imagery data into military advantage on the battlefield is somewhat problematical. This is largely because the overhead imagery itself is only one important input to a broader equation. As one military expert has noted: "Ultimately, a nation with obvious hostile intent and armed with the best satellite imagery available must still be able to convert that information into combat capability" (Grundhauser, 1998, p. 67).[11] New users of satellite imagery must first master several critical steps before they can expect to gain a significant military advantage by using commercial satellite imagery:

- Their access to satellite imagery must be reliable and timely.
- The interpretation, analysis, and dissemination of imagery data must be accurate and prompt.
- They must be able to use the information in weapon delivery or other military operations effectively.[12]

The first challenge for foreign militaries is to ensure that they will have access to the type of commercial or civilian satellite imagery that they need to support their military and intelligence missions. As noted earlier, access to commercial satellite imagery under normal peacetime conditions is not likely to be a problem for almost any potential user. Even states otherwise subject to arms embargoes are bound to find ways of acquiring the satellite imagery data they desire through purchases using third parties and front organizations. Yet the extent to which states are willing to rely on receiving satellite imagery provided by enterprises from other countries remains to be seen. Few national governments are probably willing to depend on foreign sources of satellite imagery to support their more-demanding tactical combat operations or important intelligence-monitoring missions. These governments will recognize that their access to overhead imagery could easily be denied if the satellite imagery data provider's government chooses to impose shutter controls on foreign users of commercial satellite imagery during crises or military conflicts.

A second major impediment for foreign military and intelligence establishments is the challenge of producing accurate and timely assessments after acquiring the commercial satellite imagery data. Well-trained and highly experienced imagery analysts are required to produce accurate imagery analyses with confidence. The absence of such expertise creates a greater risk that important objects captured in the satellite imagery will be overlooked or misinterpreted. Most advanced countries are likely to have the imagery analysis expertise needed to support their military establishments. A larger problem, however, will be generating timely assessments of the imagery analyses.[13]

The third step in the process is the difficult challenge of translating the resulting information into a military edge on the battlefield. This is perhaps the largest uncertainty for many foreign militaries. This military advantage could come either from accurately delivering a weapon against an enemy target or from using the information to gain an edge in maneuvering their military units into an advantageous position for combat operations. This is challenging under any circumstances but is particularly so for militaries confronting the U.S. military, which has some of most sophisticated intelligence and military capabilities in the world, including robust defenses against air and missile attacks.

NEAR-TERM THREAT POTENTIAL OF COMMERCIAL SATELLITE IMAGERY

The projected growth in the number of commercial and civilian observation satellites over the next several years, along with the worldwide diffusion of other information technologies, means that U.S. and allied military forces must increasingly expect to operate in an environment characterized by growing global transparency. Some observers have suggested that this trend will inevitably diminish U.S. military power by eroding the long-standing U.S. military and intelligence advantage in satellite reconnaissance capabilities. Others have argued that military commanders must be prepared to conduct their activities and operations assuming that they are under constant surveillance. However, a closer examination of current trends suggests that the new generation of commercial and civilian observation satellites is unlikely to present major concerns for U.S. defense planners in the near term. Rather, the security challenges arising from growing global transparency are more likely to be a long-term issue for the United States.

Several factors will diminish the national security risks that are likely to arise from the proliferation of commercial and civilian imaging satellites over next five years or so. First, as discussed earlier, these satellites have some important technical shortfalls as a reliable source of overhead data for many military missions. A key shortcoming is the uncertain timeliness of any data nonmilitary imaging satellites provide. For the existing generation of these satellites, placing satellite imagery data in the hands of the users less than 24 hours after receiving the order is the exception, rather than the rule. This constraint means that, for the near-term, commercial and civilian satellite imagery is best suited for supporting less-urgent military missions.[14]

Furthermore, even if these satellites happen to be in the right position at the right time to collect an image, adverse weather can always prevent data collection or degrade the imagery. Since most commercial and civilian imaging satellites rely on electro-optical sensors, cloud cover is a major impediment to collecting the data needed to support military combat operations or preparations. Certain conflict regions, such as Central Europe, are under cloud cover for protracted periods, which sharply decreases the number of cloud-free opportunities for collecting data on particular targets or locations. And the civilian radar imaging satellites currently available do not offer the higher resolution most military and intelligence applications require.

Second, political uncertainty is another important constraint on how much foreign adversaries are likely to count on receiving satellite imagery to support their military and intelligence activities. Few states are likely to depend on the reliability of foreign commercial sources of satellite imagery for supporting military operations during a crisis or combat. Their leaders are likely to be aware that the imagery flow could be abruptly discontinued if the government operating or regulating the satellite imagery data provider decided to restrict their access during a crisis or an armed conflict. The U.S. government has clearly declared its authority (via PDD-23) to impose shutter controls on U.S. commercial remote sensing satellite operations. These controls will temporarily restrict the collection or dissemination of imagery data for a particular area if circumstances arise that pose a risk to U.S. national security interests, international obligations, or foreign policy concerns.[15]

Of course, foreign adversaries could always circumvent these restrictions by turning to non-U.S. sources of high-resolution satellite imagery. However, at least for the near term, most foreign civilian and commercial observation satellites will be operated by countries or companies that the U.S. government can substantially influence, particularly during in a crisis or conflict situation. The projected higher-resolution imaging satellites will be operated by the United States, U.S. allies, or other countries with a major stake in preserving U.S. market access for their imagery sales or other commodities.

Finally, the capability of most foreign militaries to exploit satellite imagery as a force multiplier will probably develop slowly over time, if at all. Few of the foreign military adversaries facing the United States and its allies possess long-range, precision-guided weapon systems that could use satellite imagery data for mission planning and targeting purposes. Neither are these adversaries likely to be proficient in the use of advanced intelligence, surveillance, and reconnaissance technologies that can seamlessly integrate satellite imagery data with other forms of intelligence for supporting battlefield operations. Rather, foreign militaries will most likely use their unprecedented access to higher-resolution commercial satellite imagery as a test bed for developing experience in imagery analysis, exploitation, and timely dissemination. Over the near term, they are most likely to concentrate on supporting military missions that can be performed before hostilities break out (e.g., making military maps and identifying potential targets in neighboring states). Even these rudimentary uses of satellite imagery data are likely to be a major improvement for many foreign militaries that have had little or no access to high-resolution imagery data on their neighbors and potential adversaries in the past.

Thus, in combination, these conditions significantly limit the utility of commercial and civilian imaging satellites for providing imagery data that can satisfy the demanding information requirements of modern military operations. Yet despite major impediments that will limit the abilities of foreign adversaries, U.S. defense planners would be prudent to observe how potential adversaries gain greater experience with satellite imagery over the longer term, possibly changing a low-risk situation for the worse.

LONG-TERM THREAT POTENTIAL OF COMMERCIAL SATELLITE IMAGERY

Several long-term trends could diminish the existing impediments facing foreign militaries that might be interested in using commercial and civilian observation satellites to bolster their military capabilities vis-à-vis the United States or its allies. A key technological trend is that the potential military utility of these imaging satellites, and their supporting ground infrastructures, will steadily improve over time. Next-generation commercial and civilian observation satellites will undoubtedly offer higher-resolution imagery data, improve the accuracy with which satellite imagery is tied to ground target locations, and increase the speed with which data are delivered to users. Because radar imaging satellites are immune to the effects of weather and time of day, radar satellites capable of collecting higher-resolution images (less than 5 m) would have special military significance.

Similarly, foreign militaries could exploit advances in ground equipment and image-handling infrastructure to enhance the military uses of available satellite imagery. For example, new mobile ground receiving stations are available on the international marketplace that will help reduce the delays in receiving commercial and civilian satellite imagery data. And technically sophisticated adversaries could approach the nonmilitary imaging satellites as a virtual constellation of space-based imaging sensors instead of a series of largely unrelated satellite systems. Thus, by building up their image-handling capabilities and expertise, foreign militaries could greatly improve their chances of acquiring timely imagery data from almost any available commercial or civilian satellite that is in proper position to collect an image of the desired location.

A second important trend is that future civilian and commercial observation satellites could be owned and operated by some foreign governments and companies that are either indifferent or even hostile to U.S. national security interests. One possibility is the emergence of a state-owned enterprise or an offshore commercial firm with mercenary-like willingness to sell satellite imagery to any country or group, including aggressive states, wanting to purchase its products and services despite the potential security consequences. Another possibility is that over time some foreign military adversaries of the United States and its allies will acquire their own sources of satellite imagery. One option would be to take advantage of the technology trend toward much smaller satellite systems that is opening up the prospects for much-less-expensive imaging satellites that are also capable of producing higher-resolution imagery (see Chapter 20 in this volume). Alternatively, wealthier countries interested in acquiring satellite imagery might help fund the imaging satellite programs of other countries in exchange for receiving imagery of interest to their national security establishments. In either case, U.S. diplomatic and commercial leverage would probably be insufficient to discourage these new types of actors from providing satellite imagery to potential foreign military adversaries.

Finally, over the next several years, many foreign countries could gradually begin acquiring the types of military technologies and doctrines needed to make more effective use of commercial satellite imagery in supporting their combat operations. Few foreign military adversaries facing the United States and its allies currently rely on precision strike systems, such as long-range cruise missiles capable of delivering conventional munitions at point targets. However, the steady diffusion of dual-purpose technologies, such as Global Positioning System (GPS)–assisted guidance systems, and the proliferation of advanced military technologies is increasing the likelihood that foreign militaries will be interested in and capable of exploiting satellite imagery to obtain the geospatial data required to support their fledgling long-range strike capabilities.[16] In addition, mission-planning software and other imagery data-handling technologies required for supporting tactical operations, such as long-range air strikes or special operations, are increasingly available on the international marketplace.

Thus, U.S. national-security policymakers need to distinguish between the near-term and long-term trends that will determine whether foreign military adversaries can readily exploit commercial and civilian satellite imagery to threaten the United States and its allies. Various impediments will substantially curtail the prospects that foreign military adversaries will be able to make effective use of commercial and civilian satellite imagery in any conflict with the United States in the short run. The long-term trends, however, increase the potential—but not necessarily the likelihood—that foreign adversaries could acquire more-reliable access to high-resolution satellite imagery and could then apply the imagery data to support a wide range of military missions.

U.S. Options to Limit Adversary Access to Satellite Imagery

Unlike most other countries, the United States has a rich menu of options available for limiting the potential access of foreign military adversaries to commercial and civilian satellite imagery that could be used in supporting attacks on U.S. targets. These options range from adopting national policies that encourage U.S. commercial remote sensing satellite firms to dominate the global marketplace for satellite imagery to more-forceful options for use in a crisis or armed conflict against non-U.S. imaging satellite systems that are perceived as providing substantial assistance to a foreign military adversary. These options include the following:

- Shaping the space remote sensing environment—Peacetime U.S. national policies can play an important part in favorably positioning the United States to deal with any potential risks from foreign military adversaries gaining access to satellite imagery. Export controls can slow the diffusion of sensitive imaging technologies and know-how. However, the United States is probably better served by faithfully implementing the 1994 PDD-23 policy that encouraged U.S. commercial remote sensing satellite firms to acquire a dominant share of the global marketplace (Johnson, Pace, and Gabbard, 1998, pp. 31–32). The success of U.S. firms could discourage other countries from developing their own imaging satellites or at least encourage them to depend on U.S. manufacturers for high-quality subsystems or even turnkey imaging satellite systems.
- Impose shutter controls on U.S. firms—In emergencies, the U.S. government could restrict foreign access to U.S. commercial imaging satellites because they are likely to produce the highest resolution commercial imagery that is available for supporting military operations.[17] Under the PDD-23 guidelines, the restrictions would be imposed on collecting and/or disseminating U.S. commercial satellite imagery data for a particular region for a limited time.
- Undertake diplomatic approaches to other governments—If the situation is serious enough for the U.S. government to impose shutter controls on U.S. imaging satellite firms, other countries that are strong allies and friends are likely to impose similar restrictions on their own civilian or commercial observation satellites.[18] The United States would probably take a strong diplomatic approach to persuade or pressure other governments not to permit imaging operations that could undercut the U.S. shutter-control policy toward a foreign military adversary for the duration of the crisis or armed conflict.
- Adapt U.S. military doctrine and operations—Despite shutter controls and diplomatic approaches, U.S. military planners will prudently assume that a foreign military adversary could receive satellite imagery from various sources. Hence, planners would probably increase the tempo of U.S. military operations in a contingency to diminish the usefulness of any commercial satellite imagery available to an adversary. Given that the operational parameters of commercial and civilian imaging systems will be widely known, various concealment and deception

strategies could be used to inject substantial uncertainty about the soundness of assessments made on basis of commercial satellite imagery available to an adversary.
- Disrupt foreign imaging satellite operations—Several possible military means exist for temporarily disrupting the operations of imaging satellite systems. If needed, the United States could probably use electronic warfare means or even information warfare to degrade or temporarily disrupt the operations of an imaging satellite suspected of being a source of imagery data for a foreign military adversary. A more forceful military response would be to destroy the ground station's receiving antenna or its supporting power station, assuming the station is located on the territory of a foreign adversary.[19]
- Employ antisatellite measures—As an ultimate resort, the United States could employ direct means to degrade or permanently disable any imaging satellite that was believed to be supplying the foreign military adversary with invaluable intelligence on U.S. military troop deployments and movements. One possibility is to "dazzle" the sensitive optics and overheat the electronics of an imaging satellite using a ground-based or airborne laser system (Oberg, 1999, pp. 68–69). An alternative possibility would be some type of space-based interception using a kinetic weapon system or other mechanism to destroy the imaging satellite.

Although antisatellite attacks are the most frequently discussed space control option, they are probably the least likely given the strong incentives for the United States and most other countries to avoid setting a precedent of undertaking attacks on space systems. Space-based systems are crucial to the full range civilian, military, and commercial sectors among the technologically advanced countries. Rather the knowledge that the United States possesses the military means to disrupt ground control operations or even destroy satellite systems is likely to help persuade other governments and their commercial remote sensing firms to comply with U.S. diplomatic requests aimed at preventing their civilian or commercial satellite imagery from falling into the hands of a foreign military adversary during a crisis or conflict.

Thus, the steadily expanding access of foreign military adversaries to rapidly improving commercial and civilian satellite imagery should be a matter of concern for U.S. defense planners. However, compared to almost any other country, the United States is in the best position to recognize and effectively counter any significant threat that is likely to arise. It is easy to agree with the view of one knowledgeable observer that:

> The military impact of this proliferation [of space remote sensing] is cause for concern, but not yet catastrophic. Potential military responses include improvements in concealment and deception, direct countermeasures, and force structure that emphasize speed and concentration of destructive power. (Preston, 1994, p. 110).

Conclusion

The wartime utility of civilian observation satellites was first highlighted in the early 1990s when the United States and its coalition allies used low-resolution, multispectral Landsat and SPOT images to support their military operations against Iraqi forces. The Gulf War experience revealed that even lower-resolution civilian observation satellites could provide imagery data relevant to a wide range of military requirements. The new generation of nonmilitary imaging satellites is somewhat better suited to supporting a broad range of military tasks because these satellites offer higher resolution and more frequent coverage. Nonetheless, some important constraints still limit the potential military utility of these images, including the inability of existing commercial observation satellites to satisfy fully the timeliness requirements of many military and intelligence applications. This limitation is likely to diminish over time as the tasking and imagery data retrieval systems for these satellites improve.

U.S. defense planners face the challenge of deciding how best to take advantage of commercial observation satellites to support U.S and allied military operations while hedging against potential security threats from unprecedented global access to improved satellite imagery. However, the United States is in a better position than other countries to take advantage of this imagery source as an important supplement to national imaging capabilities. The U.S. military is also in a strong position to anticipate and counter any potential military risks that could arise from the unprecedented access foreign military adversaries have gained to commercial satellite imagery, even if these adversaries can significantly improve their ability to use nonmilitary satellite imagery over the longer term. Hence, commercial observation satellites present less of a significant military threat to U.S. military forces than a continuing challenge for the U.S. national security community to adapt its military doctrine, intelligence procedures, and institutional cultures to make the best use of what commercial observation satellites have to offer.

Notes

1. For a good survey of these imagery sources, see Williams, Heckman, and Schneeberger (1991).
2. According to Spires (1997, p. 253):
   the Defense Intelligence Agency intervened to prevent U.S. news media from obtaining Landsat data of the Kuwait-Iraq-Saudi border, which might have revealed the Coalition buildup in preparation for the "left hook" offensive maneuver at the war's start.
3. As the earlier Gulf War discussion noted, even lower-resolution satellite imagery can be exploited to support a broad range of military requirements. For an argument that 1-m imagery is not a watershed in terms of military utility compared with lower-resolution imagery, see McCue (1994).
4. Chapter 20 in this volume provides the specific imagery interpretation resolution requirements for different objects of interest to military planners.
5. Given the recurring U.S. involvement in coalition force operations, such as the Gulf War and Kosovo conflict, the ability to share commercial satellite imagery provides another useful means for developing greater situational awareness among coalitional partners.
6. Joint military doctrine is defined as "fundamental principles that guide the employment of forces of two or more Services in coordinated action toward a common objective." (DoD 1998.) Doctrine covering joint operations of U.S. military forces is found in a number of key Joint Publications. However, to date, joint space doctrine (i.e., Joint Publication 3-14) has not been approved or implemented, thus contributing to uncertainties about how information and data from space systems would be incorporated into operational planning and exercises. This has encouraged the Air Force, Navy, and Army to take somewhat different approaches to developing their space doctrine in lieu of an agreed joint doctrine.
7. Established in 1996 as a merger of eight different organizations and entities, NIMA was intended to provide a single-agency focus for users of imagery and geospatial information within the U.S. government. As a combat support agency, NIMA is dedicated to supporting U.S. military operations, as well as to supporting important national missions. Its director is the functional manager for imagery and imagery intelligence including research, development, test, and engineering and procurement within the national and military intelligence programs. (See P.L. 104-201).
8. Title 10 assigns warfighting responsibility to the theater and functional unified commands (commanders in chief) but assigns responsibility for organizing, training, and equipping interoperable forces for assignment to combatant commands to the military services. The unified commands identify the operational capabilities and strategies needed to shape and conduct operations within their particular areas of interest. (See DoD, 1995.)
9. On the potential risks of asymmetries in imagery access and exploitation capabilities, see Krepon et al. (1990), pp. 22–23 and 97–101.

10. Also note that Bowers and Keys (1998) does not discuss possible terrorist use of satellite imagery at all.
11. He goes on to warn against making assumptions that foreign militaries can make effective use satellite imagery to support their warfighters without recognizing the "profound" asymmetries that exist vis-à-vis U.S. capabilities and experience in using satellite imagery to support military operations.
12. Black (1999, pp. 110–111) clearly notes the need to avoid overestimating an adversary's ability to do these things successfully to make effective military use of commercial satellite imagery.
13. As the Gulf War experience revealed, even the United States, with unequaled intelligence resources devoted to imagery analysis, was hard pressed to produce imagery assessments that could satisfy the demanding operational requirements for up-to-date intelligence information needed to support fast-changing tactical operations.
14. One important exception, however, is Canada's RADARSAT-1 satellite, which is optimized for the rapid return of imaging data, which has been essential in supporting its sea ice–monitoring mission. This satellite can take advantage of the fact that its radar imaging sensor can penetrate cloud cover over the desired targets.
15. The U.S. National Oceanic and Atmospheric Administration, the Department of Commerce executing agency for commercial remote sensing licenses, has laid out the underlying principles for this policy; the regulations and the related implementation guidance can be found at http:www.licensing.noaa.gov/.
16. Along with a GPS-assisted guidance system, commercial space imagery would be a key enabling technology for development of a long-range cruise missile capability for many countries (see McMahon and Gormley, 1995, pp. 21–25; also see Barrie and Clark, 2000).
17. The expectation that U.S. commercial imaging satellites would be unmatched by foreign competitors in the international marketplace is consistent with the U.S. government's 1994 PDD-23 policy to encourage the U.S. commercial remote sensing satellite industry (see Baker, 1997, pp. 6–7).
18. Only the United States and Canada have publicly declared policies that would impose shutter controls on their commercial imaging satellites in emergencies. Other governments responsible for civilian or commercial observation satellites most likely have similar plans for restricting satellite imagery during crises or conflicts, even if they have not publicly announced such policies. For a description of the U.S. and Canadian policies, see U.S. Department of Commerce (no date).
19. Given their nonmilitary design, commercial imaging satellite operations are likely to be particularly vulnerable to electronic or physical attacks on their satellite control systems for telemetry, tracking, and control signals or on the power sources for the ground control and receiving stations. However, their vulnerability will also depend on the political consequences of taking action against ground installations located in countries not party to the conflict (Black, 1999, pp. 109–110).

References

Ackerman, Robert K., "Air Force Planners Exploit Commercial Space Imagery," *Signal*, June 1995.

Anson, Sir Peter, and Dennis Cummings, "The First Space War: The Contribution of Satellites to the Gulf War," in Alan D. Campen, ed., *The First Information War*, Fairfax, Va.: AFCEA International Press, 1992.

Baker, John C., *Trading Away Security? The Clinton Administration's 1994 Decision on Imaging Satellite Exports*, Washington, D.C.: Georgetown University, Institute for the Study of Diplomacy, 1997.

Barrie, Douglas, and Colin Clark, "Cruise Missile Worries Spark MTCR Action," *Defense News*, July 24, 2000, pp. 1, 58.

Black, J. Todd, "Commercial Satellites: Future Threats or Allies?," *Naval War College Review*, Vol. 52, No. 1, Winter 1999.

Bowers, Stephen R., and Kimberly R. Keys, *Technology and Terrorism: The New Threat for the Millennium*, Warwickshire, UK: Research Institute for the Study of Conflict and Terrorism, May 1998.

Caceres, Marco, "Focus Sharpens for Imaging Satellite Market," *Aerospace America*, Vol. 38, September 2000.

DoD—*See* U.S. Department of Defense.

Dunn, Lewis A., and Marjorie Robertson, *Satellite Imagery Proliferation and the Arms Control Intelligence Process*, McLean, Va.: Science Applications International Corporation, April 22, 1997.

Gordon, Michael R., and Bernard E. Trainor, *The Generals' War: The Inside Story of the Conflict in the Gulf*, Boston: Little, Brown and Company, 1995.

Grundhauser, Larry K., "Sentinels Rising: Commercial High-Resolution Satellite Imagery and Its Implications for US National Security," *Airpower Journal*, Vol. 12, Winter 1998, pp. 67–68.

Gupta, Vipin, presentation for Sandia National Laboratories, at the "Secret No More: The Security Implications of Global Transparency" conference, cosponsored by George Washington University's Space Policy Institute and the National Air and Space Museum, Washington, D.C., May 21–22, 1998.

Hoffman, Bruce, "Responding to Terrorism Across the Technological Spectrum," in John Arquilla and David Ronfeldt, eds., *In Athena's Camp: Preparing for Conflict in the Information Age*, Santa Monica, Calif.: RAND, MR-880-OSD/RC, 1997, pp. 351–359.

Johnson, Dana J., Scott Pace, and C. Bryan Gabbard, *Space: Emerging Options for National Power*, Santa Monica, Calif.: RAND, MR-517, 1998.

Krepon, Michael, Peter D. Zimmerman, Leonard S. Spector, and Mary Umberger, eds., *Commercial Observation Satellites and International Security*, New York: St. Martin's Press, 1990.

Kutyna, Donald J., "Indispensable: Space Systems in the Persian Gulf War," in R. Cargill Hall and Jacob Neufeld, eds., *The U.S. Air Force in Space*, Washington, D.C.: U.S. Government Printing Office, 1998.

Lesser, Ian O., et al., *Countering the New Terrorism*, Santa Monica, Calif.: RAND, MR-989-AF, 1999.

McCue, Brian, "The Military Utility of Civilian Remote Sensing Satellites," *Space Times*, Vol. 33, No. 1, January–February 1994, p. 11.

McDonald, Robert A., "NRO's Satellite Imaging Reconnaissance: Moving from the Cold War Threat to Post–Cold War Challenges," *Defense Intelligence Journal*, Vol. 8, No. 1, Summer 1999, pp. 81–82.

McDonnell, Janet A., *Supporting the Troops: The U.S. Army Corps of Engineers in the Persian Gulf War*, Washington, D.C.: U.S. Government Printing Office, 1996.

McMahon, K. Scott, and Dennis M. Gormley, *Controlling the Spread of Land-Attack Cruise Missiles*, Marina del Rey, Calif.: American Institute for Strategic Cooperation, January 1995.

National Reconnaissance Office, Web site, 2000. Available at http://www.nro.odci.gov/index1.html (last accessed November 28, 2000).

Noyes, Harry F., III, "Air and Space Forces: The One Endures as the Other Emerges," *Airpower Journal*, Spring 1990 (Web edition).

Oberg, James, "Spying for Dummies," *IEEE Spectrum*, Vol. 36, No. 11, November 1999.

OTA—*See* U.S. Congress, Office of Technology Assessment.

Preston, Bob, *Plowshares and Power: The Military Use of Civil Space*, Washington, D.C.: National Defense University Press, 1994.

Public Law 104-201, Title XI, National Imagery and Mapping Agency Act of 1996.

Space Warfare Center, "The Aggressor Space Applications Project and the Emerging Commercial Imagery Threat," not dated.

Spires, David N., *Beyond Horizons: A Half Century of Air Force Space Leadership*, Peterson AFB, CO: Air Force Space Command, 1997.

U.S. Air Force, Air Force Basic Doctrine, AFDD-1, September 1997.

U.S. Air Force, *Doctrine Document 2-5.2, Intelligence, Surveillance, and Reconnaissance Operations*, April 21, 1999, pp. 26, 31.

U.S. Congress, Office of Technology Assessment, *The Future of Remote Sensing From Space: Civilian Satellite Systems and Applications*, Washington, D.C.: U.S. Government Printing Office, July 1993.

U.S. Department of Commerce, NOAA Licensing of Commercial Remote Sensing Satellite Systems, no date. Available at http://www.licensing.noaa.gov/ (last accessed November 17, 2000).

U.S. Department of Defense, Joint Publication 3-14, draft, no date.

—, *Unified Action Armed Forces*, Joint Publication 0-2, February 24, 1995.

—, *DoD Dictionary of Military Terms*, Washington, D.C., Joint Publication 1-02, 1998. Available at http://www.dtic.mil/doctrine/jel/doddict (last accessed November 20, 2000).

U.S. Congress, House of Representatives, Joint Hearing of the Committee on Science, Space, and Technology and the Permanent Select Committee on Intelligence *Scientific, Military, and Commercial Applications of the Landsat Program*, 102nd Congress, 1st Session, Washington, D.C.: U.S. Government Printing Office, 1991.

U.S. Space Command, *U.S. Space Command Vision for 2020*, February 1997.

—, *Long Range Plan: Implementing USSPACECOM Vision for 2020*, March 1998.

Williams, Richard S., Jr., Joanne Heckman, and Jon Schneeberger, *Environmental Consequences of the Persian Gulf War, 1990–1991: Remote-Sensing Datasets of Kuwait and Environs*, Research & Exploration Series, Washington, D.C.: National Geographic Society, 1991.

Winnefeld, James A., Preston Niblack, and Dana J. Johnson, *A League of Airmen: U.S. Air Power in the Gulf War*, Santa Monica, Calif.: RAND, MR-343-AF, 1994.

Section II: National Remote Sensing Programs & Policies

During the Cold War, the United States and Soviet Union possessed a near monopoly on imaging satellites in both the military and civilian realms. For most of the Cold War years, the superpowers were the only countries with sufficient capabilities to develop, produce, and operate their own constellations of earth-observation satellites. Since the late 1980s, however, there has been a steady growth in the number of countries (and multinational organizations) that have fielded their own civilian remote sensing satellites. These countries include both highly industrialized countries, such as France, Japan, and Canada, and some newly developing countries, such as Israel and India. In addition, other countries, including Australia, Norway, and Singapore, have developed extensive expertise in using satellite and airborne imagery for civilian and commercial remote sensing applications without operating their own imaging satellites.

Section II presents a representative survey of the remote sensing programs of several different countries that highlights their civilian and commercial imaging satellite programs and policies. In addition to offering specific insights on particular national imaging satellite programs, these chapters serve as a reminder that the major source of global transparency is the increasing number of countries, and their commercial enterprises, that are committed to pursuing earth-observation satellite programs.

Kevin O'Connell and Gregory Hilgenberg analyze the U.S. civilian and commercial remote sensing programs and policies in Chapter 7. Their analysis highlights the extensive investment that the U.S. government has made in supporting civilian earth-observation systems since the 1950s, including a continuing series of meteorological satellites and the new Earth Observing System satellites dedicated to studying global change. They examine in some detail the U.S. Landsat program, which produced the world's first civilian remote sensing satellite in 1972 and continues with the launch of Landsat 7 in 1999. O'Connell and Hilgenberg also examine the considerations that led the U.S. government in the mid-1990s to encourage U.S. firms to invest in developing commercial observation satellites and the extensive ground-based infrastructures needed to support these satellites.

George Tahu reminds us in Chapter 8 that Moscow actually led the way in the early 1990s in relaxing Cold War restrictions on satellite imagery by permitting declassified imagery (with resolutions as sharp as 2 m) from Russian military satellites to be sold on the international marketplace. However, in marketing data, Russian enterprises have struggled with a combination of bureaucratic, financial, and technical impediments. Nonetheless, Russian enterprises are exploring new forms of imaging satellites and imagery data sales that could once again set new precedents for marketing in this field.

In assessing the French remote sensing experience in Chapter 9, Isabelle Sourbès-Verger and Xavier Pasco point out that the Satellite Pour l'Observation de la Terre (SPOT) system laboriously built much of the current market for earth-observation data and will continue to be a major market force. They note that Paris and Washington differ in their definitions of the "commercialization" of remote sensing.

Kazuto Suzuki observes in Chapter 10 that Japan's remote sensing program is at a crossroads in making a difficult transition from an orientation toward technology push driven by government agency objectives toward market pull, in which customer requirements are paramount. He also assesses the major change that occurred in the Japanese remote sensing program following the North Korean launch of a Taepodong rocket over the Japan mainland in September 1998. This event galvanized Japanese determination to acquire a series of high-resolution information-gathering satellites for monitoring purposes.

In Chapter 11, Gerald Steinberg analyzes the satellite remote sensing programs of various countries in the Middle East and the Persian Gulf. He gives substantial attention to an examination of Israel's rapidly developing imaging satellite programs, including its ambitious plans to enter the commercial satellite imagery marketplace with a series of high-resolution imaging satellites. Steinberg also offers an alternative perspective on the benefits and risks of commercial satellite imagery as a force for global transparency. He argues that, at least for countries located in regions where the risk of armed conflict with neighbors is high, such as Israel, the benefits are uncertain at best. Rather, unrestricted access to high-resolution satellite imagery may actually erode national security by impeding the ability of these countries to conceal their vulnerabilities.

A similar ambivalence toward the value of satellite imagery is found in India's remote sensing policy. India's IRS-1C and IRS-1D satellites have gained it a significant presence in the international market for satellite imagery. However, in reviewing the evolution of the Indian remote sensing program in Chapter 12, Deborah Foster notes that high-resolution satellite images also raise difficult data policy and bureaucratic issues for India as its top government officials try to balance the national development benefits associated with selling and acquiring satellite imagery against the potential national security concerns that high-resolution imagery raise.

Canada's RADARSAT program is undergoing major changes as Canadian officials seek to define their national interests in developing the next-generation radar imaging satellite. The analysis of the Canadian remote sensing program that Michel Bourbonniere and Louis Haeck provide in Chapter 13 underscores the international policy challenges associated with moving to the RADARSAT-2 program, which is focused on meeting both civil and commercial remote sensing needs.

The indepth survey of national remote sensing programs presented in Section II is necessarily less than comprehensive. A more complete inventory of the past, present, and planned land remote sensing satellite systems, as well as their key imaging characteristics, can be found in Appendix B. Given the dynamic nature of these diverse national and commercial remote sensing satellite programs (i.e., program delays, launch failures, new starts, etc.), the reader is encouraged to check the Web site of the American Society for Photogrammetry and Remote Sensing, www.asprs.org/asprs/news/satellites/, for periodic updates to the data presented in Appendix B of this volume.

7

U.S. Remote Sensing Programs and Policies

Kevin O'Connell and Greg Hilgenberg

From the outset of the space age, the United States has led in developing and fielding earth-observation satellites. In the military sector, the United States assumed the lead in the 1960s with the CORONA spy satellite program. In the civilian sector, the United States launched the first weather satellite in 1960, the first of the Television and Infrared Observation Satellite (TIROS) series. Beginning in the 1970s, the Landsat program provided the first civilian remote sensing satellites for monitoring changes on the earth's surface. And in September 1999, the United States took the lead in the commercial sector with the launch of Space Imaging's commercial high-resolution observation satellite, IKONOS.

This chapter begins with a review of U.S. civilian observation satellite programs, which helped form a strong foundation in earth observation for development of the commercial remote sensing industry. The Landsat satellite series receives special attention because of its role as a technology development program that has come to emphasize continuity of the digital record for assessing geospatial change worldwide. The chapter also focuses substantial attention on the advent of U.S. commercial remote sensing firms, beginning with a discussion of the presidential policy directive that made this trend possible. Finally, the chapter briefly reviews some of the government programs that are encouraging the development of the commercial remote sensing industry in the United States.

Current U.S. Civilian Remote Sensing Programs

The U.S. commercial remote sensing programs of the 1990s did not arise simply as a set of competing entrepreneurial ventures pursuing a new market but rather as an outgrowth of U.S. civilian and national security programs that had been operational since the 1960s. This section reviews the leading U.S. civil and commercial remote sensing satellite programs, paying special attention to the first generation of U.S. commercial observation satellite programs and the U.S. government policy that has encouraged their development.

THE CIVIL SECTOR

The civil sector encompasses the agencies, organizations, and individuals who use remote sensing to promote the general welfare and provide a public good or who use the data for exploration and discovery within a civil program. Overwhelmingly, meteorology is the primary mission for satellites in the civil sector, whether for real-time weather forecasting or short- and long-term climate studies (Table 7.1). Despite their comparatively coarse ground sample resolutions, these satellites do connect with high-resolution satellites just going into operation.[1]

NOAA'S ENVIRONMENTAL MONITORING SATELLITE PROGRAMS

The National Oceanic and Atmospheric Administration (NOAA) of the U.S. Department of Commerce (DOC) operates most U.S. civil meteorological satellites. There are two primary programs: The Geostationary Operational Environmental Satellite (GOES) and the Polar-Orbiting Operational Environmental Satellite (POES).

Table 7.1—U.S. Civil Observation Satellites

Lead Agency	Satellite	Launch	Primary Mission
NOAA	GOES	Continual[a]	Continuous, broad-area weather forecasting
	POES/DMSP	Continual[a]	Near–real-time, high-resolution weather forecasting
NASA	OrbView-2	1997	Ocean color monitoring; carries a NASA sensor but is owned and operated by a commercial firm, Orbital Sciences
	TRMM	1997	Measuring global rainfall measurement
	QuickScat	1999	Measuring global wind speed and direction patterns
	EOS/TERRA	1999	Measuring global surface and atmosphere interactions
	EOS/PM-1 (Aqua)	TBD 2000	Assessing long-term global climate changes
	EOS/CHEM	TBD 2002	Measuring the dynamics of earth's atmosphere
USGS	Landsat 7	1999	Civil land remote sensing

[a] The satellite constellations are continually replenished by successive launches.

GOES provides continuous broad-area, low-resolution coverage of the earth for general weather forecasting. From geostationary orbit, a single satellite can continually watch over an entire continent and provide a constant stream of meteorological data.

In contrast, POES provides near–real-time, high-resolution coverage for weather forecasting. Each POES satellite carries a specialized sensor, the Advanced Very High Resolution Radiometer, which measures surface land and sea temperatures in the visible and thermal infrared spectral bands. Critical applications of this technology include monitoring volcanic eruptions and detecting forest fires.

For several years, the U.S. military had a separate but similar program, the Defense Meteorological Satellite Program (DMSP). But in May 1994, POES and DMSP were combined by Executive Order to create a single national program, the National Polar-Orbiting Environmental Satellite System, with its program office within NOAA. Part of this program's mandate is to develop a new series of spacecraft in conjunction with NASA (and, eventually, the European Meteorological Operational Weather Satellite program [METOP]). The resulting constellation will consist of three highly capable weather platforms that will operate well into the 21st century, providing near–real-time coverage for U.S. civil and military agencies, as well as numerous users worldwide.

NASA'S EARTH SCIENCE ENTERPRISE PROGRAMS

Remote sensing for environmental science is another core mission for U.S. civil remote sensing. The National Aeronautics and Space Administration (NASA) often operates at the cutting edge of technology in developing and fielding high-quality sensors to view the earth for various research applications. This effort was introduced in 1989 as the Mission to Planet Earth but was renamed in 1998 as the Earth Science Enterprise (ESE) (see NASA, 2000a). This enterprise is as an umbrella program covering a large range of activities, including exploiting remote sensing data from both surveillance and meteorological satellites to support global change studies (Table 7.1). Phase I heavily emphasizes earth observations from several U.S. and international spacecraft. Phase II, the Earth Observing System (EOS), will include a number of satellites with remote sensing capabilities that are presently in operation or that should be launched in the next few years.

One of NASA's more novel remote sensing missions was the Shuttle Radar Topography Mission in February 2000. The ultimate goal was to produce a digital, three-dimensional, high-resolution map of the earth's surface between

60°N and 60°S latitude using synthetic aperture radar (SAR) and interferometry. NASA conducted this mission jointly with the National Imagery and Mapping Agency (NIMA), which had a common interest in acquiring radar imaging data for generating the relatively precise topographic maps known as digital elevation models.

LANDSAT CIVIL REMOTE SENSING PROGRAM

NASA designed the Landsat series of remote sensing satellites (see Table 7.2) to be the country's main civilian instrument for monitoring natural changes and the effects of human activities on the earth's land surface (see NASA, 1999). Over its lifetime, this program has been an occasional victim of bureaucratic rivalry and uncertain domestic political support (see Chapter 3 in this volume) but has also been a major success in many respects. Landsat has provided a solid technological foundation for developing the U.S. remote sensing sector, has made a large and growing archive of satellite images available for a broad range of scientific and practical applications, and has stimulated international interest in earth observation by helping to build a global network of remote sensing expertise and infrastructure.

The extensive archives of Landsat imagery data are vital to a broad range of scientific, civil, and other users who depend both on the images and on the continuous, consistent updates that support change detection and analysis. This is especially important in the primary applications for Landsat data, which include

- agricultural monitoring
- land-cover change, including tropical deforestation and coastal monitoring

Table 7.2—Landsat Satellite Series

No.	Launch	Mission Status	Altitude (km) [a]	Revisit Rate (days)
1	July 1972	Ended January 1978	917	18
2	January 1975	Ended February 1982	917	18
3	March 1978	Ended March 1983	917	18
4	July 1982	Standby [b]	705	16
5	March 1984	Operational [c]	705	16
6	October 1993	Launch failed	705	16
7	April 1999	Operational	705	16

[a] All missions had circular, polar, sun-synchronous orbits.
[b] Effectively decommissioned.
[c] Can only relay data via direct downlink to ground stations.

- global biophysics and biochemistry
- ecology and forestry.

Although NASA is responsible for managing the development and production of Landsat, the Department of the Interior's U.S. Geological Survey (USGS) now has the leading role in operating Landsat 7 and in managing the resulting imagery data. The USGS runs the Earth Resources Observation System (EROS) Data Center (EDC) in Sioux Falls, South Dakota. The EDC is the world's largest archive of civilian remotely sensed data covering the earth's land masses, housing millions of cartographic maps, satellite images, and aerial photographs. EDC also operates the National Satellite Land Remote Sensing Data Archive, a legislatively mandated program designed to maintain a high-quality database of space-acquired images of the earth for use in studying global change and other related issues. This archive has become an important part of NASA's ESE program and is the primary archive and distribution node for Landsat 7 data.

The Landsat program can probably be best described as providing a continuous and consistent digital record of the earth's land resources (see Table 7.2). The initial satellites emphasized demonstrating new technologies and missions, and the early program pioneered the use of civil satellites for collecting multispectral digital images for surface monitoring applications. The early success and subsequent longevity of the Landsat program has created certain requirements for the present system:

- high-volume collection of earth land-surface imagery
- an emphasis on continuity in sensor performance and calibration
- a worldwide network of ground stations
- a major imagery data archive of cloud-free, sunlit land images
- a strong focus on the scientific community and civilian agencies as the primary users of Landsat imagery data.

This legacy includes a well-developed ground infrastructure. Over 20 ground stations on six continents and the EDC in South Dakota receive Landsat data. Most importantly, Landsat, as the world's first civilian remote sensing satellite, has played a crucial role in establishing international support for the basic principles that govern the operations of earth-observation satellites today, including the concept of open skies and the principle that sensed states have a right to nondiscriminatory access to civilian satellite imagery of their territories.

TECHNOLOGY DEVELOPMENT

The Landsat program has been central to the development of space-based sensors for nonmilitary earth-observation satellites. Up to the early 1990s, the U.S. government's civil and national security space sectors completely dominated remote sensing and so pioneered the use of new and improved technologies. As foreign and commercial interests matured, the Landsat program was no longer the sole leader in cutting-edge technology for remote sensing satellites. Today, the Landsat program concentrates on filling the specific needs of its civil and scientific users, which means that sensors are chosen primarily to meet user applications rather than to advance technology. While commercial firms are competing to achieve very high resolutions, Landsat is concentrating on providing continuity in medium-resolution multispectral imagery, which is better suited to resource management and mapping needs.

Landsats have carried three primary types of sensors to complete the remote sensing mission. As Table 7.3 indicates, the main improvements occurred with the addition of new types of sensors, although smaller-scale technological improvements were implemented with each consecutive satellite. In the 1970s, the Return-Beam Vidicon (RBV) system was based on the earlier, more mature analog video technology. The RBV consisted of three coaligned television cameras and was initially designed to cover the visual and near-infrared spectral bands at a resolution of approximately 80 m (Mika, 1997).

Landsat was also a pioneer in the design and use of the multispectral scanner (MSS), a digital video scanner similar to a reflector telescope. In the MSS, a scanning mirror directs images of the earth to a fiber-optic array and

Table 7.3—Sensor Characteristics of the Landsat Series

No.	Primary Sensors	Resolution (m)	Data Rate (MBPS)
1	RBV, MSS	80, 80	15
2	RBV, MSS	80, 80	15
3	RBV, MSS	30, 80	15
4	MSS, TM	80, 30	85
5	MSS, TM	80, 30	85
6[a]	ETM	15 panchromatic, 30 multispectral	85
7	ETM+	15 panchromatic, 30 multispectral	150

[a]Landsat 6 failed to reach orbit in October 1993.

then to photomultiplier tubes. This type of sensor was the state of the art in the late 1960s, and subsequent MSSs have incorporated significant upgrades, including new materials, expanded spectral coverage, and greater resolution.

Eventually, however, a more advanced sensor, the Thematic Mapper (TM), became the staple of the Landsat sensor suite. The new sensor is called "thematic" because it produces images in spectral bands tailored to specific "themes," or missions, such as agriculture, forestry, or geology. The TM instruments on Landsats 4 and 5 represented a different and decidedly better technological approach than the MSS in all areas, including increased spectral resolution and coverage, improved scanning methods, and use of lightweight composites. The TM spectral bands used a more advanced method for collecting and distinguishing the reflective characteristics of vegetation and minerals. A 15-m panchromatic band was finally added as part of the Enhanced TM Plus (ETM+) sensor package on Landsat 7 to increase imagery resolution.[2]

Thus, each successive Landsat has benefited from a steady process of incremental improvements in resolution, calibration, sensitivity, and accuracy from adding spectral bands to increase utility. All of the Landsat sensors that have reached orbit have operated better than expected and for longer than they were designed to, producing invaluable data for terrestrial applications. Yet, over time, the Landsat series has backed away from the cutting edge of new technology. As new civil and commercial remote sensing players have entered the field, however, Landsat can no longer be called a technology pioneer. This does not detract from the Landsat mission, however, since the program plays a valuable role providing civil users with well-calibrated, multispectral digital imagery that addresses a broad range of civil, commercial, and even military requirements.

Nonetheless, the future of the Landsat program is presently uncertain, even as Landsat 7 has been operating at full capability following its successful launch in April 1999. If the Landsat follow-on is funded, whether by government or the private sector, it will probably be used to sustain the U.S. government's commitment to maintaining continuity in the flow of medium-resolution, digital imagery data.

LANDSAT DATA POLICY

As originally envisioned in 1972, Landsat data was intended to be made available to the general public at the lowest cost and by the easiest means possible, which meant that government would have to subsidize all remaining costs. But some problems intervened, including issues about the relative investments of the public and private sectors, separation of military and civilian priorities, competition versus cooperation with other nations, and the uncertainties of government subsidies.

The data policy for Landsat 7 (NASA, 1997), essentially returned Landsat imagery to the status of a public good. The key elements of the policy are to

- ensure that unenhanced data are available to all users at cost
- ensure timely and dependable delivery of unenhanced data to all civilian, national security, commercial, and foreign users and to the National Satellite Land Remote Sensing Data Archive
- ensure that the United States retains ownership of all unenhanced Landsat 7 data
- support the development of the commercial market for remote sensing data
- ensure that commercial value-added services continue to come exclusively from the private sector
- ensure that the data distribution system for Landsat 7 is as compatible as possible with the EOS Data and Information System.

It is important for potential users of Landsat data to understand what type of data they can receive from the ground stations. Landsat 7 Data Policy (NASA, 1997) mandates that the EDC will provide only limited data processing, leaving all further data enhancements to users and value-added providers.

U.S. Commercial Remote Sensing Programs

The U.S. remote sensing community received a small shock in 1986 when France launched its first Satellite Pour l'Observation de la Terre (SPOT-1). In only a few years, SPOT Image vaulted past Landsat as a seller of satellite imagery. Then, when India launched the first Indian Remote Sensing (IRS-1A) satellite in 1988, it seemed as though everyone was getting in not only on proliferation of remote sensing technologies but also on the inevitable creation of a commercial remote sensing market.

At the time, both U.S. government and industry leaders feared being left behind in an industry they felt the United States should dominate. By the early 1990s, Congress heeded the call to loosen the federal monopoly on remote sensing technology and data distribution. After weighing the priorities of commerce and competitiveness against those of national security, Congress passed the Land Remote Sensing Policy Act of 1992, which opened the door to licensing of U.S. commercial remote sensing satellites.

Presidential Decision Directive 23 on foreign access to remote sensing space capabilities further galvanized the willingness of U.S. aerospace firms to proceed with commercialization of observation satellites (White House, 1994). Building on earlier legislative directions, such as the 1992 act, this directive removed many uncertainties that had been inhibiting Executive Branch action on licensing commercial remote sensing satellite operations and set forth government guidelines on foreign sales of imagery data, sensitive technologies, and even complete turnkey observation satellite systems (Baker, 1997). More importantly, in an important post–Cold War shift in perspective on U.S. space capabilities, this policy envisions that commercialization would be a "win-win" situation for U.S. commerce and security. As stated, the goal of the U.S. policy is

> to support and to enhance US industrial competitiveness in the field of remote sensing space capabilities while at the same time protecting US national security and foreign policy interests. (White House, 1994.)

The implementation of PDD-23's policy has drawn significant criticism both inside and outside the government, but the number of licenses granted for commercial observation satellites did increase rapidly after 1994 (see Table 7.4).

Table 7.4—U.S. Commercial Remote Sensing Licensees

Company	System	Date Issued	Note
EarthWatch	EarlyBird	January 4, 1993	Satellite failed December 1997
Earth Observing Satellite Corporation (EOSAT)		June 17, 1993	Acquired by Space Imaging
Space Imaging (LMC)	IKONOS	April 22, 1994	Operational since 1999
OrbImage	OrbView-1 OrbView-3 OrbView-4	May 5, 1994	Operational since 1995 Launch due 2001 Launch due 2001
OrbImage	OrbView-2	July 1, 1994	Operational since 1997
EarthWatch	QuickBird	September 2, 1994	Launch due 2000
AstroVision		January 23, 1995	—
GDE Systems Imaging		July 14, 1995	Acquired by BAE
Motorola		August 1, 1995	—
Boeing	Resource21	May 16, 1996	—
CTA Corp.		January 9, 1997	—
RDL Space Corp.		June 16, 1998	—
STDC		March 26, 1999	Acquired by ESSI; launch tentatively due 2001

And a U.S. firm launched the first high-resolution commercial observation satellite, IKONOS, in 1999.

PDD-23 extended DOC's responsibility for granting U.S. government licenses to U.S. private remote sensing enterprises (first granted by the Remote Sensing Policy Act of 1992), with NOAA's National Environmental Satellite Data and Information Service (NESDIS) in turn being responsible for licensing commercial remote sensing satellites (NOAA, 2000). NOAA also leads interagency reviews involving the departments of Defense, State, and the Interior, as well as other relevant U.S. government agencies. In comparison, the Department of State is the lead agency for reviewing any requests by U.S. firms to export sensitive technologies or even turnkey imaging satellite systems to other countries.

The Secretary of Commerce is responsible for ensuring that U.S. firms that have license approval adhere to the U.S. government guidelines for operating commercial observation satellites, as set forth in PDD-23. Among other things, the licensee must

- maintain records of all satellite tasking for the previous year and permit U.S. government access to the record
- not change the operational characteristics of the satellite system without formal notification and U.S. approval
- obtain U.S. government approval for any encryption devices that deny unauthorized access
- provide for U.S. government review of its intent to enter into significant agreements with foreign parties
- make unenhanced imagery data available to "sensed states" as soon as the data are available and for reasonable cost, terms, and conditions
- provide the U.S. government access to and use of data when required for national-security or foreign-policy purposes
- limit imaging when national security or international obligations and/or foreign policies may be compromised.

Some of these regulations have fueled a continuing debate, particularly over the last provision, which gives the government the right to impose "shutter control"—restrictions on the collection and/or distribution of commercial satellite imagery data—during emergencies. Some U.S. firms have been concerned about the Clinton administration's unwillingness to specify what foreign policy or military problems would constitute such an emergency. This tension is largely unavoidable: On one hand, commercial firms have a strong interest in minimizing the risk of unexpected disruption of their remote sensing operations through government action. On the other, the Executive Branch has an equally strong interest in avoiding specificity that might limit its decisionmaking flexibility in this area.

Shutter control remains a most controversial element of U.S. regulatory controls over the commercial remote sensing industry. To attempt to clarify U.S. policy, the departments of Commerce, State, Defense, and Interior and the intelligence community signed a memorandum of understanding (MOU) in February 2000. According to the Fact Sheet (DOC, 2000), the purpose of the MOU was

> to establish interagency procedures concerning the process for handling remote sensing licensing actions, and consultation regarding interruption of normal commercial operations consistent with the President's policy on remote sensing.

Specifically, the MOU covered consultation during review of licensing actions and regarding interruption of normal commercial operations and coordination before release of information provided or generated by other agencies.

Even though the rules for obtaining a license have existed for several years, government agencies only began finalizing the methods of carrying out the policies that deal with commercial imaging firms in 1999, in particular following the successful launch of IKONOS. While DOC remains the focal point for licensing and shutter control actions, it is clear that the departments of Defense and State and the intelligence community have significant consultative oversight of certain commercial remote sensing matters:

> If the relevant Secretaries are unable to resolve any issues, the Secretary of Commerce will so notify the Assistant to the President for National Security Affairs, who, in coordination with the Assistant to the President for Science and Technology, will seek to achieve a consensus within the interagency, or failing that, by referral to the President. (DOC, 2000.)

In a further effort to clarify policy related to PDD-23, NOAA published a revised licensing regime for commercial remote sensing vendors in July 2000. In part using public (industry and government) comment, NOAA attempted to elaborate on the guidelines for licensing and operating commercial imaging satellites (DOC, 2000). There, NOAA provided more-precise definitions of important concepts that have historically caused some confusion and misunderstanding between government and industry, such as the distinction between *operational* and *administrative* control of space and ground assets. More importantly, NOAA has attempted to eliminate excessive and redundant regulations about notifications, technology transfer, license amendments, export controls, and levels of foreign investment. With regard to national security and foreign policy obligations, NOAA commented that any shutter control limitations would be imposed for the smallest possible area and the shortest time necessary to protect the U.S. security issues at hand.

Another debate-producing aspect of the original guidelines concerned the decision that both the Executive Branch and Congress made to assuage Israeli security concerns by prohibiting U.S. firms from either collecting or distributing high-resolution (better than 2 m) satellite imagery of Israel. U.S. firms are concerned that such precedent-setting restrictions could have a chilling effect on potential foreign investors, partners, and customers by encouraging other countries to restrict how they can be imaged by commercial satellites (Simons, 1998; Ferster, 1998). These concerns are heightened by what appears to be an ambitious Israeli commercial space program, including a constellation of high-resolution remote sensing satellites.

U.S. Commercial Remote Sensing Firms

Given the opportunity of commercial remote sensing, U.S. firms have stepped out boldly and aggressively. After the unbridled optimism and enthusiasm about the commercial market that seemed to mark the mid-1990s, U.S. firms have had to continue to adapt in the face of new risks and challenges, whether induced by government policy, their own delays and technical setbacks, or the slower emergence of the commercial market. Among the most interesting aspects of U.S. firms has been the diversity and the ever-changing nature of their business models.

Three U.S. firms currently lead the industry in providing commercial remote sensing imagery—Space Imaging, EarthWatch, and Orbimage—although this is likely to change as a function of technical and market success and the arrival of newcomers. These firms share some common characteristics: an aerospace legacy, similar technologies and approaches to satellite imaging, sophisticated digital data archives, and multinational partnerships. Although all focus on selling information products based on higher-resolution satellite imagery, their market strategies differ somewhat. In addition, several new enterprises are pursuing other imaging technologies and market segments.

SPACE IMAGING

In October 1994, Space Imaging International (Thornton, Colorado), spun off from Lockheed with the goal of creating the first commercial high-resolution satellite. Space Imaging is now a leading competitor in commercial remote sensing (see Space Imaging, 2000). This was the first commercial firm to deploy a high-resolution satellite, with its successful launch of IKONOS in 1999. Space Imaging continues a limited partnership with Lockheed, as well as with Raytheon E-Systems Inc., Mitsubishi Corp., Singapore's Van Der Horst Ltd., South Korea's Hyundai Space & Aircraft, Europe's Remote Sensing Group, Swedish Space Corp., Thailand's Loxley Public Co. Ltd., and several other international investors.

Space Imaging presently maintains key strategic alliances with firms in Europe; Japan; the Middle East; and, most recently, Latin America. These regional affiliates, operated and financed outside of Space Imaging, maintain local ground stations that can receive imagery directly from one or more satellites within the Space Imaging constellation and distribute data products throughout their respective regions in cooperation with a network of distributors and value-added resellers. The affiliates have assumed significant responsibilities for tasking the IKONOS imaging satellite within their communication cones and for marketing the imagery data. Within North America, Space Imaging has formed agreements with a large array of resellers and distributors, effectively covering all major markets in Canada, the 50 states, and parts of Mexico.

While often discussed only in the context of the high-resolution IKONOS satellite, Space Imaging continues to diversify the remote sensing sources from which it can produce output, whether in the form of raw data or value-added products. In 1996, Space Imaging made a key acquisition when it purchased EOSAT from Lockheed Martin. EOSAT began as the U.S. government partner in charge of commercializing the Landsat program by operating and selling data from Landsats 4 and 5 and grew into the largest U.S. provider of worldwide space-based imagery. Space Imaging also later gained rights to sell imagery from India's IRS-1C and IRS-1D satellites. Space Imaging also distributes imagery from the Japan Earth Remote Sensing (JERS) satellite and the European Space Agency's Remote Sensing (ERS) satellite. In 1997, the firm further increased its scope by gaining distribution rights to Canada's RADARSAT-1 imagery.

Space Imaging has also taken advantage of partnerships with the U.S. government. Like other remote sensing vendors, the firm participates in

NASA's Scientific Data Purchase program and has agreed to provide imagery products to NASA and other research organizations over several years. Space Imaging has also entered into contractual arrangements with NIMA to provide imagery and data products to the U.S. military and other federal agencies.

The firm suffered a painful blow when the first IKONOS satellite was lost soon after launch on April 27, 1999.[3] While the loss was significant, Space Imaging was able to rebound and deploy the second IKONOS satellite successfully on September 24, 1999. IKONOS is in a near-polar, sun-synchronous, circular orbit at an altitude of 682 km and is capable of providing 1-m panchromatic and 4-m multispectral imagery. IKONOS produced its first publicly distributed image on October 12, 1999—a 1-m panchromatic image of downtown Washington, D.C.

Commercially, Space Imaging has tried to become a "one-stop shopping" company, supplying both imagery and value-added services and providing the highest-quality products and services among commercial contenders. This goal was clearly demonstrated when Space Imaging acquired Pacific Meridian, a value-added firm that specializes in geographical information system (GIS) and imagery applications, in March 2000 with a view to improving end-to-end service capabilities. The general market strategy is to focus first on traditional markets and then on creating and expanding newer markets, such as real estate and insurance. Even though Space Imaging performs many of its value-added services in house, it also has extensive agreements with outside firms, especially overseas.

The firm has also developed an impressive database and distribution network to catalog and sell remote sensing imagery. Under the CARTERRA brand name, Space Imaging advertises end-to-end management and production of remote sensing products using the latest GIS techniques. Besides the IKONOS satellite, Space Imaging and CARTERRA have access to the EOSAT archive and distribution network of observation data from Landsat, ERS, JERS, RADARSAT-1, and IRS satellites, as well as aerial photography from the Digital Aerial Imagery Systems-1 [DAIS-1] operation. They have also created an agreement with TerraServer to distribute Landsat and IRS-1C and IRS-1D data.

EARTHWATCH

One of the first U.S. commercial firms to pursue commercial remote sensing specifically was EarthWatch Incorporated, formed in March 1995 as a merger between Ball Aerospace & Technologies Corp. and WorldView Imaging Corp. (see EarthWatch, 2000). In January 1993, WorldView became the first commercial organization to receive a license from DOC to operate a commercial, high-resolution, remote sensing satellite, EarlyBird 1.

EarthWatch is a privately held company whose main headquarters are in Longmont, Colorado. Ball Aerospace, based in nearby Boulder, Colorado, constructs the imaging satellites in cooperation with the Eastman Kodak Company and Fokker Space B.V. EarthWatch's investment partners include Hitachi Ltd.; Datron Systems Inc.; Nuova Telespazio; and MacDonald, Dettwiler and Associates Ltd. To catalogue data and serve customer needs, EarthWatch also maintains an on-line database called Digital Globe®, which includes a world-class mapping production facility that can use both satellite and aerial data to produce mosaics, orthorectified imagery, and digital elevation models.

Despite being first out of the licensing gate, EarthWatch had its own setback in December 1997 when the EarlyBird 1 satellite ceased communicating after a few days in orbit. The EarlyBird satellite had been designed to provide 3-m panchromatic and 15-m multispectral imagery, which would easily have exceeded the performance of any nonmilitary system to date. In reaction to the loss, EarthWatch decided in 1998 not to deploy the second EarlyBird satellite but to focus instead on the higher-resolution QuickBird imaging satellite.

Even with the critical loss of EarlyBird, EarthWatch has continued to push forward and has received contracts from NASA and NIMA to use the Digital Globe® database for image processing. Until the QuickBird satellites successfully reach orbit, however, EarthWatch's prospects will remain uncertain, given their commercial operations. As an example, EarthWatch has twice restructured its debt and equity in an attempt to take the company into full commercial operation.

The EarthWatch marketing strategy is noticeably different from those of its near competitors and reflects the more streamlined nature of the firm. Initially, it had expected to be first to market with EarlyBird but has reemphasized low-cost operations to meet customer needs at lower prices since the failure. EarthWatch does not plan to offer significant in-house capabilities for value-added products, instead relying on partnerships with outside private firms to meet user needs.

The future of EarthWatch depends heavily upon the launch and operation of the QuickBird satellites. QuickBird 1 has been designed to provide 1-m panchromatic and 4-m multispectral and near-infrared imagery. Going outside the norm, EarthWatch plans to deploy QuickBird 1 in a 600 km, circular orbit at an inclination of 66°, giving it the relatively unique ability to obtain images of the earth at different times of day under different lighting conditions. Among other benefits, this orbit will give QuickBird 1 the ability to revisit previously imaged sites much more quickly than a satellite in sun-synchronous orbit can. QuickBird 2, however, will follow the industry standard of a sun-synchronous near-polar orbit and will offer same-time observations under consistent viewing conditions. Recently, in yet another indication of a shift in business model, EarthWatch announced sales and marketing agreements with five ground station providers to facilitate the ingestion of images after launch.

ORBIMAGE

Another early entry into the commercial observation satellite race was Orbimage (see Orbimage, 2000). It was founded in 1991 as a subsidiary of Orbital Sciences Corporation, a space and information system company that designs, builds, operates, and sells a broad range of space-related products and services.

Orbimage highlights its continuing experience in building and operating earth-observation satellites for the U.S. government. One of its major successes is OrbView-2, also called SeaStar, launched in August 1997. This satellite produces daily color meteorological imagery using a 1-km resolution multispectral sensor. OrbView-2 also has a number of commercial customers, primarily fishing fleets. NASA, a primary user, sponsored most of the development of the platform to carry the Sea-Viewing Wide-Field-of-View Sensor, a component of NASA's ESE program. This sensor provides the earth-science community quantitative data on global ocean bio-optical properties.

Orbimage will enter the resolution race with planned the launches of OrbView-3 and OrbView-4 sometime in 2001. OrbView-3 will carry 1-m panchromatic and 4-m multispectral sensors providing truly high-resolution imagery. The OrbView-4 satellite will be placed in a complementary orbit to that of OrbView-3, which will give Orbimage more flexibility to meet its imagery-collection needs. In addition to 1-m panchromatic and 4-m multispectral sensors, OrbView-4 will have an 8-m hyperspectral sensor. The cutting-edge hyperspectral technology is being developed with funding from the U.S. Air Force, under the designation Warfighter-1. For security reasons, the hyperspectral products from this satellite will have limited distribution, but Orbimage will likely be able to sell degraded imagery products on the open market.

Orbimage has solidified its strength with on-orbit systems but has also made dramatic progress with imagery archives and distribution. Currently, the firm has U.S. distribution rights to earth imagery obtained from several current or expected systems, including all four OrbView satellites, the Canadian SAR platforms RADARSAT-1 and -2, and the Russian SPIN-2 satellite.[4] In September 1999, Orbimage also entered into a five-year agreement with SPOT Image to gain exclusive U.S. distribution rights to all SPOT imagery. In return, SPOT Image gained the rights to distribute imagery from both OrbView-3 and -4 in Europe and other selected nations around the world.

Orbimage maintains a central processing facility in Virginia but also operates or licenses over a dozen ground stations worldwide, which aid in distribution of imagery data. In expectation of the launches of OrbView-3 and -4, the firm contracted with NTT Data Corporation of Japan in March 1999 to build a ground station to receive and process OrbView imagery. This agreement also allows Orbimage to set up distribution rights in Japan for panchromatic and multispectral high-resolution imagery.

Finally, Orbimage has also aggressively pursued the ability to distribute observation imagery over the Internet. The company maintains the on-line Orbnet Digital Archive in Dulles, Virginia, to provide customers global access to various remote sensing data products. Orbnet contains products not only from OrbView satellites but also from the civil NOAA program and aerial photography. In September 1999, Orbimage joined with ESRI, a leading GIS company, to distribute OrbView data within a specialized archive, called OrbView Cities, which carries detailed, 1-m panchromatic layouts of a growing number of large U.S. cities.

OTHER U.S. COMMERCIAL FIRMS

Only half of the licenses on the DOC list can be attributed to these three companies. For a number of reasons, the "big three" firms are the only ones that have progressed rapidly in implementing their business plans, have ongoing or completed construction of the actual space hardware, and have launched imaging satellites or have firm plans to launch in the near term. However, other U.S. companies, such as those below, may also enter the commercial remote sensing market by developing their own imaging satellites or at least imagery distribution systems aimed at commercial users.

Of note with respect to Orbimage's WARFIGHTER, as well as some of the companies mentioned below, is the emergence of business models that envision commercialization of new phenomena, such as radar and hyperspectral data, and the U.S. policy concerns that arise therefrom (see Chapter 5 in this volume).

Research Development Laboratories Space Corporation. The only U.S. private company with a license for a SAR remote sensing satellite is the

Research Development Laboratories (RDL) Space Corporation. RDL, the parent firm, has extensive experience with defense space-based radar technologies, including contracts to process data from all Space Shuttle SAR missions and from a joint project with NASA and the Jet Propulsion Laboratory to develop LightSAR. The RDL Space Corporation was created in 1996 to commercialize high-resolution SAR imagery from space. While the majority of U.S. remote sensing firms build sensors for the visible or near-infrared spectral bands, RDL hopes to capitalize on the unique qualities of radar. RDL plans to build and launch its first satellite, RADAR-1, in 2001, to provide 1-m radar imagery from low earth orbit.

Despite the new regulatory environment, more-intense operating restraints have been placed on RDL—and radar technology in general—than on electro-optical systems, for U.S. national security reasons. The U.S. government may insist on limiting the resolution to 5 m, even though RDL contends that it needs a 1-m capability to be competitive against such foreign competitors as Canada's RADARSAT-2, which is planned to have a 3-m resolution capability.

Earth Search Sciences, Inc. Other U.S. companies expect hyperspectral imagery to be the eventual winner in the commercial remote sensing market. Earth Search Sciences Inc. (ESSI), a U.S. aerial firm, has already tested hyperspectral cameras for remote observation, in conjunction with the Department of Energy and NASA. To exploit the possibilities of the industry fully, ESSI purchased the one commercial firm developing space-based hyperspectral sensors, Space Technology Development Corporation (STDC), of Alexandria, Virginia.

STDC was established in 1993 to commercialize space-related technologies developed in government laboratories. The company had been developing the Naval EarthMap Observer (NEMO) in coordination with the Naval Research Laboratory but ran into financial difficulties in 1999. ESSI acquired STDC and hopes to restart the program, tentatively planning to launch NEMO in 2001. The satellite will carry a sensor package that includes a 5-m panchromatic sensor and, most notably, a 30-m hyperspectral sensor. ESSI expects to operate NEMO on a commercial basis for 75 percent of its tasking, and the Navy will govern the tasking priority for the remainder.

The NEMO program is another groundbreaking venture that combines commercial and military interests in a single remote sensing platform. If successful, this program would provide detailed characterizations of coastal environments around the globe to U.S. naval planners directly supporting warfighters. In an expected five years of operation, NEMO will map the majority of the earth's surface, emphasizing littoral regions, at medium resolution in addition to serving the specific needs of commercial customers.

Resource21. Another potential player in the commercial remote sensing race is Resource21, a Boeing Corporation subsidiary dedicated to developing a multispectral observation satellite system. Boeing received the necessary license from DOC in May 1996, but work has not, as yet, proceeded past the developmental stages. The company has a partnership agreement that also includes Farmland; the Institute for Technology Development; and GDE Systems Inc., a member of Marconi Integrated Systems.

Resource21 plans to focus mostly on customers in precision agriculture but will also serve the forestry, government sales, and service sectors. The emphasis on agriculture is somewhat unusual among the rising commercial firms, but Resource21 expects the need and the market for farm applications to grow. The company plans to operate at least a pair of coplanar satellites in sun-synchronous orbits at an altitude of 743 km. To maximize applicability to agriculture, Resource21 will have multispectral sensors with a resolution of at least 20 m and perhaps as good as 10 m.

TerraServer. Not all the important players in the commercial remote sensing industry build and operate satellites to produce imagery data for sale. Two developments in the 1990s have been key to the industry's more-general acceptance: a near revolution in user-friendly software for digitally manipulating imagery and the accumulation of searchable archives of vast amounts of data that allow efficient distribution to customers. TerraServer is one Internet company that has taken advantage of these trends by loading satellite and aerial imagery into a large and agile database and passing the imagery along to its consumers at retail prices. Potential customers can quickly click through a few screens to isolate the images they desire and can then order and pay for the images on line. TerraServer began in June 1998 with substantial support from Aerial Images, supplier of Russian SPIN-2 imagery, and StorageTek, a company specializing in high-volume data storage.

Other Types of Firms. Of course, it is worth noting that the commercial remote sensing industry involves much more than the most publicized commercial observation satellite firms. It also encompasses the firms that manufacture the hardware—the satellite systems and ground stations—and the firms that create the imagery processing and data archiving technology and software. The industry also includes the value-added providers—a broad range of image data-processing and analysis firms that focus on translating the satellite images into information products that are more valuable to diverse users. All these components of the remote sensing industry stand to gain from the U.S. government's decision to support the development of U.S. commercial observation satellites that will be highly competitive in the international marketplace for geospatial data and information.

THE U.S. GOVERNMENT'S USE OF COMMERCIAL REMOTE SENSING DATA

Consistent with PDD-23, U.S. government agencies have determined that a commercial remote sensing industry is important not only in an overall sense but also possibly for the agencies' individual missions. This has recently led the government to encourage industry development and advances in applications development. NASA and NIMA are prominent among the users of commercial remote sensing.

NASA tasked its Commercial Remote Sensing Project (CRSP) with accelerating the development of the U.S. remote sensing industry, creating mutually beneficial partnerships between ESE scientists working with NASA and the remote sensing industry, and making NASA a good customer for the U.S. remote sensing industry. Since 1987, this program has grown to involve over 50 U.S. firms from the remote sensing and GIS sectors (see NASA, 2000b).

NASA expects to develop and validate remote sensing applications, form trade associations, train a workforce to populate the industry in the 21st century, and conduct joint research and development. In addition to assisting development, NASA is helping out with sales. A key element of CRSP is the Science Data Purchase program, which was developed in response to administration and congressional interest in purchasing suitable remote sensing data from the private sector for NASA's scientific and research needs. This $50 million program was funded within the NASA ESE Program to provide scientific data to the science community.

Developing a much larger and more diverse market for satellite imagery is critical to expanding the commercial industry. Another program within the CRSP, the Earth Observations Commercial Applications Program, has created cooperative relationships with U.S. companies to expand the use of remote sensing technology in the geographical information industry. One of this program's central purposes is to broaden general knowledge and use of remote sensing technology in the marketplace.

Because of its role as the imagery manager for the Department of Defense and the intelligence community, NIMA also has a strong potential interest in a commercial remote sensing industry. Ideally, commercial remote sensing is a

valuable source of supplemental data for addressing U.S. national security requirements, as well as a series of new commercially driven applications for remote sensing of potential use to national security. NIMA's created a Commercial Imagery Program to seek "opportunities to leverage industry's proven capabilities" (NIMA, 2000) and to
- encourage and nurture the ongoing exchange of information with industry through visits, demonstrations, conferences, and exhibits
- promote the use of commercial solutions in NIMA's core and supporting business processes and systems
- promote constructive teamwork between NIMA and industry
- influence industry leaders to develop commercial products, services, and interoperable technologies that better match NIMA's needs.

Like NASA, NIMA began a program to buy data products from U.S. remote sensing firms and to form cooperative relationships to leverage any opportunities the private firms might offer. NIMA's Commercial Imagery Program (CIP) focuses on learning and exploiting the capabilities of all potential suppliers of remote sensing imagery. Starting in 1998, NIMA contracted multiyear purchasing agreements with such U.S. firms as EarthWatch, Orbimage, and Space Imaging to obtain imagery data and services in support of such federal agencies as the Department of Defense and the Federal Emergency Management Agency. To date, these contracts have largely focused on funding infrastructure enhancements that will expedite the firms' abilities to collect and deliver imagery data to government users.

While such contracts have provided U.S. commercial remote sensing firms with much-needed financial support, there is still controversy and debate over the utility of commercial remote sensing to satisfy security requirements. Timeliness of commercial products, competition with commercial users, and the sensitivity associated with the U.S. government's security interest in a given area are among the top concerns with the use of this imagery. Among the longer-term issues are the extent to which the industry is robust enough to reliably provide some data sets of routine import to military and security missions (e.g., mapping data), and the extent to which the national security satellite architecture should be adapted to focus on more pressing tasks.

Commercialization creates another complex problem for the security community. There is an insufficient understanding of the extent to which older classified imagery, current national technical means imagery, and commercial remote sensing data and products intersect, whether in support of national security requirements, or in terms of U.S. limits on allowing access to data. For example, the U.S. Intelligence Community in 1994 began releasing previously classified photoreconnaissance imagery from earlier (1960 to 1972) programs, such as CORONA, to the public. Although these huge sums of old data are of doubtful utility for modern commercial and civil applications, their release is of concern to U.S. firms. Private remote sensing satellite firms are concerned that the U.S. government will declassify large amounts of higher resolution imagery data in ways that will diminish the firms' prospects for success in the commercial market.

Because of the changes in the remote sensing industry, NIMA and National Reconnaissance Office (NRO) have begun to recognize that access to and the control of these assets is fundamentally different, which demands adopting new strategies for serving U.S. national security interests. Both the Director of Central Intelligence and the Deputy Secretary of Defense approved one such strategy in 1999, which recognized the new environment and represents a far-more-open approach than in the past. In this strategy, the intelligence community has determined that remote sensing is strategically important and that the industry will play a key role in pursuing overall information superiority. The primary risks include the near-term financial instability of the commercial industry, the security implications of the commercial availability of high-resolution imagery, and questions about whether commercial firms can in fact provide advertised performance. NIMA and NRO have calculated that commercial suppliers can be reasonably counted on to handle less-urgent, low- and medium-quality tasks, while the intelligence community can concentrate on more-urgent, high-end tasks. Yet the implementation of the current strategy is very uncertain, based on ambivalent implementation and uncertain funding.

Concluding Observations

The advent of commercial observation satellites has blurred the once relatively clear distinction between U.S. civilian and military programs. The growing list of satellite capabilities and data applications cuts across traditional sectors, demanding that civilian, military, and commercial players increase cooperation. This more-complex environment presents both opportunities and challenges for policymakers shaping U.S. remote sensing programs and the associated data policies. These diverse U.S. earth-observation satellite programs are focused on satisfying disparate civil, military, and commercial remote sensing requirements but also provide complementary capabilities for promoting greater global transparency. The imagery data that these satellites produce is becoming increasingly available to a wide range of users as impediments to public and commercial access to satellite imagery data decrease.

However, it is important to remember that the U.S. remote sensing effort involves much more than a series of satellites; it is rooted in a highly developed scientific and industrial infrastructure, and its influence is bolstered by a broad array of value-added firms and application-oriented research programs. Thus, the U.S. lead in remote sensing stems not only from the ability to operate imaging satellites but also from a growing body of expertise in transforming raw satellite imagery into products suitable for a broad range of users.

Yet, even with all the history and inherent advantages that the United States seems to have, it is not clear that it has yet demonstrated the ability to dominate remote sensing in the future, commercial or otherwise. The risks to U.S. private firms are substantial, including the usual technical and market risks inherent in any high-technology commercial venture that has substantial domestic and international competition. U.S. government policies and regulatory practices have in fact increased the risks for commercial firms, although more-recent developments portend a more-sophisticated understanding of how the U.S. government might be a better customer for commercial remote sensing and a better partner in regulating it. Yet the broader strategy needed to integrate civil, commercial, and national security remote sensing for overall U.S. objectives remains evasive.

And what of transparency? The role of information in statecraft and warfare, by all accounts, is growing. Information technologies, including remote sensing, are creating greater transparency in political and military affairs. Yet even as U.S. strategists push to make greater use of the information, the proliferation of these technologies and the resulting capabilities continue to cause concern. While not completely comfortable with the concept itself, the United States implicitly understood the effects the remote sensing commercialization would have on security, given the twin pillars of Cold War experience with remote sensing and the desire to expand access to information technologies and their products. Yet while PDD-23's development involved various cost-benefit analyses, official behavior since the directive's release has not reflected full comfort with the findings. Despite extensive U.S. experience with transparency—whether as a de facto condition or as a strategy—government agencies continue to be leery of its consequences. Remote sensing commercialization provides an essential rationale for the United States to understand and better deal with its consequences.

Notes

1. For a discussion of U.S. and international civil remote sensing and its relation to commercial space, see *Aviation Week* (1998).
2. The original design of Landsat 7 included an improved MSS, the High-Resolution Multispectral Stereo Imager (HRMSI), to provide 5-m panchromatic and 10-m multispectral resolution. The Air Force had become interested in Landsat after the Gulf War and wanted to have a more capable civil sensor, but the HRMSI was canceled in 1994 due to budgetary pressures and mission priority disagreements, leaving Landsat 7 as a single-sensor satellite.
3. Anomaly teams working in cooperation with Lockheed Martin, the Athena II payload fairing supplier, determined that the satellite failed to achieve orbit because the fairing failed to separate from the satellite.
4. Orbimage holds the worldwide distribution rights for RADARSAT-2 imagery products. As discussed in more detail in Chapters 13 and 20, RADARSAT-2 will be owned and operated by MacDonald Dettwiler, a Canadian firm that is another Orbital Sciences Corporation (OSC) subsidiary. Plans for OSC to build the RADARSAT-2 bus fell through when U.S. liscensing uncertainties convinced Canada to go with Alenia Aerospazio, an Italian firm.

References

Baker, John C., *Trading Away Security? The Clinton Administration's 1994 Decision on Imaging Satellite Exports*, Pew Case Studies in International Affairs, Washington, D.C.: Georgetown University, Institute for the Study of Diplomacy, 1997.

"Civil Space-Earth Imaging Poised for Commercial Revolution," *Aviation Week & Space Technology*, April 6, 1998, pp. S1–S12.

Covault, Craig, "NIMA Infortech Retools U.S. Space Recon Ops," *Aviation Week & Space Technology*, August 7, 2000, pp. 62–65.

EarthWatch, Digitalglobe.com: Your Planet Online (EarthWatch home page), 2000. Available at http://www.digitalglobe.com/ (last accessed August 9, 2000).

Ferster, Warren, "New Imaging Restrictions Stun Industry," *Space News*, (July 27–August 2, 1998), 1, 19.

Mika, Aram M., "Three Decades of Landsat Instrument," *PE&RS*, Vol. 63, No. 7, July 1997, pp. 839–852.

National Aeronautics and Space Administration, Landsat Data Policy, 1997. Available at http://geo.arc.nasa.gov/sge/landsat/l7policyn.html (last accessed August 9, 2000).

—, Landsat Program, 1999. Available at http://geo.arc.nasa.gov/sge/landsat/landsat.html (last accessed August 9, 2000).

—, Destination: Earth, 2000a. Available at http://www.earth.nasa.gov/ (last accessed August 9, 2000).

—, Earth Science Enterprise: Commercial Remote Sensing Program, 2000b. Available at http://www.crsp.ssc.nasa.gov/ (last accessed August 9, 2000).

National Imagery and Mapping Agency, NIMA Commercial Office information August 7, 2000. Available at http://www.nima.mil/ (last accessed August 9, 2000).

National Oceanographic and Atmospheric Administration, National Environmental Satellite, Data, and Information Service, 2000. Available at http://www.nesdis.noaa.gov/ (last accessed August 9, 2000).

Orbimage, home page, 2000. Available at http://www.orbimage.com/main/main.htm (last accessed August 9, 2000).

Simons, John, "U.S. Prohibits Some Satellite Imaging of Israel," *The Wall Street Journal*, July 24, 1998, A16.

Space Imaging, home page, 2000. Available at http://www.spaceimage.com/ (last accessed August 9, 2000).

The White House, Office of the Press Secretary, "Foreign Access to Remote Sensing Space Capabilities," Fact Sheet, Washington, D.C., March 10, 1994.

——, "Regarding the Memorandum of Understanding Concerning the Licensing of Private Remote Sensing Satellite Systems," Fact Sheet, Washington, D.C., February 2, 2000.

U.S. Department of Commerce, National Oceanic and Atmospheric Administration, "Licensing of Private Land Remote-Sensing Space Systems (Rules and Regulations)," *Federal Register*, Vol. 65, No. 147, July 31, 2000, 15 CFR Part 960, pp. 46,822–46,837.

——, "Licensing of Private Land Remote-Sensing Space Systems (Proposed Rules)," *Federal Register*, Vol. 62, No. 212, November 3, 1997, pp. 59,317–59,331.

Russian Remote Sensing Programs and Policies

George J. Tahu

With a long heritage in earth observation and a broad variety of sensing platforms and instruments, the Russian Federation has substantial expertise in remote sensing capabilities. Since 1965, the former Soviet Union, and now Russia, has operated a variety of meteorological and oceanographic satellite systems. In its role as geopolitical superpower, the country developed a variety of satellite programs for space-based surveillance and reconnaissance missions. Earth-observation technology was also used on the early Salyut space stations and accompanying Soyuz spacecraft missions, and a variety of sensor equipment was installed on the Mir space station. As Russia's expertise and interests evolved, it also developed dedicated environmental monitoring systems for resource management and other civilian applications.

The Russian Federation has now amassed several years' worth of archived imagery and continues to collect data from a variety of earth-observation instruments. Imagery from these systems was once restricted to government users, but the opening up of the Russian space industry over the past decade and the pervasive drive for commercial contracts with foreign partners have led to increasing availability of Russian earth-observation data and systems on the international market. Russia's remote sensing satellite operators and image processing organizations now commercially offer data and imagery from a variety of national satellite systems, including declassified imagery from military systems.

Russia set the precedent for commercially available high-resolution imagery on the international market. Commercial distribution of high-resolution images was authorized in 1987 through the Soviet trade association Soyuzkarta, which offered imagery with spatial resolutions of approximately 5 m. In 1992, two Russian firms began to sell selected images with resolutions as low as 2 m (Gupta, 1995, p. 98). A 1993 license issued to one of these organizations officially limited the sale of imagery to resolutions no better than 2 m and also restricted the sale of photographs of military significance (i.e., areas of current conflict) and imagery taken over the territory of the former Soviet Union.

Despite their nascent lead in commercial sales of high-resolution imagery, Russian organizations were unable to capitalize on their "first-to-market" advantage. A variety of technical obstacles (such as infrequent satellite operations, film-based imagery archives, and lack of digital data lines), cultural barriers, lack of business and marketing expertise, and political and economic restructuring, prevented Russian remote sensing organizations from capturing a significant share of the commercial imaging market. Nevertheless, the potential for competition stemming from Russia's willingness to provide high-resolution imagery, along with other countries' announced plans to develop high-resolution commercial imaging systems, was one of several factors leading to changes in U.S. government policy to allow private firms to develop, launch, and sell high-resolution satellite imaging services (Steinberg, 1998).

After Russian remote sensing organizations established data distribution and marketing partnerships with Western firms, the technical infrastructure and marketing expertise evolved to substantially increase global access to Russian imagery and remote sensing systems. While Russia continues to operate a variety of meteorological and advanced reconnaissance satellites, it is the commercially available land remote sensing systems that are of primary concern to the discussion of global transparency and security. To varying degrees, high-resolution panchromatic, multispectral, and radar imagery is commercially available for all of the following land remote sensing space platforms:

- Resurs-O digital scanning satellites
- Resurs-F/Kometa film-return photographic satellites
- Almaz radar imaging satellites
- Mir Space Station camera systems and sensors.

Despite the similar names of the Resurs-O and Resurs-F, the two spacecraft perform distinctly different missions. The Resurs-O series provides medium- and high-resolution multispectral digital scanning imagery for the mapping and analysis of earth resources. The Resurs-F–Kometa series is a short-duration film-return satellite system providing medium- to high-resolution panchromatic photographic imagery.[1] The Almaz spacecraft carried synthetic aperture radar (SAR), providing an all-weather imaging capability. A variety of photographic cameras and earth-observation sensors were also installed aboard the Mir space station base block, as well as the Kvant-2, Kristall, and Priroda modules. Table 8.1 outlines the Russian remote sensing instruments for which imagery is commercially available.

The majority of Russian imagery offered for commercial sale is archived from previous missions. As a result of economic crises and post-Cold War national priorities, Russian space organizations have not been able to maintain the robust pace of imagery collection of Soviet times, and the availability of newly acquired imagery is limited. Film-return missions based on the Resurs-F satellites are infrequent and unpredictable. The latest Resurs-O satellite (No. 4) failed in orbit, and Resurs-O No. 3 has partially failed. The follow-on SAR spacecraft, the Almaz-1B, sits unfinished at its manufacturing plant awaiting financing. The fate of the Mir space station—either deorbit or conversion into a commercial facility—and the potential future for its imaging systems remain uncertain.

Table 8.1—Commercially Available Russian Earth-Observation Instruments

Cameras	Platforms	Data Type	Resolution (m)	Coverage (km)
KFA-3000	Resurs-F3	Panchromatic	0.75–1.5	21 x 21
DD-5[a]	Kometa	Panchromatic	2	14 x 14
KVR-1000	Kometa	Panchromatic	2–3	40 x 40
TK-350	Kometa	Panchromatic	8–10	200 x 300
KAP-350	Mir	Panchromatic	30–40	N/A
KFA-1000	Resurs-F1, Mir	Multispectral	5–8	120 x 120
MK-4	Resurs-F2, Mir	Multispectral	6–10	120 x 120
KATE (KFA)-200	Resurs-F1	Multispectral	15–30	245 x 245
MKF-6MA	Mir, Salyut 6/7	Multispectral	25	110 x 160
KATE-140	Mir, Salyut 6/7	Panchromatic	50	450 x 450
MSU-E	Resurs-O, Mir Priroda	Multispectral	33–45	45
MSU-SK	Resurs-O, Mir Priroda	Multispectral	140–170	600 x 600
SAR	Cosmos 1970	SAR	25–30	20 x 240
SAR	Almaz-1	SAR	15–30	20 x 240

a Digital Data–5 (DD-5) is a generic term for declassified military satellite data. The DD-5 system provides 2-m–resolution black-and-white images covering an area of 14 x 14 km. These images are used to create digital maps at scales of 1:100,000 to 1:4,000 and to create thematic maps, including digital road maps.

Several system upgrades and next-generation satellites are on the drawing board; however, such future missions and next-generation designs depend on securing additional financing. With limited resources from a highly constrained federal space budget, Russian organizations are desperately seeking non-budgetary resources from commercial investors. As a result, the future competitive challenge and security implications that Russian commercial imagery poses remain at best uncertain.

Russian Remote Sensing Policy: How is Imagery Governed?

While Russia has long been involved in intergovernmental data-exchange initiatives, its approach to data policy is not well defined. At a minimum, this much is known: According to established government policy, commercial sale of Russian remote sensing data and imagery is limited to products with resolutions no better than 2 m.[2] However, Russian organizations may use degraded higher-resolution imagery to produce sellable images or products (such as maps or geographic information system data sets) within 2 m of resolution. In addition, Russian policy restricts the sale of imagery covering areas of current military conflict. Beyond these general guidelines, a clear and consistent definition of Russia's commercial data policy is not well known to the international community, and it appears that the operational norms (especially in regard to commercial distribution) have been determined more through an ad hoc development of practices by individual agencies and enterprises supported by case-by-case legislation and government decrees.[3]

According to the Russian Constitution adopted in December 1993, primary responsibility for federal policy development and implementation was assigned to the President of the Russian Federation. In August 1993, the Russian parliament adopted the law of the Russian Federation "About Space Activity," which laid a foundation for regulating space activity. However, that legislation provided only a framework for the legal process; in practice, space regulation has been performed primarily by edicts of the president, decrees and directives of the government, and ministerial or departmental regulations (Tarasenko et al., 1999). However, active participation by the president, the government cabinet, or the parliament in the development and prioritization of space programs and infrastructure is not typical Russian practice. This is largely due to a historical lack of internal support staff dedicated to and capable of preparing and implementing such programs. While there have been intermittent efforts to establish special advisors to the president to address space matters, these positions have been subject to frequent cycles of political and bureaucratic restructuring. Compounding the limited space-policymaking infrastructure at the top levels of government is the overwhelming national priority to concentrate on economic restructuring and recovery. To date, the role of these policymaking institutions in the Russian government has been mainly limited to endorsing initiatives of the Russian Aviation and Space Agency and the Ministry of Defense (Tarasenko et al., 1999). As a result, the institutional structures that have driven remote sensing policy in other spacefaring nations have been severely limited in Russia. By contrast, the United States has had an expanded institutional infrastructure—a heritage of industry lobbying; think tanks; congressional oversight; Executive Branch direction; and mature, stable regulatory agencies—to engage in an active national debate on remote sensing policy.

The Russian Aviation and Space Agency (Rossiiskoye Aviatsionnoye Kosmicheskoye Agentstvo, or Rosaviakosmos) is the executive agency responsible for implementing federal space policy and for executing the Federal Space Program of Russia.[4] Rosaviakosmos acts as a government customer for space technology and as a co-customer, along with the Ministry of Defense, for space systems used for both civilian and military purposes. It has not been in Rosaviakosmos' traditional purview to engage in regulatory or legislative initiatives to regulate commercial endeavors. Where Rosaviakosmos has had a more active concern in the past is in the efficient use and application of programs that are geared toward civil state use, such as meteorological, oceanographic, and environmental monitoring satellites. To a certain extent, this has included encouraging and licensing the operators of these systems to seek commercial distribution partnerships. Likewise, while the Ministry of Defense has close relationships with the companies offering high-resolution imagery and data for sale, its organizational mission has a limited stake in the outcome of commercial endeavors beyond limiting spatial resolution and imaging areas of current conflict. The dire state of Russia's economic condition and the desire to convert the military-industrial complex to generate commercial revenue have generally given the Russian aerospace industry a clear path to pursue commercial efforts wherever possible. Yet, Russia has a limited heritage of legislation and policymaking infrastructure to address commercial remote sensing activities. Nevertheless, there are still institutional processes though which remote sensing policy is determined.

The institutional bodies that officially hold authority to determine Russian remote sensing data policy are Rosaviakosmos, the Federal Service for

Hydrometeorology and Environmental Monitoring (RosHydromet), the Federal Service of Geodesy and Cartography (RosKartografia), the Ministry of Defense, the Ministry of Emergency Situations, and the Ministry of Ecology. In 1994, the then–Russian Space Agency reportedly asked these organizations to develop an earth-observation policy to be discussed within the space agency's scientific and technical council. Organizations and institutions with interests in remote sensing issues were also invited to participate in this process. According to officials in Rosaviakosmos, the resulting policy was approved within the council and was incorporated into the Federal Space Program up to the Year 2000, the document that serves as the official basis for all space activity in the Russian Federation.[5] Reportedly, these organizations were to then begin considering a "post-2000" policy.

The remote sensing policy incorporated in the Federal Space Program has been characterized by officials as very broad. According to officials in Rosaviakosmos, policymakers took into account several factors when developing the goals, missions, and principles of the earth-observation program: the economic situation of the country, the need to maintain and develop space and ground infrastructures, the state of the art in space technology, commercialization and trends in the world space market, the interests of information consumers and providers, and Russia's obligations and future prospects in international cooperation. The resulting program contained projects that are intended to maintain Russia's status as a space power, capture a place in the world space market, and contribute to the general economic development of the country.

The Russian Federal Space Program's remote sensing activity is funded through the annual budgetary process. The majority of budgetary funds for remote sensing are allocated through Rosaviakosmos, the Ministry of Defense, RosHydromet, and RosKartografia. Depending on the details of a program, a small proportion of funding related to remote sensing may also be allocated through the Russian Academy of Sciences or the Ministry of Science and Technical Policy. According to yearly allocated funding, individual space projects may shift the scope of work and timetables for the overall program. As a result, the annual budgetary process can effectively "amend" national policy and priorities as defined by the Federal Space Program, even though the broad official goals may remain the same.

Russian Data Acquisition and Distribution

Russian remote sensing data can be acquired one of two ways: by directly contacting a Russian data acquisition, processing, or distribution organization or by contacting a Western distributor that has developed a commercial partnership with one of these Russian organizations.

Two primary organizations acquire, process, and distribute Russian remote sensing data to domestic and foreign users. The Scientific Research Center on Space Hydrometeorology (SRC Planeta) and the State Scientific Research and Production Center (Priroda) are large scientific and production institutes involved in both the development and operation of civilian remote sensing systems and the distribution of data to users. Both are state-owned organizations, and each can operate to a greater or lesser extent as a broker of remote sensing data. Three other state-owned organizations that were specifically created and licensed to market Russian imagery commercially are Sovinformsputnik, Soyuzkarta, and the Converted Technologies Center—an organization offering newly declassified military imagery. In addition to these institutions, a variety of consortia and institutional alliances have been created between some of the distribution organizations and the technical enterprises involved in the design, manufacture, and operation of remote sensing satellites and instruments. Most of these alliances were created to market data products and services commercially.

Before recent improvements in archiving, distribution, and marketing capabilities, Russian organizations offering satellite imagery commercially followed a standard procedure: A potential customer contacted the Russian organization and requested imagery with a given set of parameters (longitude and latitude coordinates, coverage area, time, cloud cover, resolution, etc.); the Russian organization then searched its archives for imagery that might fit the customer's requirements; if the requested imagery was available, the Russian organization provided a description of the imagery and its parameters and could offer the imagery in a variety of product formats (photographic positives or negatives, digital data, etc.). If no suitable imagery was available within existing archives, the Russian organization would generally negotiate a price for acquiring new imagery from operational remote sensing systems. Russian organizations typically also offer data-processing services in addition to raw imagery. Once an agreement to purchase is made, the method of delivery is arranged between the customer and provider.

Depending on the relationships between potential customers and Russian imagery providers, this general process could be slow and cumbersome because of such traditional obstacles as distance, difficult communication, and differences in culture between Russian organizations and typically Western customers. Where improvements in infrastructure and marketing skill have not been made, these obstacles still remain. As the predominantly scientific remote sensing organizations improve their business practices and marketing skills, these obstacles can be overcome. In the meantime, Russian remote sensing data—while very accessible in principle—may not be so easy for potential customers to acquire without the help of distributors.

WESTERN DISTRIBUTION PARTNERSHIPS

In response to the lack of technological infrastructure (such as digital data lines) and the business and marketing skills needed to provide commercial services, many Russian remote sensing organizations have forged partnerships with Western distributors. Some of these partnerships have been created with well-established international remote sensing centers; other partnerships are with computerized clearinghouses or image-processing firms. As a result, the relationships between distributors and the Russian providers are unique to each partnership. Available imagery, services, and prices also vary among distributors.

Some distributors simply may act as knowledgeable liaisons between customer and provider. These distributors do not maintain Russian imagery in house but rather process requests from potential customers and contact their Russian provider or partner to determine whether the requested imagery is available. Other partnerships have been created in which imagery catalogs and archives are also maintained by distributors. In the case of the Resurs-O system, direct data reception and archival arrangements were made with European distributors. Russian providers that offer digital data are more able to provide their partners with on-site catalogs or satellite reception capabilities, while the distribution arrangement of providers that still rely on manual hard-copy archives (such as with film-return systems) will be limited until their archives can be digitized and duplicated. Delivery of Russian remote sensing products varies with what imagery format is requested. Distributors that maintain close relationships with imagery providers can usually deliver manually archived images in two weeks to two months, while digital data may be deliverable within days.

Western distributors generally fund and operate their own marketing efforts. Many of these distributors have made extensive use of the World Wide Web to market and sell Russian satellite imagery. Sample images and searchable "quick look" databases are now commonly available on the Internet for almost all remote sensing systems, including Russian satellite imagery.[6]

HIGH-RESOLUTION IMAGERY AND DISTRIBUTION

High-resolution Russian imagery is commercially available from film-return camera systems flown on the Resurs-F (also designated "Kometa" or "Kosmos") series of satellites, as well as on the Mir space station. Priroda, under the authority of the Federal Service of Geodesy and Cartography, is the primary organization responsible for acquisition, storage, processing, and application of civilian photographic images from cameras aboard Resurs-F remote sensing satellites and the Mir space station. Priroda's archives contain more than 2 million multispectral and panchromatic scenes, ranging in ground resolution from 2 to 30 m and covering most areas of the world with varying degrees of redundancy.

Development of the Resurs-F spacecraft began in 1975 at the Central Specialized Design Bureau (TsSKB) in Samara. The Resurs-F system was based on a series of military photoreconnaissance spacecraft and includes two variants (Resurs-F1 and Resurs-F2) of a Vostok capsule-derived design equipped with different photographic instruments. The Resurs-F series spacecraft are composed of a bus with a braking propulsion system, a spherical reentry module in which recoverable payloads (cameras and microgravity experiments) are located, and an on-orbit maneuvering propulsion system. Resurs-F1 satellites carried a payload of three KATE-200 cameras and two KFA-1000 cameras. The Kometa-designated satellites are essentially the same spacecraft as the Resurs-F but are equipped with KVR-1000 and TK-350 cameras. In the mid-1980s, TsSKB developed the Resurs-F2 spacecraft. Resurs-F2 carried a single MK-4 camera. Resurs-F2 satellites included two solar panels to provide the additional power needed to increase mission duration from Resurs-F1's 16 days to 30 days. A variant of the design equipped with the new KFA-3000 has been designated Resurs-F3. All products originating from the Resurs-F spacecraft are analog photographic imagery.

Commercial distribution of images taken by Resurs-F spacecraft was first authorized in 1987 through the trade association Soyuzkarta, which was established to market Priroda's photographic, cartographic, and geodetic products. Until 1991, Priroda's images were available to foreign users only through Soyuzkarta. Today, like most other space organizations seeking commercial contracts, Priroda interacts directly with foreign customers. Priroda maintains an archive of all images taken from four types of cameras: KFA-1000, MK-4, KATE-200, and KFA-3000. On September 1, 1993, Priroda received a license from the Russian Space Agency to sell imagery with resolutions up to 2 m.

The license restricted sales of photos of military significance (i.e., areas of current conflicts) and imagery taken over the territory of the former Soviet Union. Priroda degrades imagery with resolutions better than 2 m before it is released. Imagery with original resolutions of 3 to 30 m is not usually altered.

In 1991, TsSKB and other enterprises associated with the imaging satellites it designed established the Sovinformsputnik Interbranch Association to market the imagery produced by the Kometa satellite series. Sovinformsputnik is a state-owned commercial association that is licensed to sell satellite imagery from the Soviet and Russian defense archives of reconnaissance and mapping satellites and that participates in the development of commercial imagery satellite programs. Sovinformsputnik provides images from two types of cameras: the KVR-1000 2-m–resolution camera, and the TK-350 10-m–resolution camera. Imagery from the KVR-1000 and TK-350 cameras is also available through Priroda, but the original negatives are stored and distributed by Sovinformsputnik or TsSKB.

Sovinformsputnik is authorized to distribute satellite imagery commercially and to sell associated products and services, including topographic, thematic, and digital maps. Sovinformsputnik has produced high-resolution color images by overlapping images from the 10-m–resolution color European SPOT-XS camera with the 2-m–resolution black-and-white KVR-1000 camera. While government restrictions prevent distribution of images with resolutions better than 2 m, Sovinformsputnik claims to have access to images with resolutions of 0.75 m or better, which are then degraded to 2-m resolution for distribution.

In 1992, the Russian government adopted a resolution to make satellite photography acquired by spacecraft under the jurisdiction of the Ministry of Defense available for civilian use. In accordance with this decision, the Russian Space Agency was given the right to order and use high-resolution images acquired from defense satellites for commercial and scientific purposes. In June 1995, the Russian Space Agency and the Ministry of Defense established the Converted Technologies Center to disseminate raw remote sensing images and processed products to domestic and foreign users on a contractual basis. Under the aegis of the Russian Space Agency, and with the cooperation of the Ministry of Defense, the center gained access to image archives from Russian defense satellites, civil remote sensing spacecraft, and meteorological satellites. Images from defense satellites are declassified by the ministry and degraded to resolutions no better than 2 m. The center offered images and products derived from three types of cameras: the KVR-1000, TK-350, and DD-5.

A newcomer to the imagery distribution business is the state-owned armament-exporting company Rosvooruzheniye. The company began marketing space services as recently as 1997 and 1998. Rosvooruzheniye's close ties to the Ministry of Defense allow it easy access to earth-observation data acquired by military spacecraft, and it offers a wide range of satellite products ranging from high-resolution imagery to meteorological data.

WESTERN DISTRIBUTORS OF HIGH-RESOLUTION IMAGERY

As early as 1993, Russian high-resolution imagery providers were establishing partnerships with Western distributors. In that year, Priroda Center and Jebco Information Services Ltd. of England signed a contract forming a joint venture called WorldMap. WorldMap established a digitizing center at Priroda's Moscow laboratory to digitize analog imagery from the KFA-1000, KATE-200, and MK-4 cameras. After the original negatives were digitized in Moscow, the WorldMap products were shipped and archived in England and marketed through a worldwide network of WorldMap distributors, organized by Jebco Information Services. The United Kingdom's National Remote Sensing Centre Ltd. obtained a distributorship agreement from Jebco Information Services. The agreement gave the center access to the newly digitized imagery, as well as the to two million analog photographic images acquired over the past 19 years.

In July 1995, Sovinformsputnik and U.S.-based Aerial Images of Raleigh, North Carolina, along with Central Trading Systems, Inc., Huntington Bay, New York, signed an agreement to market images under a joint commercial project called Space Information, 2-m resolution (SPIN-2). The SPIN-2 partners offer images from the KVR-1000 and TK-350 cameras launched aboard a variant of the Kometa spacecraft.[7] In May 1996, the first SPIN-2 spacecraft was launched aboard a Soyuz rocket, only to be destroyed by a launch failure. A second attempt to launch a SPIN-2 satellite was successful and provided high-resolution data, which the venture now sells commercially. Space Liaison and Imaging Corporation, an international technical and marketing consulting firm based in San Diego, California, also forged an agreement with Sovinformsputnik to distribute KFA-1000, TK-350, and KVR-1000 imagery from its archives. The Earth Observing Satellite Corporation, Inc., of Lanham, Maryland, which was formed in 1984 to commercialize the U.S. Landsat remote sensing satellite program, also made agreements with Sovinformsputnik to distribute KVR-1000 imagery from Kometa and Mir camera systems.

The company that operates the Mir space station, Rocket and Space Corporation (RSC) Energia, formed a partnership with the German firm Kayser-Threde to market the Mir space station and other RSC Energia transportation and space systems. The subsidiary, Energia Deutschland GmbH (Energia-Germany), sells KFA-1000 imagery taken from cameras aboard Mir. In addition, RSC Energia has an agreement with Eurimage for distribution of Mir imagery worldwide. Mir data has also been offered commercially by the North American branch office of RSC Energia, Energia USA.

Imagery from Sovinformsputnik is also distributed through Eurimage, a consortium formed in 1999 by four leading European companies in the earth-observation business: British Aerospace (United Kingdom), Dornier (Germany), SSC-Satellus AB (Sweden), and Nuova Telespazio (Italy). Eurimage provides worldwide digital and photographic imagery through a network of some 40 distributors in 28 countries in Europe, North Africa, and the Middle East. Eurimage is the European node of the global ImageNet service from Core Software Technology. Core Software Technology has agreements with Sovinformsputnik to distribute and market imagery through its network of partners.

MULTISPECTRAL IMAGERY AND DISTRIBUTION

SRC Planeta, a large conglomeration of scientific and industrial centers with ground- and space-based assets, is the primary organization responsible for the acquisition, interpretation, and distribution of scanned remote sensing data from Russia's meteorological, oceanographic, and earth-resources satellite systems. For land remote sensing, this consists of the Resurs-O series of satellites. SRC Planeta operates under the authority of RosHydromet, the primary user of SRC Planeta's processed data. The space systems SRC Planeta uses are developed and operated under the framework of Rosaviakosmos and the Federal Space Program; the space agency is a customer of SRC Planeta's services on a contractual basis.

Resurs-O satellites provide high- and medium-resolution images for earth-resource mapping and analysis, including crop, soil, forest, water, and pollution monitoring. Resurs-O provides medium-resolution digital images with wide swathes, which bridge the gap in coverage and detail between the *Satellite Pour l'Observation de la Terre* and Landsat thematic mapper and the U.S. National Oceanic and Atmospheric Administration's Advanced Very High Resolution Radar (AVHRR). Resurs-O1 No. 1 operated for three years after its launch in 1985. The Resurs-O1 No. 2, launched in 1988, was switched off after seven years of operation. Instruments aboard the Resurs-O1 No. 3, which was launched on November 4, 1994, include a 45-m—resolution visible imager, a 170-m—resolution infrared imager, and a 600-m—resolution thermal infrared imager. In mid-1998, Resurs-O1 No. 3 suffered a failure of its X-band transmitters and is now only partially operational. Resurs-O1 No. 4 was launched on July 10, 1998, but both of the onboard transmitters have failed, and data can no longer be received from the satellite. The next-generation satellite, Resurs-O2 (also known as Resurs-Arktika), is expected to include an upgrade to a 20-m–resolution visible imager and a 200-m–resolution side-looking radar to replace the Ukrainian Okean satellite for ice-cover monitoring. Russian officials hope to launch Resurs-O2 by 2002 (Tarasenko et al., 1999).

SRC Planeta is responsible for acquisition, processing, and distribution of data from Resurs-O spacecraft. Remote sensing data from SRC Planeta's satellites is processed and interpreted into a variety of forms, such as thematic maps, radar images, and geographic information system data sets. SRC Planeta satellite data are available to users as raw data (satellite data without any processing but with annotations on the object, survey dates, and geographical coordinates), primary data (satellite data after noise elimination, structure restoration, and radiometric and geometric corrections), and thematic data (satellite data after desired thematic processing applied to ecological studies, land assessment, resource evaluation, etc.). SRC Planeta distributes these data to RosHydromet, domestic customers, and to foreign cooperative partners and customers. SRC Planeta formed cooperative arrangements with Norway and Sweden to receive and process data from Russian satellites through Satellus AB in Kiruna and Norwegian processing centers in Tromsø and Bergen.[8] These data are relayed over the Internet from Bergen to SRC Planeta and European users. However, because of the failure of the Resurs-O1 No. 3 X-band transmitters, data can now only be received by the Russian stations in Novosibirsk. The Russian organization Research and Development Center (ScanEx), which developed the ground segment data-acquisition and preprocessing systems for Resurs-O, also currently offers archived Resurs-O data, as well as ScanER ground stations for receiving Resurs-O1 data.

WESTERN DISTRIBUTION ALLIANCE: SATELLUS AB AND SOVZOND

A good example of the variety of consortia and commercial foreign partnerships that are developing is that of Sweden's Satellus AB and SOVZOND. Satellus signed a strictly commercial contract with SOVZOND, a consortium of VNII Elektromekaniki, SRC Planeta, and the Institute of Space Device Engineering to distribute data from the Resurs-O spacecraft. Before the failure of Resurs-O No. 3, Satellus regularly received raw Resurs-O data transmissions from its tracking station in Kiruna, Sweden.

According to Satellus negotiators, there are no restrictions or conditions on the use or distribution of the Resurs-O data. However, there is an understanding that Satellus AB will not market the imagery or its services within the former Soviet republics. SOVZOND receives a "commission" for the imagery Satellus sells; Satellus also purchased production software and hardware from SOVZOND. According to Satellus, the company is free to market the data and its services as it wishes. All users and customers are treated the same, as the operation is strictly commercial.

Satellus is also partnered with Eurimage, which markets imagery and services in Europe. Eurimage sells archived Resurs-O images only, while Satellus sells the near–real-time data and resource-monitoring and mapping projects. Satellus is working to establish distributors in Australia, Japan, Canada, and South America. In principle, Resurs-O imagery is also commercially available from Rosaviakosmos (via SRC Planeta), although such purchases (if one has occurred at all) would be extremely rare. "Quick looks" at Resurs-O imagery are available on the World Wide Web through ImageNet.

RADAR IMAGING SATELLITES

In July 1987, a prototype radar imaging satellite equipped with an SAR with 25 to 30 m resolution was successfully placed in orbit under the designation Cosmos-1870. The Cosmos-1870 satellite operated for more than two years and was apparently a successful demonstration. In March 1991, the follow-on Almaz-1 radar imaging spacecraft was launched—as a civil environmental monitoring satellite, after the Ministry of Defense completely refused to sponsor the system (Tarasenko et al., 1999). The Almaz-1 differed from Cosmos-1870 in its advanced radar and digital data transmission system. Almaz-1 had a planned 30-month minimum operational life span, but increased solar activity required frequent orbit boosts that depleted the fuel reserves and required a controlled reentry in October 1992 (Wilson, 1998). The Almaz-1B satellite was hoped to have been launched as a replacement, but lack of funding has delayed the proposed launch indefinitely. The Almaz-1B spacecraft is partially complete and lies in storage awaiting further funding. A joint U.S.-Russian venture called Sokol Almaz Radar Corp. (a joint venture between U.S. Sokol Group, Inc. and the satellite manufacturer, NPO Mashinostroenye) was seeking foreign investment to fund the project.

Radar imagery from Cosmos-1870 and Almaz-1 is commercially available with a ground resolution of 15 m in a 45 km-swathe that is up to 250 km long. In 1990, Space Commerce Corporation of Houston, Texas, first acquired western marketing rights for Almaz data under a joint venture agreement with the Soviet space marketing firm Glavkosmos and the Almaz manufacturer, NPO Mashinostroenye. Imagery was first available at $1,600 per 40 by 40 km scene (Wilson, 1998). By 1992, the price for available radar imagery was halved to $800 per scene. The prices applied to foreign customers; domestic users paid significantly less (Federation of American Scientists, 1995). By 1996, officials at the Almaz Scientific Center stated that Space Commerce Corp. was no longer marketing Almaz data, although the company still advertises on the World Wide Web. Other distributors that have advertised commercial sale of Almaz imagery are Spot Image and Hughes STX (now Raytheon STX) of Lanham, Maryland. Hughes STX signed an agreement with Mashinostroenye to distribute and provide processing and analysis of Almaz SAR data. Despite these multiple marketing efforts, by the time of Almaz-1's reentry, only a few hundred images had been sold (Wilson, 1998). Imagery from Almaz is also commercially available directly from NPO Mashinostroenye and its Almaz Scientific Center.

Future Russian Remote Sensing Systems

Most manufacturers of Russian remote sensing satellites have a variety of proposed designs for follow-on or next-generation versions of the satellites they have flown in the past. Despite possessing the technical ability to implement these programs, a persistent obstacle has been lack of financing—both from federal budgetary and from commercial investment sources—that has slowed or prevented the realization of many of these designs.

The developer of the Resurs-F series, TsSKB-Progress, has designed a new satellite, the Resurs-DK,[9] which has the potential to overcome many of the limitations to commercialization that earlier Russian imaging systems have faced. The Resurs-DK is designed as a civilian remote sensing satellite capable of providing digital imagery at 2- to 6-m resolution in the visual, near infrared, and ultraviolet frequency bands. The Resurs-DK would also be capable of transmitting data to ground stations in real time or through store-and-forward protocols. When the Resurs-DK was proposed to the Russian Space Agency in 1996, the agency determined that the satellite design merited support and gave it "top priority" status (Malcovski and Semeyonov, 1998). However, federal budget allocations for the project have not been forthcoming. At the same time, Ministry of Defense payments to TsSKB for other federal programs are in arrears. Consequently, survival of the enterprise has relied upon revenues generated through its commercial launch services provided in partnership with France (*Rossiyskaya Gazeta*, 1999). As a result, Resurs-DK development has progressed slowly. At a May 21, 1998, press briefing in Berlin, TsSKB Deputy General Designer Chechin claimed the first Resurs-DK would be launched at the end of 1999 or the beginning of 2000. More recent statements by officials place a first launch in 2001.

Sovinformsputnik has described plans to transfer cameras from the current Resurs-F/Kometa series of satellites to new spacecraft designs called Resurs-IS and Resurs-Spektr. The new series are reportedly based on an improved spacecraft bus designed by TsSKB for defense-related optical and electronic reconnaissance satellites. However, insufficient funding has delayed this move for several years, and officials at Sovinformsputnik have stated that Resurs-IS and Resurs-Spektr exist only on paper. The company is seeking foreign investment to support the development of these new spacecraft designs. The Nika-Kuban (Nika-K) satellite is another design proposed to replace the Resurs-F series. Apparently derived from the sixth-generation Kuban photoreconnaissance satellite, Nika-K is designed to provide high-resolution images: 1- to 5-m resolution in black and white, 6- to 8-m resolution in color. A new color camera is reportedly under development for this satellite. Nika-K supposedly would contain more film and more propellant and would be capable of 45-day missions. However, it is uncertain when the first launch of a Nika-K satellite could occur, since the satellite development has not reached the flight-test stage because of a lack of financing.

Despite its current inability to provide new Almaz imagery, NPO Mashinostroenye has received media attention in response to its recent efforts to commercialize its earth-observation capabilities. The company announced that it is marketing two remote sensing satellites based on a small universal spacecraft bus design called Condor. One design is proposed to carry electro-optical instruments to provide panchromatic imagery with 1-m resolution and multispectral imagery with better than 1-m resolution. A second design carries a radar imaging instrument capable of 2- to 15-m resolution, although some sources have claimed this design could provide 1-m radar imagery. Condor imagery data could be transmitted to ground stations operated by NPO Mashinostroenye or authorized distributors in real time. The satellites are designed for a 3- to 10-year life span. An experimental Condor satellite is reportedly being manufactured, with an intended launch date in 2001. NPO Mashinostroenye has also received attention for a reported deal to provide

Iraq with 220 high- and medium-resolution digital images of Iran, Saudi Arabia, Turkey, and Syria.

The Khrunichev State Research and Production Space Center has announced it is pursuing a constellation of commercial electro-optical infrared and/or radar imaging satellites, called Monitor, based on the Yakhta F98M spacecraft bus (*Novosti Kosmonavtiki*, 1999; Wall, 2000). According to Khrunichev's plan, the first satellite in the constellation, Monitor-E, would be launched by 2001 and would provide 8-m panchromatic and 25-m two-band multispectral imagery. A second Monitor-E would be launched in 2002. Two more satellites, designated Monitor-I, would include an additional 60- to 90-m–resolution infrared imager and are slated for launch in 2002 and 2003. In 2004, a single Monitor-S satellite would be launched carrying a dual 8-m panchromatic sensor. Khrunichev also plans to include radar satellites in the constellation with the 3- to 5-m–resolution Monitor-R3 and Monitor-R23 radar satellites in 2004 and 2005, respectively. The company hopes to launch its first 1-m–resolution satellite in 2006. Khrunichev is expected to build the ground station and data-exploitation system for the Monitor imagery and envisions that 2-m satellite dishes could allow customers to receive data directly. According to Khrunichev officials, development of the initial Monitor-E satellite was funded jointly by the Russian government and Khrunichev. Funding for the follow-on satellites is the subject of continuing negotiations. NPO Mashinostroenye's Condor system is viewed as the main Russian competitor to the Monitor system, prompting Khrunichev to intensify its development efforts. The first Monitor-E satellite is scheduled to be manufactured and tested by November 2000, with a launch in December 2000 or early 2001.

The VneshNauchPribor Innovation and Engineering Center and the Research and Development Center ScanEx have jointly proposed another unique constellation, called Transparent World. As the name implies, the goal of the system is to provide sufficient space and ground segments to ensure global daily access to remote sensing data to as great a number of users as possible in real time, as well as to provide a network of regularly updated archives. The constellation design includes six to seven small satellites based on either of two small spacecraft buses (Kolibry or Elf) designed by the Arsenal Design Bureau. Each satellite would carry an electro-optical sensor capable of 50-m resolution in 200-km swathes. The data would be transmitted continuously, received by a global network of ScanER ground stations similar to those deployed to receive Resurs-O1 imagery. The proposal envisions the possibility of equipping the satellites with higher–resolution sensors and radar.

The Program Research Center of Rosaviakosmos reportedly has an as yet unnamed new environmental monitoring and earth-resources observation system in the design project phase. The system envisages the use of both small and heavy multipurpose spacecraft buses that would carry payloads with diverse capabilities (active and passive sensors, multispectrum, varying spatial resolution). Reportedly, it includes a heavy Akron-1 satellite by NPO Lavochkin that is based on an electro-optical reconnaissance satellite and is capable of 2- to 5-m–resolution imagery with 30-km coverage in eight bands of the visible and infrared spectrum. The system is designed for government user needs and would therefore require state funding.

Conclusion

Each organization that exercises control over a Russian remote sensing platform has made commercialization efforts and several have made alliances with foreign distributors. Within the known commercial arrangements, there are no indications that foreign governments or the scientific community will be treated as a special class of customer, although this is apparently entirely at the discretion of the distributor. At present, it is likely that institutions will continue to set their own policies within the bounds of Rosaviakosmos' approval. At the same time, the Russian government, through Rosaviakosmos and funding mechanisms, retains the ability to set more-detailed policy that enterprises would be obliged to follow. In light of Russia's continuing economic crisis, it is unlikely that the government would try to restrict an institute's efforts to commercialize data sets. In this context, there appear to be few policy barriers to expanding data availability or commercialization efforts.

In response to the evolving market in commercial imagery, Russia is reportedly considering reducing the commercial sale limitation to 1-m resolution by the end of 2000. Unconfirmed reports indicate that commercial sale of 1-m imagery may already have taken place and that satellite images of some "crisis" territories have been sold to some customers. However, such deals are commercial secrets with additional sensitivity, so there are no known public examples of such sales. Russia set the precedent of putting high-resolution imagery on the commercial market and is likely to continue to set other precedents. Media reports suggest that Russian enterprises have raised the possibility of selling higher-resolution imagery or satellite imaging technology to a variety of countries in the Middle East and/or Asian Pacific that have a strong interest in imaging capabilities and products. Given the severe funding constraints on Russian national satellite capabilities and the limited domestic support for industrial enterprises associated with Russia's imaging satellite programs, it is plausible that these options are being actively explored.

As Russia's role in the commercial imagery market matures, its approach to remote sensing data policy becomes increasingly relevant to concerns about commercial competitiveness, international security, and transparency. Yet Russia remains somewhat of a wild card in this regard. Scarce federal funding limits the ability of Russian enterprises to continue to fly imaging satellites; at the same time, the possibility of commercial investment from Western partners keeps Russian designers hopeful for future programs. In addition, Russia's approach to data policy and industrial restructuring will always be influenced by immediate foreign policy and economic concerns. Amidst this uncertainty, Russia's earth-observation capabilities and its potential in the commercial market will make the Russian approach to imagery policy and remote sensing development an area to watch for the foreseeable future.

Notes

1. The Kometa spacecraft is a film-return system that uses a spacecraft bus and return capsule designs similar to those of Resurs-F. Kometa carries TK-350 and KVR-1000 cameras, while Resurs-F carries KFA-1000 and KATE-200 cameras. The designation difference is also based on historical development: Kometa was based on the military Yantar mapping and reconnaissance satellites, and Resurs-F was based on the Zenit series of military remote sensing satellites. Sovinformsputnik stores Kometa imagery, while Priroda stores Resurs-F imagery. Because of their higher resolution and stronger connection with military reconnaissance, several Kometa spacecraft bore the "Kosmos" designation; thus, the imagery from these systems is often marketed in the West as "Kosmos" data. According to Kramer (1992), the name Kosmos (or Cosmos) is used to designate any of a series of unmanned satellites that were launched starting in 1962 with Cosmos 1 (the counting in 1988 was around 1800; in 1993 it was around Cosmos-2200). The Cosmos satellite series has been used for a wide variety of purposes, including Earth observation. There are also many satellites with military payloads under the Cosmos designation.
2. In June 2000, Sovinformsputnik announced that it was ready to distribute Russian 1-m panchromatic images and that 1-m–resolution images were available from the Russian imagery archives. As this book went to press, it was unconfirmed whether this represented a departure in Russian government policy.
3. This chapter is based on review of available official documents and interviews with Russian scientists and government officials, foreign partners, and distributors of Russian remote sensing data.

4   In March 1999, Boris Yeltsin issued a presidential directive to transfer companies of the Russian aviation industry from the Ministry of Economy to the Russian Space Agency and rename the agency to the Russian Aviation and Space Agency.
5   The Federal Space Program was developed by the Russian Space Agency, the Ministry of Defense (VKS Space Forces), and the Russian Academy of Sciences in collaboration with the Ministry of Economics, Ministry of Science and Technology, and other interested ministries and agencies. The document is not readily available to those outside of the government.
6   The following is a partial list of known Russian imagery suppliers on the World Wide Web: Eurimage, http://www.eurimage.com; ImageNet, http://www.imagenet.com/; Satellus AB, http://resurs.satellus.se/; SPIN-2, http://www.spin-2.com/; and Geosys, http://www.geosys-intl.com/english/geodata/images.htm (all last accessed July 18, 2000).
7   In 1995, the designation Resurs-T ("T" for Topography, its primary mission) was an initial name for the SPIN-2 spacecraft. However, by the time of launch, the spacecraft was flown under the Kometa designation and known in the West as SPIN-2.
8   The Swedish Space Corporation's Satellitbild negotiated the original agreement to receive and process Resurs-O data. On July 1, 1999, the operations in MDC, Satellitbild, and the Remote Sensing Services department at the Swedish Space Corporation were merged into a new organization, Satellus AB.
9   The "DK" suffix is in honor of the TsSKB General Designer, Dmitry Kozlov.

References

Belgian Federal Office for Scientific, Technical, and Cultural Affairs (OSTC), *TELSAT Guide for Satellite Imagery*, Brussels, Belgium, 1995. Available at http://telsat.belspo.be (last accessed July 18, 2000).

Bodin, Boris V., "Federal Space Program to the Year 2000 and the Electric Propulsion," Moscow: Russian Space Agency, undated.

Committee on Earth Observation Satellites, *Coordination for the Next Decade: CEOS 1995 Yearbook*, Surrey, United Kingdom: Smith System Engineering Limited, 1997.

Congressional Office of Technology Assessment, *Civilian Satellite Remote Sensing: A Strategic Approach*, Washington, D.C.: U.S. Government Printing Office, OTA-ISS-607, September 1994.

——, *Remotely Sensed Data: Technology, Management, and Markets*, Washington, D.C.: U.S. Government Printing Office, OTA-ISS-604, September 1994.

Federation of American Scientists, FAS inquiry with Almaz Corp., Washington, D.C., June 1995. http://www.fas.org/ssp/guide/russia.

Gupta, Vipin, "New Satellite Images for Sale: The Opportunities and Risks Ahead," *International Security*, Vol. 20, No. 1, Summer 1995.

Johnson, Nicholas, *The Soviet Year in Space 1990*, Colorado Springs, Colo.: Teledyne Brown Engineering, 1991.

Kramer, H. J., *Earth Observation Remote Sensing*, Berlin: Springer-Verlag, 1992.

Novosti Kosmonavtiki, No. 12, November 1999.

Rossiyskaya Gazeta, August 7, 1999, pp. 1–2.

Russian Federal Space Program, Draft Budget, 1996.

Steinberg, Gerald, "Dual Use Aspects of Commercial High-Resolution Imaging Satellites," *Mideast Security and Policy Studies*, Begin-Sadat Center for Strategic Studies, Bar-Ilan University, No. 37, February 1998.

Tarasenko, Maxim, et al., Russian Space Industry, Washington, D.C., Federation of American Scientists, 1999. Available at (http://www.fas.org/spp/civil/russia/ (last accessed July 18, 2000).

Toschi, F., "Analysis of Earth Observation Data Policies," Strasbourg, France: International Space University, 1996.

Wall, Robert, "Krunichev Unveils Remote Sensing Plan," *Aviation Week & Space Technology*, March 6, 1999, p. 52.

Wilson, Andrew, ed., *Jane's Space Directory 1998–1999*, Alexandria, Va.: Jane's Information Group Inc., 1998.

9

The French Pioneering Approach to Global Transparency

Isabelle Sourbès-Verger and Xavier Pasco

In 1978, some six years after the United States had launched the Landsat-1 earth-resource satellite, France decided to take its own pioneering approach to global transparency by beginning its own earth-observation satellite program, the Satellite Pour l'Observation de la Terre (SPOT).[1] As is common with high-technology programs, the French authorities considered SPOT to be strategic as a tool for sovereignty and so decided to promote it regardless of its real or supposed commercial value. Space is considered a national strategic asset and therefore an area appropriate for state intervention. It was also important to be pragmatic, making the best of the public's investment in remote sensing —nationally or Europe-wide—without precluding possible commercial spin-offs.

But since France was not the first to enter this field, becoming a real player required adopting an original strategy. The program would need to sustain development of the French space agency—the Centre National d'Etudes Spatiales [National Center for Space Studies] (CNES). Further, the space industry, both in France and in Europe generally, needed new strategic objectives. Since the establishment of the European Space Agency (ESA) in 1975, France has had an active, leading role developing Europe's presence in space and relationships with other spacefaring countries. Commercialization appeared very early in the planning process as the most convenient way to achieve a French or European space observation capability.

The idea that "global transparency" could be a stabilizing factor in international relations lent official legitimacy to a policy of commercialization. In 1978, France became the first nation to propose establishing an international satellite monitoring agency, under the auspices of the United Nations, to aid disarmament verification. Although the prospects for commercializing remote sensing data seemed to have faded a few years later, the principle itself was unquestioned in France. The French did not consider commercialization a prerequisite for a remote sensing program, quite unlike the approach the United States took in establishing the Earth-Observing Satellite Corporation (EOSAT) in 1984.

The 1994 U.S. decision to open high-resolution remote sensing activities to private business, however, broke a de facto consensus in the remote sensing world (White House, 1994). In France, any public service has always had the latitude to earn money from its activities. The U.S. action forced France to refine its approach to be either public driven *or* truly commercial. Moreover, because high spatial resolutions parallel the performance of the French military satellite series (Helios), commercializing high-resolution data has security implications for the French establishment.

The French Historical Context
EARLY VIEWS OF POTENTIAL USERS: THE LANDSAT EXPERIENCE

The very existence of the Landsat program and scientific cooperation in data investigation led the French space community to support developing national competence in this field. This idea stemmed directly from the experience Landsat provided, which demonstrated both the innovative techniques the scientific community could use and some of the limitations of the data Landsat could provide (see Brachet, 1985).

CNES had been engaged in an exploratory remote sensing program since 1970. The first flights of specially equipped aircraft, conducted in concert with an evaluation by the Institut Géographique National [National Geographical Institute] (IGN), led to better understanding and appreciation of the uses of remote sensing.

In 1971, CNES and IGN officially created the Groupement pour le Développement de la Télédétection Aérospatiale [Association for the Development of Remote Sensing] (GDTA).[2] Initially created to test the then-new airborne infrared technology, GDTA rapidly enlarged the scope of its mission. The GDTA began thinking about facilitating the access of French users to Landsat data in 1974 and signed a contract with Telespazio in 1975 for the use of the first Landsat ground station in Europe, at Fucino, Italy.

European interest in remote sensing techniques took shape very soon after that. To assess the role these techniques might play in remote sensing development, ESA created a consultative body, the Earth Observation Program Group, in 1976. This group recommended that ESA develop a remote sensing program and that the agency begin by developing a European network, Earthnet, to receive the then-available data. ESA quickly adopted the network as an optional program in 1977. The network began with the Fucino ground station and continued with the creation of another in Kiruna, Sweden, in 1979. Another strong incentive for extending this European network of ground stations was the U.S. Seasat. The satellite's synthetic aperture radar could look through cloud cover (common weather in Europe), which made its data especially valuable. Despite Seasat's limited life (only 100 days in 1978), the satellite still provided much greater coverage of most of Europe than Landsat did in several years of operation. A Seasat receiving station was set up in Oakhanger, Great Britain, and special processing equipment was built in Oberpfaffenhofen, Germany. The network has since been constantly improved and expanded.

In pursuit of its own remote sensing capabilities, ESA decided to favor development of a radar system in 1981 (see ESA, 1984). The European Remote Sensing (ERS) mission was to take a scientific and experimental approach to development of various technologically complex instruments (a synthetic aperture radar, a radar altimeter, a scanning infrared radiometer, etc.) for studying planetary changes and the environment. This approach was specifically adapted to ESA's goals, not to be particularly commercial but to contribute to global transparency through the pursuit of science. Continuity has been one of the major characteristics of this program: ERS-1 was launched in 1991 and ERS-2 in 1995, and Envisat should be launched in 2001. A global network of ground receiving stations have also been crucial for exploiting the very rich scientific possibilities the program offers (see Duchossois, 1999); see Figure 9.1.

THE CNES APPROACH TO REMOTE SENSING

From 1973 through 1976, CNES was in disarray because of shrinking budgets and uncertainties about the future of the institution. For political reasons, it was desirable to build on the investment in ESA, which France strongly supported. As the leader in the Ariane effort, France's contribution was already 37 percent of the entire ESA budget and would climb to 52 percent in 1976.

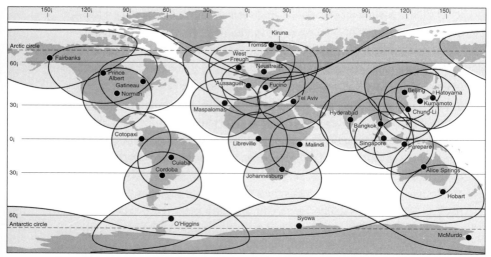

Figure 9.1—ERS Ground Receiving Stations

The increasing investment in Ariane had direct implications for CNES's space center in Toulouse. Work on the French Diamant launcher came to a halt in 1975. However, it was important not to lose the expertise the Toulouse center had gained, so CNES made it a priority to propose new programs that would help retain the center's expertise, whatever its future might be. Since ESA was in charge of scientific and experimental programs, the French team focused on satellite applications, such as telecommunications and remote sensing, which evolved into, respectively, the Symphonie and SPOT satellite programs.

The Toulouse space center did a technical study on an initial concept for a remote sensing satellite in 1974. At the time, a working group—the Groupe des Ressources Terrestres [Terrestrial Resource Group] (GRT)—was in charge of evaluating potential remote sensing applications. This group was chaired by a representative of Bureau de Recherche Géologique et Minière and included representatives of several public bodies interested in such applications as cartography, oceanography, geology, agriculture, and land management. GRT issued a report two years later, at the height of the CNES crisis, that recommended the development of a space platform for earth observation that could be launched by an Ariane rocket to guarantee European autonomy in acquiring space remote sensing data. GRT also recommended setting up methods and an organization for dealing with the raw space data collected.

CNES wanted to include a high-resolution visible (HRV) charged-coupled device (CCD) sensor and a medium-resolution visible infrared (MRVIR) instrument to increase the operational usefulness of the platform.[3] In December 1976, France officially proposed carrying out its project under European auspices to the ESA Council. CNES made several presentations on the SPOT project in different European capitals. However, with the notable exception of Belgium and Sweden, it was clear early on that most of the member states were not interested in the project. Of course, at the time, the ESA budget was almost entirely devoted to Spacelab and Ariane projects. Further, interest in optical remote sensing systems was weak in countries with often-cloudy skies. Opposition was particularly strong in Germany, which was more interested in radar techniques.

Given these circumstances, CNES decided to study the feasibility of pursuing the project on a national basis, with the participation of other interested states. In May 1977, a proposal was made to remove the MRVIR instrument and replace it with two HRV instruments, each with a panchromatic resolution of 10 m in addition to the 20-m initial multispectral performance (see Figure 9.2). The objective here was both to be original, compared with Landsat, and to stress the importance of high-resolution applications for cartographic needs. The swing capacity of the HRV entry mirror was extended to ±27 degrees to increase the stereoscopic imagery possibilities and allow repeat views of some areas of particular interest (monitoring vegetation, for example). After a failed attempt to cooperate with the United States in July 1977,[4] the SPOT project

Figure 9.2—SPOT Family Sensors

Satellite	Sensor	Resolution	Wavelength
SPOT 1-2-3	**HRV** *multispectral*	20 m	0.5, 0.59, 0.61, 0.68, 0.79, 0.89
	panchromatic	10 m	0.51 – 0.73
SPOT 4	**HRVIR** *multispectral*	20 m	0.5, 0.59, 0.61, 0.68, 0.79, 0.89; 1.58, 1.75
	monochromatic	10 m	0.61–0.68
SPOT 4-5	**Vegetation**	1150 m	0.43–0.47; 0.61, 0.68, 0.78, 0.89; 1.58, 1.75
SPOT 5	**HRG** *multispectral*	10 m	0.5, 0.59, 0.61, 0.68, 0.79, 0.89; 1.58, 1.75
	monochromatic	5 & 2.5 m	

was submitted to the French government, which formally approved it in September 1977. In this regard, Belgium and Sweden's willingness to participate in the program (at an original level of 4 percent each) eased the political decisionmaking process. Sweden had expressed interest as early 1977 and formalized an agreement in November 1978, and Belgium signed an agreement in June 1979.

By the time the Comité Economique et Social,[5] directed by Prime Minister Raymond Barre, gave the official green light for the project on February 9, 1978, the satellite design had found its final and definitive form.

THE COMMERCIALIZATION ISSUE

In 1978 and 1979, as the SPOT spacecraft was being defined in detail, numerous studies were conducted to assess possible economic spin-offs from the program. Cartographic applications were seen as particularly promising, which validated the choice of the 10-m resolution. In 1980, when the issue of funding the development of SPOT-2 was raised, the organization and the exploitation of the system became major areas of consideration. A data-distribution policy was needed, and one of the main issues was what type of policy should be adopted. Should it be similar to or different from the one prevailing in meteorology, i.e., quasi-free access?

The commercialization issue emerged naturally because of a convergence of rationales. First, the "scientificlike" approach was impractical, both because of the existence of Landsat and because ESA was in charge of all scientific programs. Second, CNES had to demonstrate SPOT's self-sustainability to the political decisionmakers, especially by stressing its applied nature. To do that, CNES attempted to gain a better knowledge of the possible French users and set up the "SPOT users' service" within GTDA.

Finally, the French agency recommended the simultaneous adoption of the following principles:

- Imagery would be distributed on a commercial and nondiscriminatory basis; pricing would be the same regardless of the customer.
- The company Spot Image would be established, starting with capital of 25 million FF. The company's mission would be to organize imagery distribution and to set up the corresponding customer support services.
- Continuity of service should be guaranteed for at least ten years after the launches of SPOT-1 and SPOT-2 by constructing SPOT-3 and SPOT-4.

The New Commercial Way
THE CREATION OF SPOT IMAGE

The formal decision to create Spot Image, SA, came in October 1981, along with the decision to develop SPOT-2, using the research and development budget, as had been done for SPOT-1. In 1983, the Belgian and Swedish governments decided to participate in SPOT-2, as they had in SPOT-1. The SA structure became effective July 1, 1982. This particular corporate structure has certain specific advantages, since it allows both public and private partners (including the Belgian and Swedish partners), preserves some of the dynamism traditionally associated with the private sector,[6] and facilitates international involvement.

Sweden endorsed this approach and created the SATIMAGE Company (Satellitbild) to satisfy Swedish industrial interests in the SPOT program (Astermo, 1985). The involvement of Belgium and Sweden in the SPOT program was directly linked to their desire to maximize the industrial cooperation already developed in the Ariane program. Table 9.1 shows the distribution of the initial capital investments between the shareholders, illustrating the weight of involvement of the respective actors.

Table 9.1—Shareholders of Spot Image, 1982

Organization	Share (%)
Centre National d'Etudes Spatiales (CNES)	39.0
Bureau de Recherche Géologiques et Minière	10.0
Institut Français du Pétrole	10.0
Institut Géographique National (IGN)	10.0
Groupe Aérospatiale Matra	7.5
Société Européenne de Propulsion	7.5
Swedish Space Corporation	6.0
Financial sector shareholders	6.0
Belgian industrials (3)	2.0
Belgian state	2.0

Table 9.2—Shareholders of Spot Image, 1985

Organization	Share (%)
CNES	39.0
IGN	10.0
Matra	8.8
Société Européenne de Propulsion	8.8
Institut Français du Pétrole	7.4
Bureau de Recherche Géologique et Minière	7.4
Swedish Space Corporation	6.0
International Company for Scientific and Industrial Services	2.6
Belgian state	2.0
Belgian industrials (3)	2.0
Institut Régional pour le Développement Industriel de la Région Midi-Pyrénées[a]	1.2
Banque de Paris et des Pays-Bas	1.2
Banque Nationale de Paris	1.2
Crédit Lyonnais	1.2
Valoring (Société Générale)	1.2

[a] Regional Institute for the Industrial Development of the Midi-Pyrénées Region.

A 1985 increase in the capital investment to allow development of a commercial distribution network led to a slightly different distribution of shares, with the industrial sector (especially Matra and the Société Européenne de Propulsion) gaining weight (see Table 9.2). This followed the guarantee Hubert Curien, then—Minister of Research, offered in June 1985 that the program would continue through 1998 with the development of SPOT-3 and 4 (Cervelle, 1989). The government decided to fund SPOT-3 in October 1986.

In 1981, the ultimate goal was to use the financial gains derived from the sale of imagery to finance the post—SPOT-4 space segment and the expenses related to exploitation. But regardless of the financial results, continuity of service and further improvements were guaranteed from the start.[7] It was originally planned to launch SPOT-2 in 1987, but it was not launched until January 1990. SPOT-3 was launched in 1993. Spot Image was now able to ensure reliable and continuous service to its future customers. The launch of SPOT-4 in 1998 and the planned launch of SPOT-5 in 2001 reflect the same commercial strategy of assured continuous service.

All of this has made it easier to adapt to the necessities and limitations of a market that has not yet matured, yet does not prevent Spot Image from becoming purely commercial, when and if that becomes possible. This particular approach was a natural outgrowth of the French habit of using public-private partnerships to manage risky programs. The French approach for SPOT was also fundamentally different from the one the United States took in 1984 with the decision to commercialize Landsat data, which reflected a very different political management culture. Interestingly, the U.S. Office of Technology Assessment (OTA, 1994, p. 94) has since cited SPOT as an example of government-private partnership:

> In this arrangement, the government and one or more private firms would enter into a partnership to build, operate and distribute data from a land remote sensing satellite. . . . The experience of the French space agency, CNES, and Spot Image provides one possible model of such an arrangement.

The ability to field satellites on a regular basis paid off for France, especially at a time when Landsat was experiencing unexpected delays and technical hurdles. If we consider the lower turnover of the year 1993 with the loss of SPOT-3 only three years after its launch, we may easily realize how important service continuity has been (see Table 9.3). This policy has allowed France to fulfill its initial mission of not leaving the United States in sole possession

Table 9.3—Turnover (in million FF)												
1986	1987	1988	1989	1990	1991	1992	1993	1994	1995	1996	1997	1998
35	55	90	123	165	204	218	178	220	208	218	224	263

SOURCE: Spot Image

of the field of acquiring and distributing data and images. The point was to be a presence in remote sensing, not simply a commercial competitor.

The use of SPOT imagery worldwide has also grown as a result of some unexpected, well-known world events in which SPOT imagery has played a role. Some incidents that were particularly demonstrative in this regard were the Chernobyl nuclear accident, which occurred two months after SPOT-1 was launched in 1986; the preparations for Desert Storm in 1991; and the UN Special Commission's multilateral verification inspections that began after the Gulf War. Moreover, SPOT-1 and SPOT-2 have functioned for much longer than expected (more than 12 years for SPOT-1 and more than eight years for SPOT-2), which has helped Spot Image maintain a reliable customer base. While there have been breakdowns in the onboard data recorders, the growing number of ground stations has largely compensated for this. Finally, Landsat's difficulties helped Spot Image penetrate the North American market. The U.S. subsidiary, Spot Image Corp., was founded in 1982; its sales have averaged 20 percent of total annual sales, making the United States the largest national market for the SPOT products.

SPOT MARKETING CHARACTERISTICS

In 1983, CNES concluded an agreement with Spot Image, which received the responsibility for distributing the data and setting pricing policy and was supposed to reimburse around 46 million FF as a contribution to the program.

One of the SPOT program's founding principles was a fundamentally open data policy. This meant that there was to be no discrimination, except in exceptional circumstances that would be determined on a case-by-case basis. However, intrinsic practical constraints, such as programmatic priorities or cloud cover, effectively limit the imaging possibilities.

The SPOT data commercialization plan has, from the beginning, relied on development of a large and widespread distribution network. The difficulty, however, has been adjusting to a market that is poorly defined, geographically and thematically. The situation here is quite different from that for meteorology, for which there are generally well-defined institutional customers. SPOT policy has favored selling licenses to local firms that have experience in exploiting and/or distributing remote sensing images. The license permits the holder to duplicate and distribute data within a specific marketing area, as long as Spot Image receives a fee for each copy sold. At the end of 1986, there were nearly 40 such resellers in 36 countries, almost half in Europe (see Brachet and Fontanel, 1985). By 1999, there were more than 70 resellers in 60 countries.

To facilitate the use of SPOT data, ground equipment for receiving the data directly has also been set up in any country asking for it, although such stations receive only the data for the geographical region in which they are located. As of early 1986, five countries (Bangladesh, Canada, India, People's Republic of China, and Saudi Arabia) had negotiated agreements for such equipment. By 1999, the network comprised 23 receiving ground stations in 20 countries (see Figure 9.3). Finally, Spot Image has set up four subsidiaries over the years, in Australia, the People's Republic of China, Singapore, and the United States. The receiving ground stations and subsidiaries are considered to reinforce the local presence because of the strong role they play in developing local markets for SPOT data.

The United States, Japan, and France rank as the principal national customers. The main regional customers from 1986 to 1995 were Europe, North America, and the Asia-Pacific region (see Figure 9.4). But since 1995, the Asia-Pacific region has been increasing its share and is now followed by Europe and

Figure 9.3—SPOT Ground Receiving Stations

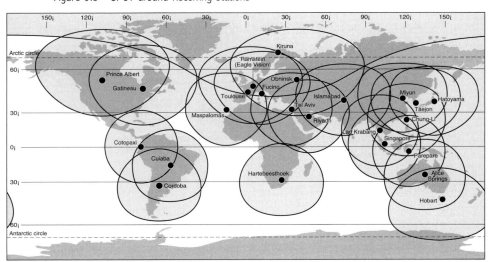

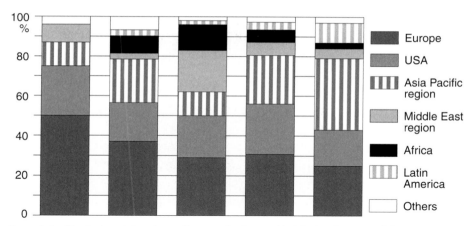

Figure 9.4—Distribution of Spot Image Revenues by Geographical Region (courtesy of Spot Image)

North America (although the Asian financial crisis may have slowed this trend).

Distribution of sales among applications seems to be linked to the space and spectral characteristics of the system, with cartography and agriculture dominating (see Figure 9.5). Exploitation of the vegetation instrument on SPOT-4 may further increase the relative weight of agricultural applications.

NEW COMMERCIALIZATION ISSUES

Setting commercialization up as a principle (indeed, a "golden rule") for Spot Image has had various consequences. The official policy is to set prices according to market potential (i.e., "what the market will bear"). Since the main customers for satellite data are public institutions, it was necessary to begin by stirring up their interest in SPOT products, which relegated developing a private-sector customer base to being a secondary objective. This was problematic for two reasons: Scientific data from the initial civilian observation satellites (e.g., Landsat, JERS, ERS, and IRS) were nearly free for the science community, and that community's budgets tend to be limited (despite efforts to introduce the uses of SPOT data in the academic arena, through the Programme d'Evaluation Préliminaire SPOT). This, in turn, limited the amount that could be charged for SPOT data to a level that would be inadequate from a commercialization point of view. (See European Science Foundation and National Research Council, 1998, p. 25.)

Whether there even is a true market for remote sensing satellite data remains largely a mystery, even to the main actors in the field. It is interesting

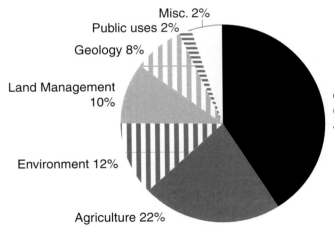

Figure 9.5—Distribution of Spot Image Revenues by Applications in 1998 (courtesy of Spot Image)

to note that, in mid-1987, the U.S. National Oceanic and Atmospheric Administration (NOAA) informally proposed to CNES that the two organizations cooperate on what was called the Joint Optical Mission Program, which was intended to develop a new generation of American-French remote sensing satellites. At the time, the market was considered to be too small and the images too expensive for private customers, meaning that competition between two systems that were more or less similar would be rather counterproductive. However, U.S. political considerations rendered the idea moot, and discussions ended early in 1988, after *Aviation Week and Space Technology* had revealed these "private" contacts. Still, the idea of cooperation was one of the three options suggested in a Congressional Budget Office (CBO) study to help finance work on remote sensing:

> A third option would be to conclude an agreement with other countries or create an international public institution that would coordinate the design, launching and operation of remote sensing satellites. This approach might include a role for the private operation of satellites and distribution of data. (CBO, 1991.)

The market for remote sensing satellite data has, in fact, never been consolidated, although it has been growing. The high degree of internationalization of SPOT distribution is indeed positive, but the dominant role that national and international public customers play as customers for satellite imagery makes it difficult to define a true commercial market (Institut de l'Audiovisuel et des Télécommunications en Europe, 1996). Spot Image has been taking this evolution into account for some time and has progressively looked for ways to diversify by providing additional services. One of the main current proposals consists of including SPOT products in a global "system offer." The market itself has been driven to greater diversification, with an increasing number of small firms offering expertise in different fields.

In September 1999, the distribution policy for different satellite imagery products (such as those of ERS or RADARSAT) took another step forward with the signing of a commercial agreement between Spot Image and Orbimage. The share capital of Spot Image increased in June 1999, up to 49 million FF, again with a new distribution of the shareholders (see Table 9.4). Note that, for the first time, the government's share of the firm is lower than the total shares owned by private companies.

Table 9.4—Shareholders of Spot Image, 1999

Organization	Share (%)
CNES	38.5
Matra[a]	35.7
IGN	7.8
Swedish Space Corporation	6.0
Alcatel Space Industry	3.4
Belgian participants	3.3
Telespazio Spa	1.6
Others	3.7

[a] Including Matra Marconi Space.

However, even though sales volume has increased, the debate over the market potential for satellite data has been stalled for 15 years (see, for instance, OTA, 1984, or OTA, 1987). The ambiguous nature of remote sensing has again led to questions about its commercialization potential. By and large, there are currently two main points of view. The first is that some societal needs must be fulfilled, which has been seen as the basic justification for the SPOT public investment. The other stresses the need to consider uses that are more classically commercial, such as providing some kind of value-added information. In practice, this may result in modes of commercialization that vary according to user type and product category.

Earth Observation and Security Issues
SECURITY CONCERNS

The French decision to sell panchromatic pictures with 10-m resolution was the first step toward closer association of civilian and military stakes, as technology improvements continue to yield increasingly better performance. Another step came in 1999 with the launch of IKONOS, the first of the 1-m class of commercial observation satellites. This trend toward constantly improving technical performance will continue with the appearance of hyperspectral and high-performance infrared sensors.

As in the debate over commercialization, the core national security issues have evolved over time (OTA, 1984, 1987). Maintaining global transparency has been seen as important because foreign competitors would simply gain market share by filling the unexplored niches. But despite that reality, a global transparency policy is likely to be met with serious internal reservations.

At the time the decision to have 10-m resolution was made, France considered the capability to be fully justified because its principal target application was cartography. Moreover, the limitations the Pentagon had imposed on Landsat (30-m resolution in panchromatic mode) created an opportunity to penetrate the market. Preliminary discussions in the government about future French military remote sensing needs led to consideration of technologies that the military could also use.[8] SPOT and Helios have such a synergy. Finally, selling images with rather good resolution demonstrated to disarmament organizations the idea that France had presented through its proposal for an international satellite monitoring agency (known in the United Nations as the ISMA proposal): the use of independent capabilities for verification and armament. The French Ministry for Cooperation (the ministry in charge of cultural and technical cooperation with foreign countries), because it favored building new types of technical relationships, also supported the idea.

However, the resolution of SPOT sensors has always been a concern for the Secrétariat Général à la Défense Nationale [the General Secretariat for National Defense] (SGDN), an "interministerial" service under the prime minister. As SPOT-1 was launched, the situation was still unclear, and SGDN was still lobbying for some encryption of data dealing with sensitive French sites or, at a minimum, that only 20-m resolution data be distributed.

Experience suggests that the 10-m limit is a benchmark for national and international security, because it has not been perceived as being particularly intrusive. A number of independent analyses have shown the notable strategic value of SPOT images,[9] and SPOT has benefited from being the first unclassified satellite able to provide information useful for crisis management. At the end of the 1980s, as the first French military observation satellite, Helios, was under construction, the idea of using of high-resolution civilian satellites for disarmament and verification purposes was not considered legitimate. In the same vein, India's launch of the 6-m resolution (according to the Indian Space Research Organization) IRS 1-C Indian satellite in December 1995 was seen as a serious breach in what appeared in France to be an international consensus about not allowing civilian systems to become involved in security matters. In this context, many have considered the U.S. presidential directive opening the market to high-resolution imagery (White House, 1994) somewhat surprising.

On the other hand, the ability to use SPOT as a diplomatic tool to enhance the French influence in the world has been considered positive, in particular in the Foreign Ministry. For example, a Western European Union (WEU) meeting in Rome in March 1990, held at French initiative, involved discussions about developing an observation satellite as a European instrument for verification and disarmament. This event led to a series of reports and conferences on the subject. The 1991 creation of the WEU Satellite Centre for satellite imagery processing and interpretation in Torrejon, Spain, was viewed as a first step in this direction, even if it was initially limited to exploiting only civilian images. The center has since acquired the technological capabilities for timely processing of Helios imagery, as well as images from other civilian sources (Indian, Russian, or American). Moreover, the center should also benefit politically from its joining with the European Union under the auspices of the formal Common Foreign and Security Policy.

In parallel, there has been some general support for developing cooperation networks (especially with French-speaking developing countries), through the education and training of foreign experts and the transfer of image processing techniques. In this respect, the second-most-important sector of remote sensing activity—the sale of receiving ground stations—demonstrated several direct and indirect benefits. It helped the distribution and commercialization of French technologies and also proved useful when the onboard recorders broke down on SPOT-1 and SPOT-2.

But SPOT's worldwide network of ground stations able to receive data directly, covering a very large area, was established at a time when the resolution level was not judged to be destabilizing. New foreign policies that favor better resolutions have today raised new challenges.

THE 1-M LIMIT AND NATIONAL AND INTERNATIONAL SECURITY

As a matter of principle, the French military defines data of 1-m resolution as a security concern. The military itself might prefer still better resolutions, even at an experimental stages, but images with 1-m resolution do fulfill certain operational requirements. For example, such imagery is suitable for classical military "identify, recognize, analyze" typology.

French decisionmakers are thus still naturally cautious about liberalizing the availability of high-resolution imagery on the international market. Helios imagery, for instance, cannot be sold commercially; even within France, its distribution is still limited to a fairly small number of military officials. As much as possible, sensitive Helios imagery is distributed through a specially endorsed institution, such as the WEU Satellite Centre, to participating member states. However, these data are processed and interpreted with special restrictions, and the images cannot be included in a report as other original information might be. Since maintaining a major role in the security decisionmaking process is important to France, it is difficult to see this evolution as a sign of increasing transparency. Rather, this sharing arrangement should clearly be viewed more as a political tool than as a military service.

The new official U.S. position (White House, 1994) has led French decisionmakers into ongoing discussions about their national interests in commercializing high-resolution imagery. This issue is particularly important given the expectation that commercial remote sensing will evolve both technically and strategically. This raises a number of different possible strategies for the French military that will depend on a higher-level political decision.

The dual-use nature of technologies constitutes one area of interest, since such technologies appear central to planning for future military and civilian imaging satellite systems. Rather than borrowing and adapting technology from one area (civilian or military) to apply to the other, dual-use technologies take into account the mission needs and technical specifications of both from the start, looking at the task as a whole. In this way, the civilian activity can gain from the military experience, and the military can have greater access to systems that, because they are developed in common, are more in line with military requirements than purely civilian systems would be. Hence, a high-resolution civilian satellite could also be better suited for meeting the military's general-purpose needs, providing additional, easily accessible information.

In some sense, this general trend perfectly illustrates the traditional broad French perspective on earth observation issues, which takes into account diplomatic, commercial, technical, and military approaches. In this regard, the 1-m issue is only one among others, which may require some of the actors to adapt their own strategies, taking the changing international background into account. This would represent a new, real challenge for French space policy.

Conclusions

The changes in U.S. remote sensing activities have obliged France to reconsider its own policy on commercial remote sensing. The related issues are still under debate, given that there are no immediate technical and commercial deadlines and that the U.S. experience will be highly instructive as France shapes its own national posture. Considering the French experience, a number of issues will have to be addressed before engaging in new developments. Indeed, among the hardest and most crucial questions to be worked out will be the continuity of new systems; the types of guarantees offered for accessing the data, considering national-security limits; and the reality of the envisioned market.

From a more-specifically French point of view, several shortfalls prohibit comparable national policy choices. The technical investments have remained limited compared to those the United States has so far made. The development of geographical information systems does not yet appear to be a driving force in the satellite imagery market. Furthermore, the French national resources for helping satellite imagery sales cannot compare to those of the United States, neither in the institutional domain, where a new agency exists with a broad mandate (National Imagery and Mapping Agency), nor in the industrial sector. Finally, the point for France is not so much to gain leadership in the remote sensing field as to consolidate its almost 15-year presence in the market.

Notes

1 SPOT was initially known as the Satellite Probatoire d'Observation de la Terre.
2 GDTA was joined by the Bureau de Recherche Géologiques et Minières [Mining and Geological Research Bureau], the Institut Français du Pétrole [French Oil Institute], and the Bureau du Développement pour l'Agriculture [Bureau of Agricultural Development] as new participants in 1973 and the Institut Français de Recherche pour l'Exploitation de la Mer [French Institute for Sea Exploration] in 1991.
3 Like the Landsat multispectral scanner, the MRVIR was intended to use a mechanical scanning technique providing 200-m resolution in the infrared band. The HRV instrument used the new CCD technique for three spectral bands with a 20-m resolution.
4 NASA had previously been indecisive about cooperation on the first Landsat multispectral scanner. The idea had been to coordinate experiments with the multispectral scanner and CCD sensors on both Landsat and SPOT.
5 The Economic and Social Committee is a limited "interministerial" committee that acted as the de facto administrative body in charge of space.
6 The Directorate for Electronics and Computers in the Ministry of Industry demanded this.
7 According to official numbers, Spot Image had revenues of $40 million in 1998. SPOT-4 development and launching cost $590 million dollars, and the expenses for the next 6 years are expected to be $50 million U.S.

8   At the time, a specific satellite, the Satellite Militaire de Reconnaissance Optique, was being considered but was ultimately dropped.
9   See, for example, Jasani (1987) or the numerous SIPRI studies on the subject.

References

Astermo, Svante, "Le rôle de la société SATIMAGE," *Société Française de Photogrammétrie et de Télédétection*, Bull. n°100, 1985-4, pp. 47–48.

Brachet, Gérard, "Le programme SPOT: Historique, objectifs, description du système et organisation générale," *Société Française de Photogrammétrie et de Télédétection*, Bull. N°100, 1985-4, pp. 13–25.

Brachet, Gérard, and André Fontanel, "Organisation du service de diffusion des images SPOT," *Société Française de Photogrammétrie et de Télédétection*, Bull. N°100, 1985-4, pp. 37–46.

Cervelle, Bernard, *SPOT: Des yeux braqués sur la Terre*, Paris: Presses du CNRS, 1989.

Congressional Budget Office, *Encouraging Private Investment in Space Activities*, Washington, D.C.: U.S. Government Printing Office, February 1991.

Duchossois, Guy, "Les programmes européens ERS et Envisat," *Société Française de Photogrammétrie et de Télédétection*, Bull. N°154, 1999-2, pp. 25–31.

The European Science Foundation and the National Research Council, *U.S.-European Collaboration in Space Science*, Washington, D.C.: National Academy Press, 1998.

European Space Agency, "20 years of European Cooperation in Space '64–'84," ESA report 1984, ESTEC, 1984.

Goudy, Philippe, "De SPOT-1 à SPOT-4," *Société Française de Photogrammétrie et de Télédétection*, Bull. N°154, 1999-2, pp. 12–15.

Institut de l'Audiovisuel et des Télécommunications en Europe, "La filière aval de l'observation de la Terre," Paris: Ministère des Postes, des Télécommunications et de l'Espace, July 1996.

Jasani, Bhupendra, and Toshibomi Sakata, *Satellites for Arms Control and Crisis Monitoring*, Oxford: Oxford University Press, 1987.

Office of Technology Assessment, *Remote Sensing and the Private Sector*, OTA Technical Memorandum, March 1984.

—, *Commercial Newsgathering from Space*, OTA Technical Memorandum, May 1987.

—, *Civilian Satellite Remote Sensing, A Strategic Approach*, OTA-ISS-607, September 1994.

Verger, Fernand, Cambridge Encyclopedia of Space, Cambridge: Cambridge University Press, 2001.

The White House, Office of the Press Secretary, "Foreign Access to Remote Sensing Space Capabilities," Fact Sheet, Washington, D.C., March 10, 1994.

10

Japanese Remote Sensing Policy at a Crossroads

Kazuto Suzuki

Japanese policies on remote sensing programs, if not space programs as a whole, have been driven by scientific and technological enthusiasm. The objectives for Japanese remote sensing policy were clearly set to "catch up" with other "space advanced countries" or to avoid being left behind with respect to international standards of space technology (Space Activities Commission [SAC], 1999a). Until the mid-1980s, remote sensing work in Japan was driven by an interest in earth sciences (e.g., meteorology and resource observation), and the focus was on technology development. Since the mid-1980s, increasing concerns about environmental protection and the implications for international cooperation have added extra value to remote sensing programs, but the major policy issue was to develop technologically advanced equipment.

The technology-biased interest in remote sensing, however, seemed to be changing in recent years. The "catch-up" strategy has been successful to some extent in remote sensing technology, and there is general recognition of the need to shift the political focus to exploiting the technology. More importantly, however, North Korea's launch of the Taepodong in 1998 shocked the Japanese space policy community into moving on to military space reconnaissance.

This chapter explores how Japan developed its remote sensing capabilities and where this development is heading. The focus is on the institutional aspects of Japanese policymaking and the objectives that have shaped a technology-driven remote sensing policy that, as a result, contributes to global transparency. The latter part of this chapter deals with recent changes in Japanese remote sensing policy, particularly in the military field, and tries to assess the effects of the new policy on remote sensing policy in general.

The Policymaking Framework for Civil Remote Sensing Policy

Japan has developed a unique style of policymaking for remote sensing programs (see Figure 10-1). The responsibilities for developing the strategy are mainly divided between two ministries: The Science and Technology Agency (STA)[1] has the policy competence to supervise overall program and technological development, and the Ministry of International Trade and Industry (MITI) responsible for developing payloads (mainly radar sensors) and for commercialization.

THE ROLES OF STA AND NASDA

STA is responsible for space policy as a whole, with the main objective of developing Japanese space technology to an international standard. Since STA's competence is limited to science and technological issues, its policies, mainly

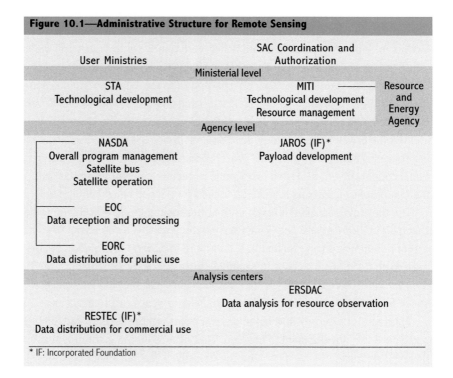

Figure 10.1—Administrative Structure for Remote Sensing

* IF: Incorporated Foundation

initiated and implemented through the National Space Development Agency (NASDA), are inclined to focus more on technological development than on the utilization of its technology.

There is no doubt that NASDA has a strong capability to initiate, design, implement, and operate remote sensing programs. It is no exaggeration to say that NASDA's Office of Earth Observation Systems is the powerhouse of remote sensing policy. This office plans, designs, and conducts the research and development for remote sensing technology and for every remote sensing satellite. The office is also responsible for disseminating imagery data; it operates the Earth Observation Center (EOC), which was established in 1978 to receive, process, and store data from Japanese and foreign (Landsat, *Satellite Pour l'Observation de la Terre* [SPOT], and European Remote Sensing [ERS]) satellites.

To distribute remote sensing data to a wider range of users, NASDA established the Remote Sensing Technology Center of Japan (RESTEC) in 1975. RESTEC is an incorporated foundation that is funded by the private sector for nonprofit activities in the public interest and that is supervised by NASDA and STA. The center is mandated to cooperate with and support EOC in developing data analysis technology and distributing remote sensing data for "general users" at a market price.[2] In the context of commercialization of data use, it is perhaps worthwhile to note that RESTEC has never used the term *commercial purpose*, preferring instead the term *general purpose*, because of the center's legal status as an incorporated foundation. Since organizations having legal status are prohibited from seeking profit, RESTEC should be regarded as a completely different kind of organization from other "spin-off" companies, such as EOSAT or SPOT Image.

The creation of Earth Observation Research Center (EORC) as a part of the Office of Earth Observation Systems in 1995 marked one aspect of NASDA's policy shift from technological development to utilization of remote sensing data. Since 1995, the number of research projects using remote sensing data has dramatically increased, and EORC has accommodated a large number of scientists outside NASDA (Committee for NASDA Evaluation, November 1998a). EORC is developing its expertise in operational meteorology, map production, land use, resource exploration, and monitoring natural disasters and has made remarkable progress in these areas.

Besides analyzing data, EORC also distributes remote sensing data. Because of EORC's legal status as part of NASDA, distribution had been limited to public and research institutions. However, NASDA recently authorized EORC to provide data not only to nonprofit organizations but also to commercial and private entities for research uses, for the cost of reproduction (EORC, 2000). This policy change reveals two important aspects of NASDA's hesitation toward commercialization of remote sensing data: On one hand, NASDA recognizes that commercial and private actors have become important users, but on the other, believes that commercial users have not yet matured enough to apply remote sensing data.

It is worth mentioning that NASDA's legal structure prevents a smooth transformation of remote sensing programs to data utilization.[3] Article 22 of the Law of Establishment of NASDA limits its activities to developing space hardware and managing "accompanying tasks," such as receiving and analyzing remote sensing data. Any activities other than these—for example, commercialization of satellite imagery data—should be authorized by the Minister for STA. But the minister is very unlikely to grant authorization, since these activities may conflict with the competence of another ministry, such as MITI, and the Minister for STA would not like to challenge the other ministry's competence. Thus, the utilization and commercialization of data are not priority tasks for NASDA, and therefore its remote sensing programs tend to be more technology oriented.

Having said that, the policy environment has been changing STA's competence in recent years. Since 1995, the government has strongly emphasized science and technology, which favors STA and NASDA. However, a series of failures in space activities in recent years, such as the malfunction of the Advanced Earth Observation Satellite (ADEOS), increased criticism of STA and NASDA. Under these circumstances, the government ordered NASDA to set up an international peer review committee to evaluate its performance (Committee for NASDA Evaluation, 1998b). At the same time, SAC set up a subcommittee to review fundamental questions that these failures posed (Committee for NASDA Evaluation, 1999). Although these committees respected the achievements NASDA has made, they recognized the need to shift from a catch-up strategy to more industry-oriented activities and the need to concentrate resources and manpower on focused strategic objectives.

It is still too early to forecast whether these recommendations will change the attitudes of STA and NASDA, since STA will be integrated into the Ministry of Education (under the supervising ministry of the Institute of Space and Astronautical Science [ISAS]) in 2001. The structural changes in the Japanese administration will certainly affect the policymaking priorities and strategies, but the blueprint for the new policymaking framework has not yet been completed.

THE ROLES OF MITI AND AFFILIATED ORGANIZATIONS

MITI's role in the remote sensing program may give the impression that Japan is ready to participate in the commercial satellite data market because of MITI's reputation as an organizer of Japanese economic success (for example, see Callon, 1995, and Johnson, 1982). However, such an assumption may lead to a misunderstanding of the Japanese remote sensing program.

Three points are worth mentioning here. First, the main objective of MITI's involvement in remote sensing is to develop technology to help stabilize the supply of natural resources and energy, since MITI supervises the Agency of Natural Resources and Energy. Japan depends on foreign countries for 98 percent of its oil and many other natural resources, and MITI is concerned about the vulnerability of the energy trade. Second, it is important to know that it is the Aerospace, Defense and *Space Industry* division, within the Machinery and Information Industry Bureau, that deals with space policy. This implies that MITI is mainly responsible for "industrial" space policy, rather than policy on services related to space activities. Third, within MITI, the term *commercialization* is associated with sales of manufactured goods. At the moment, the discussion of commercialization is dominated by the potential of H-IIA launcher, and there is very little discussion about the commercialization of remote sensing data.

These points are also true for MITI's affiliated organizations for remote sensing: the Japan Resource Observation System Organization (JAROS) and the Earth Remote Sensing Data Analysis Center (ERSDAC).[4] Although some organizations under MITI's supervision, such as the Institute for Unmanned Space Experiment Free Flyer,[5] are very keen on commercialization, JAROS and ERSDAC are mainly technology-driven organizations. JAROS and ERSDAC were established as incorporated foundations to develop hardware (particularly radar technology) and to promote data analysis technology, respectively. Because of their legal status, JAROS and ERSDAC are both less autonomous than NASDA, and their role is therefore limited to implementing MITI's decisions. In short, JAROS and ERSDAC are similar to NASDA and RESTEC in terms of their objectives and paradigm, which is to achieve technological goals.

It is thus appropriate to say that MITI's involvement does not represent a Japanese intention to compete in commercial data distribution but rather a particular interest in resource management and technology development.

OTHER ACTORS

A few other actors have certain roles in Japanese remote sensing activities. The Environment Agency's role in promoting environmental tasks in satellite remote sensing has become more significant in recent years. For STA and NASDA, close contact with the Environment Agency is important for establishing both a *raison d'être* and a financial justification for remote sensing programs. Although it does not have specific program for remote sensing as such, the Ministry of Education has encouraged universities and research institutions to use satellite data.

The Ministry of Transport (which supervises the Meteorological Agency), the Ministry of Agriculture, and the Ministry of Construction (which supervises the National Geographical Office) are major public data users with certain amount of influence on policymaking in remote sensing activities. In particular, the Ministry of Transport finances and operates the Geosynchronous Meteorological Satellite (GMS) system (better known as "Himawari" series) and the follow-on Multifunctional Transport Satellite (MTSAT) program, while NASDA is responsible for developing satellites. Generally, the relationships between these ministries and STA are cooperative and are based on a clear division of labor as developer and users.

Above these two ministries, SAC is in charge of interministerial coordination. SAC has a variety of features (e.g., it is directly affiliated with the Prime Minister; its members are independent of ministerial interests, the chairmanship of the Minister for Science and Technology is nominal; etc.) that have led to its being regarded as a strong, authoritative, and commanding committee for planning space programs. However, its actual role in remote sensing policymaking is no more than consulting and aggregating the different interests of STA, MITI, and the user industry, and authorizing program proposals from ministries. Although a reform of the administrative structure is scheduled to be completed in 2001, when STA will merge with the Ministry of Education and when MITI will be integrated with the Economic Planning Agency, the division of labor in remote sensing policy will likely remain as it is.

Data Policy

Japanese policy on data distribution and utilization has also been unique in terms of how data utilization and customer needs are conceptualized. The most distinctive feature is the multitier pricing policy. As shown in Table 10.1, the price and conditions differ according to how the data will be used. This shows that NASDA's priority is on research- or public-purpose uses and suggests that data distribution is still considered to be public service, although EORC and RESTEC are now gradually stepping into commercial data distribution.

Table 10.1—Data Distribution Schemes

Scheme	Users	Cost (Distributor)	Conditions
Science	Researchers for scientific research, data correction and verification	Free (EOC, EORC)	Open tender for particular mission organized by NASDA
Selected organization	Public organizations: Met Office (J), Agency of Fishery (J), NOAA (U.S.)	Subject to agreement (EOC)	Special agreement with NASDA
Open access	General public	Free (EORC)	Limited images on Internet
Research purpose	Researchers	Cost of reproduction (EORC)	Certain conditions are applied
Public interest	Public authorities	Cost of reproduction (EOC, EORC)	Certain conditions are applied
General distribution	General users, private companies, other researchers	Market price (RESTEC)	Distribution from archive

Source: SAC Planning Committee Report (1997).

The other unique aspect of Japanese data policy is its regulations for data use. For legal and historical reasons, satellite imagery data distributed through NASDA and RESTEC should be limited to "peaceful purposes." NASDA, however, uses a wider interpretation of "peaceful purposes," including so-called "passive military use" or "nonaggressive use," such as monitoring sensitive areas and verifying arms control treaties, which do not involve offensive actions. On this basis, NASDA constantly provides remote sensing data to the Japanese Defense Agency (JDA). Evidently, such an interpretation infringes on the Diet Resolution of 1969, "Principle of Space Development and Utilization of Japan," which is regarded as a rigidly legal binding to prohibit any space activities related to military affairs, but data policy seems to be implicitly exempted from the spirit of the resolution. A part of the reason is that satellite imagery data are already available in the global commercial market, and it is pointless for NASDA and RESTEC not to distribute images to JDA when foreign commercial sources would do so. Perhaps the issue of "peaceful purposes" would be significant when it comes to a question of distribution to foreign users. NASDA currently distributes only 6 percent of its total images (rather low-resolution imagery data) to foreign users, mostly public institutions (SAC, 1997, pp. 63–71). However, the arrival of relatively high-resolution data (2.5-m panchromatic optical data) from the Advanced Land Observation Satellite

(ALOS) in 2002 may attract some attention from foreign military and intelligence authorities.

Japanese data policy exercises could certainly contribute to global transparency. The principle of fostering research and public-purpose uses has enhanced access to satellite imagery for a wider range of users, and perhaps the wider concept of "peaceful purposes" may be helpful for organizations wishing to use satellite imagery for nonaggressive purposes. However, as the NASDA Evaluation Report pointed out, the user community in Japan is not yet fully developed because the data-distribution network is underdeveloped and because of a lack of infrastructure, software, and capable manpower (Committee for NASDA Evaluation, November 1998b, p. 17). Thus, Japan's actual contribution to global transparency, in terms of volume and accessibility to satellite imagery data, is not as great as policy means allow.

The Data Distribution Network

Although the liaisons between providers and users are still underdeveloped, interest in organizing the data distribution network has been increasing, particularly since the establishment of EORC. For example, NASDA's new program, the Pilot Project for Partnership, aims to develop a data distribution and utilization system, with both NASDA and interested users, and to get feedback from the user community to improve data processing and analysis. So far, the pilot project has been operated by local authorities, such as Gumma Prefecture Office for environment analysis and Iwate Prefecture Office for fishery resource management, but NASDA is trying to attract private companies to this scheme as well (SAC, 1997, pp. 72–80).

The Satellite Remote Sensing Promotion Commission (130 members) and the Earth Science Technology Forum (540 members) organize the demands of small groups of users of particular data and, in general, function as efficient feedback mechanisms. It is important to note, however, that the members mostly come from scientific and academic research fields, and there is little feedback from commercial applications.

Having said all that, change seems to be in the air for commercialization. For some time, Japanese trading companies (*Sogo-Syosya*) have been involved in selling remote sensing imagery data to Japanese users. Nissho-Iwai Trading had contracts with EOSAT and SPOT Image to sell their images in competition with RESTEC,[6] but these contracts were terminated in 1991. The market had not yet matured by that time, and few data analysts and users understood the value of remote sensing data. In 1998, knowing these experiences, the Mitsubishi Trading Co. signed an agreement with Space Imaging to distribute remote sensing data commercially. Whether such a private initiative can be successful is difficult to assess, but it has the potential to pressure NASDA to move toward more commercial- and user-oriented approaches.

Historical Progress

In the early period of Japanese space development, remote sensing programs were lower priority than launcher development and telecommunications satellites. The success of Landsat led to Japanese interest in remote sensing. The Japanese remote sensing strategy therefore followed the so-called "Landsat model"—developing remote sensing satellites for scientific and research purposes. The Fundamental Policy of 1978[7] outlined the strategy for developing remote sensing programs beginning with marine observation, which required a lower resolution, and moving to land observation, using radar and optical sensors. There was no doubt that Japanese interest in using remote sensing data increased during 1970s and 1980s, even as each program stressed technological development. The GMS and Marine Observation Satellites (MOS) were successful examples of this strategy. GMS was developed in cooperation with the Meteorological Agency not only to monitor cloud movements but also to learn spin satellite stabilization from the United States. The objective of MOS was primarily to develop three types of sensors (visible and near infrared, thermal infrared, and microradiation) and the three-axis stabilized satellite bus (SAC, 1997).

However, as other ministries gradually began to recognize the usefulness of remote sensing data, STA and NASDA were unable to continue indulging only their own interests and had to take users seriously. The Japanese Earth Resources Satellite (JERS-1) was a good example of meeting the demands of users and developers. The Japanese government (MITI in particular) had realized Japan's severe vulnerability during two oil crises in the 1970s. It was imperative for MITI to reduce Japan's dependence on Middle East energy resources and to develop its own resource-supply system. This situation also presented an opportunity for NASDA to develop a high-resolution radar satellite, although the responsibility for the sensor was finally allocated to MITI and JAROS. There was a bitter disagreement between STA and MITI on how the work should be shared. STA naturally demanded the right to develop the optical sensors and the SAR technology as a part of its remote sensing programs.

NASDA has already acquired technical expertise on, at least, optical sensors and considered developing the optical sensor a technological challenge. MITI demanded the right to develop the SAR and OPS, because it wanted to be fully capable of monitoring resource exploration. MITI did not have experience developing spacecraft but had created JAROS with support from space industries (Toshiba, NEC, Hitachi, Fujitsu, and Melco). Eventually, MITI was given the responsibility for developing the JERS sensors and subsequent radar sensor developments. Consequently, this conflict added an extra feature to the enduring "unpeaceful relationship" between STA and MITI.

The next large program for remote sensing, ADEOS, was also a "multi-objective" satellite. ADEOS was not at first planned to be primarily a remote sensing satellite. Its satellite bus was initially developed as a space platform, a step toward manned space station technology. Although the idea of a space platform was a keystone program in the Fundamental Policy, the program needed to provide some extra value. Therefore, two other objectives, environmental monitoring and international cooperation, which became NASDA's new strategic agenda in the mid-1990s, were added. Since NASDA completed "domestic technology development" programs, such as H-II, it sought more international cooperation and environmental projects to justify its programs to the Ministry of Finance. Therefore, it was symbolic as well as substantial to invite NASA and the French Centre National d'Etudes Spatiales to participate. The Tropical Rainfall Measuring Mission (TRMM),[8] the Advanced Spaceborne Thermal Emission and Reflection radiometer (ASTER),[9] and the ADEOS-2 programs also featured an emphasis on international cooperation.[10]

ALOS, which is planned to be launched in 2002, reflects a changing mood in Japan for remote sensing. The Fundamental Policy of 1996 gave higher priority to remote sensing with particular emphasis on "user-developer network." ALOS was therefore well-placed to set the first example of user-developer collaboration: The development agencies (NASDA and JAROS) would have to take user needs into account when specifying the design and capabilities of the satellite. Even so, the main emphasis of the program was on developing high-resolution (2.5 m) sensors.

In retrospect, the development of Japan's remote sensing program has been surprisingly constant, as outlined in the Fundamental Policy of 1978. See Table 10-2 for a list of the major satellites. Even though NASDA and STA had funding difficulties, they effectively introduced different policy objectives to secure program funding. However, the failure of ADEOS and subsequent

Table 10.2—Major Remote Sensing Satellites

Satellite System (operational years)	Main Instruments	Resolution (m)
Marine Observation Satellite (MOS-1: 1987–1995) (MOS-1b: 1990–1996)	MESSR (visible near infrared) VTIR (visible thermal infrared) MSR (microwave radiation)	50 900 (visible) 2,700 (infrared)
Japanese Earth Resources Satellite (JERS-1: 1992–1998)	SAR (synthetic aperture radar) Optical sensors (visible near infrared and shortwave infrared)	18 18
Advanced Earth Observation Satellite (ADEOS: 1996–1997)	AVNIR (visible and near infrared) OCTS (ocean color and temperature scanner)	700 8–16
Advanced Earth Observation Satellite 2 (ADEOS-2: 2000)	AMSR (microwave scanning radiometer) GLI (global imager)	6,000–60,000 250–1,000
Advanced Land Observing Satellite (ALOS: 2002)	PRISM (panchromatic stereo mapping) AVNIR-2 (visible near infrared) PALSAR (synthetic aperture radar)	2.5 2.5 10–100
Information Gathering Satellite (Optical) (IGS: 2002)	Panchromatic Sensor (based on PRISM) Multispectrum Sensor (based on AVNIR-2)	1 1
Information Gathering Satellite (Radar) (IGS: 2002)	SAR (based on PARSAR on ALOS)	1–3

criticism certainly affected the management of the remote sensing program. In 1999, NASDA began a study on the Global Change Observation Mission, which, in response to critics, will be carried out by four satellites instead of one large ADEOS-3 satellite (Kallender, 1999c).

Having said that, the roles of NASDA and STA as driving forces in technological development seem to be coming to an end, particularly since Japan is finally achieving its technological objective—high-resolution sensor technology—through the Information-Gathering Satellite (IGS) program. Japanese remote sensing program development would be, as the NASDA Evaluation report suggested (Committee for NASDA Evaluation, 1998b, pp. 18–20), carried out by the space industry for more commercial purposes. In particular, since the agreement with the United States in 1990 prevents NASDA from developing a non–research and development satellite, NASDA's role may become smaller when the current wave of commercialization takes over core remote sensing activities.

Information Gathering Satellite

For 50-odd years, Japan's rather peculiar notion of "security," based on constitutional constraints and the alliance with the United States, prevented Japan from active involvement in international security, including using outer space for anything with military implications. Although Japan has "Self-Defense Forces," they are only operational within Japanese territory and the surrounding area, so developing military space systems was considered unnecessary. As for intelligence gathering, Japan heavily depends on its alliance with the United States.

It has been said that Japan did not become interested in remote sensing for security until the Taepodong launch in September 1998 caught the attention of politicians and initiated the decision to introduce IGS. Indeed, although the launch certainly triggered national debate, the ruling Liberal Democratic Party (LDP), Ministry of Foreign Affairs (MOFA), and JDA had been considering some kind of reconnaissance program for a few years before the Taepodong launch. The idea was to increase Japanese military autonomy in the post–Cold War period and to deal with the new U.S.-Japan Security Guidelines and the introduction of theater missile defense. As early as 1996, politicians held three meetings in LDP's Council of Foreign Affairs and Security to discuss the necessity of IGS, and eventually the issue became one of the principal features of LDP's foreign policy (LDP, 1998). In 1998, following this decision, JDA began constructing the Imagery and Mapping Support System, which is planned to be completed in 2000, to use high-resolution commercial data (Keidanren, 1999a).[11] MOFA began an investigation of reconnaissance satellites of the world "to encourage the debate over introduction of Japan's own satellite" in 1998 (Obuchi, 1998).

Nevertheless, these activities were seen as a slow and lengthy process because of the Japanese Diet Resolution of 1969. However, the Taepodong launch was so shocking and frightening that the concept of "peaceful use of space" was widened to accept the idea of IGS, which LDP, JDA, and MOFA took advantage of. LDP held seven intensive meetings from September 10 to October 15, 1998; proposed that the government develop four satellites (two optical and two radar) on October 29; and sent a group of powerful politicians to the United States to study actual development and operation of reconnaissance satellites from November 9 to 13 after the Cabinet decision on November 6 (Tamura, 1999).

INITIAL CONFUSION WITHIN THE GOVERNMENT

The movement to introduce IGS was, indeed, uncharacteristic of the Japanese "bottom-up" administration culture. Usually, decisions in space development are proposed and negotiated among relevant ministries, and SAC authorizes the final outcome. But the IGS process was led by LDP, MOFA, and JDA, which had never been involved in space policymaking procedures, which led to a series of problems. First, there seemed to be little agreement on details among the ministries. On one hand, STA, NASDA, and SAC, together with MITI and industry,[12] were afraid that IGS would squeeze the budget and delay other programs, ALOS in particular (SAC, 1998). On the other hand, there was little communication between JDA and MOFA on how the satellite should be procured and how the data should be used. Both ministries approached the U.S. government to explain the Japanese position on foreign procurement and data distribution, but their propositions were somehow inconsistent and confusing (*Asahi Shimbun*, April 14, 1999). Furthermore, the policy confusion was accelerated when LDP decided to use IGS not only for reconnaissance but multiple purposes, to avoid criticism for undermining the Diet Resolution. The Ministry of Transport, the Ministry of Home Affairs, the Public Security Investigation Agency, the Fire and Disaster Management Agency, and the Environment Agency were just a few of the many groups that tried to snatch a slice of the pie.

THE RACE AGAINST TIME AND MONEY

However, the situation became calm at the end of 1998. The Committee for IGS Promotion, chaired by the Cabinet Secretary-General, took responsibility for designing the whole system and coordinating interministerial interests. The committee had to tackle several urgent problems. First, there was the question of time constraints. The government wanted to have the satellite immediately, since it was desperately seeking information about North Korea. If the government decided to develop the satellite with domestic technology, it was expected to take at least four years. There were other options, such as (1) procuring a satellite from the United States, and (2) using U.S. intelligence information (combined with commercial data). The committee decided on domestic procurement, both to ensure autonomous and concealed operation and data gathering and to satisfy the demands of Japanese satellite manufacturers (Kallender, 1999a).

The committee's decision to develop IGS using only domestic technology faced strong pressure from the U.S. government. On the surface, the U.S. government appeared to accept the idea of an autonomous Japanese intelligence capability and seemed willing to cooperate with Japanese authority. But underneath, Washington sought to sell a complete U.S. system or at least one complete satellite (Kallender and Ferster, 1999; *Nihon Keizai Shimbun*, September 28, 1999). Such pressure put the committee in a difficult position.

Since the decision to develop four IGSs using domestic technology gave the relevant ministries and agencies (STA-NASDA, MITI-JAROS, and the Ministry of Post and Telecommunications' Communication Research Laboratory) a chance to develop new technologies for a higher-resolution sensor and an encrypting signal transmitter, it was impossible to purchase a fully U.S.-made system. Hence, the committee decided to purchase some key technology from the United States that did not exist in Japan.[13] The memorandum of understanding exchanged in September 1999 satisfied all the parties involved. On the one hand, Japanese intelligence capability will continue to rely heavily on the United States for crucial technology, on the other, Japanese agencies saved face and some money by investing in new technology. But most of all, the Japanese government was relieved, for the sake of further defense and intelligence cooperation, that it avoided upsetting the U.S. government.

While the committee found a middle ground between autonomy and alliance, another equilibrium, between cost and efficiency, needs to be found. For NASDA and other relevant agencies, the decision to develop IGS with domestic technology, despite the purchase of some key technologies from the United States, was good news. It allowed them to seize an opportunity for another technological development program. Moreover, they had felt overwhelmed by the additional workload. As the NASDA Evaluation Report clearly pointed out (Committee for NASDA Evaluation, 1998b, pp. 13–15), NASDA was already understaffed, and it was evident that the extra work for IGS would exhaust its human resources. In response, the Prime Minister decided to use the so-called "Prime Minister's special expenditure framework," which is a portfolio that the prime minister can allocate without the consent of the Ministry of Finance to the IGS program. As Table 10.3 shows, the scale of the budget was unprecedented. More than 70 billion yen ($670 million U.S.) was allocated from the Prime Minister's portfolio, in addition to 11 billion yen ($103 million U.S.) for IGS development. The funding for programs related to remote sensing tripled in fiscal year 2000; this not only relieved the cost pressures on other programs but also provided an opportunity for NASDA to recruit staff using this project funding (usually staff expenses are paid from general accounts) and increase manpower.

IGS DATA DISSEMINATION AND GLOBAL TRANSPARENCY?

The third question was how the satellite data should be used. The committee set up priorities for data distribution: defense, diplomacy, and large-scale disaster. The implication is that data can be used for purposes other than defense but that the priority in satellite operation and "shutter control" should be given to security issues. However, Kensuke Ebata, a Japanese military commentator, has said that the agency that will be dealing with imagery data will have members coming from different ministries loyal to their original organizations, so it is not difficult to imagine tense conflicts within the agency (*Sankei Shimbun*, June 28, 1999).

In terms of imagery data dissemination, the committee has not yet made clear any official position on whether data will be classified or partially declassified for public purposes, such as disaster management. In the SAC subcommittee meeting, the committee hinted that the reason to choose domestic procurement was to secure "secrecy" for Japanese information-gathering activities (SAC, 1999b, p. 16). Nevertheless, researchers and remote sensing engineers (including NASDA engineers), as well as some ministries holding

Table 10.3—Budget for Remote Sensing Related Activities (million yen)

Organization	Program	1997	1998	1999	2000 (estimate)
CO[a]	IGS	—	—	—	11,252 +*71,021*
STA	ADEOS-2	9,501	15,431	3,965	3,400
	ALOS	1,818	2,108	8,469	14,559
	IGS study	—	—	6,800	—
	Other Programs	13,279	11,967	10,153	11,290 +*1,183*
MITI	ALOS (PALSAR)	1,141	1,484	1,671	1,241
	IGS (SAR)	—	—	2,188	—
	Other Programs	5,802	3,876	3,496	3,040
Ministry of Post and Telecommunications		195	197	161	35 +*200*
	IGS (Transmitter)	—	—	890	—
Environment Agency[b]		906	1,049	1,056	*1,166*
MoT[c]		3,687	3,770	3,552	2,492
MoC[d]		1,640	1,665	1,676	1,705
MoA[e]		63	93	79	66
Totals		38,032	41,640	44,156	122,650

Source: SAC.
[a] Cabinet office.
[b] Environment Agency.
[c] Ministry of Transport (Meteorological Agency) excluding MTSAT budget.
[d] Ministry of Construction (National Geographic Office).
[e] Ministry of Agriculture, Fishery and Forestry.
Italic: Budget from Prime Minister's special expenditure framework.

high expectations for the multipurpose concept, are strongly arguing for civilian use of imagery data. Since the committee decided to use U.S. technology, which is under the restriction of PDD-23, it does not seem that the imagery data will be widely disseminated through RESTEC for general use,[14] but this does not rule out the possibility of using data for civilian public activities.

The fourth question is the expertise and organization of data processing and analysis. Although NASDA (EOC, EORC, and RESTEC) and ERSDAC have accumulated expertise for data analysis to a certain extent, they are short of staff and lack the experience needed for security analysis. JDA has been analyzing commercial data, but its expertise was fragmented into small groups in the Land, Sea, and Air Self-Defense Forces (these units merged into a single JDA Information Office in 1997) (JDA, 1997). The committee considers 200 to 300 staff members necessary for analysis and has been negotiating with the U.S. government to provide training. Even if the U.S. government agrees to cooperate, it is still questionable whether the whole data processing and analysis system will be ready in 2002.[15]

So far, there is a lot of uncertainty about the IGS, but two things are certain: (1) Although IGS was an unfamiliar "top-down" decision, all parties involved now seem satisfied with the advantages the program will bring, and (2) the IGS program would not increase global transparency. The program appears to be another technological satellite at this moment, but the real question is whether Japan will be able to manage the data analysis and distribution and, more importantly, what it will do with the data. Certainly, IGS will increase Japan's capability for autonomous intelligence gathering, but imagery alone is not enough for recognizing and analyzing objects—it is the information that gives a life to imagery data and that Japan crucially lacks at this moment.

Conclusion

Japanese policy on remote sensing is at a crossroads. The catch-up strategy has successfully and efficiently achieved its perceived goal of establishing a technological base and allowing Japan to join the "club of space advanced countries." The bias toward hardware development, however, has left the data utilization, analysis, and distribution system underdeveloped. Current structural changes, such as the commercialization of satellite imagery data, a series of failures in orbit and at launch, the reform of NASDA activities in response to SAC and international peer review, reforming the ministerial structure, and developing a military-oriented satellite poses many questions about the future of Japanese remote sensing activities. What is needed is a strong political leadership to design a new strategy for remote sensing and the management of space activities. Together with a high technological standard and a record of peaceful engagement in international affairs, this is a good opportunity for Japan to break its old habits of just catching up and instead to develop new strategies for contributing to global transparency.

Acknowledgment

I am grateful to Masami Onoda of NASDA and to the editors for their helpful comments on earlier drafts of this chapter.

Notes

1. Some "agencies," such as STA, JDA, and the Environment Agency, have their own ministers and the autonomy to formulate their own budgets (and to negotiate with the Ministry of Finance). Their role is more like that of a full ministry but with less power in the government. Other agencies, such as the Resource and Energy Agency, are nonministerial and are junior to their "parent" ministries. NASDA, although the English translation uses the word "agency," has a special status as a public corporation. As such, it is strictly under the financial control of its supervising ministry but, as an expert organization, has a powerful voice in policymaking.
2. RESTEC's original purpose was to distribute Landsat data, but it gradually became a research center for data reception and analysis and a distributor of Japanese satellite imagery data.
3. I am grateful for Masami Onoda for suggesting this point.
4. Although both JAROS and ERSDAC were established to be under the joint guidance of MITI and STA, the major influence—financially as well as politically—comes from MITI. STA's role is limited to technological assistance.
5. This institute was created to operate the MITI-ISAS-NASDA codeveloped Space Flyer Unit but was spun off to become MITI's think tank for spacecraft commercialization.
6. According to one employee, Nissho-Iwai was charging less than RESTEC was for data (from the Japanese Web site http://www.geocities.co.jp/Technopolis/4025/tokumaru2.html).
7. The Fundamental Policy of Japan's Space Development is a long-term strategy document drafted by STA and approved by SAC. The original document was adopted in 1978 and was amended in 1984, 1989, and 1996. Although the document defines future projects rather vaguely, it rigidly binds future programs and therefore provides a source of stability for Japanese space development.
8. TRMM is a joint project of Japan and the United States, with Japan responsible for developing the precipitation radar instrument and launch with H-II, and the United States responsible for developing the satellite bus and four other sensors, as well as for satellite operation.

9   ASTER is a JERS-1 follow-on sensor developed by JAROS. It was integrated into NASA's EOS-AM1 with other four sensors.
10 In addition to these programs, there is an initiative from a remote sensing researcher to develop Ricesat, an international project to use a radar imaging satellite to monitor rice crops, the Asia-Pacific's staple food (Kallender, 1999b).
11 Keidanren is the most powerful employers association in Japan.
12 For the position of industry toward IGS, see Keidanren (1999b).
13 Likely technologies are those for (a) changing and adjusting the angle of optical sensors, (b) fixing an optical sensor in a precise angle, (c) transmitting data to a ground station, (d) recording imagery data, and (e) material for optical lens (*Yomiuri Shimbun*, September 23, 1999; *Mainichi Shimbun*, September 29, 1999; Usui and Ferster, 1999).
14 Although RESTEC is in charge of developing ground systems for image analysis, it is unlikely that RESTEC will be able to use imagery data from IGS (RESTEC, 1999).
15 Even some JDA officials are skeptical about the readiness in 2002.

References

Asahi Shimbun, April 14, 1999.
Callon, S., *Divided Sun: MITI and the Breakdown of Japanese High-Tech Industrial Policy, 1975–1993*, Stanford, Calif.: Stanford University Press, 1995.
Committee for NASDA Evaluation, *Evaluation Report of the First Meeting of the Subcommittee on Earth Observation*, November 1998a.
—, *NASDA Evaluation Report* (English version), November 1998b.
—, *Report of SAC Subcommittee of Fundamental Problems of Space Development*, May 1999.
Earth Observation Research Center, "What's EORC," last update February 9, 2000. Available at http://www.eorc.nasda.go.jp/EORC/Welcome/index.html (last accessed April 21, 2000).
EORC—*See* Earth Observation Research Center.
JDA—*See* Japanese Defense Agency.
Japanese Defense Agency, *Securitarian*, March 1997.
Johnson, C. A., *MITI and the Japanese Miracle: The Growth of Industrial Policy*, Stanford, Calif.: Stanford University Press, 1982.
Kallender, Paul, "Firms Vie to Win Japan's Biggest Satellite Deal," *Space News*, February 8, 1999a.
—, "Researchers Push New Satellite to Monitor Rice," *Space News*, June 28, 1999b.
—, "NASDA Details 4-Satellite Climate-Monitoring Plan," *Space News*, August 9, 1999c.
Kallender, Paul, and Warren Ferster, "U.S. Lobbies Japan for Role in Reconnaissance Satellites," *Space News*, May 3, 1999.
Keidanren, "Interview with Mr. Sato," *Keidanren Bulletin*, January 19, 1999a.
—, *Proposal for Establishing Comprehensive Space Development and Utilization Policy, Strengthening Industrial Base for Space Industry, and Promoting Industrialization*, July 6, 1999b.

LDP—*See* Liberal Democratic Party.
Liberal Democratic Party, "Proposal for Introduction of Information Gathering Satellite," *Newsletter*, October 29, 1998.
Mainichi Shimbun, September 29, 1999.
Nihon Keizai Shimbun, September 28, 1999.
Obuchi, Keizo, reply in Minutes of Upper House of Diet, April 23, 1998.
Remote Sensing Technology Center, *Annual Activity Plan 1999*, April 1999.
RESTEC—*See* Remote Sensing Technology Center.
Sankei Shimbun, June 28, 1999.
SAC—*see* Space Activities Commission.
Space Activities Commission, *Report on the Evaluation of Earth Observation*, March 1997.
—, Minutes of the Tenth Meeting of the Planning Committee, November 9, 1998.
—, *Report of SAC Subcommittee of Fundamental Problems of Space Development*, May 1999a.
—, Minutes of the Fourth Meeting of the Planning Committee, July 16, 1999b.
Tamura, Shigenobu, "*Jimin-Tou Jyouhou-Syusyu-Eisei ni Kansuru Houbei-Cyousadan-Houkoku* [Report on LDP project team on IGS visit to the United States]" in Keidanren Space Development and Utilization Promotion Conference, "For the New Satellite Era," Ucyu, No. 47, March 1999.
Usui, Naoaki, and Warren Ferster "Agreement Opens Japanese Program to U.S. Firms," *Space News*, October 11, 1999.
Yomiuri Shimbun, June 7, 1999.
Yomiuri Shimbun, September 23, 1999.

11

Commercial Observation Satellites in the Middle East and the Persian Gulf

Gerald M. Steinberg

The Middle East and Persian Gulf will be among the regions most directly affected by the proliferation of commercial high-resolution satellite imaging. In these areas, which continue to be characterized by high levels of conflict and tension, the dual-use nature of this technology and the data it provides are most pronounced.

Militarily, the instability in the region—particularly among revisionist and pariah states, such as Iraq, Iran, Libya, and Syria, and the terrorist groups, such as Hamas, Hizbollah, and the Bin Laden network—highlight the potential strategic effects of the widespread availability of high-resolution images and data. Satellite imaging is generally seen as a means of obtaining military intelligence and, depending on the format (resolution, spectral details, and timing), is likely to have major influences on regional stability and balance of power.

In this chapter, we will consider the implications of increased transparency resulting from the commercial availability of timely high-resolution satellite imaging for the Middle East. On this basis, the chapter will describe and analyze the activities and capabilities of the countries in the region with respect to the launch and operation of imaging satellite and the analysis and distribution of their data.

The Double-Edged Sword of Transparency

In discussions and analysis of high-resolution commercial satellite imaging, beyond the economic benefits, advocates frequently point to the increased transparency this technology provides. Space imaging will make it much easier to identify and respond to unusual troop movements, preparations for surprise attacks, or missile tests. Verifying arms control agreements will also become more reliable, thereby increasing their functionality. With wide access to high-resolution satellite images, nongovernmental organizations and journalists will also be able to provide independent checks on governments, making it harder for states to hide their activities from the public (see, for example, Gupta, 1995; Florini and Dehqanzada, 1999).

However, in the high-conflict environment of the Middle East, the potential advantages of transparency are open to question, and in some ways, the costs may be greater than the benefits. With the legacy of overlapping conflict zones (Arab-Israeli, Persian Gulf, Turkey-Syria, North Africa, etc.) and the resulting wars and terrorism, this region is characterized by a high level of instability.

Under these circumstances, the increased availability of the timely information that high-resolution satellite imaging can provide will, first and foremost, be used for military intelligence. In the Middle East, efforts to negotiate arms control agreements and various confidence-and-security-building measures to increase stability have not progressed significantly. The meetings of the Multilateral Working Group on Arms Control and Regional Security between 1992–1994 ended in an impasse without any accomplishments. In the Middle East, the United Nations Arms Registry has made little headway, with most states either ignoring this voluntary measure toward transparency in conventional weapons or only providing information that has been made available from other sources. Thus, for the foreseeable future, the use of satellites for verifying formal regional arms control agreements is likely to be limited.[1]

This environment highlights the dual-use nature of high-resolution satellites, with both military and civil applications. During the Cold War, satellite imaging was a major tool for military intelligence (see Steinberg, 1983; Gaddis, 1988). Now, this technology is becoming available in regional contexts, through both dedicated military programs and commercial systems (Florini, 1998; Tahu, Baker, and O'Connell, 1998). Similarly, the images provided by the French Satellite Pour l'Observation de la Terre (SPOT) system and by Russian and Indian high-resolution imaging satellites are also frequently used for military purposes (Krepon, 1990, p. 21; Richelson, 1990, p. 55). There are indications that Iraq and Iran used SPOT images during their eight-year war (Krepon, 1990 p. 23). SPOT has also been used to obtain information regarding the Dimona nuclear reactor complex in Israel (which is closed to all but essential personnel with the highest level security clearances), sites in Iran, Iraq, Serbia, and Bosnia (Zimmerman, 1989).

Among the states in the region, Israel is probably the most sensitive to the dual-use nature of high-resolution commercial imaging satellites. Israeli defense officials expressed their sensitivity to this issue in the early 1990s through their concern about the effects on Israel's national security of the U.S. government's decision to remove the limitations on high-resolution commercial imaging. Israel's very small territory allows detailed and repeated coverage with a relatively limited number of images, making the nation vulnerable to attacks based on data accessible through commercial high-resolution imaging satellites. Israel's deterrence posture and strategy are based on maintaining a high degree of uncertainty in the eyes of potential enemies, particularly with respect to the nuclear- and ballistic-missile potential. Israeli policymakers feared that Arab states and Iran, as well as terrorist groups, would be able to exploit these high-resolution images to obtain very detailed intelligence on Israeli capabilities and deployments. Their ability to target Israeli sites with a high degree of precision would alter the balance of power fundamentally, particularly if these images were combined with Global Positioning System (GPS) data and used to target accurate cruise or ballistic missiles (Steinberg, 1995, 1998b).

The military applications of commercial satellite imaging, and the potential destabilizing effects, are not confined to the Middle East. Indeed, the ability to receive detailed images from the battlefield and distribute them (or analyses based on them) directly and immediately to the commanders in the theater is increasingly being recognized as a major innovation in conventional warfare. In 1995, U.S. Secretary of Defense William Perry noted the central role that space forces play, stemming from their "exceptional capabilities" for collecting, processing and distributing data. Similarly, Russian military have analysts noted that

> The collection, processing, storage, transmission and display of information in them . . . has been made the basis of operation. It permitted formalizing the process of assessing events that are occurring and coming up with preferable decisions. . . . (Menshikov and Rodionov, 1996.)

Regional powers, rogue states, and terrorist groups that gain access to the data and the capability would be able to exploit it for the same purposes. Eliot Cohen (1996, pp. 44–45) has noted that

> A military cliché has it that what can be seen on the modern battlefield can be hit, and what can be hit will be destroyed. Whereas at the beginning

of this century this applied with deadly certainty only to front line infantrymen, it now holds not only for machines on the front lines but for supporting forces in the rear. . . . As all countries gain access to the new forms of airpower [space based reconnaissance and unmanned aerial vehicles], hiding large scale armored movements or building up safe rear areas chock-a-block with ammunition dumps and truck convoys will gradually become impossible.

U.S. government officials have increasingly recognized the potential impact of the proliferation of this technology. Keith Hall, Director of the U.S. National Reconnaissance Office, has stated that "Real-time imagery capabilities provided by E-O [electro-optical] and other technologies is causing a revolution in warfare." (Hall, 1998.) In analyzing the 1991 Gulf War, he noted that "Satellite reconnaissance was a major factor in the rapid US victory." However, Hall's public statements do not address the effects on U.S. interests and allies or on regional stability in such areas as the Middle East.

In contrast, other analysts emphasize the destabilizing impact of these capabilities on a global basis. For example,

> Islamic Jihad could get its hands on a one-meter resolution picture of . . . a US Air Force General's headquarters in Turkey, convert the shot to a precise three-dimensional image, combine it with data from a GPS device...and transmit it to Baghdad, where a primitive cruise missile, purchased secretly from China could await its targeting coordinates. (Lane, 1996, p. 24.)

Vipin Gupta (1995, p. 115) has noted that the effect

> depends on how the new remote sensing services will be distributed through the political landscape, how belligerent states will use the high-resolution images, and how observed states will respond to routine overhead imaging by their neighbors.

He warns that unlimited sales of high-resolution imaging could disrupt "delicate balances of power," complicate the containment of international crises, and fuel developments in offensive weapons capability (Gupta, 1995, p. 117). Similarly, former Central Intelligence Agency Director James Woolsey concludes, "This very comfortable world people have been living in where fixed target installations on land are safe" will vanish with the proliferation of high-resolution commercial imaging (Woolsey, 1996).

As noted above, the Middle East is characterized by a high level of conflict, and instability is increased by a number of revisionist powers (Iraq, Iran,

Syria, Libya), as well as numerous terrorist groups with access to funds and advanced technology, including the Bin Laden network. As a result, high-resolution satellite imaging, whether dedicated military platforms or commercial systems, is viewed primarily in the context of its dual-use nature. As will be seen in the following discussion, security concerns are the foundation for Israel's efforts to limit the distribution of high-resolution images.

Commercial Imaging Capabilities in the Middle East

Israel has the most active and advanced space program in the region, including a dedicated military imaging system, Ofeq, and a commercial program, the Earth Remote Observation System (EROS). The military establishments of other countries, which do not have the resources for dedicated reconnaissance satellites, such as Egypt, Syria, Iraq, and Iran, are likely to be major customers for commercial imaging products, while applying the data to military intelligence applications.

However, the market for civil applications of high-resolution satellite imaging systems in the Middle East and Persian Gulf is also significant. As in other regions, space-based imaging has long been used for urban planning; locating and identifying natural resources, including water (particularly in large and remote deserts); agriculture; and environmental monitoring. In some of these applications, commercial high-resolution imaging can be used to improve efficiency. Egypt and Saudi Arabia have developed an advanced infrastructure in civil applications of remote sensing, and the United Arab Emirates (UAE) has created a commercial center to provide remote sensing services (both military and civil) to the Arab and Islamic world (all states in the region except Israel).

THE ISRAELI PROGRAM

Israel has had an active and growing space program for two decades. As in the case of the United States and USSR during the Cold War, the environment of conflict led Israel to develop imaging satellites capable of real-time intelligence for reconnaissance.

The centrality of intelligence and early warning was emphasized as the countries in the region began to acquire ballistic missiles and weapons of mass destruction. On occasion, the United States shared strategic and space-based intelligence information with Israel. However, despite the close defense cooperation with the American government, Israel did not have routine access to real-time satellite intelligence data. A Ministry of Defense (MOD) official was quoted saying

> For years we have been begging the Americans for more detailed pictures from their satellites and often got refusals—even when Iraqi Scud missiles were falling on Tel Aviv . . . . (Rotem, 1992; see also Koulik and Kokoski, 1994, p. 199.)

At times, the Israeli Defense Forces have turned to other sources, including Russia, which reportedly sold hundreds of satellite pictures of Syria, Iran and Iraq for about $1 million, as part of a secret cooperation agreement (Melman, 1998).

The Middle East peace process and the transfer of territory to Egypt and the Palestinian Authority have reduced Israel's ability to rely on ground-based early warning and intelligence installations, increasing the reliance on space-based systems. This dependence will increase if Israel relinquishes the Golan Heights to Syrian control, including the intelligence gathering station on Mt. Hermon. Following an agreement with Syria, Israel will need systems to provide early warning of any Syrian military activity. As former U.S. Department of Defense official Dov Zakheim has noted, "space-based support could buttress Israel's employment of unmanned stations that could be placed on the Golan to monitor Syrian military activity." (Zakheim, 1999.)

Ofeq. As a result of these factors, Israel launched the Ofeq-1 (Horizon) test satellite in 1988, using the three-stage Israeli-designed and manufactured Shavit launcher (see Table 11.1). The launch site is located on the Mediterranean coast. To avoid flying over hostile countries, a highly unusual flight path was used (northwest, over the Mediterranean), placing the satellite into a

Table 11.1—Israeli Military Satellite Imaging Systems

Payload	Launch Date	Description	Technical Details
Ofeq-1	September 1988	Test and proof of concept	156-kg package; spin stabilized;' reentered after 4 months
Ofeq-2	April 1990	Advanced test	156-kg package; spin stabilized; reentered after 3 months
Ofeq-3	April 5, 1995	Dedicated reconnaissance satellite	255-kg package; thruster-based three-axis stabilization; still orbiting as of January 2000
Ofeq-4	January 22, 1998 (launch failed)	Dedicated reconnaissance satellite (?)	

NOTE: All Ofeq payloads are classified and were launched by the Israeli Shavit launcher.

retrograde orbit at an inclination of 143 degrees (Congressional Research Service [CRS], 1994, p. 161).[2] Ofeq-1 was reported to be a test vehicle designed to lead to the development of an orbital reconnaissance capability, and it reentered the earth's atmosphere in January 1989. Ofeq-2 was launched in April 1990, was similar in weight and technical characteristics to Ofeq-1, and had an orbital lifetime of three months (CRS, 1994, p. 161).

Ofeq-3, launched in April 5, 1995, was apparently the first operational reconnaissance system, with a payload containing ultraviolet and high-resolution imaging sensors. Its higher perigee (369 km) and orbital maneuvering capability allow a longer life time. According to press reports, this version of the Shavit launcher included a small new Israel Aircraft Industry (IAI) third-stage rocket engine with 674 lbs. of thrust (*Aviation Week*, 1994b).[3] The satellite's orbit takes it over sites in the Middle East, including Iraq.

On January 22, 1998, the attempted launch of Ofeq-4 (reportedly equipped with an advanced imaging system) ended in failure when the booster malfunctioned (Reuters, 1998a; Rodan, 1998). Had the last launch been successful, Israel would have had two imaging systems in operation simultaneously, significantly enhancing capability. The Israeli military now hopes to attempt another launch in 2000, before Ofeq-3 reenters or is no longer operational (Schiff, 1998a).

EROS. As in the United States (where the firms and individuals involved in the military reconnaissance satellite program are centrally involved in the commercial efforts), the Israeli military and commercial high-resolution imaging satellite programs are closely interrelated. Using the technology developed for dedicated military reconnaissance systems, IAI has been planning commercial space ventures since the early 1980s. The establishment of the Space Technology Division at IAI in 1984 marked a major step toward developing this capability.

Although a government-owned firm formally under the control of the MOD, IAI enjoys a significant amount of autonomy, particularly with respect to new commercial ventures. As direct government subsidies are reduced, the firm's directors are responsible for finding independent resources for research and development and for ensuring employment in Israel's largest industrial firm. In the past two decades, the efforts to increase exports and joint ventures with foreign firms have accelerated (Steinberg, 1998a; Klieman and Pedatzur, 1991).

IAI submitted its first proposal for the EROS program in 1993. Because of the limits on government funds and the high costs of development, outside

Table 11.2—Israeli Commercial Satellite Imaging Systems

Satellite Designation	Launch Date	Resolution (m)	Type	Altitude (km)
EROS A1	Dec. 2000	1.80	Panchromatic	480
EROS A2	2001	1.80	Panchromatic	480
EROS B1	2002	0.82	Panchromatic	600
EROS B2	2002	0.82	Panchromatic	
		3.68	Multispectral	
EROS B3	2003	0.82	Panchromatic	600
		3.68	Multispectral	
EROS B4	2003	0.82	Panchromatic	600
		3.68	Multispectral	
EROS B5	2004	0.82	Panchromatic	600
		3.68	Multispectral	
EROS B6	2004	0.82	Panchromatic	
		3.68	Multispectral	

NOTE: The first EROS satellite is scheduled to be launched by the Russian START booster. No decisions have been made regarding subsequent launches.

investors were sought. The major Israeli defense technology firms, including IAI, have often sought external partners in the private sector, particularly in the United States, to expand marketing opportunities and access to development capital. Thus, the effort to find foreign partners or investors in this case was not particularly unusual. To remain the senior partner and to ensure control over technology, operations, and data, IAI rejected joint programs with the major foreign commercial firms, such as Space Imaging (Lockheed-Martin and Raytheon), OrbImage, and Earthwatch in the United States and SPOT Image in France.

In 1996, IAI agreed to work with Core Software Technology (CST), based in Pasadena, California, in the development of the EROS satellite system.[4] Core reportedly invested $150 million in the program to provide real-time imaging to ground stations around the world (Cole and Marcus, 1996; Rodan, 1995; Lardier, 1997; IAI, 1996; ImageSat International, 2000).

Although press reports claimed that the first EROS launch was scheduled for 1997, this date was never realistic, in part because the Israeli MOD refused to approve the license (see discussion below). However, planning continued, and by 1998, a program for a constellation of eight commercial lightweight, low-earth orbit, high-resolution imaging satellites was completed. This system is designed to provide potential customers, including the Israeli military, with very frequent coverage of any point on the globe, compared to the more sporadic coverage that would be available from a system of one or two satellites.

For this very ambitious program, IAI, CST, and Electro Optical Industries, Ltd. (Israel), created West Indian Space (WIS), renamed ImageSate International in 2000, incorporated in the Cayman Islands,[5] to manage the development, launch, and operations of the EROS project. The failure of the Ofeq-4 launch increased the Israeli government's interest in EROS as a means of lowering the cost of a satellite observation system (Schiff, 1998a).

EROS is advertised as a low-weight "light-sat" incorporating breakthroughs in system concepts that allow a 200-kg satellite to match the performance characteristics of satellites weighing several tons. Similarly, IAI's Amos communication satellite was the lightest communications satellite built at the time of launch, in 1996. Customers can purchase "turnkey" earth-observation systems, based on "vertical integration of IAI's satellite technology and CST's archiving, distribution, image management and exploitation technology." WIS is offering to retrofit existing ground stations, used for accessing SPOT and LANDSAT images, at a cost of $10 million (de Selding, 1999).

Israeli launch capacity is currently insufficient for the orbital altitude and weight of the EROS payload; as a result, Russian launchers are scheduled to boost the first satellites into orbit (in itself a major source of controversy).[6] The first stage of the program consists of two enhanced EROS A+ satellites, to be launched in 2000 and 2001, followed by the EROS B satellites. By 2004, the full system of eight satellites is scheduled to be fully operational. This constellation will provide data to customers and future satellite operating partners (SOPs) in geographical information system–ready form. The satellites will be launched into low-earth orbits, at altitudes between 480 and 600 km.

EROS A+ will deliver imagery at a resolution of 1.8 m and, using the "push broom sweep method" of imaging, will cover a 12.5-km^2 swathe with panchromatic imagery, provided to customers in near real time. EROS B1 will deliver 0.82-m resolution from an altitude of 600 km, covering swaths of 16.5 km^2, followed by EROS B2 through B6 (five satellites), with 0.82-m panchromatic and 3.68-m multispectral imagery. The EROS A+ sensors will consist of 10,000 detector elements in arrays that provide day-only or day-and-night performance, with time delay integration. The EROS-B series will incorporate over 15,000-element modules on a single focal plane, to provide for such different sources as an infrared layer over conventional (visible light) images (Esche, 1999).

The SOPs in the EROS network will receive worldwide access to high-resolution imaging, archiving, and distribution, using existing remote sensing ground infrastructure and additional installations, as necessary. The SOPs will receive "a dedicated regional satellite with local customer tasking to meet national and civilian satellite program requirements." In 1999, WIS and the Israeli government reportedly signed an agreement giving Israel exclusive access to images of the Middle East region obtained by the first three EROS satellites for eight years (de Selding, 1999).

Customers can also purchase priority acquisition service (PAS) programs to receive priority tasking of EROS satellites worldwide. Daily coverage (at least one revisit per day) will be provided for North and South America, Europe, Asia, the Middle East, Africa, and Australia. WIS is offering imaging services for use in civilian mapping, infrastructure management, and military intelligence via long-term contracts at $10 to 20 million annually. Company officials are basing their marketing plans on the expectation that states will opt for a system that is less expensive than it would be to develop their own military surveillance programs, which could cost five to ten times more.

In addition to selling a range of services to investors and customers, WIS is also planning to issue bonds via a private offering to help finance the $250 million needed to launch and operate the first three satellites. The full cost of the eight-satellite system is estimated at $750 million, and WIS is seeking to raise most of this sum through private investors (Rosenberg, 1999).

Details regarding possible SOPs and customers have not been published, but there has been some speculation in press reports, particularly with respect to Turkey (Davis and Rudge, 1998) and India (Davis, 1998). Asian nations that lack their own satellite capability but that need high-resolution intelligence and other data are considered to be likely customers. WIS officials declared that they will "respect the wishes of the U.S. and Israeli governments" by not providing data to states that are subject to export restrictions, such as Libya, Iraq, and Iran (de Selding, 1999). There are also reports that Israel rejected requests from other states to purchase EROS and Ofeq-type platforms (Parnes, 1995).

U.S.-Israeli Negotiations. As noted above, Israel is particularly vulnerable to intelligence data that can be made widely accessible through commercial high-resolution imaging satellites.[7] In the early 1990s, the changes in the U.S. policy in this area (Presidential Decision Directive 23 [White House, 1994], in particular), disturbed and surprised Israeli military planners, and they sought to reduce the repercussions for Israeli security concerns. In addition, the close strategic relationship between Jerusalem and Washington and the degree to which the peace process and Israeli concessions have depended on U.S. security guarantees and pledges to prevent the degradation of Israel's qualitative edge were also important factors in the Israeli efforts to limit the damage caused by what was seen as a radical unilateral change in U.S. policy.

The Israeli government first raised this issue in 1992, shortly after the United States adopted the Land Remote Sensing Policy Act (P.L. 102-555), and the UAE submitted an application to purchase an imaging satellite from Litton/Itek. Israeli officials protested, charging that the United States was planning "to supply the Arab countries with binoculars that will enable them to see every military movement here." (Rotem, 1992; see also Koulik and Kokoski, 1994.) The U.S. Department of State ultimately blocked the application (in part, in response to Israeli objections).

Israeli concerns increased after a Saudi company known as EIRAD, owned by Prince Fahd Bin Salman Bin Abdulaziz, sought to acquire a major interest in Eyeglass (now OrbImage), in return for an agreement to build a ground station in Riyadh and exclusive rights of coverage in the Middle East. The main customer was said to have been the Saudi Defense Ministry (Lane, 1996). In response, the Israeli government expressed concern that this would give the Arab States, including Iraq, access to highly accurate intelligence information and would threaten Israeli security and vital interests. The Clinton administration asked Israel not to object to the OrbImage agreement with Saudi Arabia. However, Israel cited the Saudis' negative role in regional instability, potential contribution to conflict, and support for radical Islamic groups (*Aviation Week*, 1994a).[8] Congress also took up the issue (Bingaman, 1996), and OrbImage subsequently announced that it would "exclude the territory of Israel from its viewing area and to put a technical fix on the satellite that would prevent such viewing." (See also Ferster, 1994.)

The Israeli development of the commercial EROS system complicated these discussions. As noted above, IAI's plans for EROS were largely motivated by commercial factors and were not coordinated with the MOD. This appeared to be in sharp contradiction with Israeli policy, as political and defense officials sought to prevent commercial satellite imaging of Israel, while the defense industries were planning their own high-resolution commercial imaging system (Steinberg, 1998b).

U.S. government and industry officials noted this contradiction and argued that, while the United States was limiting its own programs to satisfy Israeli demands, and at considerable cost, Israel was preparing to establish a major market presence before the U.S. firms were able to launch their own first commercial satellites. Had the initial schedule for a launch in 1997 been met, EROS would have been in orbit 18 months before the first U.S. commercial system (Dailey, 1996). U.S. industry officials also argued that Israeli commercial imaging was unfairly subsidized by the military, although the U.S. government also effectively subsidized the development of U.S. commercial imaging systems (Broad, 1997; Cole and Marcus, 1996).

In contrast, some Israeli industry officials complained that the MOD's policy of negotiating limits with the United States, while refusing to license the EROS program, would result in losing an opportunity to establish a major presence in the market before the United States gained dominance. They argued that the Americans were using the negotiations to delay the EROS program until U.S. firms are able to launch their own commercial imaging satellites (MOD, 1996).

In 1995, after considerable negotiations, the Israeli and U.S. governments agreed to coordinate policies. Washington denied the Saudis control of the satellite track from the ground and also limited the sale of state-of-the-art software for image enhancement. On March 4, 1996, the Israeli MOD issued the first formal public statement of Israeli policy, which included a ban on the use of Ofeq-3 images for commercial purposes and keeping security-related technologies and data separate from civil activities. "Any possible future commercial track" would require licensing from the MOD.[9]

In June 1996, the U.S. Senate passed an amendment to the 1997 Defense Authorization Act entitled, "Prohibition on Collection and Release of Detailed Satellite Imagery Relating to Israel and Other Countries and Areas." The amendment would have prohibited any agency or department of the U.S. government from licensing the collection or dissemination, declassification, or release by any nonfederal entity of satellite imagery with respect to Israel, or any other designated area, unless the imagery is no more detailed or precise than imagery produced by that country's indigenous satellites (Bingaman, 1996).

The final version of the amendment prohibited the sale of imaging data over Israel with a better resolution than other "commercially available" sources. The precise definition and implementation of this language was unclear, and the Israelis agreed to a 2-m limit, based on the availability of Russian KVR-1000 images, although there are questions regarding the degree to which these can be called "commercially available." In 1998, the United States set the limit at 1 m. Israel protested, and an agreement was reached in July 1998 that blacked out Israel at resolutions below 2 m. Israel also accepted this limit on the sale of EROS images, and the MOD agreed to license this program (Associated Press, 1998a; Schiff, 1998b).

During the negotiations, the positions and policies of the two governments began to converge. U.S. government officials gradually began to understand and accept the Israeli concerns regarding the security implications of the proliferation of high-resolution commercial imaging systems, not only for Israel, but primarily for U.S. security interests. At the same time, the Israelis came to understand the difficulties inherent in blackouts of specific areas, shutter control, and other limitations.[10] Officials from both countries have adopted a more flexible and pragmatic approach, seeking to balance their security and commercial interests.

Israel's Data Policy. As the EROS program progressed, the Israeli government began to develop a licensing policy for a commercial imaging system. After considerable pressure for a definitive policy statement, the MOD issued a short document, which began with the straightforward acknowledgment that "The State of Israel is developing national remote sensing capabilities." At the same time,

> The MOD prohibits the use of its remote sensing [i.e., Ofeq] defense capabilities for commercial purposes. It is the policy of the MOD to strictly differentiate between the national defense technological projects and any possible future commercial track.

Thus, like the United States and France, Israel has created a clear partition between dedicated military reconnaissance satellites and commercial ventures, even though, at some point, the commercial systems, such as EROS, may be used for national security objectives. Regarding licensing, the announcement proclaimed that

> Any industry that will express an interest in entering into the commercial remote sensing field will have to receive appropriate permits. Any industry in Israel, including IAI, wishing to enter the track of commercial remote sensing, will have to receive final approval of the MOD after a careful and thorough review of all aspects. (MOD, 1996.)

This statement acknowledges that Israel faces the same combination of military and commercial incentives that exists in the United States and other countries. In the spirit of "if you can't beat them, join them," Israeli policy had shifted, and the MOD agreed to provide operating licenses and initial contractual commitments for the EROS program. Thus, although the military factors are still dominant, commercial factors have become increasingly important. In Israel, as in the United States, France, Russia, India, and other states, sales of high-resolution imaging services are increasingly viewed as a means of offsetting the cost of developing and operating a vital military capability.

THE ARAB STATES AND IRAN

Saudi Arabia. Saudi Arabia has invested considerable resources in creating a remote sensing infrastructure, including an advanced center located in Riyadh. In late 1994, a Saudi company called EIRAD, owned by Prince Fahd Bin Salman Bin Abdulaziz, sought to acquire a major interest in Eyeglass (now OrbImage), in return for an agreement to build a ground station in Riyadh and exclusive rights to receive and distribute OrbView satellite (launched and operated by OrbImage) images in the Middle East. (EIRAD acquired a 20-percent interest in the company.) The main customer is expected to be the Saudi Defense Ministry (Lane, 1996). As noted above, this involvement raised fears in Israel regarding the use of this system to gather military intelligence information that could be used by various Arab states and terrorist groups against Israel.

In addition, the Saudi Center for Remote Sensing (SDRS), located in Riyadh, was established in 1983 and is developing an advanced capability for data analysis. In 1999, SDRS signed an agreement with RADARSAT International (Canada) for exclusive ordering, scheduling, reception, and product generation of RADARSAT-1 (8-m resolution) data for the Middle East.

Other Saudi investors have also invested heavily in different space-based commercial technologies. Prince Alwaleed Bin Talal invested $200 million in Teledesic, which is seeking to manufacture and operate a global, broadband space-based Internet communications system. Another Saudi investor has reportedly agreed to finance the development of a commercial launch system (*Aviation Week Space Business Online*, 1999). Thus, Saudi involvement and access to data in costly commercial ventures, including high-resolution satellite imaging, is part of the pattern.

UAE. As noted, the UAE sought to purchase a commercial imaging satellite from a U.S. manufacturer in 1992. Although the rules had not yet been changed, the U.S. Department of Commerce seriously considered and favored this offer before rejecting it on political and military grounds.

In 1997, Dubai signed an agreement for establishing an imagery receiving station, to be operated by Dubai Space Imaging (DSI), a joint venture formed by the U.S. company Space Imaging EOSAT and a group of UAE investors. This station will provide customers with real-time (within 20 minutes of collection) access to detailed imagery in an area within a 2,300-km radius (including all of Iraq and Iran) from the receiving station (EOSAT, 1997). In the first stage, the Dubai receiving station will use a ground imagery processing system leased from the U.S. company Datron World Communications and will receive data from the 5-m Indian Remote Sensing (IRS) earth imaging satellites IRS-1C and IRS-1D. After the launch of the IKONOS satellites, the Dubai station will be upgraded with Raytheon E-Systems technology, to receive and distribute 1-m–resolution imagery. Company officials and sales material explicitly noted that this system "is easily capable of detecting and identifying individual vehicle movements, mobile missile launchers and other military activities under clear weather conditions." (Jane's, 1998.) DSI will sell information products and services, as well as imagery, and will provide training in imagery analysis and geographical information system tools and applications. The contract with DSI is not exclusive, and EOSAT officials have stated that "we are already talking to several other countries in this region who have an interest in establishing a national ground station to exploit our imagery." (Jane's, 1998.)

Egypt. Although Egypt is not known to be investing in development or purchase of a dedicated imaging satellite (civil or military), Egypt has broad and advanced capabilities applicable to remote sensing. Egyptian analysts, both civil and military, have significant experience in processing and interpreting high-altitude aerial and satellite-based imaging data. As a result, Egyptian technicians and analysts are likely to be centrally involved in many of the application programs based on the use of commercial high-resolution satellite imaging in the Middle East.

Iraq. In December 1989, a few months before the invasion of Kuwait, Iraq launched a three-stage missile (the Al-Abid), and the Iraqi government declared that this was a test of an independent space launching capability (CRS, 1994, p. 135). During the 1980s, Iraq was also involved in the China-Brazil Earth Research Satellite imaging development project.

After the Gulf War, and the United Nations' imposition of sanctions, these projects were frozen. However, with the availability of commercial systems, the Iraqi regime will have the same access as other entities to the data and images that are produced. In late 1999, a Russian firm (NPO Mashinostroenye) reportedly delivered the first 70 (out of a total of 220) digital high- and medium-resolution images satellite images of the Gulf region to Iraq. News reports noted that "Defense analysts believe the photographs will greatly improve the ability of the Iraqi armed forces to target neighboring countries." (Coughlin, 1999; Reuters, 1999.)

The Iraqis may be barred from directly entering into contractual arrangements to receive data from U.S. and perhaps European or Japanese firms. However, as in the case of other dual-use technologies, agents and contractors will be able to purchase the data via third-party transactions. Coupled with the missiles and weapons of mass destruction that Saddam Hussein retains and an ability to launch terror operations throughout the world, intelligence information from real-time commercial imaging systems will greatly increase Iraq's capabilities.

Iran. Following Israel, Iran is the most active country in the Middle East with respect to indigenous space and satellite development and launch capabilities. In 1998, Iran announced plans to build a telecommunications satellite, to be launched in 2001 using a Shehab-4 rocket. Iranian Defense Minister, Adm Ali Shamkhani (quoted in Beal, 1999) announced that Shehab 4 is now in production as a space launcher: "Shehab 4 is not for military purposes but for launching a satellite." Although there have been no official announcements, Israeli intelligence analysts believe that Iran is seeking to acquire a reconnaissance satellite (Associated Press, 1998). Iran is also a participant in a $20 million multinational project to develop and launch a satellite for use in telecommunications and monitoring. Other participants include China, Pakistan, Mongolia, Thailand, and South Korea, with the target launch date of 2001. The official press release did not provide details on the potential monitoring activities of the satellite (Reuters, 1998b). However, given the involvement of South Korea and Thailand, whose interests are primarily economic and technological, rather than military and strategic, this could also provide the basis for a commercial imaging satellite program.

Rules of the Game?

The dual-use nature of commercial satellite imaging systems and the data that these systems can supply remain primary factors in future military scenarios involving the Middle East. Concerns regarding their potential destabilizing effects were central to the Israeli government's efforts to prevent the sale or release of high-resolution commercial imaging of its territory. In addition to the legislation in the U.S. Congress, Israeli officials also held talks with policymakers in other states, including Russia and France, to persuade them to impose limits or prevent the release of images that would compromise vital Israeli security interests.

However, as the number of satellites with high-resolution imaging capabilities increases (including the successful launch of IKONOS in September

1999), Israeli officials are also increasingly aware of the difficulties of maintaining the blackout of high-resolution images over its territory. Although the U.S. commercial technology will provide the most advanced and, therefore, the most militarily significant sources of intelligence, satellites operated by Russian, French, Canadian, Indian, and other firms and by consortia will also provide such data. Under current political conditions, the obstacles remain formidable to the negotiation of agreed "rules of the game" and a satellite imaging regime to prevent proliferation of destabilizing systems and information.

Acknowledgment
I would like to thank Aharon Entengoff for providing research assistance in the preparation of this paper.

Notes
1. In the context of bilateral arms control, it is possible that a change in the political environment could lead to a peace treaty between Israel and Syria. Such an agreement would include Israeli withdrawal from the Golan Heights and establishment of demilitarized and limited conventional deployment zones on both sides of the border. In this context, high-resolution satellite imaging might be able to supplement on-site, airborne, and other forms of verification, but the use of commercial systems, as opposed to dedicated national technical means, for this purpose will be limited.
2. Since additional thrust was needed to achieve the extra 1,200 mph of velocity required to escape into orbit from a westward launch, this restricted the size of the payload (Simpson, Acton, and Crowe, 1989).
3. Rafael, the Israeli Arms Development Authority, improved engine performance by using a sophisticated system of gauges inside the fuel tank to measure the movement of gas. When small satellite thrusters are fired, the fuel tends to slosh, which affects the satellite's attitude and orbit. The Rafael system dampens the effect of this sloshing (Barzilai, 1998).
4. CST is headed by Steve Wilson. Founded in 1991, the company developed a global on-line visual indexing and distribution infrastructure, known as ImageNet, for users of geospatial information products (i.e. satellite and aerial imagery and computer cartographic and demographic products). ImageNet claims to be "the single largest on-line source of electronically indexed satellite imagery archives" with "over 20,000 registered subscribers through franchised Primary Service and Data Providers in North America, Europe, Russia, the Middle East, Australia and the Southern Pacific Rim." (See also CST, 1999.)
5. Establishing an offshore corporation for a joint venture is unusual and probably unique in the case of IAI. The Cayman Islands was apparently chosen as a result of its unique tax status with respect to both Israel and the United States.

6. Israel has been very supportive of U.S. efforts to use sanctions against Russian firms and research centers involved in the sale of missile technology to Iran. If Israel were to contract for Russian launch services, it would be undermining its policy with respect to Iran (see Steinberg, 1999, and Schiff, 1999). Israeli industry (Modular Space Transportation and IAI) is attempting to develop a heavier launch capability, known as the Star-460, but its prospects are uncertain (Barzilai, 1999).
7. Gupta (1995, p. 117) has noted that states with long-range weapons may be able to create highly accurate maps of enemy territory and will thus be encouraged to "develop or import new guidance systems capable of directing weapons to the designated point at a comparably high accuracy."
8. On August 7, the *New York Times* reported that Muhammed Hilawi, the Saudi diplomat who defected in May 1993, claimed the Saudis were interested in nuclear weapons (Lewis, 1994).
9. Lane (1996) argues that it is in the interest of the United States to continue to lead the regulation in remote sensing, and claims that the loss of U.S. international leadership in launches has contributed to lower disincentives for missile proliferation.
10. The U.S. decision to license high-resolution commercial imagery (White House, 1994) was based on a trade-off between commercial interests, which sought minimal limitations, and national security and foreign policy interests, which argued for greater intervention controls. This resulted in the establishment of shutter controls to allow the U.S. government, at the Cabinet level, to limit the collection or distribution of data "in the event of a diplomatic or military emergency." (See Baker, 1997, pp. 6–7.) This definition was widened subsequently in the licenses the U.S. National Oceanographic and Atmospheric Administration (NOAA) issued, which include the provision that

> During periods when national security or international obligations and/or foreign policies may be compromised, as defined by the secretary of defense or the secretary of state, respectively, the secretary of commerce may, after consultation with the appropriate agency(ies), require the licensee to limit data collection and/or distribution by the system to the extent necessary by the given situation. See Florini and Dehqanzada (1999), and Chapter 7 in this volume.

References

Associated Press, "U.S. Bans Some Satellite Images of Israel, July 25, 1998a.
—, "Israeli Intelligence Suspects Developing Spy Satellite," September 5, 1998b.
Aviation Week and Space Technology, August 2, 1994a.
—, October 17, 1994b.
Aviation Week Space Business Online, "Kistler Receives Saudi Investment," July 28, 1999.
Baker, John C., "Trading Away Security?: The Clinton Administration's 1994 Decision on Satellite Export Imaging Exports," Pew Case Studies in International Affairs No. 222, Institute for the Study of Diplomacy, School of Foreign Service,

Georgetown University, Washington D.C., 1997.

Barzilai, Amnon, "Israel to Cooperate with European Space Agency on New Dutch Satellite," *Haaretz*, September 10, 1998.

——, "Launcher Experts Say Israeli Super-Rocket Possible," *Haaretz*, March 11, 1999.

Beal, Clifford, "Iran's Shehab 4 is Soviet SS-4, Says US Intelligence," *Jane's Defence Weekly*, Vol. 31, No. 7, February 17, 1999.

Bingaman, Jeff, Statement of Senator Bingaman, Hearings on National Defense Authorization Act for Fiscal Year 1997, Amendment No. 4321, "Purpose: To Prohibit the Collection and Release of Detailed Satellite Imagery with Respect to Israel and Other Countries and Areas," *Congressional Record*, U.S. Senate, Washington D.C., 26 June 1996, p. S6924–5. Available at http://www.fas.org/eye/1_01.htm (last accessed May 15, 2000).

Broad, William J., "Commercial Use of Spy Satellites to Begin; Private Ventures Hope for Profits," *New York Times*, February 10, 1997, p. 1.

Cohen, Eliot A. "A Revolution in Warfare," *Foreign Affairs*, Vol. 75, No. 2, March/April 1996, pp. 37–54.

Cole, Jeff, and Amy Docker Marcus, "Israeli-Led Venture in Satellite Imaging Poses Challenge to US Aerospace Firms," *Wall Street Journal*, February 28, 1996.

Congressional Research Service, "Space Activities of the United States, Soviet Union and Other Launching Countries/Organizations: 1957–1993," Congressional Research Report to Committee on Science, Space Technology, House of Representatives, 103rd Congress, Washington, D.C.: U.S. Government Printing Office, 1994.

Core Software Technology, Web site, 1999; see http://www.coresw.com (last accessed June 28, 2000).

Coughlin, Con, "Russian Space Pictures Enable Saddam to Target Gulf States," *Sunday Telegraph*, October 10, 1999.

CRS—*see* Congressional Research Service.

CST—*see* Core Software Technology.

Dailey, Brian, Testimony of Brian Dailey, Vice President of Space and Strategic Missiles Sector, Lockheed Martin Corporation, before the House Subcommittee on Space and Aeronautics, U.S. Congress, Washington D.C., 31 July 1996 (http://www.fas.org/spp/civil/congress/1996_h/h960731_spac_com_wit.htm).

Davis, Douglas, "'Foreign Report': Israel Giving Satellite Data to India," *Jerusalem Post*, June 11, 1998

Davis, Douglas, and David Rudge, "Turkey Wants Israeli Intelligence on Syria," *Jerusalem Post*, October 8, 1998.

de Selding, Peter B., "U.S.-Israeli Venture Plans Bold Offer," *Space News*, March 15, 1999.

EOSAT, "Space Imaging EOSAT Forms United Arab Emirates Partnership: Five Meter Resolution Imagery Will Now Be Available for Entire Region," News Release, November 18, 1997. Available from *Florida Today Space Online*, at http://www.flatoday.com/space/explore/stories/1997b/111897c.htm (last accessed May 15, 2000).

Esche, Tamir, "El-Op Aims High for Space Business" ; "U.S.-Israeli Venture Plans to Launch Eight High-Resolution Satellites," *Imaging Notes*, Vol. 14, No. 3, May/June 1999, p. 6.

Ferster, Warren, "Eyeglass to Refrain from Photographing Israel," *Space News*, No. 7, November 13, 1994.

Florini, Ann, "The End of Secrecy," *Foreign Policy*, No. 111, Summer 1998.

Florini, Ann M., and Yahya Dehqanzada, "Commercial Satellite Imagery Comes of Age," *Issues in Science and Technology*, Fall 1999.

Gaddis, John Lewis, "The Evolution of a Reconnaissance Satellite Regime," in George Farley Dallin, ed., *US-Soviet Security Cooperation*, New York: Oxford University Press, 1988, pp. 353–363.

Gupta, Vipin, "New Satellite Images for Sale: The Opportunities and Risks Ahead," *International Security* 20, Summer 1995.

Hall, Keith, Director of the National Reconnaissance Office, Remarks to the National Network of Electro-Optical Manufacturing Technologies Conference, February 9, 1998, http://www.nro.odci.gov.

IAI Electronics Group and Core Software Technologies, press release, "Israel Aircraft Industries and Core Software Technology Announce Formation of a Joint Venture Company to Enter High-Resolution Satellite Imagery Market," Tel Aviv, February 28, 1996.

ImageSat International, Web site, 2000; see http://imagesatinternational.com (last accessed December 20, 2000).

Israeli Ministry of Defense, Statement, Tel Aviv, March 4, 1996.

Jane's, "Middle East Customers to Obtain Detailed Space Imagery," *Jane's International Defence Review*, Vol. 31, No. 1, January 1, 1998, p. 8.

Klieman, Aharon, and Reuven Pedatzur, *Rearming Israel: Israeli Defence Procurement Through the 1990s*, Jerusalem: *Jerusalem Post* Publication, 1991.

Koulik, Sergey, and Richard Kokoski, "Verification Lessons of the Persian Gulf War," *Conventional Arms Control*, New York: Oxford University Press, 1994.

Krepon, Michael, "The New Hierarchy in Space," in Krepon et al. (1990).

Krepon, Michael, Peter Zimmerman, Leonard Spector, and Mary Umberger, eds., *Commercial Observation Satellites and International Security*, New York: St. Martin's Press, 1990.

Lane, Charles, "The Satellite Revolution," *The New Republic*, August 12, 1996, p. 24.

Lardier, Christian, "Proliferation of Remote Sensing Satellites as Market Expands," *Air & Cosmos/Aviation International*, April 4, 1997, pp. 34–35.

Lewis, Paul, "Defector Says Saudis Sought Nuclear Arms," *New York Times*, August 7, 1994.

Melman, Yossi, "Russia Sold Israel Hundreds of Satellite Pictures," *Haaretz*, March 30, 1998.

Menshikov, Valeriy, and Boris Rodionov, "Opinion on the Subject," Russian National Information Service, Moscow Armeyskiy Sbornik (in Russian), No. 10 (October 1996), pp. 88–90 (reprinted in FBIS Daily Report, 30 January 1997, UMA-97-018-S, US Government World News Connection, Washington DC (http://wnc.fedworld.gov/).

MOD—*see* Israeli Ministry of Defense.

Parnes, Sharone, "Israeli Officials Decline to Discuss Role of Latest Ofeq," *Space News*, April 10–16, 1995.

Reuters, "Israel to Launch New Satellite Before Old One Expires," July 7, 1998a.

——, "Iran to Build $20 Million Satellite with Asian States," August 4, 1998b.
——, "Russia Sells Iraq Satellite Photos," October 12, 1999.
Richelson, Jeffrey, "Implications for Nations Without Space-Based Intelligence Collection Capabilities," in Krepon et al. (1990).
Rodan, Steve, "Space Wars," *Jerusalem Post Magazine*, March 10, 1995.
——, "Ofek Satellite Budget Approved," *Jerusalem Post*, June 26, 1998.
Rotem, Michael, "Spy Satellite for Arab Emirates 'Serious Threat'," *Jerusalem Post*, November 19, 1992.
Rosenberg, David, U.S.-Israel Satellite-Imaging Firm Plans Bond Issue, Reuters, February 28, 1999
Schiff, Ze-ev, "A Serious Satellite Situation," *Haaretz*, January 25, 1998a.
——, "U.S. Bans High-Resolution Satellite Sales in Bow to Israeli Pressure," *Haaretz*, August 17, 1998b.
——, "Israeli Military Satellite to Ride Russian Rocket," *Haaretz*, December 3, 1999
Simpson, John, Philip Acton, and Simon Crowe, "The Israeli Satellite Launch," *Space Policy*, May 1989.
Steinberg, Gerald M., *Satellite Reconnaissance: The Role of Informal Bargaining*, New York: Praeger, 1983.
——, "Middle East Space Race Gathers Pace," *International Defence Review*, October 1995.
——, "Defence Procurement Decision Making in Israel," in Ravi Singh, ed., *Arms Procurement Decision Making*, Vol. 1, Oxford University Press and SIPRI, 1998a.
——, *Dual Use Aspects of Commercial High-Resolution Imaging Satellites*, Security and Policy Series Paper No. 17, BESA Center for Strategic Studies, Bar Ilan University, 1998b.
——, "News Analysis—Israel to use Russian launcher," *Jerusalem Post*, December 6, 1999.
Tahu, George J., John C. Baker, Kevin O'Connell, "Expanding Global Access to Civilian and Commercial Remote Sensing Data: Implications and Policy Issues," *Space Policy*, 14, 1998, pp. 179–188.
West Indian Space, Web site, 1999; see http://westindianspace.com/home.html (last accessed June 28, 2000).
The White House, Office of the Press Secretary, "Foreign Access to Remote Sensing Space Capabilities," Fact Sheet, Washington, D.C., March 10, 1994.
WIS—West Indian Space—see ImageSat International
Woolsey, R. James, quoted by Charles Lane in "The Satellite Revolution," *The New Republic*, August 12, 1996.
Zakheim, Dov S., "Hi-Tech Eyes and Ears," *Jerusalem Post*, July 30, 1999.
Zimmerman, Peter, "From the SPOT Files: Evidence of Spying," *The Bulletin of Atomic Scientists*, Vol. 45, No. 7, July 1989, pp. 24–25.

12

The Indian Space Program

Deborah Foster

Although it began humbly, with laboratories set up in a church and a cow shed, the Indian space program has achieved significant successes since its inception in 1962. The program has evolved extensive remote sensing, weather monitoring, telecommunications, and launch capabilities. The space program's objectives have remained relatively constant in its 38-year history, its achievements in space technology tied closely with development objectives. Although military use of the space capabilities has been a subtle theme since the inception of the space program, spinning off the space launch vehicle (SLV) program to establish the ballistic missile program has been the most visible connection between the space program and the military. To date, the military appears to be primarily an end user of space services, as opposed to an active participant in defining program requirements and objectives.

The Indian Space Research Organization (ISRO), which is responsible for research, development, production, and management of Indian space capabilities, is nearing an important threshold. ISRO is in the final stages of development for its geosynchronous launch vehicle (GSLV), and has planned the first test launch for 2001. Once the system is operational (more than likely requiring several additional years of development), India will have achieved a long-standing program objective: an independent space launch capability. When ISRO is no longer reliant on Russian or French launch services to reach geosynchronous orbits, it will be staged to enter the commercial satellite communications market, providing services, systems, and launches. India has capitalized on its remote sensing capabilities since the early 1990s, providing the highest-resolution panchromatic imagery (5.8 m) regularly available in the commercial market. However, India was upstaged in September 1999 when Space Imaging Corp. successfully launched its IKONOS satellite, which provides 1-m panchromatic images.

This chapter will provide a brief overview of the Indian space program as a whole and then will discuss the remote sensing program in detail, reviewing the motivation behind and management of the program, as well as addressing its military utility.

Space Program Early Years

In response to the increasing use of satellites for studying the earth's environment and the creation of the Committee for Space Research (COSPAR) by the International Council of Scientific Unions, the Indian government created the Indian National Committee for Space Research (INCOSPAR) in 1962 to represent India to the COSPAR (Pandya, 1992).[1] This enabled India to participate in an important international scientific body and served as an impetus for coordinating space research within India. This organization was located administratively under the auspices of the Department of Atomic Energy (DAE).

Shortly after the committee was set up, the Thumba Equatorial Rocket Launching Station (TERLS) was established with the objective of developing the knowledge base and infrastructure to support an Indian rocketry program. India sought and received United Nations (UN) sponsorship of the facility from the UN Committee on the Peaceful Uses of Outer Space (COPUOS).[2] In addition to prestige, this provided India with an important means of participating in the international space faring community. With this UN accreditation, India provided other states with access to research and launch facilities while gaining exposure to foreign technology and technical expertise. In fact, collaboration with the French, Soviets, and Americans at TERLS established the foundation for India's SLV and ballistic missile programs.[3]

The space program's objectives were oriented toward serving national development needs. India has a large and growing population that is dispersed over a large area. The country's communications infrastructure is immature, and there is little awareness of the natural resources within the country's boundaries. These facts made it clear to the national leadership that space-based communications and remote sensing capabilities could contribute significantly to development efforts.

Bureaucratic Formalization: The Department of Space

Once the rocketry program was well established, with hundreds of launches occurring every year at TERLS, efforts were made to house space research activities under one roof. In 1969, with the initiation of the SLV program, INCOSPAR was reconstituted into the ISRO. Then in 1972, the Department of Space (DOS) was created, replacing DAE as the overview agency for ISRO.[4] It was at this time that actual development programs for telecommunications, weather, and remote sensing satellites began.

Although the space program was considered an important component of India's development strategy, it was not until the bureaucratic consolidation of the space research efforts that the connection between the two was formally articulated. In 1972, ISRO defined the two primary space-program objectives:

1. the application of space science and technology to further national development objectives in mass communications and education via satellites.
2. the survey and management of natural resources through remote sensing technology from space platforms.

As has been a consistent theme in Indian high-technology projects, these objectives were to be met by maintaining the maximum degree of self-reliance possible: In its own words, DOS pursued a vigorous policy of collaborative activity with other scientific and industrial agencies in the country to utilize the expertise and infrastructure in these agencies in areas of relevance to the programs of the Department (DOS, 1980, p. 38).

Any technology that was developed in the process of constructing and managing India's space architecture was to be leveraged to nurture India's nascent high-technology sector.

Remote Sensing Experimentation

There was a great deal of interest in mapping the country's natural resources from space in the early 1970s. Initially, remote sensing efforts were conducted either via aircraft or balloon. When the opportunity arose, ISRO conducted extensive experiments with "borrowed" space systems or with data from foreign space systems. For example, in 1975 and 1976, ISRO used the ATS-6 to conduct the SITE program, an educational experiment involving the broadcast of educational television programs to 2,400 remote villages over the period of a year (DOS, 1975), p. 31).

ISRO also used aerial platforms to test satellite sensors. Scientists gained experience in the development and management of imagery sensors, as well as in the art of multispectral data fusion and analysis.

Foreign imagery also had an important place in India's remote sensing efforts. Imagery obtained from foreign satellites was used consistently to augment Indian land classification and utilization efforts, water surveys, geological feature mapping, geographical classification, and weather monitoring activities (DOS, 1976, pp. 34–35).

In 1975, the National Remote Sensing Agency (NRSA) was established under the auspices of the Department of Science and Technology. This organization was created to take advantage of remote sensing technology and techniques for the survey, planning, and management of India's natural resources (DOS, 1981, pp. 37, 14; DOS, 1974, pp. 35–36). NRSA served as ISRO's liaison to users of remote sensing data products. Through NRSA, ISRO accomplished several important objectives of the remote sensing program (DOS, 1974, p. 37):

- undertake, aid, promote, guide, and coordinate research in the field of remote sensing
- provide consultancy services and airborne survey facilities to user agencies
- carry out surveys by using remote sensing technology for various applications, such as agriculture, hydrology, meteorology, fisheries, geology, environmental monitoring, forestry, oceanography, topography, and resource and crop-disease surveillance
- establish, maintain, and manage databases for acquisition, storage, retrieval, dissemination, evaluation, scrutiny, and interpretation of information relating to remote sensing technology

The SITE Program

The Satellite Instructional Television Experiment Program was aimed at demonstrating the efficacy of satellite-broadcast television as a development medium, particularly in infrastructure-poor and remote areas. ISRO provided antennas, front-end converters, and ruggedized community-viewing television sets for direct reception of the television programs, and managed the uplink of programs. Not only did this experience provide insight into the role satellite capabilities could play in achieving national development objectives, but this and follow-on communications experiments provided valuable technical exposure and aided the requirement-definition process for India's multipurpose communication and weather satellites—the Indian National Satellite series (INSAT).

- cooperate and collaborate with other national and/or foreign institutions and international organizations in the field of remote sensing and allied sciences.

Experimentation continued after NRSA's establishment. Targets were renewable and nonrenewable resources, such as food crops, cash crops, forests, water, mineral deposits, and oil reserves (DOS, 1974, pp. 35–36). The data these experiments generated contributed to ISRO's objective of establishing a signature database of crops and earth-surface features in the visible and infrared spectra.

In collaboration with ISRO, NRSA established a downlink station near Hyderabad to facilitate the receipt, processing, and integration of foreign data into NRSA's imagery products.

Remote Sensing Program Development

In 1979, ISRO initiated an experimentation program with the primary objective of educating ministries, state governments, and national agencies involved in natural resource survey and management about the potential uses of the country's nascent remote sensing capabilities. Through the Joint Experiments Program (JEP), ISRO conducted a variety of experiments in an effort to demonstrate possible uses of remote sensing techniques for natural resource survey and management efforts (DOS, 1981, p. 14).[5] In addition to raising awareness about the remote sensing program, ISRO garnered important feedback to help determine performance requirements for the IRS-1A sensor suite.

In 1980, shortly after the JEP was initiated, NRSA was administratively transferred to DOS from the Department of Science and Technology. This bureaucratic reorganization served to focus remote sensing activities under one organization, enabling more effective management of financial resources and coordination of products and also provided a central point of contact for organizations interested in using remote sensing data. This identified DOS as the organization responsible for "the generation of integrated data, mainly based on remote sensing, and the analysis and dissemination of information derived from the data" (DOS, 1986, p. 20).

To meet the growing demand for remote sensing products and to expand efforts to generate awareness of the availability of remote sensing products, ISRO established the National Natural Resources Management System (NNRMS) in 1985. The NNRMS had two primary objectives: Facilitate the use of remote sensing data for resource management, and train the necessary personnel to fuse remote sensing and collateral data used in resource management. The program established five remote sensing service centers (RRSSCs) that were dispersed throughout the country and linked to NRSA's data center (NDC) outside of Hyderabad (DOS, 1985, pp. 16; DOS, 1988, p. 28).[6]

Although spearheaded by DOS, the program also received funding from several government agencies, including the Department of Science and Technology, the Geological Survey of India, the Department of Mines, the Indian Council of Agricultural Research, and the Oil and Natural Gas Commission.

In 1992, NRSA expanded this effort to encourage imagery use at the local level. The program, called the Integrated Mission for Sustainable Development (IMSD), provides a mechanism by which NRSA and universities, local entities, private entrepreneurs, and nongovernmental organizations can collaborate to generate locale-specific action plans to meet sustainable development aims (DOS, 1998, p. 37). State Remote Sensing Applications Centers, managed by RRSSCs, help local agents fuse satellite remote sensing data with socioeconomic data (DOS, 1988, pp. 24, 28; DOS, 1984, pp. 14–15).

Remote Sensing Capabilities: Technical Evolution

While generating important data about India's natural resource base, remote sensing experimentation also contributed significantly to ISRO's larger goal of creating a space-based remote sensing architecture—to include not only the development of remote sensing satellites but also the necessary launch and satellite control infrastructure. The first step in the satellite development process was the Satellite for Earth Observations (SEO) program.

The SEO program was experimental, with the dual goals of gaining experience in defining requirements for the construction and management of a remote sensing satellite. Once launched, the SEO satellites would also provide the scientists with the opportunity to process and apply the data that the satellites generated.

This research resulted in twin satellites: the Bhaskara-I and -II. These satellites had similar payloads of 1-km resolution television cameras and microwave radiometers. These sensors provided data relevant to hydrology, forestry and meteorology applications (DOS, 1976, p. 22). With assistance from foreign suppliers of some of the onboard sensors (such as cameras), as well as launch assistance from the former Soviet Union, the Bhaskara-I was launched in June 1979. The Bhaskara-II was launched in November 1981 (DOS, 1979, pp. 3, 6).

The experience gained by developing the SEO satellites was vital for the development of the Indian Remote Sensing satellite (IRS) program. The principal objective of the IRS program was to develop and deploy, in about four years, a fully qualified, integrated spacecraft with a three-year on-orbit life. An important secondary objective was to construct data-product facilities to ensure that data users had ready access to the satellite data. Although the first IRS satellite launch was planned for 1984, the first satellite in the series, IRS-1A, was not launched until 1988. Several years later, IRS-1B was launched with a similar payload.

A notable improvement in the constellation occurred when IRS-1C was launched with a panchromatic camera that provided imagery at 5.8-m resolution. This capability was further enhanced when the IRS-1D was launched with an onboard data recorder. This enabled the satellite to image sections of the earth's surface beyond the range of satellite downlink stations, leading to greater coverage by the satellite and the likelihood of generating income from the sale of data products.

India developed the P series of satellites as lower-cost, application-specific satellites with fast turnarounds from design to launch (DOS, 1996, p. 29). Several of these satellites were used in test launches of India's Polar SLV (PSLV). India's most recent IRS launch was of the Oceansat (IRS-P4) in May 1999. This is the first Indian satellite specifically intended for surveying ocean resources. Table 12.1 on the following page provides details of India's remote sensing satellites; planned systems are in italics.

Remote Sensing Capability Shortfalls

Even as ISRO incrementally improves the resolution of the IRS sensor suites, the constellation has sensors that can detect only in the visible and near-visible spectra. The remote sensing program lacks advanced capabilities, such as the ability to generate high-resolution imagery (PAN and multispectral) and the ability to image in all weather conditions (i.e., through cloud cover).

HIGH-RESOLUTION IMAGERY

IRS-P5, or Cartosat-1 is planned for launch sometime in 2001. This satellite was originally intended for cartographic applications and will have sensors similar to those of IRS-1D. However, the higher resolution will be achieved by launching the satellite into a lower orbit than was originally intended, rather than by improving the sensor itself, likely shortening the satellite's life span. Some analysts speculate that the mission of Cartosat-1 was modified after IRS coverage lapses were highlighted by Pakistan's 1998 Gharui missile test, although no information is available to confirm this assertion (*Jane's Defence Weekly*, 1998). ISRO intends to launch a 1-m resolution PAN satellite, Cartosat-2, but only in 2003 (Raj, 1999).

Table 12.1—Indian Remote Sensing Satellites (IRS Series)

Name (launch date)	Revisit Rate (days)	Sensor Type[a]	Resolution
IRS-1A[b] (3/88)	22	LISS-I	72.5 m
		LISS-II	36.25 m
IRS-1B (8/91)	22	LISS-I	72.5 m
		LISS-II	36.25 m
IRS-1C (12/95)	5	PAN	5.8 m
	24	LISS-III	23.5 m visible, near-IR
			70.5 m shortwave IR
	3	WiFS	188 m
IRS-P2[b] (10/94)	24	LISS-II	32 x 37 m
IRS-P3 (3/96)	24	WiFS	188 m
	24	MOS A	1,569 x 1,395
		MOS B	523 x 523
		MOS C	523 x 644
IRS-1D (9/97)[c]	3	PAN	5.8 m
	25	LISS-III	23.7 m visible, near-IR
			70.5 m short-wave IR
	3	WiFS	188 m
IRS-P4 (Oceansat-1) (5/99)	2	OCM	236–360 km
		MSMR	1,360 km
IRS-P5 (Cartosat-1) (2001)	Unknown	PAN	2.5 m
IRS-P6 (Resourcesat-1) (2001)	Unknown	LISS-III	23 m
		LISS-IV	<6 m
		WiFS	70 m
IRS-P7 (Oceansat-2) (2002-03)	Unknown	Unknown[d]	
Cartosat-2 (2003)	Unknown	PAN	1 m

Sources: Bharat Rakshak (2000); Deccan Herald (1999); TAG's Broadcasting Services (2000); Morgan (1998), pp. 99–100; the IRSO Web page (http://www.isro.org); and the NRSA Web page (http://202.54.32.164).

[a] Definitions: Linear Self-Scanning Sensor (LISS), Modular Optoelectronics Scanner (MOS), Multifrequency Scanning Microwave Radiometer (MSMR), Ocean Color Monitor (OCM), Panchromatic (PAN), Wide Field Sensor (WiFS).
[b] These satellites are now out of service.
[c] Similar to IRS-1C with the addition of an onboard data recorder.
[d] The payload is unknown but is expected to relate to oceanography, atmospheric sciences, and climate and meteorological studies.

SYNTHETIC APERTURE RADAR

All-weather capability is achieved through the use of synthetic aperture radar (SAR). ISRO has been collaborating with the European Space Agency since the late 1970s on various SAR-related projects, including training in SAR data analysis. ISRO established a downlink station for ESA's ERS-1 in 1991 (DOS, 1992, p. 26; DOS, 1978, pp. 72–73).

Despite 15 years of research and development, ISRO has yet to announce the intention to field a satellite with SAR. This may be related to the fact that a SAR sensor would require a satellite bus and a power source different from those on the current constellation of IRS sensors. In addition, a different satellite bus might require the use of a launch vehicle with a better lift capability than the PSLV.

The need for a high-resolution, all-weather remote sensing capability was evident to Indian rescue workers in October 1999 when a "supercyclone" hit the coast of the Indian state of Orissa. Cloud cover over the area prevented the use of up-to-date IRS data for the relief effort, necessitating ISRO's purchase of SAR data from Canada's RADARSAT International. ISRO used this data in combination with archived IRS data to identify flooded areas.[7]

Globalization of IRS

In 1992, DOS established Antrix, Ltd., to market Indian space products and space services abroad. In 1994, Antrix had negotiated an agreement with the U.S. company EOSAT to market IRS imagery products globally. The contract was renegotiated with Space Imaging in 1997, after Space Imaging Group purchased EOSAT (Ahmed, 1997). There are data reception stations in the United States, Germany, Thailand, Japan, Dubai, and South Korea that not only receive IRS data, but work with Indian-supplied hardware. By 2001, additional ground stations in Australia, Saudi Arabia, Ecuador, and Alaska will be augmented with Indian hardware to receive IRS signals (*The Telegraph*, 1999).

Data Control and Distribution

NRSA's NDC is the focal point for distribution of data products within India (DOS, 1984, p. 37). In addition to IRS data, NRSA maintains downlink terminals for many commercial foreign remote sensing sources, including LANDSAT, SPOT, ERS, and NOAA satellites.

As the central dissemination point for all satellite imagery, NRSA is able to implement effectively the data control policies that India's Ministry of Defense (MOD) has established. NRSA prohibits the distribution of images of military facilities and restricts the availability of geographical imagery, applying these restrictions to an 80-km belt along border areas and coastal zones (including gravity maps and high-resolution maps that depict geological formations and rocks, as well as topographical maps of Jammu and Kashmir) (Bagla, 1999; Jayaraman, 2000a, p. 3). Recently, the NRSA director acknowledged that the agency alters IRS imagery products to black out or erase areas deemed sensitive to national security, such as military installations and nuclear facilities (Jayaraman, 2000b).

An official with the Geological Society of India, the Indian government's official mapmaker, has reported that maps covering more than 60 percent of the Indian land mass are not available to the public because of security restrictions

(Jayaraman, 2000a). Gaining access to this information requires navigation through an interministerial bureaucratic procedure requiring a minimum of 15 stages of negotiations (Bagla, 1999). This restrictive policy is in place for users within and outside India. Users outside of India interested in purchasing IRS imagery of India must make a request to Space Imaging—which then requests the relevant imagery from NRSA. NRSA then removes sensitive features from the data product and forwards it to Space Imaging.

Recently, India's efforts at controlling access to high-resolution imagery were thwarted by the government's own hand. Without a data-control policy dealing with the distribution of commercial high-resolution data in place, the Indian state of Andhra Pradesh negotiated a $1.3 million U.S. deal with Space Imaging to purchase IKONOS imagery of the state's coastline in the fall of 1999. Assisted by the World Bank, the state government intended to use the imagery for disaster mitigation planning in coastal areas (Jayaraman, 2000a). The project involved mapping a 20-km swath along the state's coastline to facilitate the development of a cyclone hazard mitigation plan.

The deal spurred a national data policy review —not only because 14 percent of India's entire coastline is in Andhra Pradesh, but also because many military bases are located in the state. As of May 2000, the Indian government was drafting a data-control policy that would provide NRSA with sole rights to purchase high-resolution imagery (Jayaraman, 2000b). If this policy is implemented, any users within India interested in obtaining high-resolution commercial imagery will have to request the imagery through NRSA, enabling NRSA to remove sensitive features.

Military Involvement in the Space Program

As can be inferred from the lack of mention of India's national security apparatus in the preceding pages, little information is available on its involvement in space in general or the remote sensing program in particular. Certainly, the

> **From SLV to Ballistic Missile**
>
> It is ironic that the United States played a key role in India's development of a ballistic missile capability—and that this assistance was in support of an effort to improve India's weather monitoring capabilities. In the mid-1960s, a key figure in both the space program and the Indian missile program, Dr. A. P. J. Abdul Kalam, received training at NASA's Wallops Island rocketry facility. After his course at NASA, Kalam asked for and received the blueprints for NASA's Scout rocket. These engineering plans facilitated the Indian construction of its first SLV—the SLV-3 —and has since been integrated into follow-on launch vehicles, including India's most current SLVs and its ballistic missiles (Milhollin, 1989, p. 32). Of course, this is mentioned not to charge the United States with supporting a missile program that will soon be integrating nuclear warheads into its arsenal. Rather, the intention is to highlight the dual nature of rocketry and space technology—to be used to advance our understanding of the earth environment and our impact upon it as well as to advance our national security objectives (Kalam and Tiwari, 1999, p. 61; Milhollin, 1989).

relevance of multispectral imagery data fusion to the national security community's concerns and missions was clear in the early days of the space program. The emerging cold war between the United States and the former Soviet Union provided constant reinforcement to the Indian defense community of the importance of space capabilities to military operations. It is difficult to imagine that India's close collaboration with the Soviet space program did not have a military element to it, particularly since the Soviet space program served military ends.

In a 1970 white paper, Dr. Vikram Sarabhai, the visionary and driving force behind the space program in its early years, noted that

> The defense organization and the Services could utilize the services provided by INSAT for establishing communications in the remote regions . . . Naval ships equipped with small dishes could establish communication from ship to shore and from ship to ship using the satellite . . . The Air Force can utilize [satellite] communications . . . to interconnect all the operational airports . . . (Sarabhai et al., 1970, p. 11)[8]

Other than this document, there is a paucity of information relating to the involvement of the national security community in the requirement-definition process for remote sensing systems. This may mean that the national security community is primarily a user of space services rather than a participant in the program definition and management processes.

Military Use of Remote Sensing Capabilities

The remote sensing program is not likely to satisfy more than limited intelligence functions with the current (and planned) constellation. The reasons for this are discussed below.

LACK OF TIMELY REVISIT RATES

Currently, the Indian subcontinent is imaged by IRS satellites every five days. Effective integration of satellite imagery into tactical military operations would require situation updates in a matter of hours, rather than days. Not only does this mean that satellites would need to travel over the area of operations more frequently than current IRS satellites do, but the time required to downlink, process, and disseminate imagery to the appropriate users would have to be very rapid. NRSA reports that imagery requests are processed between three and ten days after submission (DOS, 1995, p. 26).

LACK OF ADVANCED REMOTE SENSING CAPABILITIES (HIGH-RESOLUTION MULTISPECTRAL DATA, SAR DATA)

It is estimated that, in some of the military's operational areas, only two days of clear data can be transmitted in a year because of cloud cover (Sarin, 1999). Space-based SAR provides two important advantages over aerial SAR: It is less vulnerable to attack, and satellites are not subject to airspace restrictions.

The Indian Armed Forces were made very aware of the lack of this capability in the late spring of 1999, when the infiltration of insurgents into Kashmir was detected through nonspace means. Imagery analysts from the Defense Image Processing Center reported in the postconflict analysis that they were unable to gain any tactically useful information from the IRS satellites during the conflict (Sarin, 1999).

Further, many of the targets of interest to the Indian Armed Forces during this most recent conflict were likely smaller than 3 m.[9] Even if there had been no cloud cover, IRS 5.8-m PAN cameras would have been of little use to operational commanders.

If the MOD determines that SAR capability is a requirement to support military operations, ISRO could purchase the capability from a commercial provider or an ally —either in the form of sensors to integrate into Indian satellite busses or in the form of complete satellites. That this has not yet happened could be related to the government's desire to avoid technological dependence on another state.

LACK OF AN INTEGRATED, REAL-TIME WIDEBAND DATA PROCESSING AND DISTRIBUTION SYSTEM

Although NRSA has a National Resources Information System linking its RRSSCs with NRSA, the function of this service is primarily to store NNRMS data products. As indicated above, NRSA's data analysis and distribution processes are not responsive to time-sensitive requirements.

If pressed, the Indian military might be able to augment the Very Small Aperture Terminal (VSAT) communication network currently used by the National Informatics Center to facilitate real-time transmission of data from NRSA. However, the INSAT constellation may require more advanced capabilities to support such a scheme.

Perhaps the Indian military could use the Internet or augment the army tactical communication network to facilitate such data exchange. Anecdotal evidence suggests that the Indian military is also taking advantage of the Internet for communication (Raj, 1999). The use of the Internet or a network with a similar protocol could facilitate the transfer of processed imagery data.

Conclusion

India's space program as a whole has, since its inception, been linked closely with the management of natural resources. Although the national security apparatus likely appreciated the utility of remote sensing data, India's remote sensing efforts, in conception and implementation, have been focused on facilitating sustainable development (Nicholson, 1997). However, given India's new security position as a nuclear weapons power and its increased interests beyond its borders, this situation is likely to change.

The increasing importance of space capabilities to enable terrestrial warfighting operations has been demonstrated in several recent conflicts, most notably the 1991 Gulf War and later in the Bosnia and Kosovo operations. It was likely that the humiliation of the 1999 Kargil infiltration brought home to the Indians the failure of their space constellation to enable military operations. Proponents of the military are increasing their calls to integrate current space assets more effectively into military operations and to develop military-specific systems. However, the fact that the government is considering the implementation of such a restrictive data-control and data-dissemination policy suggests that government leaders are more concerned about restricting access to information than with using that information to enhance both civilian and military operations.

In a country uncomfortable with transparency, it is unclear how the Indian government will respond to calls for improved, military-specific remote sensing capabilities. Technical barriers exist to developing a space architecture that enhances military operations, but the more challenging barrier to overcome will likely be the culture of mistrust between the civilian strategic planning apparatus and the military.

Notes

1. For information on COSPAR, see http://www.icsu.org/ and http://www.icsu.org/Structure/cospar.html. See also DOS (1974), p. 45.
2. India was a charter member of COPUOS, which the UN established in 1959. With the mission of facilitating international cooperation in space research, COPUOS encouraged members to establish sounding-rocket launch facilities under UN sponsorship. TERLS received UN sponsorship shortly after its establishment. Resolutions 1472, 1721, and 1963 are available at http://www.unesco.org/webworld/com/compendium/2306.html (Section 2.3, Space Law).
3. Through collaboration at TERSL, the Indians gained familiarity with both the French Viking rocket engine and NASA's Scout sounding rocket. Both of these technologies have come to play important roles in India's SLVs and ballistic missiles. For a more detailed discussion of India's efforts at leveraging foreign technology in its

space and missile programs, as well as the crossover between the space program and the Ministry of Defense, see Milhollin (1989).
4   In the Indian government, a ministry is the overview unit for a department. The Department of Atomic Energy and the Department of Space are unique in that they do not come under the purview of any Ministry. This means that the director of each department reports directly to the prime minister. It is notable that DAE and DOS have successfully pursued high-priority endeavors for the government.
5   Some of the participants in this program were the Ministry of Agriculture, Irrigation, and Mines; the Ministry of Chemicals and Petroleum; the Department of Science and Technology; and the governments of Karnataka and Gujarat. Experiments demonstrated remote sensing techniques in such areas as agriculture monitoring, forestry, and mineral and petroleum exploration.
6   The centers were located in Dehra Dun (funded by DOS); Nagpur (funded by Indian Council of Agricultural Research), Kharagpur (funded by the Department of Science and Technology), Bangalore (funded by the Department of Mines), and Jodhpur (DOS/DST). The Geological Survey of India also contributed funding to these centers.
7   See ISRO's Web site (http://www.isro.org) for a discussion of the effort.
8   The INSAT program is India's telecommunications satellite development program.
9   For a discussion of ground resolution requirements for military applications, see Table 3 of Steinberg (1998).

References

Ahmed, M., "ISRO in Space Imagery Deal with U.S. Firm," *Delhi Business Standard*, February 3, 1997 (FBIS-NES-97-024).

Bagla, Pallava, "Indian Scientists Question Government Grip on Data," *Science Magazine*, Vol. 285, No. 5428, July 30, 1999, p. 659.

Bharat Rakshak, "Indian Space Program Launch History," 2000. Available at http://www.bharat-rakshak.com/SPACE/space-history2.html (last accessed May 23, 2000).

Deccan Herald, "Indian Oceansat-1 Satellite to be launched after 25 May," April 30, 1999 (FBIS-NES-1999-0430).

Department of Space, Government of India, DOS Annual Report 1973–1974, Bangalore: Publications and Public Relations Unit, 1974.

—, DOS Annual Report 1975–1976, Bangalore: Publications and Public Relations Unit, 1976.

—, DOS Annual Report 1977–1978, Bangalore: Publications and Public Relations Unit, 1978.

—, DOS Annual Report 1978–1979, Bangalore: Publications and Public Relations Unit, 1979.

—, DOS Annual Report 1979–1980, Bangalore: Publications and Public Relations Unit, 1980.

—, DOS Annual Report 1980–1981, Bangalore: Publications and Public Relations Unit, 1981.

—, DOS Annual Report 1983–1984, Bangalore: Publications and Public Relations Unit, 1984.

—, DOS Annual Report 1984–1985, Bangalore: Publications and Public Relations Unit, 1985.
—, DOS Annual Report 1985–1986, Bangalore: Publications and Public Relations Unit, 1986.
—, DOS Annual Report 1987–1988, Bangalore: Publications and Public Relations Unit, 1988.
—, DOS Annual Report 1991–1992, Bangalore: Publications and Public Relations Unit, 1991.
—, DOS Annual Report 1994–1995, Bangalore: Publications and Public Relations Unit, 1995.
—, DOS Annual Report 1995–1996, Bangalore: Publications and Public Relations Unit, 1996.
—, DOS Annual Report 1997–1998, Bangalore: Publications and Public Relations Unit, 1998.
Jane's Defence Weekly, "New Delhi Plans Missile-Monitoring Satellite for '99," April 29, 1998, p. 14.
Jayaraman, K. S., "Imagery Deal Spurs Indian Policy Review," *Space News*, January 17, 2000a, p. 3.
—, " New Delhi Considers Strict Policy on Imagery," *Space News*, May 1, 2000b, p. 3.
DOS—see Department of Space.
Kalam, A. P. J. Abdul, and Arun Tiwari, *Wings of Fire: An Autobiography*, Hyderabad, India: Universities Press (India) Limited, 1999.
Milhollin, Gary, "India's Missiles—With a Little Help From our Friends," *The Bulletin of the Atomic Scientists*, November 1989, pp. 31–35.
Morgan, Tom, ed., *Jane's Space Directory*, 14th ed., 1998–99, London: Jane's Information Group Inc., 1998.
Nicholson, Mark, "India has high hopes for satellite launch," *Financial Times*, September 12, 1997, p. 6.
Pandya, S. P., "The Physicist," in Padmanabh K. Joshied, ed., *Vikram Sarabhai: The Man and the Vision*, Ahmedabad: Mapin Publishing Pvt. Ltd., 1992.
Raj, Gopal, "One Metre Resolution Satellite Launch in 2003?" *The Hindu*, November 18, 1999.
Raj, Yashwant, "The Old War Horse of Communication," *The Times of India*, July 27, 1999.
Sarabhai, Vikram, E. V. Chitnis, B. S. Rao, P. P. Kale, and K. S. Karnik, "INSAT—A National Satellite for Television and Telecommunication," paper presented the National Conference on Electronics, Bombay, March 24–28, 1970, Ahmedabad: Rajratan Press, 1970.
Sarin, Ritu, "Our Satellites Not Good Enough for Military Pictures—Experts," *The Indian Express*, September 26, 1999.
Steinberg, Gerald, "Dual Use Aspects of Commercial High-Resolution Imaging Satellites," Mideast Security and Policy Studies, No. 37, February 1998. Available at http://www.biu.ac.il/SOC/besa/books/37pub.html.
TAG's Broadcasting Services, Satellite Fact Sheet: IRS-P2, The Satellite Encyclopedia, February 29, 2000. Available at http://www.tbs-satellite.com/tse/online/sat_irs_p2.html (last accessed May 23, 2000).
The Telegraph, "India's Strides in Space Technology Viewed," May 16, 1999 (FBIS-NES-1999-0517).

13

Canada's Remote Sensing Program and Policies

Michel Bourbonniere and Louis Haeck

The ability to gather and use information has become crucial for modern nations, both for reasons of internal and external security and for industrial development. Well aware of this new reality, the Canadian government has taken steps to ensure that Canada achieves an information-based economy. The RADARSAT-1 and RADARSAT-2 projects are part of the effort to strengthen Canada's position in this area (see Abiodun, 1998).

This new emphasis on information presupposes that information itself or the use of the necessary technology will be regulated. Both nationally and internationally, developing a regulatory structure for an information-based economy requires balancing different human interests. One concept, however, is fundamental to democratic governments: Laws must not regulate the technology itself. The object of legal norms must be the human relations and competing interests associated with these new technologies.

This chapter will describe the two RADARSAT projects, their technology and uses; the policies, laws, and regulations affecting them; and the human relations, public and private, they have, in turn, affected.

General Historical Review of Canadian Remote Sensing Development

Canada is one of the largest countries in the world, with a variety of landscapes and climactic conditions; with vast and relatively remote, sometimes inaccessible, land and sea areas; and with a large continental shelf. In the late 1960s and early 1970s, the Canadian federal government discovered the value of airborne observation both for managing resources and the environment and for meeting its national security needs.[1] Although bureaucratic rivalries negated early efforts to set up a coherent Canadian remote sensing program, the Federal Cabinet finally approved the creation of the Canada Centre for Remote Sensing (CCRS) on February 1971 (CCRS, 1998b). CCRS has played a key role in the development of the remote sensing expertise in Canada and is mandated to provide a national service for receiving, processing, archiving, and disseminating remotely sensed data (CCRS, 1998a). The CCRS also works closely with private industry to develop geospatial information application.

In 1994, the Canadian government announced its Long-Term Space Plan (LTSP), which established four new earth-observation programs, the center of which was the RADARSAT-1 project. Canada's first earth-observation satellite, RADARSAT-1, was launched on November 4, 1995. Commercial operations began on April 1, 1996 (Evans, no date). In its first year of operation, RADARSAT-1 fulfilled over 11,000 image requests (Evans, no date). The Canadian government considers aerospace to be crucial for the transition to a world-class knowledge-based economy (Manley, 1999, p. 1), and RADARSAT-1 is a key part of that transition. The RADARSAT-1 satellite now provides the Canadian government an economical and practical tool for managing an extremely large and diverse territory.

The RADARSAT Joint Venture

From an international perspective, RADARSAT-1 is a joint project between Canada, through the Canadian Space Agency (CSA), and the United States, through the National Aeronautics and Space Administration (NASA) and the National Oceanic and Atmospheric Administration (NOAA). From an internal Canadian perspective, the RADARSAT-1 project is a joint venture involving both the public and private sectors. CSA managed the development of the satellite. The public-sector participants included the Canadian federal government and the provincial governments of Québec, Ontario, Saskatchewan, and British Columbia. The private sector participated through RADARSAT International, Inc. (RSI). RSI was established to market, process, and distribute RADARSAT-1 data (CSA, 1999b) and was originally a consortium that included MacDonald Dettwiler, COM-DEV Inc., Spar Aerospace (RADARSAT's prime contractor, now called EMS Technologies), and Lockheed Martin Astronautics.

The project cost an estimated total of $620 million Canadian (CDN), of which $500 million CDN came from the Canadian government; a total of $57 million CDN came from the four provinces; and $63 million CDN came from the private sector. The contributors and participants are entitled to data allocations (Government of Canada, no date, p. 2).

RADARSAT-1's International Steering Committee is its supreme governing body. Its members represent CSA, NASA, and NOAA. This body works through consensus to supervise the operations of RADARSAT-1 and resolve possible conflicts.

In May 1999, RSI became a wholly owned subsidiary of MacDonald Dettwiler. RSI is the exclusive distributor of RADARSAT data worldwide. RSI pays a royalty to the CSA for all data and derivative products sold. According to an April 10, 2000 article in the *National Post*, MacDonald Dettwiler is preparing itself for a public offering to make the company Canadian owned. If an ownership change does occur, RADARSAT-2 could be exempt from certain U.S. licensing requirements.

Intellectual Property Rights

CSA owns the RADARSAT-1 data and has negotiated a master license agreement with RSI for the distribution of the data. This gave RSI the exclusive right to distribute and market worldwide all RADARSAT SAR data and the data products described in the pricing section of the agreement. RSI also collects all royalties on behalf of CSA for commercial distribution of RADARSAT-1 data and data products and any value-added products from which RADARSAT-1 data can be retrieved, visually or digitally (Government of Canada, no date, p. 2). RSI establishes the prices of RADARSAT-1 data and data products.

The RADARSAT-1 project participants receive their data allocation for the cost of processing and distribution, as agreed upon in their respective memoranda of understanding (MOUs). The terms of these MOUs apply to the Canadian government, its departments, and to the other participating governments or agencies, as listed in Schedules I and I.1 of the Financial Administration Act (Government of Canada, no date, p. 3). According to the MOUs (Government of Canada, no date, p. 3), the governments participants may

1. receive data for their own use in fulfillment of their departmental mandates
2. share data among their various government departments in support of their mandates

3. redistribute data only for research and development (R&D) purposes under signed agreements which assign the use of data and data products to named investigators for research projects which are not commercial in nature
4. allow the development and distribution of value-added products on their behalf within their governmental mandates and without competing with commercial interests of RSI
5. purchase data at commercial prices for the data needs that exceed their data allocations.

Technical Characteristics

RADARSAT-1 was designed to last five years and three months but is expected to at least seven years because of its operational success. It is equipped with a synthetic aperture radar (SAR), a powerful microwave instrument that can transmit and receive signals through clouds, haze, smoke, and darkness. The satellite is thus operational regardless of weather or illumination conditions. RADARSAT-1 uses a single C-band frequency and has the ability to shape the SAR beam by varying its width, thus affecting the desired resolution, and the ability to steer the beam over a 925-km range. The position selections among the seven beam modes allow the image's swath width to vary from 50 to 500 km, the resolution to vary from 8 to 100 m, and the incidence angle to range from 10 to 60 degrees (see CSA, 1999a).

RADARSAT-1 is in a near-polar sun-synchronous earth orbit, each orbit lasting approximately 100.7 minutes.[2] RADARSAT-1 is rarely in eclipse, which minimizes the influence of diurnal variation effects. RADARSAT-1 can thus obtain multitemporal data sets, which are useful for crop prediction (see CSA, 1999a). RADARSAT-1 can obtain 28 minutes of data per orbit. The data are downlinked in real time or, if the satellite is out of range of a ground station, are stored onboard. The North American ground stations are located in Prince Albert, Saskatchewan; Gatineau, Québec; and Fairbanks, Alaska. Currently, there are certified foreign stations in the United Kingdom, Norway, Singapore, China, Australia, South Korea, Japan, Saudi Arabia, and Puerto Rico. Thailand has signed a reception agreement and will be certified in mid-2000. RSI holds the license rights for exclusive worldwide distribution. All other distribution and data-reception agreements are negotiated jointly by the CSA and RSI (CSA, 1999c).[3]

Civilian Applications
COASTAL MONITORING, GEOLOGY, AND LAND USE

Canada has an extremely long coastline, much of which is in the Arctic. RADARSAT-1's ability to monitor and map the distribution of sea ice has proven very valuable. SAR is particularly useful here because the polar areas are in darkness for extended periods, and RADARSAT-1's SAR sensor can clearly differentiate boundaries between land, ice, and water. The daily charts of the ice fields produced using the SAR data increase the safety of maritime navigation.

RADARSAT-1 data from coastal observations can be used to monitor erosion, aquaculture, intertidal seaweed, and shipping and fishing operations. Over the open ocean, RADARSAT-1 can monitor changes in waves and winds and can even promote commercial fishing by locating fertile fishing areas.

Cartography and land use studies can be done effectively using RADARSAT-1 data because the satellite provides a choice of viewing angles, including stereoscopic views. Thus, digital elevation models can be created with RADARSAT-1 data for use in engineering and mapping applications. Geological and other ground features can be readily mapped using RADARSAT-1 data. This can be particularly useful in determining such features as fault lines (see Massonnet, 1997); folds; groundwater distribution; and mineral deposits, such as oil and gas reserves. Other civil applications include environmental monitoring, hydrology, agriculture, forestry, and Antarctic mapping (Mahmood et al., 1996).

MARITIME

RADARSAT-1 data provide an excellent tool for operational maritime surveillance, for several reasons. First, RADARSAT-1 data are reliable because imaging is not contingent upon climactic conditions, such as cloud cover. Second, the multiple imaging modes help to minimize the time needed to reimage an area. Third, RADARSAT-1's ground infrastructure allows rapid delivery of data. Finally, space-based imaging is an economical way to view large ocean areas.

Ice mapping is one of RADARSAT's most important contributions. Because large areas of the ocean can be imaged, safe shipping routes can be determined from the extracted information. In fact, near-real-time data acquired from both the north and south poles have been delivered directly to ships operating at sea.

The data are of great help in responding to oil spills because the data can be used to evaluate the extent and movement of the oil on the ocean surface.

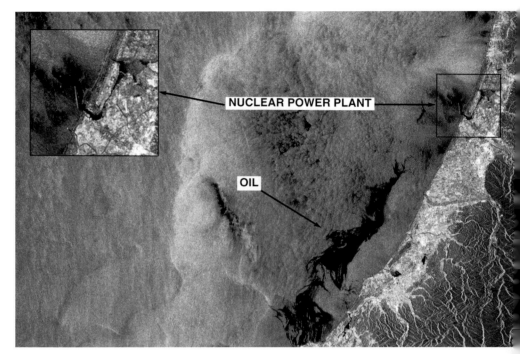

Figure 13.1—Oil Spill off Japan's Coast, January 11, 1997

First, RADARSAT-1 data are very sensitive to oil spills because of the damping effect the oil has on capillary waves. Second, the backscatter signature in open water can be used to measure wind speed and direction accurately, even in remote areas where there are no other sources of information. When the *Sea Empress* struck a rock in Wales in 1996, RADARSAT-1 acquired numerous images over several days that showed the extent of the oil slick. Similarly, the Russian oil tanker *Nakhodka* broke apart and sank in the Sea of Japan in 1997, creating a major spill of heavy fuel oil that covered 285 miles of Japan's western coastline, including the potential threat of closing down Japanese nuclear power plants in the affected region. Despite the weather, RADARSAT-1 images delivered within a few hours helped to measure the extent of the spill. Figure 13.1 is one of the images used to monitor the location of the oil spill and supporting containment and cleanup activities.

Another application is the detection of environmental pollution caused by ships illegally dumping foreign substances (such as oil-contaminated water) into the sea. Numerous cases have been reported of ships flushing bilges or empty ballast tanks. Many coast guards and navies around the world attempt to monitor coastal waters, territorial seas, contiguous zones, and exclusive economic zones extending up to 200 nautical miles from their coastlines for such illegal activities.

RADARSAT-1 can also be used to detect illegal fishing operations; in calm seas, even small (10- to 20-m long) fishing boats can be monitored. Similarly, RADARSAT-1 allows monitoring of maritime traffic and the identification of ships engaged in illegal activities or carrying illegal cargoes. One advantage here is that RADARSAT-1 can be used to determine the speed and direction of a ship under way by detecting its wake and measuring the wake's Doppler shift. Finally, port traffic can be monitored regularly, because radar guarantees data acquisition. Figure 13.2 is a RADARSAT image of the harbor and port facilities of San Diego, California. The image clearly indicates major maritime-related features, as well as a range of larger ships located at the main dock areas and even smaller boats moored in the yacht basin.

Thus, RADARSAT-1 technology makes it easier to address such hemispheric security concerns as enforcing fishing regulations or drug trafficking laws by detecting the presence of ships operating within a country's territorial waters.

Figure 13.2—San Diego Harbor Area, March 20, 1996

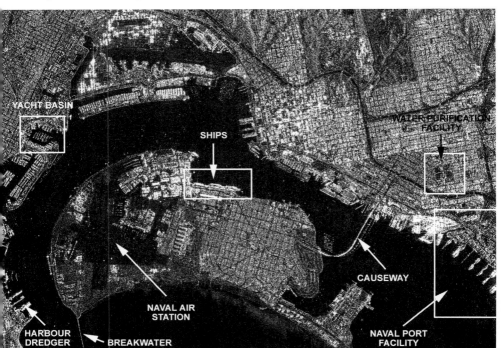

DISASTER MANAGEMENT

RADARSAT-1 images have been useful in evaluating damage in national and international emergencies and determining the need for humanitarian aid. Examples include determining the extent of such recent disasters as Manitoba's Red River flood in 1997 (see Figure 13.3) and the southern Mozambique's Limpopo River flood in 2000. RADARSAT-1's frequent revisits, beam flexibility, emergency programming, rapid data processing, and electronic delivery have proven to be extremely useful for continuously monitoring emergency situations. In the Mozambique case, the Canadian department of National Defense helped produce maps using RADARSAT-1 imagery, which in turn were supplied to the World Food Programme and the United Nations (UN) Office for the Coordination of Humanitarian Affairs and to other organizations that needed such maps for humanitarian missions (CSA, 2000).

RADARSAT Data Acquisition

RADARSAT-1 data acquisition is open and nondiscriminatory. Generally speaking, it is done in accordance with the UN principles pertaining to state-sponsored remote sensing activities (UN, 1986).[4] The following priorities have been established for acquisition:
1. spacecraft health and safety
2. emergency data acquisition
3. calibration
4. time-critical data
5. non–time-critical data.

Priority 2 refers to unforeseen natural or man-made catastrophic events. Priority 4 applies when failure to obtain the data at a precise moment would mean that the phenomenon or activity in question would not be remotely sensed. Priority 5 encompasses the vast majority of the data requests, which do not meet the criteria for higher priorities.

The RADARSAT Mission Management Office (MMO), located at CSA headquarters, near Montreal, is responsible for data acquisition. The MMO cannot guarantee acquisition of an image that has not been requested at least 14 days in advance and that does not have a priority level of 1, 2, or 3, although the MMO will make a "best effort" to fulfill such requests (Government of Canada, no date, p. 6). Moreover, a request having a priority level of 1 or 2 can take precedence over a confirmed request up to 29 hours before the scheduled imaging, when the data acquisition plan is frozen in preparation for uplink to

13 Canada's Remote Sensing Program and Policies

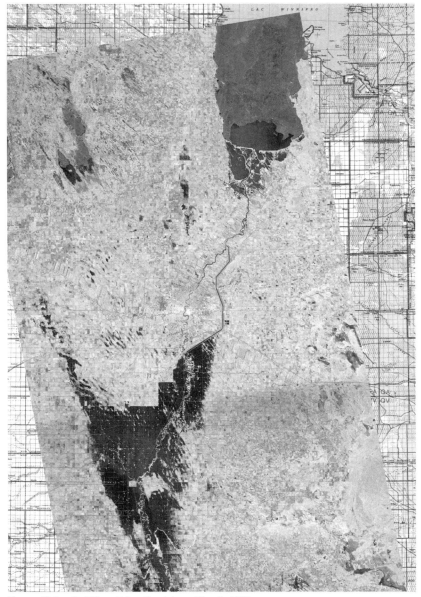

Figure 13.3—Red River Valley Flood Area, May 6, 1997

the satellite (Government of Canada, no date, p. 6; Mahmood, 1996).

Acquired data must normally be downlinked in real time to the closest functioning ground station. However, when the United States has used RADARSAT-1 to obtain data over China, the data were not transmitted to the Chinese RADARSAT ground station but were stored onboard to be downloaded to the ground station in Alaska. Although RADARSAT-1's International Steering Committee requested that this practice end by June 1999, the practice still continues for technical reasons.

RADARSAT-1 Data and Value-Added Products

Value-added service providers can purchase data to produce various products. However, the use of any data for a purpose other than those originally specified in the data purchase agreement requires RSI's consent. Data obtained by governments or agencies that are RADARSAT-1 participants and for use in a value-added process are governed slightly differently. In this case, value-added products can be developed and distributed noncommercially on behalf of the government or agency. Basically, government-obtained data must be used within the limits of governmental activity and operations.

To promote the use of RADARSAT-1 data, CCRS began the Radar Data Development Program in 1987. This program consists of competitive contracts totaling $4.5 million CDN per year for up to 15 years (CCRS, 1998c).

RADARSAT as a Dual-Use Space Asset

Space has become the proverbial "high ground" from which military planners can observe potential adversaries (Pircher, 1999). Space-based imagery is now important in military intelligence in itself and can also provide visual confirmation of intelligence reports obtained through other sources.[5] Satellite imagery is also very useful for arms control and verification (Centre for Research of Air and Space Law, 1987). Space-based earth imaging is thus a cost-effective way to ascertain international threats. In fact, the increased transparency satellites provide can reduce tensions.

But military space assets are extremely expensive, and Canada has a small military with a very limited budget.[6] The Canadian armed forces do not have the financial resources necessary for developing and deploying space systems exclusively for their own use. However, a synergy between civilian and military needs for space-based imagery has been created within the RADARSAT-1 project.

RADARSAT-1 technology can detect training activity levels and capacities, changes in training routines, military exercises, construction and/or destruction activities on military bases, transportation activities, ammunition and fuel storage sites, and vehicle convoys on roadways (representing possible unit deployments); even the absence of activity can be militarily significant. RADARSAT-1 data can be used to provide images of air installations and infrastructure and to detect the arrival or departure of large numbers of aircraft, analyze aircraft parking patterns, determine types of aircraft, and observe airport construction ("The Role of RADARSAT," no date). RADARSAT data acquisition is not constrained by time of day, climactic conditions, or obscurants (such as the smoke the massive oil well fires in Kuwait created during the 1990–1991 Persian Gulf War). All this gives military planners great flexibility in gathering intelligence.

Information is fluid and volatile, even more so in battle situations, in which conditions can change very quickly. A key issue in military intelligence is providing real-time information to military personnel in command centers and on the battlefield. Space-based data collection can help achieve this, but a transportable ground station would be needed. To this end, IOSAT, Inc., developed Sentry (see Figure 13.4), a mobile multisatellite receiving and processing station providing near–real-time image processing and management of large volumes of data.[7]

Sentry was used to support operational maritime surveillance during Maritime Combined Operational Training (MARCOT)–UNIFIED SPIRIT 1998 (see also CSA, 1999d). In this military training exercise, one of the largest Canada and other NATO countries have participated in, a large amphibious force

Figure 13.4—Sentry Mobile Ground Receiving Station

(15,000 personnel, 40 warships, and 100 combat airplanes from nine countries) landed on a hypothetical "enemy" shore. In this case, Newfoundland was "invaded" from a base in Nova Scotia (IOSAT, 1998). RADARSAT-1 was used to monitor the exact positions and courses of both enemy and friendly ships. RADARSAT-1 imagery also helped identify the exact location of the landing forces. The results were conclusive. Once RADARSAT-1 collected the data, they were immediately communicated to the Sentry receiving station. The data were then quickly processed to create an image, which was then interpreted, extracting information that was rapidly transmitted to field commanders in near real time. A mere 25 minutes elapsed between the time RADARSAT-1 collected the imagery and the time the commanders used the results.

RADARSAT-1 has also been very useful in Canadian peacekeeping operations, either for preparing maps or monitoring large internal refugee displacements. The satellite's cartographic capacity is also very important to the Canadian military, which is often deployed to locations for which updated maps are not readily available. RADARSAT-1 images give these forces the

Figure 13.5—Airfield at Goma, Zaire, November 12, 1996

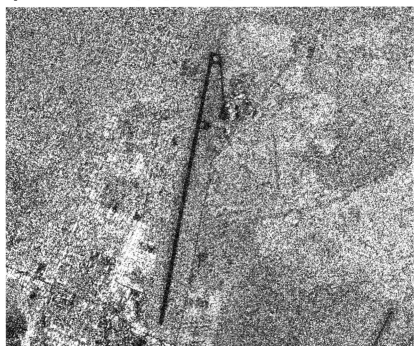

13 Canada's Remote Sensing Program and Policies

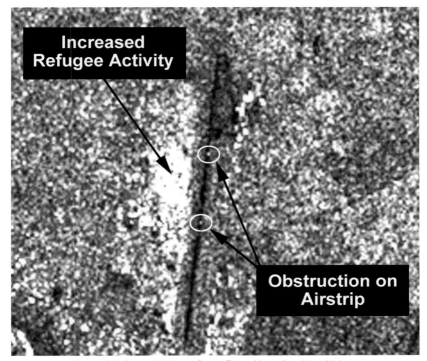

Figure 13.6—Signs of Refugee Activity at Goma, Zaire, November 22, 1996

capacity to create their own maps, thus facilitating peacekeeping and other humanitarian operations.

Figures 13.5 and 13.6 provide an example of how RADARSAT has been used to gather information about refugee displacement, in this case, the one that occurred during the humanitarian crisis in Eastern Zaire in late 1996. A comparison of the two images makes it evident that a large number of refugees arrived at the airfield in Goma, Zaire, between the times the two images were acquired. Although the resolution of the RADARSAT images is not good enough to detect individuals, the images do capture the effects the large group of people has had on area surrounding the airfields and the possibility that impediments have been placed on the runway.

RADARSAT-1 data can also be used to locate elements of infrastructure, such as seaports and pipelines, which can be legitimate military targets during conflicts. The data can also be used to evaluate oil reserves, because oil reservoirs have different radar signatures depending on the amount of oil they contain.

RADARSAT-1 imagery can also be very useful to Canada's foreign allies. The U.S. Department of Defense, for example, has moved to acquire imagery from commercial vendors to supplement and complement its existing capabilities (Johnson-Freese and Handberg, 1997, p. 193; Ferster, 1999d; Singer, 2000).[8]

National governments are of course concerned that regional rivals or terrorist organizations might make use of these images, given their military value. And some interesting images are already available on the World Wide Web.[9] Military planners must now grapple with the possibility of having to deny an enemy the use of commercial space-based imagery during conflicts, possibly by neutralizing a satellite that is the property of a neutral state.[10] In some cases, regulatory mechanisms, especially those that would restrict the operation of privately owned remote sensing satellites, must also take constitutional issues into account. Such issues as freedom of information, expression, and the press are, however, beyond the scope of this discussion; see Kramer (1989, p. 339); Martin (1982); Merges et al. (1989).

Hemispheric security concerns, including drug trafficking, can also be addressed through RADARSAT-1 technology. Crops have distinctive radar signatures through which they can be identified. In some areas, drug trafficking is a large-scale operation; the ability to detect the associated infrastructure, such as airfields and warehouses, helps law enforcement agencies accomplish their missions more efficiently.

The U.S. Connection

American-Canadian cooperation has been an important and ongoing issue in the development of the Canadian remote sensing expertise. Canada was interested in participating in NASA's Earth Resources Orbiting Satellite project (EROS). The Prince Albert Radar Laboratory, with an 84-foot diameter parabolic tracking dish, was offered as a receiving station for the EROS satellite (CCRS, 1998b). Unfortunately, the project was canceled. NASA established an alternative program, the Earth Resources Technology Satellite (ERTS), in 1969. Canada attempted to get an arrangement on the ERTS project similar to the one that had been agreed upon for the EROS project. However, despite an appeal to President Nixon's scientific advisor, foreign participation in this project was refused (CCRS, 1998b).

An agreement between NASA, NOAA, and CSA provided for the launch of RADARSAT-1. In return for NASA's deployment of the satellite, 15.8 percent of the RADARSAT-1 SAR on-time is allocated to U.S. federal departments and agencies. One of NASA's main requirements in this agreement is mapping Antarctica, to be completed within the satellite's five-year life span. This mission, now complete, yielded the first complete mosaic of the continent and its ice shelf. RADARSAT-1 has also yielded valuable data on glaciology and has provided NOAA with useful data for monitoring the northeast U.S. coast.

NATIONAL SECURITY ISSUES

As noted above, space-based earth-imaging satellites have considerable implications for national security. These assets can provide data on a state's military resources, its natural resources (e.g., water and minerals), its use of its resources, the environmental effects of its activities, its agricultural production, among other things (Doyle, 1993, p. 70).

National security concepts represent one of the most sensitive issues in Canadian-U.S. relations on space-based commercial remote sensing. Canada's announcement of a 3-m resolution capability for its RADARSAT-2 remote sensing satellite caught U.S. attention. The U.S. defense and intelligence establishments repeatedly expressed their concerns, which led to a series of discussions between the two countries.

RADARSAT-2

Unlike RADARSAT, which the Canadian government owned, MacDonald Dettwiler will own RADARSAT-2 privately. In a construction agreement, reached on December 18, 1998, CSA agreed to invest $225 million CDN, which MacDonald Dettwiler will repay by providing RADARSAT-2 data. MacDonald Dettwiler itself will contribute $80 million CDN. Another important difference between RADARSAT-1 and RADARSAT-2 is that Orbital Imaging Corp. (OrbImage), which purchased the exclusive worldwide marketing rights from MacDonald Dettwiler, will commercialize the data from RADARSAT-2 (Ferster, 1999a). (It is interesting to note that Orbital Sciences is the majority shareholder of OrbImage, that MacDonald Dettwiler is a subsidiary of Orbital Sciences, and that MacDonald Dettwiler is a minority shareholder in RADARSAT International.)

In planning the RADARSAT-2 project, CSA initially believed that NASA would be willing to barter launch services in return for data. Although the CSA-NASA Enhanced Cooperation Arrangement of 1994 stated that NASA would participate in the RADARSAT-2 satellite under conditions similar to those established for the RADARSAT-1 satellite, several obstacles to have arisen (Ferster, 1999b).

DATA POLICY

One of the most significant of these obstacles is data policy resulting from concerns about national security. The U.S. government has set the minimum allowable resolution for SAR satellites at 5 m. The 3-m resolution planned for RADARSAT-2 is clearly well below that minimum. Understandably, NASA does not want to participate in any project that does not conform to the U.S. government's national security standards. This obstacle appears to have been resolved by the intergovernmental agreement recently signed between Ottawa and Washington. This agreement, however, is not a panacea, because the ownership issue still remains.

OWNERSHIP

The second obstacle, and probably the principal issue, has to do with the ownership of the RADARSAT-2. Since RADARSAT-1 was the property of the Canadian government, bartering data for launch services between governments was easy. But the private ownership of RADARSAT-2 creates certain legal hurdles that prevent a similar barter arrangement with NASA. The ownership change is a substantial problem because NASA cannot justify a partnership with competitors to the U.S. earth-observation industry (White House, 1996). RADARSAT-2 will now be launched by Boeing on a Delta-II class booster, probably from the launch facility at Vandenberg Air Force Base, at an estimated cost of approximately $100 million CDN. This represents the increment cost of not launching RADARSAT-2 with NASA and includes, among other things, engineering costs, programming costs, and program delay costs.

REGULATION

The third, and most complex, obstacle is the U.S. regulatory structure for space-based earth imaging. Shutter control and export restrictions are only two of the many aspects of this multidimensional issue (Caddey, 2001).[11]

In the case of RADARSAT-2, a modified U.S. regulatory structure that, among other things, required export-control permits for Canadian companies created havoc. These new legal structures created a slow and uncertain export-control process, which resulted in delays for Canadian companies dealing with U.S. contractors. Since RADARSAT-2 must be ready in time to replace RADARSAT-1 when it reaches its scheduled end of service in 2003 (assuming the mission is indeed extended), these delays were perceived as creating unacceptable business risks. This belief, unfortunately, lead the Canadian government to change its main U.S. contractor. Given the close business relationships involved, this decision was indeed very difficult and probably somewhat awkward. In executing this change, the force majeure clause had to be invoked. The Canadian government and the contractor will both bear the costs of the cancellation (although the Canadian government will assume the lion's share).

It is important to stress that this decision was not based on the contractor's technical expertise but rather on a slow and uncertain regulatory structure that requires interagency consultation and, in some cases, the involvement of congressional committees. This situation, unfortunately, reduces the international competitiveness of the U.S. space industry by creating uncertain delays (for more on this, see Loeb, 2000). The U.S. government was unable to guarantee that bureaucratic obstructions would not prevent a timely launch of the RADARSAT-2 satellite. RADARSAT-2's primary contractor, MacDonald Dettwiler, was thus forced to withdraw its contract from Orbital Sciences, deciding instead to do business with a European contractor. In December 1999, MacDonald Dettwiler awarded a contract worth $74 million CDN to Alenia Aerospazio of Rome, Italy, to build the satellite platform for RADARSAT-2 (de Selding, 2000a). U.S. authorities argued that their refusal to grant the necessary export licenses was in fact Canada's fault, since the Canadian government had refused to acquiesce to U.S. demands to restrict sale of high-resolution RADARSAT-2 imagery (de Selding, 2000a).

The effect of this decision is unquestionably of greater magnitude than the mere loss of the monetary value of the contract. The European contractor will certainly gain valuable and marketable experience and expertise in developing commercial radar satellites.[12] Furthermore, as a result of this decision, RADARSAT-2's star tracker will now most probably be built in Canada rather than the United States. The noncompetitive nature of the U.S. regulatory structures could force other foreign satellite operators to imitate the Canadian decision. In 1994, showing foresight on these issues, Gen Charles Horner, former commander of U.S. Space Command, observed in reference to high-resolution imaging satellites that "[t]he free market is going to take over. . . . If they don't build them in Europe, and we don't build them in the United States, Russia is certainly going to build them." (Johnson-Freese and Handberg, 1997, p. 192.)

The loss of this contract did little to help to protect the perceived national security concerns of the United States. Given the close historical relations between Canada and the United States, the Canadian government considered this change in contractors most unfortunate but necessary. As the Canadian Industry Minister, John Manley, pointed out, for the Canadian space industry to remain competitive, it must be able to rely on leading-edge technologies and commercial revenues (Manly 1999). Orbital Sciences, after 10 months of trying, failed to get the necessary licenses for the RADARSAT-2 project.

CANADA-U.S. TRADE RELATIONS

> **Views in the United States**
>
> Air Force Gen Richard Myers, Commander in Chief, U.S. Space Command, has articulated concerns that his nation's crackdown on satellite exports may actually reduce the United States' ability to protect U.S. ground forces (Myers, 1999). So have others. The following quotations are excerpts from testimony prepared for the Senate Committee on Foreign Relations Subcommittee on International Economic policy, Export and Trade Promotion.
>
> John Hamre, Deputy Secretary of Defense: My personal conclusion after nearly three years as Deputy Secretary of Defense is that our export controls are increasingly ineffective as a tool of foreign policy or a tool of technology security. Indeed I believe we are coming perilously close to the point where export controls in their current form are causing more problems than they solve... In my judgment...the current system of export controls is broken....I consider the system broken because it fails to be an efficient form of government oversight, and is increasingly creating negative developments that are having an impact on our national security.
>
> William Reinsch, Under Secretary of Commerce for Export Administration: To sum up, we find ourselves in the ironic situation where denial or delay of exports under the rubric of national security has done more harm than good to our nation's military and economic strength.
>
> Clayton Mowry, Executive Director, Satellite Industry Association: In 1998 America racked up a 73% market share, however U.S. companies

Canada and the United States reached a tentative accord concerning the RADARSAT-2 project during President Clinton's visit on October 8, 1999. The Canadian government then announced that it was developing a data-distribution policy attuned to the concerns of its southern neighbor (Ferster, 1999e, p.1).[13] However, even with approval of the general framework of a Canadian space imaging policy, the actual regulations cannot be in place before late 2000 (Rains et al., 1999). Unlike the American regulations, the focus of the Canadian regulatory norms would not be on technological transfers but on limiting imaging distribution in certain countries. Nonetheless, Canadian regulations will most probably be heavily influenced by the American regulatory structure, as these new rules will be designed to protect both Canada's national security concerns and those of its military allies.[14] Although the U.S. government appeared satisfied with this development, several important issues remained unsettled between the two neighbors and military allies. Unfortunately, the subsequent talks between Canada and the U.S. progressed rather slowly (Travers, 2000). Nonetheless, one of the key points contained in the Canada-U.S. agreement is a statement that the Canadian regulatory structure complies with NSC/PDD-23, a policy directive President Clinton issued in March 1994 that was intended to encourage the development of the U.S. remote sensing industry.

Besides the national security, the resolution of the satellite, and ownership concerns, certain other economic issues are also hindering the RADARSAT-2 program. First, The United States has a policy of encouraging the development of U.S. commercial remote sensing satellites. PDD-23 established a delicate balance between industrial development, U.S. national security, and foreign policy interests. Thus, the secretaries of Defense and State could now exercise a shutter control policy, limiting data collection and/or distribution when the national security or foreign policy obligations of the United States are under threat.

Second, export licenses for remote sensing systems are hindering RADARSAT-2 development. U.S. national security laws have extraterritorial effects upon Canada through trade restrictions.[15] Export licenses for remote sensing equipment are contingent upon the proposed foreign recipient's willingness and ability to accept commitments to the U.S. government concerning sharing, protection, and denial of products and data and concerning constraints on resolution, geographic coverage, timeliness, spectral coverage, data processing and exploitation techniques, tasking capabilities, and ground architecture.

On March 18, 1998, the White House issued implementing guidance for NSC/PDD23. This established a standing interagency working group for remote sensing, whose purpose is to streamline the interagency review of potential exports of advanced remote sensing systems. This group will determine agreements and assurances necessary for technology exports and will thus heavily influence the RADARSAT-2 project.

Third, there is the issue of the commercial space-based imaging regulatory structure.[16] The U.S. remote sensing industry has been very critical of the regulatory structure governing its operations. The following issues are considered particularly important for the Canadian remote sensing industry.

> share of announced orders has dropped to 52% . . . Our relationship with Canada has caused particular concern for U.S. satellite suppliers. The Canadian government's decision to purchase a satellite bus from a European after waiting ten months for a U.S. license is a clear signal that we have a continuing problem. Even more troubling are press accounts as recent as last week indicating that commercial satellites may not be included in the list of products on the U.S. Munitions List under negotiation between the United States and Canada that will be eligible to receive the renewed "exemption" from State Department licensing.
>
> John Hollum, Department of State, Senior Advisor for Arms Control, outlined possible forthcoming concessions for U.S. allies:
>
> State will be moving to use high-volume export licenses for components and technical data which will be valid for four years for multiple shipment to any NATO or major non-NATO allies. Another important feature is that these new regulations will not require that details of purchase orders, contracts or end use certificates in advance of the licensing being given.
>
> On May 10, Representative Sam Gejdenson (D-Conn), the ranking Democrat on the House International Relations Committee, introduced HR 4417. This proposed legislation would reverse the March 1998 measure that transferred jurisdiction from the Department of Commerce to the Department of State.

- A U.S. license is required for a non-U.S. system that will have substantial connections with the United States (Bourbonniere, 1997). Unfortunately, the term *substantial connections* is ill-defined and, if broadly interpreted, can have sweeping applications. However, the term most probably includes launches using U.S. vehicles, U.S. cooperation in the operation of the spacecraft, tasking of satellite sensors from U.S. control stations, real-time direct access to unenhanced data, equity interest in licenses above 15 percent, distributorship arrangements involving receipt of high volumes of data, and data or image sales offices.
- At times of national security (which is not defined), the licensee must restrict the collection and/or distribution of data as directed and make unenhanced data available exclusively to the U.S. government by means of the U.S. government–furnished rekeyable encryption on the downlink (and equipping a satellite to do so is a licensing precondition). A definitional problem is also present here. There is no precise definition of what could constitute a "situation of national security." The fear is that satellite operations could be arbitrarily shut down for perceived security risks rather than for objective security concerns.
- There are severe limitations on foreign participation, either in joint ventures or as owners in proposed U.S. commercial remote sensing system operators. Furthermore, there are strict requirements for identifying non-U.S. officers and/or shareholders.

Fourth, the relationship between NASA and the U.S. remote sensing industry is important for the Canadian industry as well. Rather than building its own spacecraft or contributing to non-U.S. space ventures, NASA must procure scientific data whenever possible. NASA must justify its involvement with a foreign competitor to U.S. national industries, by appealing to the unique capabilities that it offers, in this case, those of the RADARSAT-2 satellite.

U.S. policy on commercial remote sensing aims to create a strong indigenous remote sensing industry. At the same time, the policy attempts to control the proliferation of this technology by restricting exports of the underlying technologies, which are themselves regulated by export controls applied to munitions (Johnson, Pace, and Gabbard, 1998).

INTERNATIONAL TRAFFIC IN ARMS REGULATIONS

During the 1998 Summit of the Americas, held in Santiago, Chile, President Clinton discussed with the leaders at the summit the issue of transnational threats in our hemisphere, such as the narcotics trade, weapons, and terrorism.

President Clinton then decided to implement regulations on commercial arms transfers. In response to this initiative, the secretaries of State, Commerce, and the Treasury were required to implement certain regulatory modifications. The result of this new policy is an amendment to the International Traffic in Arms Regulations (ITARs) (*Federal Register*, 1999, p. 17,532).

Before these regulatory amendments, Canadian companies were exempted from the obligation to obtain export licenses for certain high-technology materials that were on the U.S. Munitions List (USML).[17] Although these amendments were made in cooperation with the Canadian government, they did cause certain problems, raising concern in Canada.[18]

The first area of concern about the ITAR modifications is their effects on Canada's effort to transition to a knowledge-based economy. Achieving this vision requires (1) R&D, (2) access to an environment that supports the free flow of goods, and (3) timely access to advanced technology (Manley, 1999, p. 2).

The ITAR amendments are considered as an irritant to Canadian R&D efforts, to the free flow of goods between Canada and the United States, and to Canadian access to certain technologies. This is so because all the following require export licenses regardless of the category (U.S. Department of State, no date):

1. any transaction that requires congressional notification in accordance with the United States Arms Export Control Act
2. all controlled USML items and the related technical data
3. any defense service covered by Part 124 of the ITAR
4. any item, data, or service on the USML that is for end use in Canada by a person who is also a national of another foreign country
5. all U.S. government classified items, data, and services on the USML
6. the defense articles, data, and defense services identified in Part 126.5 of the ITAR. Part II presents guidelines for identification of these items as well as the USML items that do not require a license prior to export unless they meet the criteria in 1 through 5 above.

It is important to stress that the irritant is not simply the requirement for export licenses but rather the lack of timeliness in the licensing procedures. Thus, as the ITARs are presently structured, obtaining a license can cause unpredictable delays for a Canadian company. The ITAR modifications have thus complicated the situation for the RADARSAT-2 project. Export licenses are now required for certain components of RADARSAT-2, such as solar cells and the SAR payload, which will probably now be built elsewhere that the United States, and the power amplifier and switch matrix, which might now be built in Canada. Multiple licenses are sometimes required for one part, such as a license for the technology itself, a license for the export of the part itself, and a reexport license when the satellite is brought back to the United States for launching (de Selding, 2000b). Such export licensing procedures could cause endless delays for the RADARSAT-2 project. The uncertain timeliness of these licenses increases the business risks associated with the project.

The perceived need for the ITARs is certainly understandable. They are designed to control exports of sensitive technologies, the spread of which can threaten the security of the northern hemisphere. For example, both Canadian and U.S. nuclear research laboratories have been subjected to Chinese espionage efforts. China has stolen nuclear weapon secrets from government facilities at Los Alamos, New Mexico, and has essentially stolen the technology for the Canadian Slowpoke nuclear reactor.[19]

Since Canada is a close ally of the United States, the ITARs also indirectly protect Canada's own national interests. Canada both understands the U.S. concerns and agrees with the fundamental goals of the ITARs.[20] Canada has also taken steps to tighten its own security regulations. The Canadian government values the integrated defense relationship that it has built with the United States over the years. As a matter of fact, it is estimated that Canadian defense exports to the United States amount to approximately $1 billion CDN each year.[21] This being said, however, from a Canadian perspective, the ITAR rules must be streamlined to reduce their negative effect on Canada's evolution to a knowledge-based economy.

A second important point of concern about the ITARs is the delicate issue of Canadian citizens who possess multiple nationalities. Canadian constitutional laws, and more specifically the Charter of Rights and Freedoms (Canada Act, 1982), are very clear on this topic: They prevent discrimination based on nationality. Canadian courts could very well annul a regulatory structure founded on such discrimination as an *ultra vires* act of government. In an attempt to respect these constitutional restrictions, the October agreement took aim at regulating known security risks rather than risks based on nationality alone. The distinction is subtle and could prove difficult to promulgate effectively. Unfortunately, and regrettably, the modified ITARs place restrictions on Canadians who have other nationalities and have on-the-job access to U.S. technologies. Again, this creates an extraterritorial application of

U.S. regulations, a sensitive issue for Canadians. Fortunately, the Canadian Foreign Affairs Minister, Lloyd Axworthy, has met with the U.S. Secretary of State Madeleine Albright in an attempt to resolve these issues.[22]

A third concern about the ITARs comes from Canadian industry: that the U.S. regulatory structure might strike a lethal blow to the Canadian defense and space industries. There are two dimensions to this concern.

The global market of the space and defense industries is highly competitive. Many Canadian space industries have U.S. partners or use components that are manufactured in the United States. If regulatory structures reduce the competitiveness of U.S. industries, there will be a ripple effect on the Canadian industries that depend on U.S. products. Furthermore, export licensing delay adds uncertainty, which is necessarily passed on to clients. A purchaser of space products may choose not to do business with a Canadian supplier that is bogged down in an U.S. licensing quagmire.[23] In the past, Canadian companies have shipped approximately $1 billion CDN worth defense products to the U.S. market annually. The consequences of the ITARs could be disastrous for the Canadian defense industry, placing thousands of Canadian jobs in danger (Pugliese, 1999).

In addition, ITAR rules could force Canadian companies to relocate to the United States in an attempt to avoid the need for export licenses. This possibility is very real, considering that 60 percent of the Canadian defense industry is composed of subsidiaries of U.S. companies.[24]

U.S. industry has also voiced concerns about the negative effects of export licensing as it is presently structured. Unfortunately, policymakers appear to be caught under the sword of Damocles. A delicate regulatory balance must be established between objective national security needs (as opposed to perceived security needs) and a healthy and competitive space industry (see *Space Policy*, 1994). An overprotective regulatory structure could cost billions, not only to Canadian defense and space industries but also to their U.S. counterparts. A report on export trade restrictions to China notes that

> The industry is chafing under the current satellite export restrictions imposed by concerns over technology transfers to China. If the series of exports curbs is too tight, the industry could be damaged, thus harming the Pentagon's space plans. (Sietzen, 1999a.)

Sietzen then quotes Joan Johnson-Freese: "If the U.S. aerospace industry begins to suffer, so too could U.S. military space capabilities." Commenting on the effect of export trade restrictions, Sietzen (1999b) has also noted that tough new language passed by Congress last year to guard against possible illegal technology transfers in space system exports is also being blamed for the export crunch, which some have suggested may cost the U.S. commercial space industry billions in lost sales.

Fortunately, the U.S. Department of State is aware of these licensing dilemmas and is taking steps to alleviate the problems. Sietzen (1999b) reported that Congress was expected to add "$11 million to the State Department budget next fiscal year to ease the current bottleneck in issuing export licenses for satellites."

Shutter Control

Shutter control was not an issue for RADARSAT. Because RADARSAT-1 was a government-owned space asset, the Canadian government controlled the imaging process directly. The situation is quite different for RADARSAT-2, which is privately owned. Canada does not presently have a regulatory structure comparable to that of the U.S. regulations to establish licensing conditions for private operators of remote sensing satellites, and so lacks shutter control rules. Nonetheless, for RADARSAT-2, the Canadian government established itself contractually as the "custodian" of the data policy (see Bourbonniere and Haeck, 1999). The government thus can exercise shutter control without enacting a regulatory or legislative tool. However, this situation will probably change soon as a result of bilateral agreements between Canada and the United States and with the promulgation in the near future of a Canadian regulatory structure dealing with this issue.

Conclusion

Canada has developed a strong expertise in space-based earth imaging. The CCRS played an important role in developing Canadian SAR expertise, creating a strong base for the RADARSAT project. Canada's leap into space with RADARSAT-1 was very successful and displayed Canadian ingenuity in structuring a joint public-private space venture. RADARSAT-2 is a logical evolution from the RADARSAT-1 project, in which the private sector will own the next Canadian high-resolution SAR satellite. As markets develop, Canada will have a profitable niche in radar space-based earth imaging. Canada's regulatory structure will now have to follow Canada's technological advances into space and will most probably be influenced by the U.S. regulatory experience. The challenge for Canadian regulators will be to address U.S. security concerns while simultaneously protecting Canada's international competitive position. Canada will thus have to establish a delicate balance between the concerns of its southern neighbor and the highly competitive international arena within which RADARSAT-2 will have to operate. Thus, these new regulations must not impose upon Canadian commercial remote sensing, restrictions that Canada's competitors, including other NATO allies, are not burdened with. As Scott Pace (1999, p. 5) has insightfully argued, there are two regulatory cultures—guardians and merchants—and an effective Canadian regulatory structure for commercial space-based remote sensing will have to posses aspects of both paradigms.

Furthermore, Canada is an associate member of the European Space Agency (ESA) and must maintain the latitude to collaborate with its European allies. These new Canadian regulations should also deal with multispectral and hyperspectral commercial imagery. Canada's venture into satellite imaging has been marked by close technical, military, and commercial cooperation with the United States, and its future will also be contingent on not only maintaining but on improving these very important ties.

Notes

1. During the October Crisis of 1970, one of the actions the Front de Liberation de Québec took was to kidnap Québec government minister Pierre Laporte. The Department of National Defence used remote sensing technology to search for cottages in the Laurentians where he may have been taken for hiding. These cottages were generally closed for the winter in November; any that were not would show a heat target and could be a suspected hideout. (See CCRS, 1998b.)
2. RADARSAT-1 has an altitude of 798 km and an orbital plane inclination of 98.6 degrees and completes 14 orbits every 24 hours.
3. For a more detailed review of the technical capabilities of RADARSAT-1, see Mahmood et al. (1998); for examples of specific applications, see Mahmood et al. (1999a).
4. An interesting legal debate surrounds the determination of the legal values of these principles. Professor Gabrynowicz has argued that these principles are now customary international law (Gabrynowicz, 1998). We, however, disagree with her position.
5. Espionage is not illegal under international law (Haeck and Bourbonniere, 1998).
6. The CSA must also deal with a very limited budget and reduced financing in the immediate future (Berger, 1999).
7. Sentry can be used with ESA's ERS-1, ERS-2, and SPOT data and can also be very useful in environmental monitoring and emergency or disaster management operations. IOSAT's contract was announced on June 11, 1996. This system-engineering

company provides solutions for rapid data acquisition and imagery assessment from earth-observation satellites. IOSAT is based in Halifax, Nova Scotia, and is jointly owned by Satlantic Inc., Halifax; SSiG, Montreal; and Matra Systèmes & Information, Vélizy, France (IOSAT, 1996).

8. It is interesting to note that RADARSAT-1 did not have any Y2K problems. On the other hand, the *Chicago Tribune* and others reported in January 2000 that American reconnaissance satellites had a Y2K software glitch. The software glitch occurred at a ground station used to receive data from optical and radar spy satellites. The ability to process imaging data was severely impaired, reducing the ability of the Department of Defense to monitor hot spots around the world for up to three days. Data from the satellites themselves were not affected; however, data had to be rerouted to a backup ground station, where the analysis was done manually. As a result, the flow of information was greatly reduced. (*SpaceViews*, 2000.)

9. In the past year, the Federation of American Scientists has published pictures of classified facilities on its Web site and has also recently published the first privately obtained images of North Korea's ballistic missile launch pad. This is done within a project called Public Eye, the goal of which is to support the application of imagery intelligence to public policy. (Sietzen, 2000.)

10. During U.S. Army war games held February 14–24, 1999, commercial satellites played an important role in supplying the enemy with valuable intelligence. After suffering high casualties, commanders decided to disable the commercial satellite. Subsequently, space assets on both sides were disabled. (Ferster, 1999c; *Space News*, 1999.)

11. Regulatory risks are an important part of commercial ventures and must be properly evaluated, since they affect financing structures. Neither telecommunication nor earth-imaging satellites are exempt from this commercial reality. For an interesting analysis of regulatory risks pertaining to satellites, see Meredith and Trinder (1998).

12. As *Space News* (2000, p. 12) has observed: "By building the basic satellite systems for RADARSAR-2, Italy, not the United States, will gain technical experience building a small commercially focused radar satellite."

13. See also Canada Department of Foreign Affairs and International Trade (1999) for an outline of Canada's policy.

14. Canadian remote sensing operators would most probably be subject to registration and licensing, export restrictions, ownership transfer restrictions, uplink and downlink controls, and government-approved encryption and would be required to notify the Canadian government of substantial foreign agreements (Government of Canada, 1999).

15. By extraterritorial, we mean the asserted right of a state, in this case the United States, to impose its national laws on the conduct of persons, either physical or juridical, who are not the residents or nationals of that state. In this case, the United States is attempting to regulate the conduct of Canadian citizens or Canadian corporations, including Canadian subsidiaries of U.S. multinational corporations,

located in Canada but carrying on international trade activities. On this issue, see Kindred et al. (1993), p. 491; Maier (1982); Lowe (1985); and *France v. Turkey* (1972).

16 For an analysis of these regulations, see Bourbonniere (1997) and Bourbonniere and Haeck (1999).

17 These modifications were justified on the basis that

Recent escalation in defense companies' voluntary disclosures, discussions with the Canadian government and the use of the Canadian exemption by unregistered companies provides evidence that unauthorized exports are taking place both from the United States and by Canadian persons in possession of U.S. defense articles and defense services. (*Federal Register*, 1999, p. 17,532.)

According to Pugliese (1999), p. 19:

A Canadian businessman was arrested by U.S. Customs agents in Boston Feb. 11 after allegedly planning to buy gyroscopes that could be used in weapon guidance systems and attempting to ship them to China via Montreal. In May 1998, two Iranian-born Canadians were arrested in Vancouver and accused of trying to arrange the purchase of guidance equipment for Hawk missiles. The missile parts, according to the U.S. Justice Department, were to be shipped to Iran via Canada.

18 Section 126.5(b)(9) (*Federal Register*, 1999, p. 17,534) of the ITARs broadens the applicability of the export restrictions to Canada to include "Spacecraft, Remote Sensing Satellites and Military Communications Satellites in Category XV(a), (b) and (c)."

19 The latter theft has caused a severe economic loss for the Canadian nuclear power program, which has lost sales because the Chinese are now selling cheap copies. According to the Canadian Security Intelligence Service, the counterintelligence agency, "Beijing sent one of its best spies, Bu Chaomin, to Canada posing as a correspondent for a State-controlled newspaper, *People's Daily* in the late 60's." (Sallot and Mitrovica, 2000.) Furthermore, the *People's Daily* reported on January 14, 2000, that China's Ministry of Science and Technology is planning to participate in the International Space Station project. Wang Shaoqi, head of the ministry's international scientific cooperation division, says that China is discussing a role on ISS with the United States, Germany, France, Canada, Japan, and Russia (*People's Daily*, 2000). It will be interesting to see if the above-cited espionage cases affect these negotiations.

20 While Canada's Industry Minister has expressed sympathy with these concerns, he has also outlined the problems caused by tightening the ITARs (Manley, 1999, pp. 3–4):
- the inability of Canadian companies to obtain the technical data necessary to bid on U.S. contracts
- restrictions on Canadian companies attending meetings and briefings for ongoing contracts

- difficulties the Canadian Department of National Defence has experienced in maintaining operational readiness for specific systems
- substantial administrative delays as a result of the additional licensing burden.

21 It is interesting to note that the Clinton administration has recently revised its policy on encryption export restrictions. The rationale for this trade restriction was that illegal activities could be more difficult to detect if encryption technology were readily available. U.S. industry vehemently lobbied against this policy. (See Sung, 2000.)

22 While attending the ASEAN Regional Forum in Singapore on July 26, 1999, Secretary of State Madeleine K. Albright and Canadian Foreign Affairs Minister Lloyd Axworthy reviewed the bilateral consultations thus far on ITARs. A 120-day review period had been previously agreed to on April 22, 1999, to
> implement the ITAR regulations in such a way as to mitigate the effects on the North American defense industry and to determine what those effects are. Both (were) committed to evaluate and revisit the core issues of technology transfers and safeguards by the end of this period. (See U.S. Department of State, 1999.)

23 Japan shares Canada's concerns; Tomifumi Godai, Executive Vice President of NASDA has called the export controls troubling (see Lipman, 1999a). It is, however, important to note that Dan Goldin, the NASA administrator, has questioned these views (quoted in Lipman, 1999b): "Fundamental to global space cooperation and international business development is adherence to the Missile Technology Control Regime and related nonproliferation norms."

24 Bob Fisher, President of the Canadian Defence Industries Association has articulated these concerns (quoted in Travers, 2000).

References

Abiodun, Adigun Ade, "Remote Sensing in the Information Age," *Space Policy*, 14, 1998. pp. 229–238.

Berger, Brian, "CSA Avoids Steep Declines, Spending Still Scheduled to Drop 14 Percent by 2002," *Space News*, Vol. 10, No. 8, March 1999, pp. 1, 18.

Bourbonniere, Michel, "A Critical Review of American Regulations Pertaining to Commercial Remote Sensing Market Structures," *Annals of Air and Space Law*, Vol. XXII-I, 1997, p. 455.

Bourbonniere, Michel, and Louis Haeck, "Space Imaging Data Policy: A Canadian Perspective," *Annals of Air and Space Law*, Vol. XXIV, 1999, p. 33.

Caddey, Dave, "Radarsat-2: A Cautionary Tale," *Aerospace America*, Vol. 39, January 2001, pp 4–5.

Canada Act 1982 c. 11 (U.K.) in R.S.C., App. II, no. 44., arts. 1–34.

Canada Centre for Remote Sensing, "National Advisory Groups," May 20, 1998a. Available at http://www.ccrs.nrcan.gc.ca/ccrs/comvnts/advgrps/refere.html (last accessed May 30, 2000).

—, "Remote Sensing Then and Now: Early Beginnings 1960–66," July 7, 1998b. Available at http://www.ccrs.nrcan.gc.ca/ccrs/org/history/hist1e.html (last accessed May 25, 2000).

—, "Remote Sensing Then and Now: The Airborne Operation," July 7, 1998c. Available at http://www.ccrs.nrcan.gc.ca/ccrs/org/history/hist3e.html (last accessed May 25, 2000).

Canada Department of Foreign Affairs and International Trade (DFAIT), "Canada to Control Imaging Satellites, news release no 134, June 9, 1999. Available at http://www.dfait-maeci.gc.ca (last accessed July 25, 2000).

Canadian Space Agency, "Description of RADARSAT," February 25, 1999a. Available at http://radarsat.space.gc.ca/info/description/ (last accessed May 30, 2000).

—, "RADARSAT Background," February 25, 1999b. Available at http://radarsat.space.gc.ca/info/background.html (last accessed May 15, 2000).

—, "RADARSAT Data Handling and Distribution," February 25, 1999c. Available at http://radarsat.space.gc.ca/info/distribution.html (last accessed May 15, 2000).

—, "RADARSAT User Development Program," February 25, 1999d. Available at http://radarsat.space.gc.ca/info/programs/rudp.html (last accessed May 15, 2000).

—, press release, March 13, 2000.

CCRS—*see* Canada Centre for Remote Sensing,

Centre for Research of Air and Space Law, Space Surveillance for Arms Control and Verification Options, Proceedings of the Symposium held McGill University, Montreal, October 1987.

CSA—see Canadian Space Agency.

Department of the Army, Headquarters, *Space Support to Army Operations*, Washington, D.C., FM 100-18, July 1995.

de Selding, Peter B., "Alenia Wins Radarsat 2 Contract," *Space News*, January 10, 2000a, p. 3.

—, "U.S. Government Waives Monitoring of Ariane Launches," *Space News*, Vol. 11, May 29, 2000b, pp. 1, 19.

Doyle, Stephen E., *Civil Space Systems: Implications for International security*, Dartmouth: United Nations Institute for Disarmament Research, 1993.

Evans, W. M., "Canada Takes the Lead in Earth Observation," Message from W.M. (Mac) Evans, President of the Canadian Space Agency. Available at http://www.space.gc.ca/ENG/Publications/Radarsat/publ_radarsat_p1_e.html (last accessed 1999).

Federal Register, "Rules and Regulations," Vol. 64, No. 69, April 12, 1999.

Ferster, Warren, "Orbimage Gets Right to RADARSAT-2 Data; RSI's Future Unclear," *Space News*, January 25, 1999a.

—, "NASA Questions Wisdom of Launching Radarsat-2," *Space News*, February 22, 1999b, p. 1.

—, "Satellite Role Critical in Conflict, Commercial Spacecraft Pose Military Dilemma in War Game," *Space News*, March 15, 1999c.

—, "U.S. to Buy Private Imagery for Intelligence," *Space News*, Vol. 10, April 12, 1999d, p. 1.

—, "U.S., Canada Reach Radarsat 2 Accord," *Space News*, October 18, 1999e, p. 1.

France v. Turkey, The Steamship Lotus, P.C.I.J. Ser. A., No. 10, 1972.

Gabrynowicz, Joanne I., "Defining Data Availability for Commercial Remote Sensing Systems: Under United States Federal Law," Annals of Air and Space Law, Vol. XXIII, 1998, p. 93.

Garvin, James B., et al., "Taking a Clear Look at Cloud-Covered Oceanic Islands on a Seasonal Basis," *EOS, Transactions, American Geophysical Union*, Vol. 80, No. 5, February 2, 1999a, pp. 49–54

—, "Satellite Radar Images Capture a Subglacial Volcanic Eruption in Iceland," *EOS, Transactions, American Geophysical Union*, Vol. 80, No. 18, May 4, 1999b, pp. 205–207.

Government of Canada, RADARSAT-1 Data Use and Access, unclassified internal working document on file with the authors, RSCSA-ML0016-N/C, no date.

—, Communiqué No. 134, June 9, 1999.

Haeck, Louis, and Michel Bourbonniere, "Overhead Imagery and Espionage: International Law Implications," *The Caribbean Law Review*, Vol. 8, No. 2, December 1998, p. 287–298.

IOSAT, Inc., "IOSAT Wins Contract from the Canadian Space Agency to Supply Transportable RADARSAT Station," June 11, 1996. Available at http://www.iosat.com/11jun96.html (last accessed May 30, 2000).

—, "Satellite Surveillance of NATO Exercise Successful," June 28, 1998. Available at http://www.iosat.com/28jun98.html (last accessed May 30, 2000).

Johnson, Dana J., Scott Pace, and C. Bryan Gabbard, *Space: Emerging Options for National Power*, Santa Monica, Calif.: RAND, MR-517, 1998, p. 31.

Johnson-Freese, Joan, and Roger Handberg, *Space, the Dormant Frontier: Changing the Paradigm for the 21st Century*, Westport, Conn.: Praeger, 1997.

Kindred, Hugh M., et al., *International Law, Chiefly as Interpreted and Applied in Canada*, 5th ed., Emond Montgomery Publications Limited, 1993.

Kramer, G. M., "The First Amendment Viewed from Space: National Security Versus Freedom of the Press," *ADAS*, 1989.

Lipman, Jonathan, "Businesses Fear U.S. Policy Is Threat to Satellite Industry," November 4, 1999a. Available at http://www.space.com/space/business/isba_991104.html (last accessed May 30, 2000).

—, "Goldin Backs Export Controls," *Space.com*, November 4, 1999b. Available at http://www.space.com/space/business/isba_goldin_991104.html (last accessed May 30, 2000).

Loeb, Vernon, "U.S. Satellite Sales Lag Since Regulatory Shift," *The Washington Post*, March 28, 2000.

Lowe, A.V., "The Problem of Extraterritorial Jurisdiction: Economic Sovereignty and the Search for a Solution," 34 Int. & Comp. L.Q. 725, 1985.

Mahmood, Ahmed, "Conflict Management in the RADARSAT Data Acquisition Planning Process," *Proceedings of the 26th International Symposium on Remote Sensing of Environment and 18th Annual Symposium of the Canadian Remote Sensing So-

ciety, March 25–29, 1996, Vancouver, B.C., pp. 15–18.

Mahmood, Ahmed, et al., "Potential Use of Radarsat in Geological Remote Sensing," *Proceedings of the Eleventh Thematic Conference Geologic Remote Sensing*, February 27–29, 1996, Las Vegas, Nevada, p. I-475–484.

—, "Mapping the World with Remote Sensing," *EOS*, Vol. 79, No. 2, January 13, 1998, p. 17.

—, "RADARSAT Data Applications Radar Backscatter of Granitic Facies, the Zaer Pluton," Morocco Journal of Geochemical Exploration, 66 (1999a), pp. 413–420.

—, "RADARSAT-1 Background Mission Data for Flood Monitoring," International Geoscience and Remote Sensing Symposium, June 28–July 2 1999b, Hamburg, Germany.

Manley, John, Speaking notes, Honourable John Manley, Minister of Industry, August 10, 1999. Available at http://www.space.gc.ca/whatsnew/releases/speeches/1999/990810.asp (last accessed May 23, 2000).

Martin, Thomas S., "National Security and the First Amendment: A Change in Perspective," *A.B.A.J.*, Vol. 68, June 1982, pp. 680, 683.

Massonnet, Didier, "Satellite Radar Interferometry," *Scientific American*, February 1997. Available at http://www.sciam.com/0297issue/0297massonnet.html (last accessed May 25, 2000).

Maier, H. G., "Extraterritorial Jurisdiction at a Crossroads: An Intersection Between Public and Private International Law," *Am. J. Int. L.*, Vol. 76, 1982, p. 280.

McGill University, Centre for Research of Air and Space Law, Space Surveillance for Arms Control and Verification Options, Proceedings of the Symposium held at McGill University, Montreal, October 1987.

Meredith, Pamela L., and Rachel B. Trinder, "Financing a Communications Satellite Venture: Assessing Regulatory Risk is Key," *Satellite Finance*, June 1998. Available at http://www.zsrlaw.com/publications/articles/rbt980611.htm (last accessed May 30, 2000)..

Merges, Robert P., et al., "News Media Satellites and the First Amendment: A Case Study in the Treatment of New Technologies," *High Technology Law Journal*, Vol. 1, 1989.

Myers, Richard, General, USAF, "Export Controls Can Backfire," Commentary, *Space News*, April 26, 1999, p. 20.

Pace, Scott, "Merchants and Guardians: Balancing U.S. Commercial Interests in Space Commerce," in *Balancing National Interests in Space Development*, Space Policy Institute, George Washington University, 1999. (Available as a RAND reprint, RP-787.)

Pircher, Marc, "L'Observation militaire spatiale, un multiplicateur de forces," *CNES Magazine*, No. 7, November 1999, pp. 12–18;

Pugliese, David, "Export Rules Worry Canada, Industry Study Warns U.S. Changes May Cost Canada Jobs," *Space News*, March 22, 1999.

RADARSAT International, "The Role of RADARSAT for Military Intelligence," unpublished document on file with the author.

Rains, Lon, et al., "Canada Might Drop Orbital as Builder of Radarsat 2 bus," *Space News*, August 2, 1999, p. 1.

Sallot, Jeff, and Andrew Mitrovica, "Beijing's Spies Stole Canadian Nuclear Secrets," *The Globe and Mail*, January 24, 2000, p. 1.

Sietzen, Frank, "New Defense Space Policy Carries Risk, Falls Short, Experts Say," *Space.com*, July 26, 1999. Available at www.space.com/space/milspace_update.html (last accessed May 23, 2000).

—, "Satellite Makers Say Export Controls Threatens Sales," *Space.com*, August 4, 1999. Available at http://www.space.com/businesstechnology/business/satellite_exports.html (last accessed May 23, 2000).

—, "Images Long-Sought by Spies Now Online," *Space.com*, January 19, 2000. Available at http://www.space.com/space/spy_pics_000118.html (last accessed May 23, 2000).

Singer, Jeremy, "U.S. Spending Plan for Boosting Spy Imagery Called Inadequate," *Space News*, Vol. 11, No. 18, May 8, 2000, p. 1.

Space News, "A Threat that is No Games," editorial, March 22, 1999, p. 12.

—, Commentary, January 17, 2000a.

Space Policy, "Foreign Access to Remote Sensing Capabilities," No. 10, 1994, p. 243.

SpaceViews, "Y2K Glitch Reexamined," January 17, 2000. Available at http://www.spaceviews.com/2000/0117/othernews.html (last accessed May 30, 2000).

Sung, Ellen, "White House Reverses Crypto Policy," *Policy.com*, May 11, 2000. Available at http://policy.com/news/dbrief/dbriefarc330.asp (last accessed May 30, 2000).

Travers, Jim, Editorial, *The Guardian* (Charlottetown), January 27, 2000.

United Nations, General Assembly, *Principles Relating to Remote Sensing of the Earth from Outer Space*, resolution 41/65, 1986. Available at http://www.un.or.at/OOSA/treat/rs/rstxt.html (last accessed May 31, 2000).

U.S. Department of State, Guidelines for Determining Licensing Requirements for Canada, no date. Available at http://www.pmdtc.org/Canguide.PDF (last accessed July 20, 2000).

—, Office of the Spokesman, Canada: International Traffic in Arms, press statement, July 27, 1999. Available at http://secretary.state.gov/www/briefings/statements/1999/ps990727a.html (last accessed May 30, 2000.)

USAF Scientific Advisory Board, *New World Vistas: Air and Space Power for the 21st Century, Space Applications Volume*, 1999.

The White House, Presidential Decision Directive 23 (PDD-23), March 10, 1994.

—, National Science and Technology Council, Fact Sheet National Policy, Commercial Space Guidelines, paragraph (2) September 19, 1996. Available at http://www2.whitehouse.gov/WH/EOP/OSTP/NSTC/html/fs/fs-5.html (last accessed May 25, 2000).

Section III: Remote Sensing Applications to International Problems

The usefulness of commercial observation satellites depends in large part on how well the imagery data they produce can be translated into information products that address pressing human needs. Section III presents several chapters that demonstrate the challenges and benefits of applying commercial and civilian imagery to a broad range of international problems. These case studies are good examples of how openly available imagery collected from space can be used to help settle territorial disputes, enhance regional transparency among rival states, monitor nuclear proliferation, and support humanitarian relief operations. Some chapters illustrate how governments can make use of commercial and civilian satellite imagery, while others focus on the growing interest of nongovernmental organizations and multinational agencies in taking advantage of these data to address specific international problems. Thus, Section III presents practical applications of commercial observation satellite imagery in an era of growing global transparency as both governments and nonstate actors use these data to produce the geospatial information required for dealing with human conflicts and natural disasters.

Although the scientific and military utility of satellite imagery is well known, commercial and civilian observation satellites also offer an unanticipated instrument for supporting peace negotiations aimed at resolving territorial disputes. In Chapter 14, Richard Johnson assesses the role that satellite imagery and three-dimensional visualization technologies played in helping U.S. negotiators at the 1995 Dayton peace talks to overcome complicated territorial disputes that were obstructing the final peace settlement for Bosnia. Chapter 15 by John Gates and John Weikel offers a case study on the U.S. contribution of civilian satellite imagery and mapping support in securing a peace agreement between Ecuador and Peru in 1998. This ended a long-standing territorial conflict that had periodically erupted into border fighting. Both cases demonstrate how highly accurate geospatial information and advanced visualization technologies were used to help the negotiations reach a successful diplomatic outcome.

The case study by Vipin Gupta and Adam Bernstein, presented in Chapter 16, explores the opportunities for using commercial satellite imagery to help mitigate the armed conflict arising over geographical flashpoints, such as the disputes over the Spratly Islands in the South China Sea. The authors assess the potential utility of commercial satellite imagery at both moderate and high resolutions for detecting and identifying politically provocative activities, such as constructing or significantly expanding military outposts, among the disputed islands and reefs in the South China Sea.

The final three case studies highlight the potential for nonstate actors to take advantage of high-resolution commercial and civilian satellite imagery in addressing international political concerns. David Albright and Corey Hinderstein investigate in Chapter 17 how satellite imagery data could be used in conjunction with other open sources to help understand what transpired at the nuclear test sites at which India and Pakistan conducted a series of nuclear detonations in 1998. In Chapter 18, Albright and Hinderstein demonstrate how nongovernmental experts on nuclear facilities can use commercial satellite imagery to detect and identify, despite highly restricted external access, suspicious facilities that could be part of a nuclear weapon program.

Finally, many nongovernmental organizations and multinational agencies are in the early stages of developing the capability and expertise to take advantage of commercial satellite imagery to support their international activities. In Chapter 19, Einar Bjorgo examines how civilian and commercial satellite imagery can be used to help aid organizations deal with humanitarian emergencies occurring around the world. His analysis reveals that high-resolution imagery is not always necessary or desirable for producing the type of information needed to support humanitarian relief operations.

14

Supporting the Dayton Peace Talks

Richard G. Johnson

This chapter focuses on the role played by digital maps, imagery, and related technologies in helping overcome the negotiating obstacles in the Dayton peace talks of late 1995. It documents the equipment and procedures used, and identifies findings and lessons for the future.

The Context for Negotiations at Dayton

From 1991 to 1995, the former country of Yugoslavia was wracked by fighting between the Bosnian Serb, Croat, and Muslim ethnic factions that took more than 300,000 lives. The period was marked by a number of unsuccessful "peacekeeping" and negotiating initiatives. International frustration peaked in July 1995 with revelations in the press about ethnic cleansing in Muslim enclaves that had been overrun by the Serbs, prompting the United States to launch an all-out negotiating effort to end the fighting. The conflict in Bosnia-Herzegovina had just become overripe for solution.

Former Ambassador Richard Holbrooke led the U.S. negotiating team that worked with and through the United Nations Contact Group for Bosnia-Herzegovina to bring representatives of the warring factions eventually to Wright-Patterson Air Force Base at Dayton, Ohio.[1] The road toward negotiations at Dayton was complex and involved a mix of aggressive and creative diplomacy as well as demonstrated willingness by the United States and NATO to use military force.

Intense negotiations took place at Dayton from November 1 to 21, 1995, resulting in an initialing ceremony for a peace accord that was formalized in Paris in mid-December. These negotiations were called "proximity peace talks" because they were conducted in the same fashion as the shuttle diplomacy that first brought the warring factions to the table. Because significant territorial and military defensibility issues had to be resolved at Dayton, maps and imagery were a significant (but not determinant) part of the currency of negotiation.

Digital technology had matured enough by late 1995 that the peace talks at Dayton marked the first significant appearance of digital maps and automated terrain visualization in diplomatic negotiations. *Digital map* (or *digital mapping*) is a generic term that includes automated cartography (the use of computers to assist in map creation), computer-assisted map tailoring (the use of computers to modify maps or adapt them to specific uses), spatial statistical analysis (the computation of distances, directions, areas, volumes, and other relationships), and geographic information systems (GISs, which support management and use of spatially referenced data for solving complex planning and management problems). All of these disciplines actively integrate imagery or depend heavily on information derived from imagery. Terrain visualization involves interpreting mission space in visual terms, particularly by generating interactive three-dimensional depictions. One of the most powerful forms of terrain visualization involves draping imagery over a three-dimensional elevation model so that roads, vegetated areas, and other features visible on imagery appear in their correct locations in realistic landscape views on computer screens.

A number of organizations provided mapping and terrain visualization support for the negotiations leading to Dayton, for the formal Proximity Peace Talks, and for operations subsequent to the peace accord. Ambassador Holbrooke's initial diplomatic shuttles were supported by a few people working primarily with paper maps and manual measuring devices. During September 1995, the U.S. Army Topographic Engineering Center (TEC) contributed several teams of analysts equipped with portable computer systems that could perform simple area computations and compose and print maps in limited quantities; these teams accompanied Ambassador Holbrooke and his team on several of their shuttle missions.

The mapping support team deployed to Dayton started small but grew, eventually involving 55 people and over $4 million worth of mission equipment. The Defense Mapping Agency (DMA, now a component of the National Imagery and Mapping Agency [NIMA][2]) contributed senior leadership (including the agency director, Maj Gen Philip Nuber), production management personnel, library and map distribution personnel, and a variety of production equipment with operators. The U.S. Army, through TEC and the 30th Engineer Battalion, contributed technical and managerial personnel, as well as digitizing stations, map composition workstations, and hard-copy output devices. Contractors from Cambridge Research Associates (developers of the PowerScene™ terrain visualization system), Camber Corporation, ERDAS, and 3-M Corporation (developers of the Remote Replication System) augmented the DMA and Army assets at Dayton.

Subsequent mapping support for implementing the peace accords came from DMA and from U.S.-based and forward-deployed topographic assets of the military services.

These organizations used the full range of map and imagery offerings from both military and commercial providers. This included paper products,[3] raster-scanned variants of some paper products, elevation data,[4] 10-m panchromatic Satellite Pour l'Observation de la Terre (SPOT) commercial imagery, and classified imagery from national systems. The Department of State obtained special permission to use classified imagery directly in negotiations with the warring parties, indirectly providing a unique opportunity also to evaluate the potential of future high-resolution commercial satellite imagery in support of diplomatic negotiations.

Major equipment items[5] used at Dayton for digital mapping support included digitizing tables and a large-format scanner for data input from hard copy, terrain visualization systems for fly-throughs, GISs for rapid computation of areas, and plotters for precise hard-copy output. These were supplemented by large-format color copiers for low-volume jobs and a replication system for medium-volume jobs.

The technology deployed to Dayton was on the cutting edge at the time, but still had significant limitations affecting support with maps and imagery. First and foremost, the deployed systems had been developed for a variety of different purposes, and establishing compatible interfaces was a serious integration problem. Large-format color printers did not yet have the right combination of speed and high resolution; the Canon Bubblejet™, although capable of sufficient resolution and internally a digital device itself, had not yet been released with a much-needed external digital interface; and the resolution of the large-format Hewlett-Packard 650C plotters was only around

300 dots per inch. Computational capabilities in the GIS packages were two-dimensional, precluding true surface-area computations. Terrain visualization systems treated vector lines as primarily visual rather than logical entities, meaning that lines drawn during fly-throughs would not work computationally in a GIS package. Workarounds ultimately got the support teams through most of these problems; advancing technology will eventually overcome these and others to simplify future support of diplomatic negotiations greatly.

The negotiators being supported were used to paper maps, the crisp appearance of printed detail, and the flexibility of drawing on their map copies where and when they wished. Their demands for what they perceived as "quality" of product and familiarity of support inhibited the pace of weaning to the better, faster, and more flexible support possible with the digital technology provided. Despite these inhibitors, digital mapping and automated terrain visualization became core tools, used wisely, that contributed significantly to the success of negotiations at Dayton.

The Dayton Experience

The mapping support strategy for Dayton was very straightforward and came directly from Ambassador Holbrooke and LTG Wesley Clark, the senior military member of the negotiating team. The strategy was to flood the negotiation site with U.S. military maps from DMA so that there would be no question about the authenticity, quality, or source of maps used as a foundation for discussions; to keep modified maps and recomputations of territorial areas coming as fast as the negotiators needed them; to help make sure that no contending side gained a marked territorial military advantage; and not to make any mistakes that might derail the talks.

Implementing that strategy involved a complex mix of traditional and digital mapping. Figure 14.1 illustrates the process flow used and refined at Dayton. Portions of this flow, as discussed subsequently, were never fully exercised.

Over 100,000 copies of printed DMA maps were shipped to Dayton and augmented by the modest on-site production of nearly 5,000 map sheets; of these, 30,000 were actually distributed and used. Information from these maps (such as the outer boundary of Bosnia-Herzegovina) or from other sources (such as battlefield maps from the front) could be digitized and correlated to a common geometric foundation. Digitized map information (points, lines, and areas in vector form), name data, elevation data, scanned map images, and imagery could be pulled into the PowerScene™ terrain visualization systems

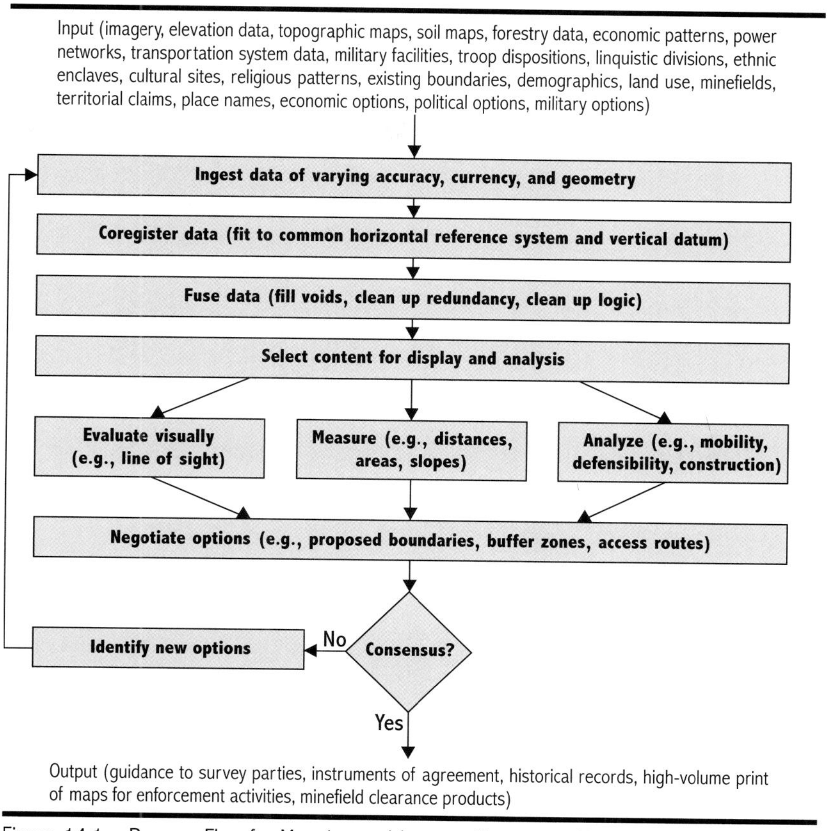

14 Supporting the Dayton Peace Talks

Input (imagery, elevation data, topographic maps, soil maps, forestry data, economic patterns, power networks, transportation system data, military facilities, troop dispositions, linguistic divisions, ethnic enclaves, cultural sites, religious patterns, existing boundaries, demographics, land use, minefields, territorial claims, place names, economic options, political options, military options)

↓

Ingest data of varying accuracy, currency, and geometry

↓

Coregister data (fit to common horizontal reference system and vertical datum)

↓

Fuse data (fill voids, clean up redundancy, clean up logic)

↓

Select content for display and analysis

↓

Evaluate visually (e.g., line of sight) | **Measure (e.g., distances, areas, slopes)** | **Analyze (e.g., mobility, defensibility, construction)**

↓

Negotiate options (e.g., proposed boundaries, buffer zones, access routes)

↓

Consensus? — No → **Identify new options** (loop back)

Yes ↓

Output (guidance to survey parties, instruments of agreement, historical records, high-volume print of maps for enforcement activities, minefield clearance products)

Figure 14.1—Process Flow for Mapping and Imagery Support to Negotiations

and presented to negotiators as still screen shots, fly-through videos, or dynamic fly-throughs under joystick control. The PowerScene™ system also supported dynamic annotation and visual assists, such as flooding, slope computations, and intervisibility exploration. Figure 14.2 illustrates what a negotiator might see on the screen while "flying" across the terrain to study options; it shows a portion of Bosnia-Herzegovina with black-and-white imagery from the SPOT commercial system draped over an elevation matrix, place names, and one of the negotiated boundary lines flanked by the limits of the associated buffer zone.

Any realignments of real or negotiated boundaries between the factions could be reflected in automated recomputations of areas and in adjustments of buffer zones. These computations were particularly challenging at Dayton because negotiators were working within and were very sensitive to a formal

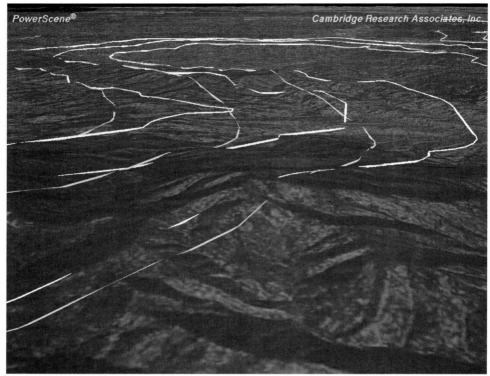

Figure 14.2—Perspective View Screen Shot on PowerScene™ (Courtesy Cambridge Research Associates)

1994 agreement to a 51-49 percent split of Bosnian territory between the Muslim-Croat Federation and the Bosnian Serbs. Further analyses, such as those for construction feasibility of the proposed corridor to the Muslim enclave of Gorazde, could be sent out to other supporting agencies and organizations.

Any map changes made at Dayton could be printed out for use in further negotiation or could be transferred electronically to DMA for use in printing maps for implementation forces, minefield clearance operations, boundary marking, or other follow-on activities. Figure 14.3 illustrates how the cease-fire and interentity boundaries negotiated at Dayton appeared in maps DMA printed from boundary tapes made in Dayton and then issued to the Implementation Force (IFOR) for use in Bosnia-Herzegovina.

The quality of map information and imagery used at Dayton was superb, although the master copy of the UN Protection Force (UNPROFOR) road map, a favorite for the negotiators because it provided "one over the country" coverage on a piece of paper that two people could work on over a coffee table, was found to have a 1-percent stretch in one direction that confounded almost all fixes. Level of detail probably had the single biggest impact. Finding one's own house depicted on a tactical map or in imagery at an air force base in a foreign country can be sobering, and knowing that anything on such a map or image can instantly become a target is even more unsettling. Seeing that same information indexed precisely in a three-dimensional terrain visualization system used for mission planning and rehearsal by combat aviators, with fly-throughs

Figure 14.3—Printed Map with Boundaries and Buffer Zones (Courtesy Cambridge Research Associates)

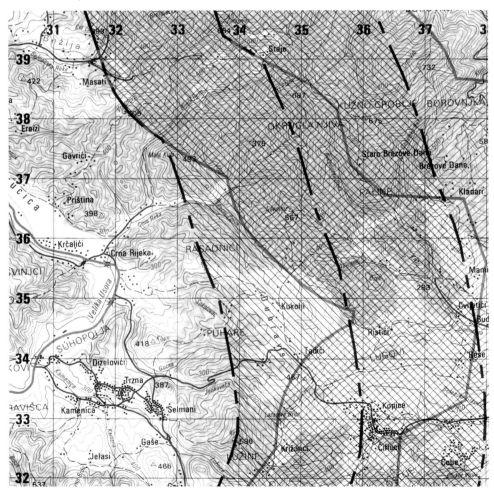

showing individual rocks and trees, ends all bravado. The first major contribution of digital mapping and imagery at Dayton, then, was putting a note of harsh reality, or intimidation, into the proceedings. In effect, the technology said: "We know about you; we've done damage before, and could do it again."

The second major contribution of digital mapping and automated terrain visualization at Dayton was flattery. Some of the most advanced equipment of the time had been brought to Dayton, with a hand-picked team from government and industry and with commitments of backup from DMA, the U.S. Army Corps of Engineers, and TEC, and was placed totally at the disposal of the negotiation effort around the clock, seven days a week, for however long the talks would take. In effect, this commitment of resources said: "This is the best equipment in the world, manned by the best team in the world, processing the finest imagery and information available. It can be an instrument of war, but we'll use it for peace because you are willing to come to the table."

A third major contribution of digital mapping and automated terrain visualization, and something not possible with any manual processes, was assurance of absolute consistency. Once information was entered and registered to a common geometry, it could be propagated uniformly across all support systems, would not change without intervention, and would not be affected by subsequent viewing or use. Negotiators could be assured: "Once you draw a line, it will stay in its correct place across the full range of scales, portrayals, or uses, such as PowerScene™, maps, computations, subsequent boundary surveys, or field use with Global Positioning System navigation."

A fourth major contribution of digital mapping and automated terrain visualization, and another not possible with analog procedures, was flexibility and responsiveness of support. The range of possibilities for support went from simple transcription of handwritten or drawn information from paper maps onto digitally rendered map displays or terrain fly-throughs to interactive adjustment of negotiation lines in response to guidance from negotiators. Terrain visualization scenes responded directly to joystick input from operators or negotiators. Buffer zones could be adjusted very rapidly when alignments changed, and terrain could then be inspected with the visualization systems to see how the changes affected military defensibility. Grades could be rapidly computed for proposed roadways. Operators could take a map with newly drawn lines from a negotiator, register and digitize the changes, and turn around precise new computations about territorial percentages within about ten minutes. Six minutes after that, under ideal conditions, a fresh transparency for the revised map could be flowing from a printer. Under the best conditions (everything working and no competing tasks), within a total elapsed time of 18 minutes from turning over the hand-drawn input or screen entries in the visualization system, the negotiator could have new computations, a fresh map showing all changes, and assurance that the changes would appear consistently in all subsequent uses. These capabilities said, in essence: "The support tools are so powerful, you can negotiate at your pace and not have to wait much for your revised maps."

The final major contribution of mapping and imagery support at Dayton may have been more evident to the mappers than to the negotiators. Many activities in mapping are tedious and require diligence and attentiveness that are difficult to maintain through interruptions and changing guidance. Digital mapping and automated terrain visualization delivered an incredible relief from drudgery. With it, operators could render the same objects at any scale. They could work on a foundation of either maps or imagery and could transfer everything they did directly to the other. They could work with extremes of detail, knowing that whatever they did would be rendered faithfully in all subsequent work. They could insert buffer zones around curving, irregular negotiated lines with assurance of accuracy whatever the geometry. They could enter only local changes to negotiation lines and recompute instantly for the whole area of conflict, saving great effort. The message of digital mapping was, in effect: "However time-consuming and trivial your task might seem, we can probably set up something that can do the task endlessly, without complaint, with confidence that no gross blunders will occur because somebody got tired."

Digital mapping support at Dayton was marked by some unrealized expectations, some capabilities that were present but never adequately used, and some tantalizing possibilities for which technology was not yet ready.

The mapping support team took more advanced capabilities to Dayton than the negotiators were ready or willing to use. For the most part, these capabilities would have required that negotiators change how they do their job; the constraints were more behavioral than technical.

For instance, TEC imported a great deal of cultural and economic data, information on political boundaries internal to Bosnia-Herzegovina, and other thematic data considered relevant to the peace talks and suitable for analysis and display by GISs. Negotiators, however, tended to focus on one thematic problem at a time and resolved the ethnic and political issues independently with their own information sources before turning to the supporting mappers later in the negotiations for thorny military defensibility issues.

The systems taken to Dayton had significant capability for preparing image-based maps; perhaps because of unfamiliarity with these types of products, negotiators consistently requested that negotiation lines be superimposed on traditional symbolized maps and did not want to hear alternatives.

It was clear from the beginning that the pace of negotiations could be greatly accelerated by getting negotiators to point to things they wanted moved on a computer screen and then flashing the results to other parties for comment and counterproposal. The technology to support this was available, but the process never got off the ground. Negotiators may have been intimidated by the technology, felt a loss of control and privacy by having to work with a computer operator not on their negotiating team, distrusted computers generally, or simply enjoyed the opportunities to consult at their own pace and deliver their proposals at the most propitious times. As a case in point, when President Slobodan Milosevic of the Serb Republic was engaged directly by the U.S. team on the night of November 16 over the access corridor to Gorazde, discussions started somewhat formally with U.S. operators of the PowerScene™ terrain visualization system showing the tactical limitations of a narrow access corridor and ended with the principals taking turns drawing with felt-tip pens on acetate over maps. The only simultaneous use representatives of the contending parties made of the PowerScene™ system in projection mode in the Command Center of the Air Force Materiel Command at Dayton occurred in a marathon adjustment session of the negotiated lines after the agreement was initialed on November 21, and the principals had departed.

Findings and Lessons for the Future

New technology often requires new behavior and will be met with resistance unless it is clearly superior to the old ways. Sometimes this resistance can be overcome by brute force, as was done at Dayton. The support team of 55 people was able to mount a continuous operation with two overlapping, 13-hour shifts that could stay ahead of negotiators who were often one deep and could not sustain much more than 18-hour work days. The technology was similarly brute-forced. The combination of visualization fly-throughs, transfer of decisions to paper maps and digitization back, and direct computation support on area percentages was forced together at Dayton; greatly aided the negotiations; and can now be considered a mature capability. The next wave of capability available to support negotiations may include three-dimensional GISs; true surficial and volumetric computations; higher-resolution

Figure 14.4—Factors in Choosing Technology for Mapping and Imagery Tasks

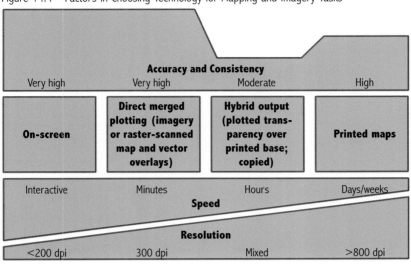

geospatial information and imagery from increasingly unclassified sources, such as new commercial imaging satellites; more-sophisticated and comprehensive meta-data (data about data) to simplify data fusion; Internetted support capabilities for advanced analyses; increased linkage between decision tools and navigation and positioning devices; greatly improved large-format printing devices with output approaching photographic quality; and perhaps even holography and immersive virtual reality.

The general work flow (Figure 14.1) worked out and used at Dayton functioned well, and should be considered for use in future negotiations. In each support task, however, providers of digital mapping and terrain visualization support should be sensitive to the negotiating styles of the people they are supporting, to the power and limitations of the technologies available, and to the time available for the task requested. Figure 14.4 summarizes some of the factors that need to be traded off in approaching negotiators who may be more open to interactive, on-screen approaches, with technology options.

Digital maps and imagery helped end the war in Bosnia-Herzegovina by permitting interactive fly-throughs of terrain on imagery and near-real-time hard-copy maps that faithfully replicated all decisions made on the imagery (for example, see Figure 14.5). The NIMA team supporting resolution of the border dispute between Peru and Ecuador duplicated this success in 1998 (see Chapter 15 in this volume). These successes may well lead to similar

Figure 14.5—Negotiated Lines and Buffer Zones, Gorazde Corridor, Bosnia-Herzegovina (see color plate)

support of future diplomatic negotiations as well as to dramatically increased demands for high-resolution, unclassified imagery from commercial sources. The lessons from the peace talks at Dayton, understood in context and applied judiciously, may help end future wars.

Summary

The Proximity Peace Talks at Dayton, Ohio, in November 1995 occurred at a time when the warring parties in the Balkans were ready to negotiate and close a deal. The U.S. Department of Defense established a digital mapping support structure at Wright Patterson Air Force Base that was, for its time, state of the art. Capabilities included automated cartography, computer-assisted map tailoring, spatial statistical analysis, and terrain visualization, all using databases of raster maps and national and commercial imagery. Negotiators took full advantage of the power of coregistering information from multiple sources, presenting detailed views of the areas of contention, analyzing intervisibility for its military implications, measuring and computing areas, and linking to other analytic support. Negotiators were less successful in using multifactor analysis in GISs, using image-based maps, interacting directly with information on computer screens, or collaborating electronically. The resolution limits of coupled printers provided great technical challenges, and negotiators were generally unwilling to adjust their negotiating style to take full advantage of available technology. Technological possibilities have improved considerably since the Dayton talks. Dramatically greater resolution of unclassified imagery is just around the corner from commercial remote sensing satellites; GISs are moving toward three-dimensional presentations and analyses; photographic-quality printing and plotting devices are proliferating; navigation and positioning technologies are increasingly being linked with digital maps and imagery; and quality grading of data will greatly simplify future data fusion. These technical advances will greatly enhance the credibility, speed, and transparency of digital mapping and imagery support to future peace negotiations.

Notes

1. Ambassador Holbrooke (1998, pp. 203–204) has noted that Wright-Patterson Air Force Base in Dayton, Ohio was selected to provide a site large enough to hold nine delegations (one for each Balkan country, the five contact group nations, and the European Union representative) that could also be sealed off from the press and other outsiders. The site had to be close enough to Washington so that senior administration officials could visit, yet far enough away from the television studios of New York and Washington to keep Balkan warlords from running off to the press whenever negotiations hit a snag.
2. NIMA was formed on October 1, 1996 from DMA, the Central Imagery Office, and elements of the Central Intelligence Agency, Defense Intelligence Agency, National Photographic Interpretation Center, and National Reconnaissance Office.
3. The primary paper products were the 1:600,000-scale UNPROFOR road map, the 1:250,000-scale joint operations graphic, and the 1:50,000-scale topographic line map.
4. Elevation data used during the negotiations were DMA's Digital Terrain Elevation Data Level 1, which has an elevation post spacing of 3 arc-seconds or about 90 m. Much higher-resolution elevation data were obtained of portions of Bosnia-Herzegovina subsequent to the peace agreement through an Interferometric Synthetic Aperture Radar for Elevation mission called Bosnia Eagle.
5. Equipment used at Dayton included two digitizing tables; two PowerScene™ terrain visualization systems, one of which was rigged for large-screen projection; three Sun workstations equipped with ARC/INFO™ GIS; four Digital Topographic Support

System—Multispectral Image Processor workstations equipped with ERDAS Imagine™ software; ten Hewlett-Packard 650C plotters; a Remote Replication System; three Canon Bubblejet™ A-1 copiers; one large-format Canon Bubblejet™ 2436 copier; and two more Canon Bubblejet™ copiers available off base for surge replication.

References

Holbrooke, Richard, *To End a War*, New York: Random House, Inc., 1998.

Johnson, Richard, "Drawing the Lines in Bosnia," *The Military Engineer*, Vol. 88, No. 576, 1996.

National Research Council, Rediscovering Geography Committee, *Rediscovering Geography: New Relevance for Science and Society*, Washington, D.C.: National Academy Press, 1997.

15 Imagery and Mapping Support to the Ecuador-Peru Peace Process

John Gates and John Weikel

On October 26, 1998, Peru and Ecuador signed an historic peace accord effectively ending the long-standing dispute over their 1,420-km border. The controversy began during the formation of these two republics in the early 1800s following the Spanish colonial period. Since then, there have been many attempts to negotiate a settlement and several armed conflicts—most notably in 1941, and more recently in 1981, 1991, and 1995. After the 1995 conflict, the Guarantor nations of the peace process established by the 1942 Rio Protocol (Argentina, Brazil, Chile, and the United States) facilitated diplomatic negotiations and provided for the establishment of the Military Observer Mission Ecuador-Peru (MOMEP), which oversaw the demilitarization of the conflict area. This chapter describes the technical support that the U.S. National Imagery and Mapping Agency (NIMA) provided to bring about a successful conclusion to the peace process.

Chronology

Luigi Einaudi, U.S. Special Envoy to the talks, has provided excellent insight to the diplomatic process during and around the negotiations. The following chronology was adapted from Einaudi (1999):

- 1809–1830: After 300 years of Spanish colonial rule, Peru and Ecuador become independent, but their common border is poorly defined.

- 1830–1938: Conflicting territorial claims remain unresolved despite frequent attempts to negotiate.
- 1941: Tensions escalate into war, and Peru quickly overruns southern Ecuador.
- January 29, 1942: The Rio Protocol is signed, with participation from Argentina, Brazil, Chile, and the United States as guarantor nations.
- 1942–1950: Ecuador and Peru demarcate 90 percent of their common border, but Ecuador halts the fieldwork.
- 1960: Ecuador formally abandons demarcation and declares the protocol "inexecutable."
- 1960–1995: Armed conflicts break out in 1981 and 1991 in undemarcated sections of the Cordillera del Condor.
- January 1995: There is a renewed outbreak in the Upper Cenepa River area.
- March–October 1995: MOMEP separates the combatants.
- January 1996–October 1998: Numerous rounds of diplomatic negotiations result in a comprehensive agreement on the border, navigation, commerce, and security.
- November 1999: The map signing ceremony is held in St. Louis, Missouri.

Historical Background

The modern history of the conflict began with the ten-day border war in July 1941.[1] Fighting broke out all along the de facto border, but Peru quickly gained the upper hand with its superior numbers, equipment, and preparation. Within two days, Peru overran portions of the southern Ecuadoran provinces of El Oro and Loja and threatened further advances. Hostilities waned, however, and the military forces slowly disengaged. With the war in Europe and the ensuing involvement in the Far East, solidarity was essential to the countries of the Western Hemisphere. Diplomatic pressure was brought to bear on both countries to find a definitive solution. On the margins of the January 1942 meeting of Foreign Ministers of American Republics in Rio de Janeiro, Brazil's Foreign Minister, Oswaldo Aranha, took the lead to broker a settlement, known as the "Protocol for Peace, Friendship and Borders between Peru and Ecuador," or the Rio Protocol. The 1942 Rio Protocol provided for withdrawal of Peruvian forces, allowed Ecuador free navigation of the Amazon River, and demarcated the border via a general geographic description of the 1,600-km border. The protocol was signed by the foreign ministers of Ecuador,

Peru, and the four guarantor nations (Argentina, Brazil, Chile, and the United States).

Later in 1942, an Ecuador-Peru mixed commission was formed, and fieldwork began to delimit the border and mark it with concrete pillars. The guarantor nations acted as mediators when the two sides could not agree. Brazil was most active in this vein, especially in the work of the Brazilian arbiter, Captain Braz Dias de Aguiar. The United States also provided invaluable assistance through our technical advisor, George McBride. For example, McBride requested aircraft from the U.S. Army Air Forces to take aerial photographs over isolated border areas. This effort was not without a price though: Two U.S. aircraft and 14 American airmen were lost in separate accidents during the 1943–1946 deployment.

In 1946, after completing the photographic campaign, the U.S. Army Air Forces finished a map of the previously unmapped area known as the Cordillera del Condor. This is an isolated and rugged area of triple-canopy jungle on the eastern flanks of the Andes Mountains—almost constantly cloud-covered or shrouded in mist. The border in this area is described in the 1942 Rio Protocol as running along the watershed line in the Cordillera del Condor between the Zamora and Santiago Rivers (see Figure 15.1). Unknown in 1942, but shown on the 1946 map, is the true extent of the Cenepa River, which intrudes into the Zamora-Santiago watershed. Figures 15.2 and 15.3 illustrate the actual configuration of the rivers in the area. In 1950, when Ecuador realized the implications of this geographic reality, it halted all work on the border and eventually declared the Rio Protocol "inexecutable" in 1960. The protocol had always been extremely controversial in Ecuador, so this decision had great popular appeal. Approximately 150 km of the border were left undemarcated and became the crucial point of argument and controversy in later conflicts and negotiations. Ultimately, both sides accepted the recommendation of the guarantors that the border in this area should not be significantly modified and best reflects the intentions of the Rio Protocol.

Following years of uneasy peace and sporadic minor incidents, armed conflicts broke out in 1981, 1991, and January 1995 in the Cordillera del Condor. In the latest conflict, at the end of five weeks of fighting, hundreds had been killed or wounded. Ecuador and Peru ended hostilities with the signing of a cease-fire agreement on February 14, 1995. The agreement called for the withdrawal of troops to the "Coangos" observation post for the Ecuadorans and to the "PV1" observation post for the Peruvians. It provided

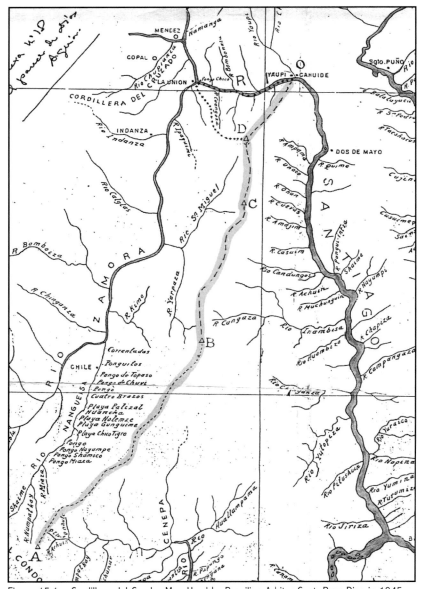

Figure 15.1—Cordillera del Condor Map Used by Brazilian Arbiter Capt. Braz Dias in 1945

specific latitude and longitude positions for the two locations and called for general demilitarization as a prerequisite for the permanent solution to the border dispute. The guarantor nations provided observers to report and verify implementation of and compliance with the cease-fire agreement.

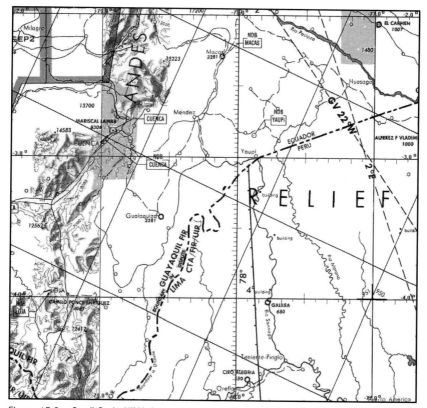
Figure 15.2—Small-Scale NIMA Aeronautical Chart (see color plate)

The renewed border fighting spurred the four guarantor nations, including the United States, to undertake diplomatic steps to defuse the hostilities and steer a course toward an overall solution to the border. The guarantors provided MOMEP to help separate the combatants and later to monitor the peace. The Chairman of the Joint Chiefs of Staff issued an order on March 11, 1995, for U.S. participation in the Guarantor Observer Group as the U.S. part in MOMEP. Operating from Patuca, a small Ecuadoran military base north of the conflict area, the group's mission was to oversee the separation of forces, demobilization, and vigilance of the demilitarized zone established in the upper Cenepa and Coangos river valleys.

Mapping Situation

Immediately after the 1995 conflict, the Commander in Chief of the U.S. Southern Command (USSOUTHCOM), was particularly concerned about the

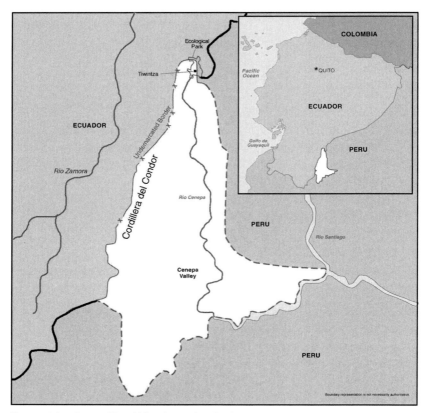

Figure 15.3—Cenepa River Valley (see color plate)

many void areas in the existing maps of the area. Defense Mapping Agency (DMA) standard products had limited coverage. No large-scale products were available, and small- and medium-scale maps had areas with no data or with severe limitations (see Figure 15.2). These gaps in map coverage were due primarily to the perennial cloud cover in the area that made complete collection of traditional imagery virtually impossible. This area is extremely rugged, covered by dense jungle vegetation, and is sparsely inhabited.

Existing Ecuadoran and Peruvian maps were the primary source of geospatial information. However, the large-scale maps also had significant void areas and were built on the 1956 Preliminary South American Datum, not on the World Geodetic System 1984 (WGS 84) datum, which is the current standard horizontal datum on all NIMA products.[2] Additionally, these products were not what our forces were accustomed to training with. In spite of their shortcomings, the maps Peru and Ecuador had produced were rescaled and merged together to provide continuous cross-border coverage. This provided a short-term solution to the mapping problem.

Early in 1995, the first requirements for developing new mapping products were spelled out. Medium-scale maps were needed as briefing graphics. As observer force deployment was considered, 1:50,000-scale topographic line maps (TLMs) were needed. When operating in specifically defined areas, the deployed forces used 1:25,000-scale TLMs. To support this operation, the DMA provided a special series of 1:50,000 scale TLMs to USSOUTHCOM that merged 30-m resolution Landsat imagery with existing TLMs to fill about 75 percent of the gaps. However, critical gaps remained.

In September 1995, the U.S. commander of the observer mission (Joint Task Force Safe Border) in Patuca, Ecuador, requested a common set of reference maps for MOMEP to coordinate tactical activities such as cease-fire verifications and separation of forces. These maps were to be devoid of politically sensitive border boundaries and marginal information. Existing border markers were to remain as unconnected points. Maps were to have no areas left uncharted, and the use of Landsat imagery was to be minimized, to provide the most complete topographic picture to the observer missions—particularly to the pilots who depended on the maps for flight operations.

An assessment of the available imagery source material of the area revealed that only 75 to 90 percent of the required areas were covered. DMA was limited to using native-edition maps from Peru and Ecuador, georectified Landsat imagery to fill gaps where possible, and DMA joint operational graphics (1:250,000 scale). Without Landsat, coverage, quality, and accuracy could not be improved unless new sources became available. The native maps were produced in the 1956 Provisional South America Horizontal Datum rather than WGS 84. Unfortunately, the geographic coordinates for the monuments established during the original border surveys in the 1940s were calculated by astronomical observations and traverses and were not tied to any modern datum.

Imagery and Mapping Support to the Negotiations

In November 1995, Canada launched a civilian synthetic aperture radar satellite called RADARSAT. The radar sensor has the unique advantage of being able to penetrate clouds, smoke, and all but the most severe weather. In an attempt to fill the voids in map coverage, DMA obtained test images to evaluate the satellite's usefulness. The test proved successful and led to the development, prototyping, and production of RADARSAT image maps over areas of ground never seen before. More than 30 images were taken of the Cordillera del Condor over the undemarcated border and scene of previous conflicts.

In the winter of 1996, in preparation for technical border negotiations to begin in April 1997, USSOUTHCOM J-5 (Plans and Policy Office) coordinated with the newly formed NIMA to produce a new suite of unclassified image maps. Hard-copy products included 1:100,000- and 1:50,000-scale RADARSAT image maps over the Cordillera del Condor and a large-format 1:150,000-scale Landsat image map of the entire Cordillera del Condor. The Landsat images over the area were nearly cloud free, but critical sections over the border were cloud covered.

In February 1997, NIMA, in separate testing unrelated to the MOMEP project, developed the capability to produce a digital elevation model (DEM) from stereo (overlapping) RADARSAT images. The DEM consisted of a matrix of elevation postings spaced approximately 100 m. This enabled NIMA to develop a three-dimensional fly-through of the Cordillera del Condor area using the DEM and RADARSAT imagery. NIMA used a deployable terrain visualization platform called PowerScene™, developed by Cambridge Research Associates, which had been used during the Bosnia negotiations in Dayton with great success (see Chapter 14 in this volume). It provided a realistic, three-dimensional visualization of the terrain for the negotiators to overfly and conceptualize possible border alignments. The PowerScene™ covered the Cordillera del Condor from approximately 4°50′ S northward to 2°50′ S. The entire scene was an average of approximately 40 km wide, and elevations ranged from approximately 200 to 3,200 m above sea level. The total area covered was approximately 10,000 km^2. The fly-through was enhanced with flags highlighting existing border monuments and significant geographical features.

Now that a capability existed to produce standard TLMs from RADARSAT imagery, USSOUTHCOM J-5 requested maps without coverage gaps in February 1997 for use in the upcoming negotiations. NIMA produced nine 1:100,000-scale sheets that straddled the border to standard specifications, except for the missing international border.

Additionally, the NIMA production facility in Panama produced hundreds of special maps to support the negotiations and MOMEP. These were largely one-of-a-kind maps and imagery products that were assembled on computers in the production facility and printed on large ink jet and/or electrostatic plotters. The quick turnaround times to produce these maps were critical for keeping pace with the rigorous schedule of the negotiations. Significantly, the only map in the final 1998 diplomatic agreement, which depicted the new border, was made in this facility.

The Negotiations

After long and arduous diplomatic negotiations that began in early 1996, the two countries exchanged six "impasses" or points of disagreement concerning the border. These impasses were then to be discussed in subsequent technical talks in Brasilia, which began in April 1997. This was the first time in more than four decades the two sides engaged in substantive direct dialog over their border. Given that this was the first round of technical talks, there was a high-level presence from all the participating nations, including the foreign ministers from Ecuador and Peru and special envoys of ambassadorial rank from the guarantor nations. The United States was represented throughout the negotiations by special envoy Ambassador Luigi Einaudi.

The high-level diplomats from the guarantor nations met with the heads of the Ecuadoran and Peruvian delegations to discuss ground rules and to "officialize" the order in which the impasses would be addressed—from easiest to most difficult. The six impasses were discussed in a series of four rounds of talks over the next six months, in the following order:

1. geodesic line intersections with river junctions and line in Napo River
2. the Lagartococha-Gueppi River area
3. the Zarumilla Canal
4. the area between the Cusumaza-Bumbuiza monument and the confluence of the Yaupi and Santiago rivers
5. the area between the Cunhuime Sur and 20 de Noviembre border monuments
6. access to the Marañon and Amazon rivers and the partial inexecutability of Rio Protocol.

It was during the series of talks on the impasses that the terrain visualization system, PowerScene™, was sent from the NIMA Outreach Office in Washington, D.C., to Brasilia. The data set included 32 RADARSAT scenes of the entire Cordillera del Condor area overlaid on the DEM generated from RADARSAT (approximate 100-m post spacing—see Figure 15.4; some areas in the mosaic were previously unmapped because of perpetual cloud cover). During the talks, the participants were allowed to "fly" through the contested area using the computer simulation and to generate hard-copy screen prints (see Figure 15.5). This proved to be an important, but not decisive, tool for the

Figure 15.4—Georectified Mosaic of RADARSAT Images in the Cordillera del Condor

Ecuadoran and Peruvian negotiators. The chief Ecuadoran negotiator, Edgar Teran, was lavish in his praise of the terrain visualization system and made the following comment in his closing statement: "The computers, the radar and the processes of virtual reality mark an abysmal difference between what was in the past and what awaits us in the future."

The radar imagery did provide an unobstructed view of the Cordillera del Condor. The imagery and resulting DEM ended any doubts as to the true "lay of the land," but this new information did not effectively change the maximalist and uncompromising negotiating tactics of the two parties at this juncture in the talks. However, it was an essential tool for the guarantor nations to assess the technical evidence the two parties presented and as an additional source of geographic information when called upon to recommend the border configuration in this area.

On November 26, 1997, Peru and Ecuador signed the Brasilia Declaration, committing each to examine in good faith a four-part proposal. On January 19, 1998, Ecuador, Peru, and the four guarantor nations signed the work plan that formalized the Brasilia Declaration. It contained brief descriptions of the work of binational commissions that were to be formed to deal with each of four proposals:

1. Treaty on Commerce and Navigation. This addressed Ecuador's desire for access to the Marañon and Amazon rivers.
2. Border Integration. The principal thrust of this commission was developing cross-border projects, such as roads, water control structures, parks, and ecological reserves. Key to the work of this commission was obtaining financing from international finance and lending institutions.
3. Border Placement. The mission of this commission was the "long pole in the tent." It was the most critical, sensitive, and complex of the four. This commission began its work in Brasilia by defining the areas to be studied by two work groups. Each group was made up of geographers and jurists from the guarantor nations and diplomats from Ecuador and Peru. One group addressed the commission's questions in the Cordillera del Condor, while the other worked on Lagartococha. The work groups responded with nonbinding "opinions." The commission accepted the opinions, then indicated the border monument locations on

Figure 15.5—PowerScene Screen Capture of the Northern End of the Cordillera del Condor (see color plate)

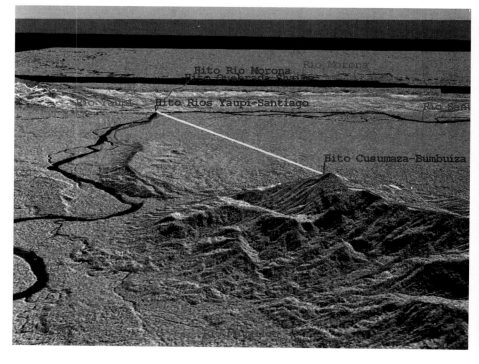

a map with the help of the work groups. The fieldwork of erecting and formalizing (measuring with the Global Positioning System [GPS] and documenting) the monuments was to be done by Ecuadoran and Peruvian teams of technicians. No aspect of the work plan was validated until the last monument was placed and accepted.

4. Mutual Confidence and Security. The establishment of a binational commission to promote mutual confidence building and security.

After renewed tensions in 1998, all remaining impasses were resolved through intense and creative diplomatic negotiations. On October 26, 1998, Peru and Ecuador signed the peace accord in Brasilia. The treaty sets the boundary in the contested sections of the border in the Cordillera del Condor and Lagartococha. Under the pact, the disputed section of the border follows the watershed line of the Cordillera del Condor mountain range, as Peru had put forward, but makes some allowances for Ecuadoran demands in the northern end of the range in the vicinity of the confluence of the Yaupi and Santiago Rivers. Additionally, 1 km^2 of Peruvian territory surrounding a former Ecuadoran outpost is granted to Ecuador's government as private property, even though it will remain under Peruvian sovereignty. Tiwintza has special significance for Ecuador because its troops successfully defended it

Figure 15.6—Members of Ecuador-Peru Border Commission and a Border Monument

against repeated assaults by Peruvian troops during the 1995 border war. The Peruvian government has 30 months to build a road from Tiwintza to the Ecuador border. The agreement also calls for two contiguous national "peace parks" on either side of the border in upper Cenepa River valley. At a combined 55 km², the new park encompasses the conflict area and will be administered binationally as an ecological reserve.

Border Demarcation

Technicians from the Ecuador-Peru border commission placed 28 border points in the Cordillera del Condor conflict area during the first five months of 1999. Since the terrain was difficult, covered by dense jungle vegetation, and potentially mined, engineers of the two countries rappelled from helicopters to the monument sites to clear and demine a 50-m radius. The helicopters could then land safely and bring in the materials to construct the monuments (see Figure 16.6). Precise geographic positions of the monuments were determined using differential GPS measurements from reference stations in the area. Sadly, seven Ecuadorans died when their Super Puma helicopter exploded and crashed over the Cordillera del Condor near the "Condor Mirador" outpost while on a flight supporting the demarcation.

Since aerial photography was still the preferred imagery for the area, Peru placed men on the ground to watch the weather. Whenever the weather cleared, they called in a Peruvian Learjet or an Ecuadoran Cessna to take photographs with standard mapping cameras and panchromatic film.

On May 13, 1999, at the Puesto Cahuide monument, the leaders of Peru and Ecuador formally ended the decades-old border dispute, sealing the peace treaty. At this jungle border spot, President Alberto K. Fujimori of Peru and President Jamil Mahuad of Ecuador dedicated an orange-painted boundary stone, the last to be laid on the disputed stretch of Amazon jungle. "We are putting an end to disputes," President Mahuad said, "closing wounds to start a new, healthy life."

Final Mapping

On June 7, 1999, the U.S. Department of State asked NIMA to participate in the final official mapping of the newly defined Peru-Ecuador border. Both countries had sent official letters of request to the Department of State asking for NIMA support. The Department of State endorsed this request for the geospatial and technical support.

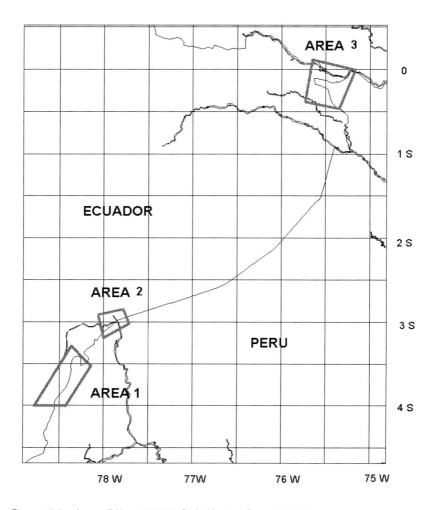

Figure 15.7—Areas of New 1:25,000-Scale Mapping Done in 1999

NIMA was to map and plot the new border monuments that were part of the demarcation agreement between the two countries. Only three areas in dispute prior to the agreement required NIMA assistance (see Figure 15.7):
1. Area 1 consisted of five map sheets that NIMA would compile from scratch at 1:25,000 scale.
2. Area 2 was compiled by Ecuador. NIMA was to add the border monuments to the 1:25,000-scale compilation.
3. Area 3 was compiled by Peru. This also only needed addition of the border monuments to the 1:25,000-scale compilation.

The project officially began with the arrival of the two delegations at NIMA in St. Louis on August 30, 1999. Ecuador and Peru supplied all the source materials and guidance for the project. Delegations from each country reviewed the work and monument placement as the project progressed. A map signing ceremony took place October 19, 1999, at NIMA in St. Louis. Delegations from Ecuador and Peru attended, together with senior officials from NIMA.

Ultimately, the border dispute was rooted in political differences and was settled through a Herculean diplomatic effort. A key element in the successful conclusion to the negotiations was the availability of accurate, complete, and unbiased imagery and mapping products. Emerging commercial radar-imaging technologies peered through the clouds and helped fill the gaps on the deficient topographic maps that existed before the 1995 conflict. The resulting imagery products and improved topographic maps proved indispensable for the final border determination. Compiling larger-scale maps that accurately depict the border will be crucial for the demining operations and the administration of these newly pacified areas.

Notes

1. One of most comprehensive English-language overviews of the history of the border conflict is Krieg (1986). Another significant contribution was McBride (1949). George McBride served as U.S. technical advisor from 1942 to 1948, leading to a unique insight into the history of the conflict and the demarcation process. This report has never been published formally but has been available in a Spanish translation.
2. WGS 84 datum is the standard horizontal frame of reference for defining latitude and longitude on all NIMA products. Included in the datum are parameters that define the size and shape of the earth, the points of origin of coordinate systems, and the gravity constant of the earth. Geographic coordinates derived from the Preliminary South American Datum 1956, a local horizontal datum covering the northern portion of South America, can differ from WGS 84 coordinates by several hundred meters. The vertical datum in both cases is mean sea level.

References

Einaudi, Luigi, "The Ecuador-Peru Peace Process," in Chester Crocker et al., eds., *Herding Cats: Multiparty Mediation in a Complex World*, Washington, D.C.: United States Institute of Peace Press, 1999, pp. 407–429.

Krieg, William L., *Ecuadorian-Peruvian Rivalry in the Upper Amazon*, 2nd ed., Washington: Department of State External Research Program, 1986.

McBride, George, *Ecuador-Peru Boundary Settlement*, report to the Secretary of State by United States Technical Advisor, unpublished, 1949.

16
Keeping an Eye on the Islands: Cooperative Remote Monitoring in the South China Sea

Vipin Gupta and Adam Bernstein

The Spratly Islands in the South China Sea have long been a source of conflict and ill will among states in the region. Brunei, China, Malaysia, the Philippines, Taiwan, and Vietnam have argued for decades about sovereignty over the area's islands, islets, reefs, and waters. China, Taiwan, and Vietnam each claim ownership of the entire archipelago, while Brunei, Malaysia, and the Philippines lay claim to parts of it (CIA, 1999). Some of these states have had serious naval skirmishes in and around the South China Sea. More often, these countries rely on less-direct but still provocative means of controlling islands of interest. The methods range from occupying the islands and constructing landing strips, markers, and other structures to more legalistic stratagems, such as selling exploratory oil drilling rights, releasing maps indicating ownership, and announcing plans to develop certain islands.

Much of the conflict stems from the ambiguity of the claims. Countries often physically occupy or build structures on the islands without ever explicitly laying legal claim to the island or reef in question. This appears to be an attempt to establish a de facto presence, which the countries hope may become de jure over time (Snyder, 1996, p. 8). Such tactics raise tensions when competitors discover the unannounced occupations and have provoked naval incidents, diplomatic protests, and the mass arrest of fishermen operating in disputed areas.

Unilateral efforts to patrol or otherwise locally monitor the islands are sporadic and are made difficult by the sparse distribution of the islands, bad weather, and the limited naval strength that most of the countries in the area have at their disposal. Multilateral cooperation is largely confined to Indonesia-sponsored annual workshops on managing conflict in the South China Sea. The workshops, which began in 1990, aim to ease tensions and increase cooperation among the contesting states. In the last several years, attendees have formed working groups devoted to scientific research, resource development, environmental protection, legal issues, and navigational and communication safety (Thomas and Dzurek, 1996, p. 301). The workshops have been useful in sustaining active communication between the parties but have not yet resolved the dispute or prevented continued unilateral attempts to take control of islands.

In this context, in which diplomatic efforts move slowly while unilateral occupation and development of the islands continues apace, it is worth considering new ways of monitoring this conflict. Sensor technologies, if used to observe the South China Sea systematically and nonintrusively, might provide the necessary technical underpinnings for treaties or agreed codes of conduct that could defuse the multilateral conflict. While not alone the guarantor of success, nonintrusive sensing technologies can at least provide the technical basis for pursuing such initiatives.

One monitoring approach that has been suggested for use in the South China Sea is overhead imaging (Snyder, 1996, p. 17; Dobson and Fravel, 1997). If changes could be seen remotely on islands and if the information could readily be shared between the interested countries, the naval skirmishes and reprisals that have occurred in the past might be avoided or might become less frequent. The states' knowledge that the region is under rigorous, multilateral remote surveillance could in itself discourage further unilateral development or takeover of islands.

This chapter explores the technical feasibility and usefulness of cooperative monitoring of the Spratly Islands with aerial and commercial satellite images. The chapter begins with a short résumé of the claims and conflicts in the territory, then discusses how we selected islands for analysis based on their political significance and the availability of archived overhead imagery. We analyze the images to determine which features can be identified and examine the technical constraints placed on the regime by the small size and widespread distribution of the islands. The chapter concludes with an assessment of both the risks and benefits of remote sensing in the context of the sovereignty disputes in the region.

Claims and Conflicts in the Disputed Territory

The Spratly Island chain (Figure 16.1) consists of about 100 islands, reefs, and sea mounts, with a total land area of less than 5 km^2. The small islands are scattered across approximately 800,000 km^2 of ocean. Fish and guano (used for fertilizer) are the principal natural resources (CIA, 1999). Some studies claim that the region has large oil and gas reserves, although the predictions vary widely and appear to be based on few hard data.[1] Despite the uncertainty of the estimates, most of the countries in the region are interested in the Spratlys because of their potential for oil and natural gas production. The islands' geostrategic value is significant because of their proximity to major shipping lanes and to the main territories of the Philippines, Brunei, Vietnam, and Malaysia. In addition, perhaps due to suspicions and

Figure 16.1—Islands, Reefs, and Shoals in the Spratly Island Region (see color plate)

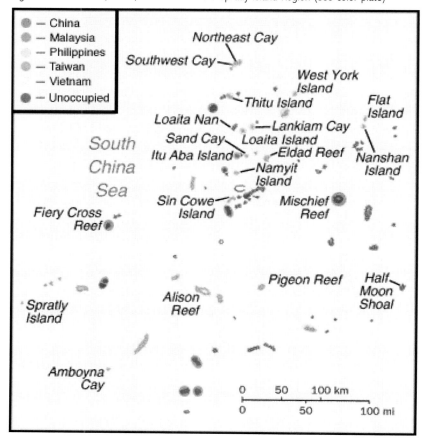

nationalist sentiments raised by decades of low-level hostilities, the islands have acquired an import beyond their current economic and strategic value, turning them into emblems of sovereignty to many states in the region (Valencia, 1997).

Actual claims to the Spratlys are variously based on historical evidence, existing occupation, and limits defined by treaties and unilateral decrees. China and Taiwan claim all features in the Spratly Islands, both above and below sea level. China occupies or has placed markers on seven to ten reefs in the Spratlys, while Taiwan occupies one of the largest islands, known as Taiping or Itu Aba. Vietnam claims everything above sea level in the Spratlys as its own; it has outposts on about 23 islands and reefs. The Philippines occupies eight islands or reefs in an area it refers to as the Kalayaan region, or "Freedomland." Malaysia claims all six islands within a continental-shelf limit defined in 1979 and occupies three of these; the other three are occupied by Vietnam and the Philippines. Brunei's claim is the most conservative, restricted to the seas surrounding Louisa Reef, apparently not even including the reef itself (Thomas and Dzurek, 1996, p. 306).

Over the past few decades, China, Malaysia, the Philippines, Taiwan, and Vietnam have all fought over sites in contested areas. Some of the salient disputes have occurred within the past ten years. The recent incidents show distinct patterns of confrontation in 1988, early 1995, and late 1998 separated by periods of tense calm:

- In 1988, China began construction on Fiery Cross Reef, also claimed by Vietnam. Vietnam then engaged in a battle with China near the reef, in which three Vietnamese ships were sunk and 72 to 75 Vietnamese were killed (*Christian Science Monitor*, 1988).
- In 1992, China announced an oil-exploration deal near the Spratlys with a U.S. company, Crestone. In 1994, Vietnam built a drilling rig in the Crestone area, even as China claimed sovereignty over Vietnam's nearby "Blue Dragon" exploration area (Thomas and Dzurek, 1996, p. 307).
- In early 1995, China began constructing several buildings on Mischief Reef that the Philippines described as "military structures." The Philippines reacted by removing Chinese markers from the island and bringing journalists to the reef to observe China's activities (Branigin, 1995).
- In March 1995, the Philippine navy arrested Chinese fisherman near Alicia Reef, which both the Philippines and China claim (Pastor, 1995). Filipino

forces also arrested 62 Chinese fishermen at Half Moon Shoal (Botbol, 1995).
- In late March 1995, Taiwanese forces fired warning shots at a Vietnamese cargo vessel that violated a Taiwan-declared exclusion zone around Itu Aba, an island it occupies in the Spratlys (Lin, 1977, p. 2). In the same week, Filipino forces fired warning shots at Chinese fishing boats that came within one mile of its occupied position on Thitu Island (Botbol, 1995, p. 78).
- On September 8, 1998, China formally protested Vietnam's sudden occupation of Orleana Shoal and Kingston Shoal. China claims sovereignty over both of these shoals (*International Boundary Monitor*, 1998).
- On November 6, 1998, the Philippines released a public statement calling for China to terminate intensified construction activities at Mischief Reef. The Philippines accused China of building more-permanent, fortified structures on Philippine territory (Philippines Department of Foreign Affairs, 1998). Chinese officials claimed that they were merely repairing fishing shelters that were damaged by storms (Associated Press, 1998; Aguinaldo, 1998).

Islands Selected for Analysis

We chose specific islands for image analysis through an iterative process. After selecting a set of islands for study based on their political significance and topographic features, we searched for suitable commercial aerial and satellite images of the islands.

The searches of the satellite archives concentrated on high-resolution (<10 m), panchromatic imagery. Since published aerial photographs and media reports indicated that island activities and infrastructures were small-scale and rudimentary, the imagery with the highest spatial detail had the best chance of providing observable evidence of island occupations. The need for high spatial resolution focused the data search on four satellite archives: declassified U.S. CORONA images, declassified Russian KVR-1000 images, Indian IRS-1C and IRS-1D images, and Canadian RADARSAT-1 images.[2] We later ordered IKONOS imagery collections of two Spratly locations.

Table 16.1 lists the set of islands selected for detailed examination; the initial group consisted of Mischief Reef, Itu Aba, Southwest Cay, Northeast Cay, Fiery Cross Reef, and Spratly Island. The table gives the geographic location of each island and explains why it was selected for image analysis. Every one

Table 16.1—Selection of Islands in the South China Sea for Detailed Image Analysis

Island and Location	Occupants	Claimants	Significance
Mischief Reef 9°55′ N, 115°32′ E	China	China Philippines Taiwan Vietnam	China took control of the reef in early 1995, triggering an ASEAN protest of the surprise occupation. Aerial photographs indicate that China constructed several octagonal structures on the reef. China added larger, more permanent structures around 1998.
Thitu Island 11°03′ N, 114°17′ E	Philippines	Philippines	Thitu Island is the largest Philippine-occupied island. The island reportedly has an airstrip, lighthouse, and power plant (Chung, 1994, p. 15). China, Taiwan, and Vietnam also have claims to the island.
Commodore Reef 8°21′ N, 115°14′ E	Philippines	China Malaysia Philippines Vietnam	Located 170 km SSW of Mischief Reef.
Alicoa Annie[1] 9°22′ N, 115°27′ E	None known	China Philippines	Located 56 km SSW of Chinese-occupied Mischief Reef, this reef was near the site of a Philippine navy confrontation with Chinese fisherman in March 1995.
Yuan Anha 8°08′ N, 114°40′ E	None known		Located 80 km WSW of Commodore Reef, Yuan Anha is comparable in size. Consequently, it could become a target for occupation in the future.
Itu Aba 10°23′ N, 114°21′ E	Taiwan	China Taiwan Philippines Vietnam	This island is one of the larger islands in the area. It reportedly has a marine garrison, helicopter pad, meteorological center, and power plant. A communications facility is reportedly under construction, and a 1,800-m airstrip is apparently under consideration (Lin, 1997, pp. 324, 330).
Southwest Cay 11°26′ N, 114°20′ E	Vietnam		Occupied by different countries, these two islands are only 4 km apart and illustrate how two countries that do not border each other can create a potential front line in the South China Sea.
Northeast Cay 11°27′ N, 114°21′ E	Philippines		
Fiery Cross Reef 9°40′ N, 113°02′ E	China	China	China began construction on this reef in early March 1988. Vietnam tried to disrupt this construction, which resulted in an armed conflict. The reef reportedly holds a large Chinese observation post, along with a helicopter pad and pier (Valencia, 1996, p. 39).
Spratly Island 8°39′ N, 111°55′ E	Vietnam	China Taiwan Vietnam	Vietnam has reportedly built a lighthouse, 600-m airstrip, harbor, power plant, and radio station (Chung, 1994, Table 6; Hanoi Voice of Vietnam Network, 1996, p. 70).
Subi Reef 10°54′ N, 114°06′ E	China	China	Located approximately 26 km SW of Thitu, this island reportedly has a blockhouse, supply platform, helicopter parking apron, and 100-W HTIW412 radio communicator. A Chinese garrison force reportedly occupies the reef (Contreras et al., 1998).

[1] The island was selected for final analysis. The others were considered but rejected for such reasons as lack of suitable imagery.

of these was either a significant base of operations or a flashpoint with a documented history of threats and conflict between two or more countries. Our subsequent search found high-resolution satellite images for only two of the six selected islands: Mischief Reef and Thitu Island. Aerial imagery was found for four of the six selected islands, but only 12 photographs were publicly available. This result was quite revealing. It indicated that declassified and commercial satellite imagery have not been used on a sustained, systematic basis to monitor the principal conflict areas in the South China Sea. It also indicated that aerial imagery of these same areas was either limited or generally kept secret by the respective conflicting parties.

As a result, the selected set of islands was revised to accommodate the limited archive of suitable images of the study area. The final group of islands analyzed with aerial and commercial satellite imagery consisted of Mischief Reef, Thitu Island, Subi Reef, Commodore Reef, Alicoa Annie, and Yuan Anha. Each island was selected by considering its relative size and significance and the availability of archived images. Preference was given to islands imaged by commercial satellite platforms because this particular technology has not been tested for South China Sea monitoring.

Analysis of Aerial and Satellite Imagery

We analyzed images of each of the final set of islands for infrastructure, ongoing activities, and natural phenomena, searching for manmade and natural changes, as well as for evidence that could confirm or refute accounts in published articles about specific islands. The image analysis enabled us to assess the technical capabilities and limitations of the various imaging systems used and to gauge the likely utility of imaging satellites planned to be deployed within the next few years.

MISCHIEF REEF

Aerial Imagery. In early 1995, the appropriately named Mischief Reef emerged as a major flashpoint in the South China Sea when China began constructing and occupying buildings on the reef. Filipino fishermen reportedly discovered the Chinese occupation, were detained by Chinese personnel at Mischief Reef, and notified Philippine authorities after being released (Branigin, 1995). Philippine naval and air force reconnaissance confirmed that the Chinese had taken control of the reef, which is located 215 km west of undisputed Philippine territory (Reuters, 1995). The Philippine government perceived the Chinese action as a direct challenge to its own sovereignty claim over Mischief Reef.

The Philippine Department of National Defense reported that several Chinese naval ships were sighted in close proximity to the reef, including a Yukan-class amphibious warfare ship and a Dazhi-class submarine support ship (Branigin, 1995).[3] As evidence, the Philippines released aerial photographs showing Chinese activities in the area. Figure 16.2 is an aerial photograph of an armed Chinese ship that was reportedly at Mischief Reef in early 1995 (note that the annotations on all images are based on our image interpretation, unless otherwise noted). The spatial resolution of the image is 0.2 m. Our analysis revealed this to be a Chinese Yannan-class ship, an armed survey and research vessel. The inset is a reference image of a Chinese Yannan-class ship from Sharpe (1995, p. 140). The connected circles show the features in the overhead image that match the reference image. Because of the overhead view, neither the flag nor the identification marks along the bow could be seen. Figure 16.3 presents aerial photographs of two armed ships that the Philippines identified as Chinese. In the large photograph, the flag is not identifiable, but the ship's pennant number ("420") is legible. The class is

Figure 16.2—February 1, 1995, Philippine Aerial Photographs Showing a Ship Reportedly Located at Mischief Reef (overhead aerial image courtesy Agence France-Presse; inset reproduced with permission from Jane's Information Group)

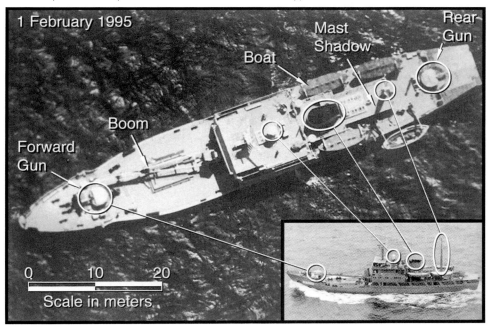

Figure 16.3—Spring 1997 Philippine Aerial Photographs Showing Two Ships Reportedly Photographed in the South China Sea and Identified as Chinese (courtesy Agence France-Presse)

not known for certain, although the ship may be a variant of Yenlai-class. The inset shows a Chinese Yannan-class ship with a pennant number of "24" or "124" (see the reference image of a Yannan-class ship in the inset of Figure 16.2). Both ships were photographed in spring 1997 near Philippine-occupied areas in the South China Sea.

In addition to the ship activity, the Philippines presented aerial photographs of four permanent building clusters on Mischief Reef that had been built by China (Branigin, 1995). Each cluster was constructed on top of steel pylons because the reef submerges at high tide. Figures 16.4 and 16.5 show two of the four building clusters in great detail. Acquired a few months after China took control of the reef in 1995, the photographs show the personnel at the site carefully watching the observation aircraft as it flies by. Both photographs also show the deployment of armor shields along the periphery of the building clusters. The inset of Figure 16.4 shows a possibly armed individual behind one of the armor shields, facing the observation aircraft.

The importance of high-resolution imagery was concretely demonstrated in late 1998, when the Philippine government announced renewed Chinese construction on Mischief Reef. The Philippines accused China of a substantial military buildup. China responded that the activity merely involved the repair of existing structures damaged by heavy storms. The conflicting accounts

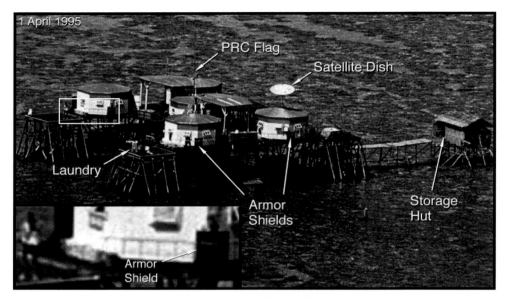

Figure 16.4—Aerial Photograph, Taken a Few Months After China Took Control of Mischief Reef, Showing One of Four Building Clusters China Constructed and Occupied (courtesy Agence France-Presse) (see color plate)

Figure 16.5—Aerial Photograph Showing Another Chinese Building Cluster on Mischief Reef (courtesy Agence France-Presse) (see color plate)

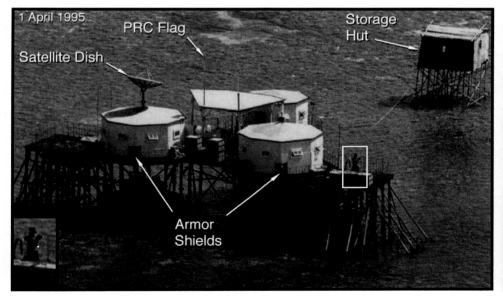

could have lingered on as an unsubstantiated factual dispute, but high-resolution photographic evidence helped to clarify the type of activity that was taking place.

Philippine aerial reconnaissance photographs show significant construction activity around the existing building clusters in 1998. Figure 16.6 shows approximately 40 construction workers laying out the skeletal frame for a new, large concrete structure. The Figure 16.6 inset shows the five-story concrete structure nearly completed, just two months later. Figure 16.7 shows approximately 65 workers building a truss structure adjacent to the existing platforms. The photograph also shows shrouds covering portions of the platform and a probable antiaircraft or antiship system surrounded by sandbags.

The Philippines also released aerial photographs of Chinese naval ships that supported the construction operation. Figure 16.8 shows two Chinese amphibious warships at Mischief Reef on November 7, 1998. Both are large, Yuting-class ships equipped with bow doors and helicopter decks.

Figure 16.9 shows three Chinese naval ships positioned next to each other at Mischief Reef on December 10, 1998. Our analysis of this photograph revealed that the ship in the middle is a Dayun-class support ship with Yuting-class ships on either side of it. These ships are armed and are designed for amphibious warfare. With a full load displacement of 4,800 tons and a length

Figure 16.6—Aerial Photograph of Construction at a Chinese-Occupied Area on Mischief Reef (courtesy Agence France-Presse) (see color plate)

Figure 16.7—Aerial Photograph Showing New Construction Activity at a Second Chinese-Occupied Area on Mischief Reef (courtesy Agence France-Presse) (see color plate)

Figure 16.8—November 7, 1998, Philippine Aerial Photograph Showing Two Ships Reportedly Anchored at Mischief Reef (aerial image courtesy Agence France-Presse; inset inset reproduced with permission from Jane's Information Group)

of 130 m, each ship can carry 250 troops and 10 tanks (Sharpe, 1997, p. 129). The inset is a reference image of a Chinese Yuting-class ship (Sharpe, 1995, p. 131). Interestingly, the ship in the foreground appears to be exactly the

16 Keeping an Eye on the Islands: Cooperative Remote Monitoring in the South China Sea

Figure 16.9— December 10, 1998, Philippine Aerial Photograph Showing Three Ships Reportedly Anchored at Mischief Reef (aerial image courtesy Agence France-Presse; inset courtesy Ships of the World)

same as the one shown in the reference image—both are marked with pennant number 991. The ship in the background is marked with pennant number 934. Ships 991 and 934 were the first and second Yuting-class ships to be commissioned into the Chinese PLA navy (on September 1992 and September 1995, respectively). Both were built at the Zhonghua Shipyard (Sharpe, 1997, p. 129).

The high-resolution aerial images of the 1998 construction and ship activity at Mischief Reef clearly show a rapid buildup of fortified structures with logistical support and protection from the Chinese PLA navy. The construction of new multistory concrete buildings on the reef indicates that the activity went well beyond the repair of existing shelters.

The high-resolution aerial photographs show a marked increase in Chinese capability on Mischief Reef. China established an expanded, fortified presence that improves the ability of deployed personnel to withstand harsh weather and resist attack. The fortifications can also be used to project naval power

Figure 16.10—September 4, 1997, IRS-1C Satellite Image of Mischief Reef (courtesy Space Imaging, Inc.)

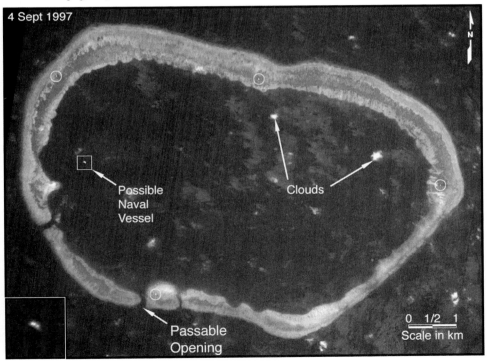

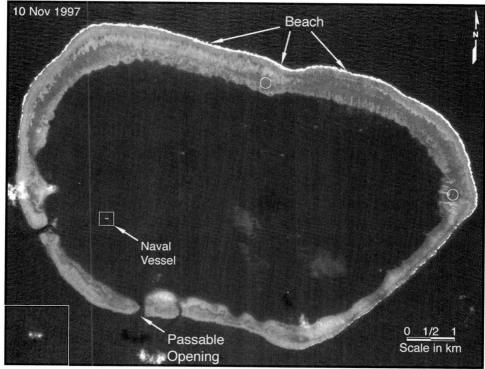

Figure 16.11—November 10, 1997, IRS-1C Satellite Image of Mischief Reef (courtesy Space Imaging, Inc.)

into the surrounding area and serve as a sanctuary for Chinese fishing vessels operating in disputed waters.[4] The naval forces that have been observed at the reef can quickly reinforce China's occupation with more personnel, material, and heavy weaponry.

IRS-1C Satellite Imagery. IRS-1C satellite images of Mischief Reef show the entire landscape and the surrounding sea. In Figures 16.10 and 16.11, the reef structure, lagoon with a passable opening to the sea, small reef islets inside the lagoon, and the ephemeral beach along the northern periphery are easily visible. Acquired 2.5 years after the Chinese occupation, these panchromatic images show significantly less detail than the aerial photographs.

Figure 16.10 is an IRS-1C panchromatic sensor image acquired at a resolution of 6 m, which was then resampled so that each pixel covers a ground area of 5 by 5 m. The image shows the reef, with four noticeable blips along the north, south, east, and west edges of the perimeter. These blips may be part of the four building clusters described in the media reports on the Chinese occupation. However, it is not possible to be certain about this because the individual buildings within each cluster are not visible. In addition, a 68±5-m long feature inside the lagoon could be a naval vessel but could not be conclusively identified because its shape was blurred, and its shadow was not visible in the lagoon (see inset).[5]

In Figure 16.11, also an IRS-1C panchromatic sensor image, two of the four blips along the reef periphery could not be seen, but a feature inside the lagoon could be positively identified as a naval vessel (see the inset for a close-up). Although the satellite image was too blurred to allow identification of the vessel's class with certainty, the feature's measured length of 71±5 m did eliminate some ship classes from consideration (Sharpe, 1995, pp. 114–142). If the vessel was Chinese, it was too small to be a destroyer, frigate, or ballistic missile submarine and too large to be a coastal patrol boat. The vessel was also too small to be either a Yukan-class supply ship (120 m) or a Dazhi-class support ship (107 m). Both ship classes have been observed at Mischief Reef on earlier occasions. Thus, based on a process of elimination, the imaged ship was probably either a replenishment ship (e.g., Fulin class), a survey and research ship (e.g., Yannan class), a cable ship (Yudian class), or a civilian transport and/or fishing ship.

Moderate-resolution panchromatic satellite imagery provided a synoptic view of Mischief Reef, as well as conclusive evidence of a ship deployment. The initial results suggest that this type of commercial imagery can be used under favorable weather conditions to detect large warships and supply ships. If compared with an overhead imagery database of known ships, the class of the imaged ships may also be identifiable.

However, the same imagery failed to reveal the building clusters shown in the aerial photographs. This suggests that higher-resolution imagery is required to detect smaller-scale features, such as the manned outposts shown in Figures 16.5 and 16.6. Such high-resolution imagery is also likely to be required to determine the nature and purpose of specific activities. Thus, we ordered the highest resolution commercial imagery available at the time, using Space Imaging's IKONOS satellite.

IKONOS Satellite Imagery. The image in Figure 16.12 was acquired by the IKONOS satellite on November 26, 1999, and is panchromatic with 1-m resolution. Its 95 km^2 field of view just covers the entire island. In this section, we provide an initial analysis of the image, describing its most evident features. As expected, the 1-m resolution imagery reveals much more detail than the

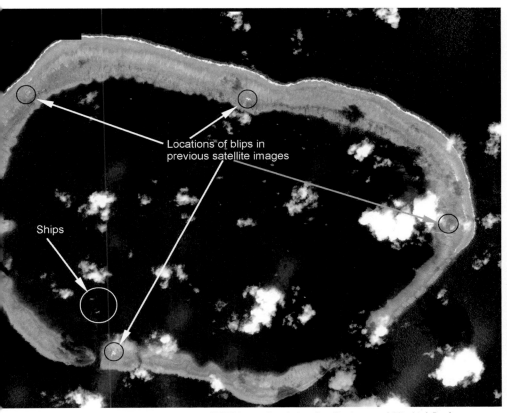

Figure 16.12—November 26, 1999, IKONOS 1-m Resolution Satellite Image of Mischief Reef (courtesy Space Imaging, Inc.)

previously acquired 6-m resolution satellite images. As the analysis below will show, high-resolution satellite imagery is a significant addition to the set of tools available for cooperative monitoring purposes; it can remove much of the inherent ambiguity of medium resolution.

Figure 16.12 is the IKONOS 1-m resolution image of the entire reef. The areas on the reef where activity was detected with medium-resolution IRS satellite imagery are labeled as are ships in the waters enclosed by the reefs.

Figure 16.13 is a close-up that reveals man-made structures at two of these sites, on the northern and southern sides of the island. Individual buildings can be seen, in contrast with the earlier images, in which only amorphous blips could be detected. The positions of the two built-up sites match those seen in the IRS-1C medium-resolution imagery. The scale of these two installations is about 100 m, somewhat larger than the estimated 20-m scale of the blips at

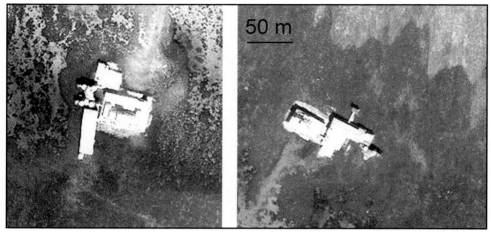

Figure 16.13—Details of Figure 16.12: Southern and Northern Sites (courtesy Space Imaging, Inc.)

these locations in the previous images. This apparent change in scale is consistent with the detailed records of activity shown in the aerial photographs and by independent reports (Fisher, 1999, pp. 5ff).

Details of the eastern and western features are shown in Figure 16.14. In contrast to the northern and southern sites, these features appear grainier. However, the features highlighted here are consistent with published reports that these two sites have been dismantled (Fisher, 1999, pp. 5ff).

Figure 16.14—Details of Figure 16.12: Western and Eastern Sites (courtesy Space Imaging, Inc.)

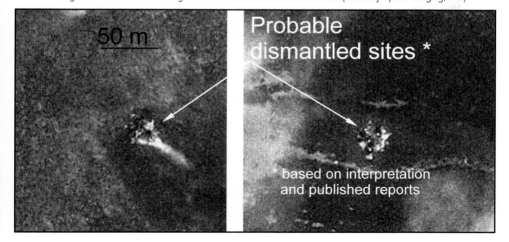

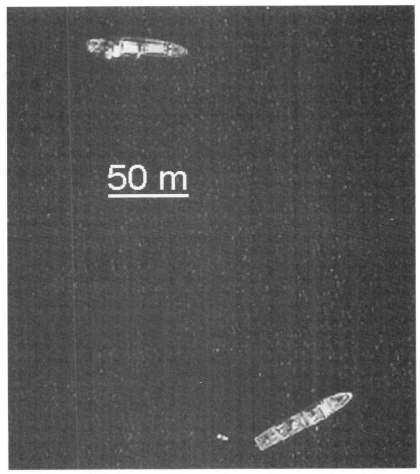

Figure 16.15—Close-up of Ships in the Interior Waters at Mischief Reef (courtesy Space Imaging, Inc.)

All four areas should be compared with the blips seen in Figures 16.10 and 16.11. The absence of new activity at the eastern and western sites is consistent with published reports, and the presence of activity at the northern and southern sites is confirmed both by published reports (Fisher, 1999, pp. 5ff) and by aerial photography (Figures 16.10 and 16.11).

Figure 16.15 shows a blowup of the two ships seen in the Mischief Reef lagoon. Both are 64 to 68 m long and thus are smaller than the Yuting class ships shown in the aerial photographs (Figures 16.4 and 16.5).

Summary. The images of Chinese activities at Mischief Reef provide some working experience on what can and cannot be done with overhead imagery to monitor the South China Sea as a whole. The lessons drawn from such experience can be expanded further by analyzing images of other areas and activities in the South China Sea.

THITU ISLAND

Thitu Island is the English name for the largest Philippine-occupied island in the South China Sea. Its Filipino name is Pagasa. Philippine forces took control of the island in the 1970s (Dzurek, 1996, p. 21). Since China, Taiwan, and Vietnam also claim this relatively large island, the Philippines established a substantial military presence there to reinforce its claim, deter attempts by others to take over the island, and defend itself in case of attack. The approximately 100-man military force on Thitu Island can also be used to project power to other parts of the South China Sea—particularly air power (Jimenez, 1999a).

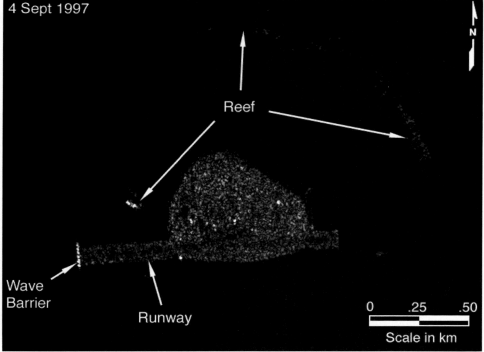

Figure 16.16—September 4, 1997, RADARSAT-1 Image of Thitu Island

16 Keeping an Eye on the Islands: Cooperative Remote Monitoring in the South China Sea

Figure 16.17—February 1998 Aerial Photograph of Thitu Island (see color plate)

In contrast with the fortified outposts at Mischief Reef, Thitu Island is one of the few islands in the South China Sea with an airfield (known as Rancudo Airfield). Figure 16.16 is a RADARSAT-1 synthetic aperture radar image of Thitu Island, acquired at C-band with a resolution of 8 m. Although the radar image has the characteristic speckled appearance, the runway is clearly visible and measurable ($1,294 \pm 5$ m by and 96 ± 5 m). The wave barrier at the end of the runway and portions of the reef can also be seen. The radar image provided information on the large-scale infrastructure, but little detail on such small-scale features as aircraft, vehicles, and buildings. The runway is clearly evident. Built on top of the island's coral base, it is substantially wider than the island itself. The image also shows the wave barrier at the end of the runway, as well as portions of the island's surrounding reef structure. Scattered throughout the island are a few bright blips that could not be identified, but are most likely vehicles, buildings, and other parts of the island infrastructure.

Aerial photographs of Thitu Island show the runway surface composition and significantly more detail on the island vegetation and support infrastructure. Figure 16.17 shows several buildings surrounded by natural vegetation, as well as an observation tower and aircraft parking area. The photograph shows that the island's coral foundation extends well beyond the beaches. Thus, there is available space to enlarge the operations at the island. Indeed, that is reportedly what the Philippines intends to do.

In early 1999, Philippine government officials announced plans to expand and upgrade the infrastructure at Thitu Island. To accommodate a wider array of military and civilian aircraft, the government plans to cement the entire airstrip. There are also unconfirmed reports of extending the runway (Villaviray, 1999), although that may not be necessary because its current length is capable of handling large transport aircraft.

The air operations are planned to be upgraded further with the construction of "temporary" hangars to house Air Force airplanes (Jimenez, 1999b), reportedly OV-10 aircraft, needed for increased surveillance activities and increased visibility of the Philippine Air Force in the South China Sea (Jimenez, 1999b).[6] In addition to the expansion of the air operations, the Philippines also plan to construct a pier, causeway, and base operations center (Villaviray, 1999).

The Philippines reportedly plans to open the island for tourism as well (*Asia Pulse*, 1999). To attract visitors to the remote area, the government has proposed construction of at least ten housing units for the use of fishermen and tourists. Civilian use of the island could help defray the cost of the occupation and solidify the Philippines' claim of sovereignty over the islands in this area.

The facilities at Thitu Island and the announced plans to improve the infrastructure make this site ripe for continued aerial and satellite monitoring. Overhead imagery could be used to determine whether such plans ever get implemented and how the changes would affect existing agreements and declarations. As illustrated in Figures 16.13 and 16.14, the moderate-resolution satellite imagery will most likely be useful for detecting and tracking large-scale changes, while 1-m resolution aerial and satellite imagery will probably be most useful for analyzing new construction activity and monitoring the air traffic to and from the island.

SUBI REEF

Just 26 km southwest of Thitu Island, Subi Reef is occupied by China, which took control of the reef in 1988 (Valencia, 1988). Although it is only above water at low tide, it is centrally located within the archipelago (see Figure 16.1). It is the northernmost Chinese-fortified position in the Spratlys and is relatively close to other strategic islands, such as Taiwan-occupied Itu Aba and Philippine-occupied Thitu Island. Consequently, Subi Reef is well suited as a way station and as a monitoring post for observing Taiwanese, Philippine, and Vietnamese activities in the immediate area (see Figure 16.1).

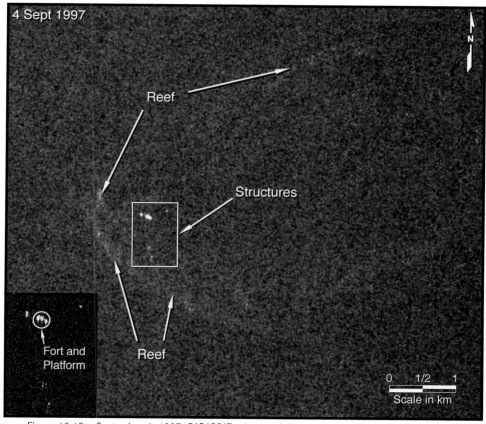

Figure 16.18—September 4, 1997, RADARSAT-1 Image of Chinese-Occupied Subi Reef.

Figure 16.18 is a RADARSAT-1 image of Subi Reef, excerpted from the full scene that shows Thitu Island (see Figure 16.16), which shows the faint outline of the submerged reef. While analyzing the full RADARSAT scene of Thitu Island, we accidentally discovered that the satellite had also imaged China's occupied position on Subi Reef. The discovery was made from the observation of several bright blips inside the reef perimeter (circled in Figure 16.18 inset). The blips were too coarse to be identifiable, but their bright appearance indicated that the features were distinctively shaped and probably metallic. The high radar backscatter from these objects suggested that these were manmade structures.

Aerial images of Subi Reef confirmed what RADARSAT-1 detected. Figure 16.19 shows two aerial photographs of the fortified Chinese position there. The Philippine Air Force acquired both photographs and released them in May

Figure 16.19—Aerial Photographs of the Fortified Chinese Structures on Subi Reef (courtesy Philippine Armed Forces)

1997. Filipino image interpreters provided the annotations in the overhead photograph, which appear to be accurate. The inset is a perspective view of the same structures shown in the overhead aerial image. The perspective view was acquired using an aerial camera that was pointing south. The photographs show a round platform, a connecting bridge, and a three-story concrete fort built on top of a raised foundation. The overhead photograph also shows four naval guns and a communications dish.

These specific features could not be identified in the RADARSAT image because the resolution was too coarse. However, using the aerial photographs, it was possible to identify the blips that corresponded with the fort, platform, and bridge. This was done by comparing the orientation of the blips with respect to north with the orientation shown in the overhead aerial photograph. From this comparison, the structures shown in the aerial photographs were matched with the corresponding blips in the RADARSAT image (see Figure 16.18 inset).

The analysis of Subi Reef demonstrated how image information from different sensor platforms could be combined to provide a clearer overall picture. The satellite image showed the layout and level of activity at the reef, while the aerial photographs provided details on the fortifications. The satellite image was useful for detecting the structures as part of a wide-area sweep, while the aerial photographs were useful for identifying the "dots" that were detected. Each sensor platform compensated for the principal technical limitations of the other sensor platform.

COMMODORE REEF, ALICOA ANNIE, AND YUAN ANHA

This subsection provides an analysis of IRS-1C medium-resolution panchromatic images from these three islands in the Spratly chain. The images illustrate some strengths and weaknesses of this type of imagery. The geographic layout of a reef can be seen in more detail than existing maps. Blips can be detected that may indicate human occupation, and the images can also provide some assurance that such occupation of a given island is missing or only on a small scale. On the other hand, medium-resolution imagery cannot, in itself, provide conclusive evidence of a small-scale military presence, and it cannot be used to search for features obscured by clouds.

Commodore Reef. Commodore Reef is located in the southeastern quadrant of the Spratly archipelago (see Figure 16.1), 110 km west of the Philippines' Palawaan island and 125 km northwest of Malaysia. Commodore Reef is claimed by China, Malaysia, the Philippines, and Vietnam. The Philippines took control of the reef in the 1970s, although one report states that they deserted it in the mid-1980s (Valencia, van Dyke, and Ludwig, 1997, p. 228).

Commodore Reef was the site of one major incident. In April 1988, the Malaysian navy detained Philippine fishermen operating near the reef. The Philippine government protested the detention, and Malaysia ultimately released the fishermen as a gesture of goodwill. However, Malaysia explicitly noted that the release was done without prejudice to its claim over Commodore Reef (Dzurek, 1996, p. 25).

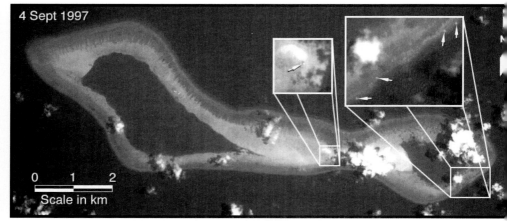

Figure 16.20—September 4, 1997, IRS-1C Satellite Image of Commodore Reef (courtesy Space Imaging, Inc.)

Figure 16.20 is an IRS-1C panchromatic image of Commodore Reef with a resolution of 6 m. The reef is shaped like a figure eight, consisting of two separate lagoons. It is one of the larger reefs in the South China Sea, and portions of it remain above water at high tide (Valencia, van Dyke, and Ludwig, 1997, p. 228). The satellite image in Figure 16.20 shows little detail on human occupation of the reef. Careful examination of the image revealed five distinct blips connected by linear features that could be evidence of manmade structures (see insets). However, because of the image's coarse resolution, it was not possible to verify the reported Philippine presence on the reef with a high degree of confidence. It was also not possible to determine whether there were any inhabited sites on the portions of the reef that were obscured by cloud cover (see Figure 16.20).

The IRS-1C satellite image of Commodore Reef illustrates two strengths and two weaknesses of moderate-resolution panchromatic imagery. It can show the geographic layout of a reef in more detail than existing maps. It can also detect blips that may be evidence of human occupation. On the other hand, it cannot, in itself, provide conclusive evidence of a small-scale military presence and cannot be used to search for features obscured by clouds.

Alicoa Annie and Yuan Anha. Alicoa Annie (also known as Alicia Annie) and Yuan Anha are two large reefs in the South China Sea that have no known occupants. However, both are in relatively close proximity to other reefs and islands occupied by various claimant states. Alicoa Annie is just 56 km south-southwest of Chinese-occupied Mischief Reef. Yuan Anha is less than 75 km from reefs occupied by the Philippines, Malaysia, and Vietnam (see Figure 16.1).

Figure 16.21—March 15, 1998, IRS-1C Satellite Image of Alicoa Annie (courtesy Space Imaging, Inc.)

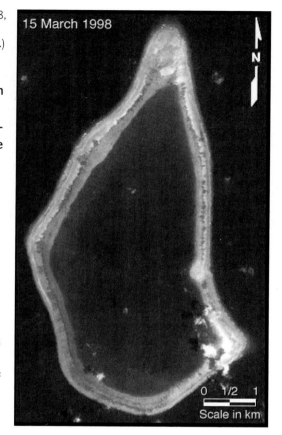

Although the Spratly conflict has sometimes been intense, Alicoa Annie and Yuan Anha have been relatively peaceful. There have been no documented confrontations at Yuan Anha; one at Alicoa Annie. On March 25, 1995, the Philippine navy captured four Chinese fishing boats near Alicoa Annie and arrested 62 fishermen (Pastor, 1995). The detention of the fishermen took place just two months after China took control of Philippine-claimed Mischief Reef. The arrests were consistent with a pattern of seizing fishing vessels of a claimant state shortly after it occupies another Spratly reef or island.

Figures 16.21 and 16.22 are IRS-1C panchromatic images of Alicoa Annie and Yuan Anha, respectively, each with a resolution of 6 m. Both were acquired in March 1998 and both are almost completely cloud free. Alicoa Annie and Yuan Anha are appreciably larger than the other selected islands and show no evidence of human occupation whatsoever. Alicoa Annie is about the same size as Mischief Reef, and Yuan Anha is the largest of the six islands and reefs selected for detailed study.

Careful examination of Figures 16.21 and 16.22 failed to reveal any blips, paths, or structures that could be attributed to the physical occupation of the reefs. This suggests that moderate-resolution images may have limited utility for detecting possible inhabited structures and for verifying the absence of such structures on disputed reefs.[7]

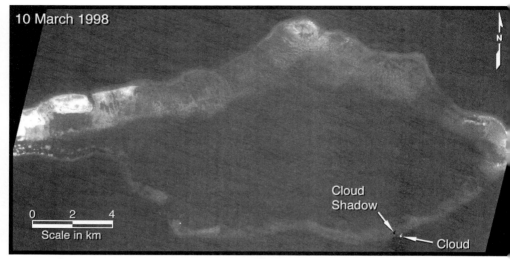

Figure 16.22—March 10, 1998, IRS-1C Satellite Image of Yuan Anha (courtesy Space Imaging, Inc.)

Conclusions

Our technical analysis used a variety of aerial and satellite sensors to analyze the selected islands and reefs in the South China Sea. Each study site revealed at least one capability or limitation associated with overhead imaging. The analysis of Mischief Reef demonstrated how high-resolution aerial images were needed to detect and assess the controversial activities there. Moderate-resolution panchromatic imagery from satellites proved to be of limited use in monitoring specific activities at Mischief Reef, but it was useful for ship detection and synoptic analysis. The study of Thitu Island demonstrated the value of radar imagery for detecting large-scale infrastructure independent of light conditions or cloud coverage. The aerial photographs showed smaller structures, such as the individual buildings and observation tower. The accidental discovery of Chinese-occupied Subi Reef in the RADARSAT-1 image was a realistic example of how radar imagery could be used as a detection tool. In addition, the analysis showed how coarse-resolution features from satellite imagery could be correlated with high-resolution aerial images of the same feature.

The examination of Commodore Reef highlighted the limitations associated with the exclusive use of moderate-resolution panchromatic imagery. The IRS-1C satellite's bird's-eye view was partially obscured by cloud cover and did not offer enough spatial detail to verify human occupation of the reef. But that view did, however, show a few "blips" of possible human occupation that could be examined more closely with higher-resolution imagery. Alicoa Annie and Yuan Anha served as controls for the study. The image analysis of these two large reefs with no known occupants came up with no evidence of physical occupation. This suggests that such imagery may be useful for dispelling false allegations of physical occupation and for verifying the absence of structures on disputed reefs.

In addition to obtaining empirical results from the study of selected islands and reefs, we also derived results from the different types of remote sensing systems that were used. Table 16.2 charts the strengths and weaknesses of four general types of aerial and satellite imaging sensors. The assessment is based on the technical and operational utility of each sensor for monitoring the South China Sea specifically. The fourth type, high-resolution satellite imaging, was used in this study in a very limited way, but its strengths and weaknesses can also be extrapolated from high-resolution aerial imaging and moderate-resolution satellite imaging.

From Table 16.2, we can devise an imaging strategy optimized for monitoring the South China Sea. The overall technical objective would be to monitor all civilian and military activity in the area without provoking an armed incident. To do this, the imaging strategy would have to rely on satellite imaging as much as possible; it is the only platform that can scan the entire South China Sea routinely for ship and island activity. And in contrast with aircraft, imaging satellites can operate from a safe altitude without the risk of being misinterpreted as fighter-bombers.

Moderate-resolution radar imaging satellites are the best space-based system for performing routine search missions over the South China Sea. Unaffected by adverse weather, radar imaging satellites are the only systems capable of searching for controversial activities that may deliberately use cloud cover as a screen. Radar images from satellites, such as RADARSAT-1, are well suited for detecting the large ships and outposts that have been observed in the South China Sea. If the images are delivered 8 to 72 hours after their acquisition, they could provide early warning of controversial or prohibited activity, whether civilian or military.

Any activity detected in a radar image would most likely appear as a fairly coarse blip. Consequently, higher-resolution images would be needed to investigate further. To facilitate the creation of a stable monitoring regime, commercial imaging satellites would be the preferred "close look" platform

Table 16.2—The Strengths and Weaknesses of Aerial and Satellite Imaging for Monitoring the South China Sea

Remote Imaging Technology	Strengths	Weaknesses
Low-altitude (<5,000 ft) aerial imaging	• Can acquire perspective views of ships, structures, and activity with very high spatial detail, approximately 0.2- to 0.5-m resolution (see Figures 16.3–16.17) • Can be cued to examine suspect sites at any time of day	• Acquires images with very small fields of view and thus do not show the surrounding areaCannot be used effectively to search for activity throughout the entire South China Sea • Cannot be used in adverse weather conditions • May be interpreted as hostile; highly vulnerable to antiaircraft fire
High-altitude (>5,000 ft) aerial imaging	• Can acquire overhead images of ships, structures, and activity at high spatial resolution, approximately 0.5- to 1-m resolution (see Figures 16.2 and 16.19) • Can acquire perspective views at 1- to 5-m resolution • Can be used to monitor a limited set of specific islands and reefs • Can be cued to examine suspect sites at any time of day	• Acquires images with small fields of view and thus do not show much of the surrounding area • Cannot be used efficiently to search for activity throughout the entire South China Sea • Cannot be used in adverse weather conditions • May be interpreted as hostile; highly vulnerable to antiaircraft fire
Moderate-resolution (5–10 m) satellite imaging	• Can use radar imagery to search for ships and structures throughout the South China Sea independent of weather or lighting conditions (see Figures 16.16 and 16.18) • Can use panchromatic or multispectral imagery to search for possible occupied sites and cue aerial sensors to suspect sites (see Figures 16.10, 16.11, and 16.20) • Operates at an altitude of 500 to 1,000 km; the observed do not see the satellite that is imaging them • Is invulnerable to antiaircraft weaponry	• Cannot provide enough spatial detail to determine the nature of observed activity • Cannot acquire panchromatic or multispectral images of islands at night or during bad weather conditions • Can acquire images only at fixed times each day
High-resolution (1–2 m) satellite imaging	• In addition to all the strengths of moderate-resolution satellite imaging, can show high spatial detail of ships, structures, and activity (see Figures 16.13 and 16.15)	• Cannot acquire panchromatic or multispectral images of islands at night or during bad weather conditions • Can acquire images only at fixed times each day • Covers a smaller area than moderate-resolution satellite imaging

because the images could be acquired from a safe, remote vantage point, transmitted to a central image archive, and shared with all claimant states and other interested parties.

If such images are not available in a timely fashion because of the orbital position of the satellite, the tasking load, or cloud coverage, aerial imaging would be the next viable option for investigating any suspect features found in the radar satellite sweep. To minimize the risk of vulnerability and armed confrontation, the aerial imaging should be done using visibly unarmed aircraft flying around the suspect site at maximum possible standoff distances. To maximize the degree of transparency in the South China Sea, the aerial data should be shared cooperatively so that claimant states and interested parties can independently assess the observed activity. Such sharing could build on the precedent established by the Philippines' routine release of aerial images of South China Sea sites.

Our remote sensing study shows that aerial and satellite imaging can provide timely, substantive information on ships, structures, and activities in the South China Sea. With the devised optimal monitoring strategy, future studies can now consider verification provisions for specific political agreements tailored to the capabilities and limitations of aerial and satellite imaging. With the September 1999 launch of the IKONOS satellite, all the necessary aerial and satellite imaging platforms for South China Sea monitoring are in place. From that point onward, a formal monitoring regime optimized for the South China Sea will be technically feasible. Whether such a regime will be realized will ultimately depend on multilateral agreement to an interim or permanent solution to the Spratly conflict.

Even this initial look at high-resolution imagery reveals the wealth of new information it can provide. The size and possible functionality of installations can be identified, and individual buildings only a few tens of meters in size can be discerned. Individual ships can also be detected and in some cases identified. In contrast, classes of ships could be ruled out in the medium-resolution case, but conclusive identification could not be made as a rule.

In a cooperative monitoring context, high-resolution imagery can be used to demonstrate with high confidence that significant unilateral development is or is not taking place. In the particular case of monitoring the Spratly Islands, the improvement in detail makes the images both easier to analyze and easier to present and explain to adjudicating bodies and to the states party to the monitoring regime.

Acknowledgments

The authors would like to thank John Baker, Larry Brandt, George Harris, Frank Pabian, Mark Valencia, and David Wiencek for their assistance and comments on previous drafts of this chapter. The authors would also like to thank Tom Budge, John Olsen, and Arian Pregenzer for their guidance and support during the course of this research project.

This research on which this chapter was based was supported in part by the Cooperative Monitoring Center at Sandia National Laboratories. The analysis, views, and opinions expressed herein are those of the authors alone.

Notes

1. A news report (*International Gas Report*, 1995) cites one Russian estimate of 1 billion tons of crude oil. The same report notes that China and the Philippines disputed this estimate, presumably as being too low. Chinese geology and resources minister Song Ruixang's September 1994 estimate of 30 billion tons of crude is cited in *International Gas Report* (1994). Conversely, E. F. Durkee, an exploration geologist with expertise in Asia, says that referring to the Spratlys as "oil-rich" is "largely rubbish" and notes that few hard data exist for the area (Durkee, 1995).
2. RADARSAT archives were included in the image search because the imagery offered both relatively high spatial detail and structural information on the imaged features (surface roughness, metallic composition, etc.).
3. It is possible the Yukan-class ship was actually a Yuting-class amphibious warship. Both types of ships have similar profiles, and both operate in the South China Sea.
4. There has already been one publicized case in which a Chinese fishing vessel reportedly fled to Mischief Reef to elude the Philippine Navy (*Manila Times*, 1998).
5. The feature's shadow could not be seen because it blended with the dark lagoon water. The cloud shadows could not be seen in the lagoon either. The high sun elevation (69 degrees) also reduced the shadow's size.
6. Rockwell built the OV-10 aircraft for the U.S. Marine Corps. It is widely used for reconnaissance but can also be configured as a fighter-bomber (Jane's, 1986, pp. 492–493).
7. If camouflage is used to blend the appearance of structures with the surrounding reef, high-resolution imagery or radar imagery would most likely be required to detect such structures or verify the absence of such structures.

References

Aguinaldo, Sandra, "RP to China: Dismantle Structures," *The Manila Times*, November 9, 1998.
Asia Pulse, "Philippines Plan to Open Spratly Island to Tourism," February 18, 1999.
Associated Press, "Chinese Building on Disputed Reef," November 9, 1998.

Botbol, David, "Reportage on PRC Presence on Spratly Islands: AFP Journalists Confirm Presence," FBIS-EAS-95-063, April 3, 1995, pp. 77–78.

Branigin, William, "China Takes Over Philippine-Claimed Area of Disputed Island Group," *The Washington Post*, February 11, 1995, p. A18.

Central Intelligence Agency, World Fact Book, Spratly Islands, 1999, http://www.cia.gov/cia/publications/factbook/.

Christian Science Monitor, "Conflict in Spratlys Spurs Rift Between Hanoi and Moscow," June 14, 1988.

Lin, Cheng-Yi, "Taiwan's South China Sea Policy," *Asian Survey*, Vol. 37, No. 4, April 1997.

Chung, Chien, "Economic Development of the Islets in the South China Sea," paper delivered at the S. China Sea Conference, American Enterprise Institute, Washington D.C., September 7–9, 1994.

Contreras, Volt, Cynthia Balana, Juliet Javellana, and Martin Marfil, "Chinese Installations in Spratlys Detailed," FBIS-EAS-98-320, November 16, 1998.

Dahlby, Tracy, "South China Sea: Crossroads of Asia," *National Geographic*, Vol. 194, No. 6, December 1998.

Dobson, William, and M. Taylor Fravel, "Red Herring Hegemon: China in the South China Sea," *Current History*, September 1997, pp. 258–263.

Durkee, E. F. "Oily Claims," letter to the editor, *Far East Economic Review*, March 30, 1995.

Dzurek, Daniel, "The Spratly Islands Dispute: Who's On First?" *Maritime Briefing*, Vol. 2, No. 1, 1996.

Hanoi Voice of Vietnam Network, "SRV: MT Institute Installs Radio Station on Truong Sa, other islands," FBIS-EAS-96-111, June 7, 1996, p. 70.

Fisher, Richard D., "Heritage Foundation Special Report: China's Expansion in the Spratly Islands," 1999. Available at http://www.heritage.org/exclusive/spratly/page5.html (last accessed June 9, 2000).

International Boundary Monitor, "FOCUS: Vietnam Occupies Two More South China Reefs," September 15, 1998, pp. 7–10.

International Gas Report, "Disputed Nansha Islands Rich in Oil and Gas," September 16, 1994.

——, "Spratly Island Reserves," No. 270, March 3, 1995, p. 32.

Jane's, *Jane's All the World's Aircraft 1986–87*, London: Jane's Publishing Company Limited, 1986.

Jimenez, Raffy, "Coping in Sweet, Sad Spratlys," *The Manila Times*, January 13, 1999a.

——, "China Aircraft on Frequent Mischief Flights," *The Manila Times*, December 19, 1999b.

Manila Times, "China Demands Fishers' Release," December 2, 1998.

Pastor, Rene, "Philippines Holds Four Chinese boats in Spratlys," Reuters World Service, March 25, 1995.

Philippines Department of Foreign Affairs, Press Release on Mischief Reef, November 6, 1998.

Reuters World Service, "Philippine Navy Sends Ship to Disputed Shoal," February 2, 1995.

Sabangan, Annie, Johnna Villaviray, and Raffy Jimenez, "China Fortifies Hold on Spratlys," *The Manila Times*, January 21, 1999

Sharpe, Richard (Capt. RN), ed., *Jane's Fighting Ships: 1995–1996*, London: Butler and Tanner Limited, 1995.

——, ed., *Jane's Fighting Ships: 1997–1998*, London: Butler and Tanner Limited, 1997.

Snyder, Scott, *The South China Sea Dispute: Prospects for Preventive Diplomacy*, Washington D.C.: U.S. Institute for Peace, August 1996.

Thomas, Bradford L., and Daniel J. Dzurek, "The Spratly Islands Dispute," *Geopolitics and International Boundaries*, Vol. 1, No. 3, Winter 1996.

Valencia, Mark, *China and the South China Sea Disputes*, London: IISS, Adelphi Paper 298, 1996.

——, "Troubled Waters," *Bulletin of Atomic Scientists*, January-February 1997, p. 50.

——, "The Spratly Islands: Dangerous Ground in the South China Sea," *Pacific Review*, Vol. 3, No. 2, 1988, pp. 438-443.

Valencia, Mark J., Jon M. van Dyke, and Noel Ludwig, *Sharing the Resources of the South China Sea*, The Hague: Martinus Nijhoff Publishers, 1997.

Villaviray, Johnna "RP Eyes Expansion of Spratlys Facilities," *The Manila Times*, February 25, 1999.

17 The Role of Commercial Satellite Imagery in Locating South Asian Nuclear Test Sites

David Albright and Corey Gay Hinderstein

Commercial observation satellites offer an important new instrument for monitoring nuclear proliferation, one that is available to both governments and nongovernmental organizations. The underground nuclear test explosions India and Pakistan conducted in May 1998 highlight the benefits and limitations of using higher-resolution commercial satellite imagery to locate and characterize underground nuclear weapon test sites on an unclassified level.

In May 1998, India and Pakistan each conducted two series of underground nuclear weapon test explosions. The tests, whose precise timing had been largely unanticipated, heightened awareness of the potential for a nuclear arms race in the region and the benefits to the region of swift implementation and enforcement of the Comprehensive Test Ban Treaty (CTBT). In India's case, the effort to learn more about the May 1998 tests was facilitated by earlier open-source information that accurately located and characterized the Indian nuclear test site using a variety of high-resolution commercial satellite images together with open-source reporting. In the case of Pakistan, a lack of similar advance information about the test site, beginning with the correct geographical location, complicated the efforts of nongovernmental experts to determine the yield and other key data about the tests and associated nuclear devices. Nonetheless, once seismic data and other open-source information became available after the tests, the first Pakistani nuclear test site was also identified and characterized using high-resolution commercial satellite imagery. Post-test analysis of the test sites also revealed evidence of concealment efforts both India and Pakistan made to hide test-site preparations from the very types of satellites used to document their activities.

This chapter examines two cases in which commercial satellite imagery has been used to identify and characterize South Asian nuclear test sites. The first focuses on the Indian nuclear test site in the Rajasthan Desert in western India. It reviews how an earlier study (Gupta and Pabian, 1996) used commercial satellite imagery to assess allegations of Indian nuclear test preparations in December 1995. We also assess how this earlier analysis provided an open-source means of tentatively identifying the Indian nuclear test site in light of the actual nuclear tests India conducted in May 1998. The second case (Albright, Gay, and Pabian, 1999) focuses on the Pakistani nuclear tests. It details how a combination of commercial satellite imagery and other open-source material was used to identify and characterize Pakistan's first nuclear test site, despite Pakistan's efforts to hide its nuclear test preparations.

Case Study I: Rajasthan Desert Test Site, India[1]

Commercial satellite imagery was used in conjunction with media reports and scientific journal articles to investigate the December 1995 allegations that India was making preparations for a nuclear test in the Rajasthan Desert (which the Indian government officially denied at the time). Taking the allegations against India as an example of a possible future CTBT compliance dispute, a study was conducted to test the utility of commercial satellite imagery as a potential tool for CTBT monitoring and verification (Gupta and Pabian, 1996). Imagery-derived information was used to assess the credibility of media reports regarding possible nuclear test–related activity in that area. The study led to the first public identification of the exact location of the 1974 nuclear test and a detailed characterization of the adjacent, then–still clandestine, nuclear test site that the Indians used to conduct their nuclear test series in 1998. The Indians have since admitted that they did originally intend to conduct nuclear tests in late 1995, but postponed their plans until 1998, "because news of the preparations leaked out and then Prime Minister P.V. Narishmha Rao backed down in the face of international pressure." (Chengappa, 1998.) In contrast to previous remote sensing studies of known nuclear test sites after long periods of testing, this investigation demonstrated the utility of commercial satellite imagery to evaluate nuclear test preparations at a little–known, undeclared, site *before* a nuclear test.

LOCATING THE 1974 NUCLEAR TEST CRATER

The first task of this exercise was to find the crater formed during that lone nuclear test, because all the media reports describing Indian nuclear test preparations (prior to the detonations of May 1998) located the alleged preparatory activity in the immediate vicinity of the 1974 ground zero.

Media Survey of the May 18, 1974, Nuclear Test. On May 18, 1974, India detonated a "peaceful nuclear explosion" in the Rajasthan Desert. After the test, some information was released regarding the location and physical description of the test area. India's official Bhabha Atomic Research Center reported that the test device was placed in a 107-m shaft and had a yield of approximately 12 kilotons. A helicopter photo of the subsidence crater was published in several publications, but the scale of the photo and the precise location were not released. The location of the test site was reportedly closest to the towns of Pokhran, Loharki, and Khetoli, but the distances varied by report.

The International Seismological Center (ISC) reported that the test took place at approximately 26.99 degrees north, 71.80 degrees east. Because seismic estimates are limited in the degree of precision they can achieve, and include significant error margins, this information proved inadequate for filtering the conflicting media reports on the test location.

In 1992, one researcher claimed to have located the 1974 crater on an IRS-1A image from October 26, 1998 (Jasani, 1992, pp. 68, 96). However, the resolution of that particular image (36.25 m) was insufficient to correlate positively with the published pictures from the time of the test. (It has since been possible, using higher-resolution satellite imagery, to determine that the crater-like feature is only a natural depression).

Correlating the Media Information with Commercial Satellite Imagery. Two archived satellite images were used in the search for the 1974 crater (see Figure 17.1). Both images are panchromatic, and they cover large portions of the error circle associated with the ISC seismic location estimate of the May 18, 1974, test. One was a March 25, 1995, 10-m Satellite Pour l'Observation de la Terre–3 (SPOT-3) image, and the other was a May 24, 1992, 2-m KVR-1000 image. The images were selected for their coverage of the area identified as closest to the test site by locating the towns repeatedly mentioned in the press reports.

The SPOT image showed agricultural and desert areas, towns, a rail line, and roads. No feature was found that could be positively identified as the

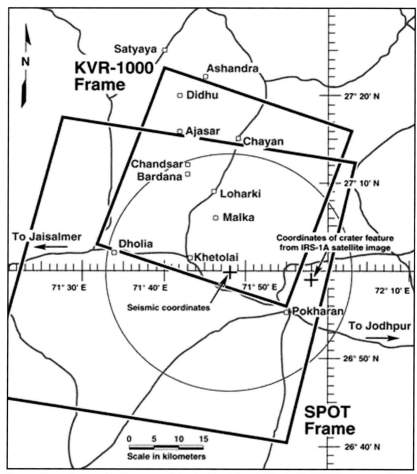

Figure 17.1—Geographic Coverage of the Archived March 25, 1995, SPOT Image and May 24, 1992, KVR-1000 Image

crater. The KVR-1000 image, however, revealed a large, circular depression located about 6 km south of Loharki and about 12.5 km north of the ISC seismic coordinates. The overhead view of the approximately 165-m-diameter depression appeared similar to that of the 1974 crater depicted in the low oblique helicopter photograph published in the 1975 Indian scientific journal (see Figure 17.2). To determine conclusively whether or not the depression was indeed the 1974 crater, the KVR-1000 image was digitally rendered as a perspective view to replicate the viewing angle of the aerial photo. Comparison of the two images revealed multiple correlative features (see Figures

17.2 and 17.3), confirming that the crater resulting from the 1974 nuclear test explosion had been located on the KVR-1000 imagery.

REPORTS OF RENEWED NUCLEAR TEST PREPARATIONS: 1981–1995

In 1981, U.S. officials and the media accused India of undertaking extensive nuclear test preparations near the site of the 1974 test. Reports included accounts of surface excavation, possible village evacuation, and erection of fencing. A meeting was apparently scheduled to discuss a four-day evacuation of nine named villages; however, the meeting never took place, and there is no evidence that a nuclear test was ever conducted.

In 1982, more information surfaced regarding activity at the alleged nuclear test site. According to media reports, the Army cordoned off an area from which lights and noise were observed during the night. Reports about town evacuation plans were sketchy. P. K. Iyengar, former Indian Atomic Energy Commission Chairman, is now quoted as confirming that "Indira Gandhi

Figure 17.2—Annotated Helicopter Photo of the Crater. The four physical features labeled on the photo were found in the KVR-1000 image of the crater. Note how the bushes in the foreground cast shadows toward the camera.

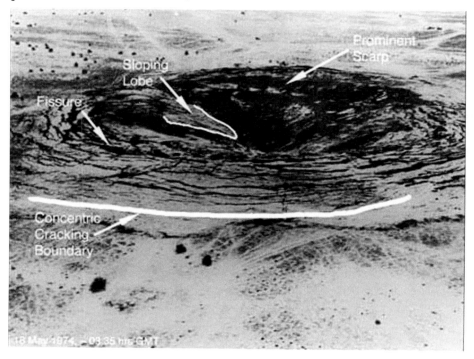

okayed more nuclear tests in 1982–1983. Three shafts were dug—the same used for the current round of tests—but at the last minute she told the scientists to call it off." (Chengappa, 1998.)

In late 1995, U.S. newspapers reported another round of apparent test-site preparations. Reported activity included cleaning out a test shaft and work on instrumentation for data collection. Apparently, a new fence was erected around the test site, and residents reported an increase in military convoys and troop movements to and from the area. At the same time, more detailed information emerged on the layout of the test-site area. Regional press described four testing ranges, only one of which was dedicated to nuclear activities. The reports stated that conventional artillery and missile testing also take place in the same general area, which is reserved for military testing and training activities.

LOCATING THE NEW NUCLEAR TEST SITE

With the 1974 test crater having been positively located, satellite imagery of the region surrounding the crater was then analyzed to determine (1) whether a site suitable for nuclear testing did exist and, if so, (2) whether there was

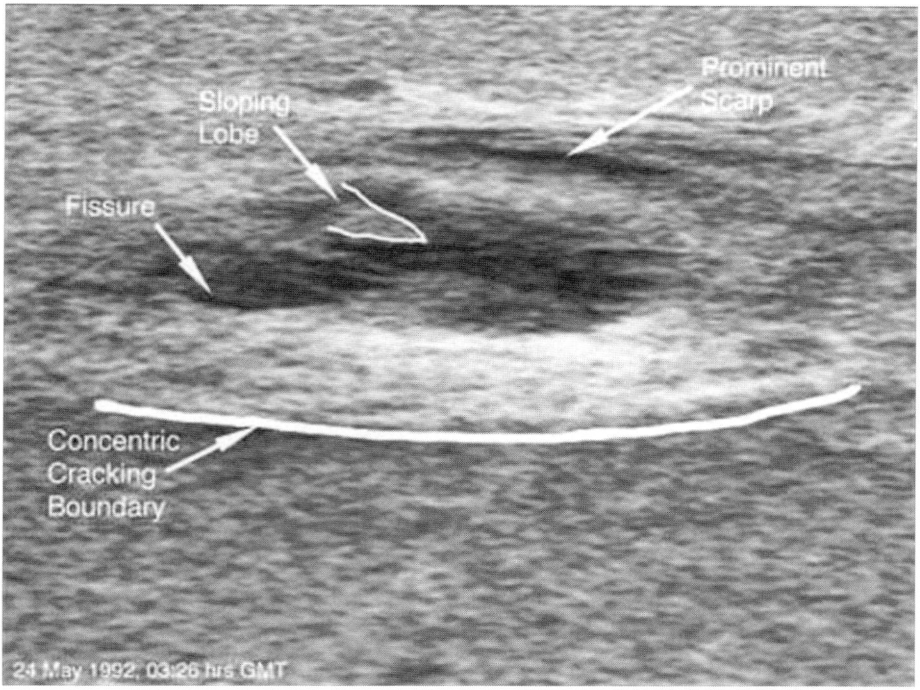

Figure 17.3—KVR-1000 Perspective View of the Large Circular Depression

any sign of recent activity in the vicinity of the site. That search revealed the presence of a handful of military sites, including one very highly secured unconventional military test area, surrounded by open desert, that exhibited discernible features that were most consistent with nuclear testing, mobile missile testing, or a combination of both. New activity was also discernible at the site that could be best explained as being associated with preparations for future nuclear weapon testing.

Gupta and Pabian (1996) concluded that

India may indeed have taken the necessary steps in this area to be capable of conducting up to two underground nuclear tests with relatively short notice as part of a contingency plan for testing at some future date.

Although, according to Chengappa (1998, p. 31),

when news of the impending 1995 tests leaked out, [Indian Prime Minister] Rao backed down under U.S. pressure but he left instructions with the team: be prepared to carry out the tests at a month's notice.

Gupta and Pabian (1996) further concluded that

Although the commercial satellite imagery derived evidence cannot by itself be considered as proof of such activity (even when combined with other reporting), it can certainly be viewed as sufficient to raise specific questions concerning the possibility of future Indian nuclear testing at this site.

FAST FORWARD TO 1998: CONCEALMENT AND CAMOUFLAGE

The events of May 1998 confirmed the test-site locations identified in Gupta and Pabian (1996). Post-test images of the Indian nuclear test site used to conduct the two largest underground nuclear weapon tests on May 11 were broadcast on Indian television and published in the Indian press. These images showed that the two test locations were the same as those identified in the study as the two well-secured probable test-shaft locations.

It should also be noted, however, that a third shaft, also reported to have been used on May 11, was not identified in Gupta and Pabian (1996). This third test shaft location is evidently in the area of sand dunes in the western portion of the test site. The authors may have been unable to identify the third location (which Joshi, 1998, states was dug in the early 1980s, at the same time as the two others) because the shaft was too small to be readily identifiable on the best resolution commercial satellite imagery then available to the authors. It is also possible that Joshi (1998) was incorrect on this point and that the shaft was dug some time after April 1996 (the date of the latest imagery available to the authors). In any case, such problems should be avoidable in the future, given that commercial imaging satellites are now becoming available with significantly improved resolution, both spatial and temporal.

Post-test Indian media imagery (some of which was rebroadcast on CNN) also showed that India took steps to conceal their 1998 nuclear test activities from overhead observation consistent with a 1998 editorial in the *Hindustan Times*, which reported that U.S. imagery analysts "evidently misread the telltale signs of a nuclear test site for the dummy villages which cleverly camouflaged the subterranean goings on." Such a deception effort is borne out by the fact that the Indians tried to disguise the site by constructing several barracks-type buildings clustered near the vertical emplacement shafts. The buildings were further disguised through the use of disruptive camouflage painting. Other pre- and post-test imagery of the test locations published after the tests also showed that the vertical emplacement shafts lacked the head-frame structures commonly used to lower a nuclear device below ground. Instead, the Indians used a single simple pulley on transverse supports. The presence of camouflage netting around the shaft heads both before and after the tests showed that the pulley structures were very likely always concealed beneath the netting prior to the tests to inhibit detection and overhead monitoring further (Chengappa, 1998, pp. 26–32).

Following the May 1998 Indian nuclear weapon tests, Indian government officials took particular pride in the fact that they were apparently able to fool the world's intelligence collection and analysis agencies in their efforts to uncover India's nuclear test preparations. Among the other precautions they said they took to avoid detection included having scientists travel in disguise (as military officers) and, perhaps most importantly, having Indian government officials issue misinformation to foreign envoys, stating that no tests were imminent. Specifically regarding efforts to thwart satellite surveillance, the Indians claim that they operated "mostly at night to avoid detection by snooping satellites" (Chengappa, 1998, p. 31) and routinely maintained a relatively high level of innocuous activity as a cover for real test activity. The Indians also feigned missile test activity on the Indian east coast to divert the attention of overhead satellites (see also Joshi, 1998; Chengappa, 1998; Bedi, 1998).

Case Study II: Ras Koh Nuclear Test Site, Pakistan[2]

Immediately following the May 28, 1998, nuclear weapon tests, Pakistan released detailed videotape images of their mountain test site during and after the tests.[3] Albright, Gay, and Pabian (1999) used multisource data integration techniques and analysis of the videos and other available open-source data, such as media reports, seismic data, mapping and geologic references, and commercial satellite imagery, to determine the precise geographic location, geologic setting, operational features, and layout of the May 28, 1998, nuclear test site. The analysis also exposed some of the techniques the Pakistanis used before the tests to attempt to thwart overhead detection and monitoring of nuclear test activities prior to testing. Comparing pre- and post-test imagery also revealed the extent of some of the physical test effects, evident as recent rock slides on mountain slopes resulting from the underground detonations.

Figure 17.4—Before, During, and After the Tests on May 28, 1998 (see color plate)

LOCATING THE MAY 28 TEST SITE

The day after the initial nuclear test explosions (five simultaneous tests were claimed), Pakistan Television released terrestrial footage of the blasts. The videos showed dramatic surface disturbances, including rising yellowish-white-greyish-white dust clouds and falling rock slides. The light-toned coloration of the dust contrasted sharply with the darker tones of the mountain moments before the test. Other than a simple reference to Baluchistan Province, no more detailed geographic locational information was initially released with the videos (see Figure 17.4).

Most media reporting, both before and after the Pakistani nuclear tests, accurately located the May 28, 1998, test site as being within the Chagai District of Baluchistan Province in western Pakistan (see also Chapter 23 in this volume). Nonetheless, the media frequently, and mistakenly, referred to a more precise location for the site, within an oval-shaped upland area known as the Chagai Hills, near the Afghan border just north of the town of Dalbandin. Such erroneous reports even claimed to have found the nuclear test sites in the Chagai Hills on commercial satellite imagery (see, for example, Weiner, 1998; Windrem, 1998; Hough, 1998). However, just before the May 28 test, a *Times of India* report suggested that the test site was instead located south of the Chagai Hills in the 200-km–long Ras Koh range (Gannon, 1998). That linear mountain range, trending north east-south west and having greater relief than the Chagai Hills, is located south of the town of Dalbandin (see Figure 17.5). Shortly after the May 28 tests, a Reuters reporter provided possible coordinates for the test site, obtained via unnamed sources, that also placed the test site within the Ras Koh range southeast of Dalbandin.[4] Final determination of the true location of the test site would depend on post-test correlation of seismic data, other post-test media reporting, and any available terrestrial and satellite imagery.

Seismic data alone also proved inconclusive for locating the test site. Both the U.S. Geological Survey (USGS) and the Prototype International Data Center (PIDC), the organization established to coordinate remote monitoring for the CTBT Organization, placed the area of the May 28 seismic signals in a lowland plain between the Ras Koh range and Chagai Hills east of Dalbandin. While the associated uncertainties of these estimates included at least a portion of the Ras Koh range, they did not cover any area having the topographic features shown in the videos the Pakistanis distributed.[5] The preliminary seismic estimates did, however, narrow down the area of the Ras Koh range in which the site was most likely located. More importantly, the videos suggested that the test site was actually situated on the southern flank of steeply sloping mountain terrain. Given that the videos were taken at the time of the test, at 3:16 p.m. local time (see Figure 17.4) and that the sun was directly behind the camera (as there were no shadows facing the camera), the camera had to

17 The Role of Commercial Satellite Imagery in Locating South Asian Nuclear Test Sites

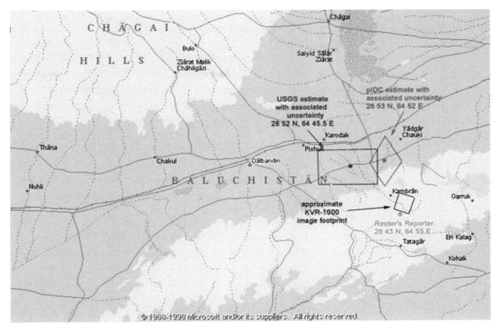

Figure 17.5—Geographic Location of Pakistan's May 28 Nuclear Test Site (see color plate)

have been looking easterly. The mountain shown rising abruptly from right to left, must, therefore, have been rising from the south. The significant break in slope at the base of the mountain was also consistent with geologic faulting.

Media reports stated that a very small population in the area of the test site was evacuated and also provided geographical references. However, the villages mentioned in these reports could either not be found on available maps or were too far away from the area of the seismically derived coordinates to be credible (see, for example, Reuters, 1998; *Dawn*, 1998b).

An archival search of geologic maps and one SPOT satellite browser image (freely accessible via the Internet) narrowed down the most probable location of the test site to a mountainous area east-southeast of the seismic estimates, known as Koh Kambaran. That area was also roughly midway between the calculated seismic coordinates and those provided by Reuters. The southwestern flank of that mountain was also found to be bounded by a significant geologic fault, giving added assurance that this was the most promising area in which to begin a more detailed imagery search. Such a search, however, required purchasing high-resolution imagery.

Due to the high cost per area of image coverage, only a relatively small portion of a Russian KVR-1000 satellite image was obtained (covering only

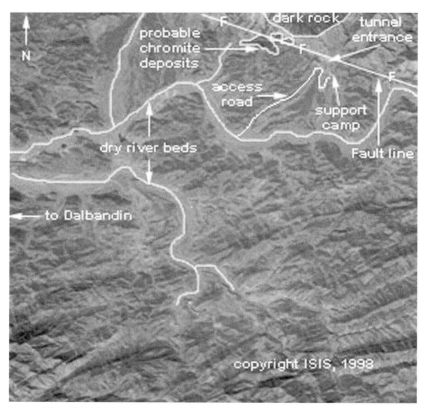

Figure 17.6—KVR-1000, 2-m Resolution Satellite Image taken February 25, 1998. It shows Pakistan's first nuclear test site in the upper right corner. Annotated by ISIS and Frank Pabian. (Distributed by SPIN-2)

25 km^2) that was limited to the extreme southern flank of the Koh Kambaran mountain area. Despite having missed inclusion of the most prominent readily identifiable feature (the mountain summit) shown in the televised terrestrial videos, that image did show both the geologic fault and, more significantly, evidence of some human activity at the base of that mountain.[6]

A CLOSER LOOK: CONFIRMATION AND CAMOUFLAGE

The Russian satellite image (see Figures 17.6 and 17.7) showed a complex of roads and buildings in the upper right-hand (northeastern) corner, consistent with a small test-support camp. Confirmation that these features were indeed associated with activity to support Pakistan's first nuclear test was made through correlation with additional terrestrial images taken on the ground by Pakistan

Television during Prime Minister Nawaz Sharif's June 19, 1998, visit to the test site. A Pakistani media report at the time also specified that the test site the prime minister visited was in the "Ras Koh mountain range in the Chagai district." (*Dawn*, 1998b.) Frame captures from that video are included here as Figures 17.8 through 17.10.[7] Figure 17.8 is a side-looking aerial view of the south face of the mountain along with a few of the larger site-support structures. The site layout consists of a small collection of buildings on an upper terrace, including

Figure 17.7—Closer View from KVR-1000 Image of Pakistan's First Nuclear Test Site (Distributed by SPIN-2)

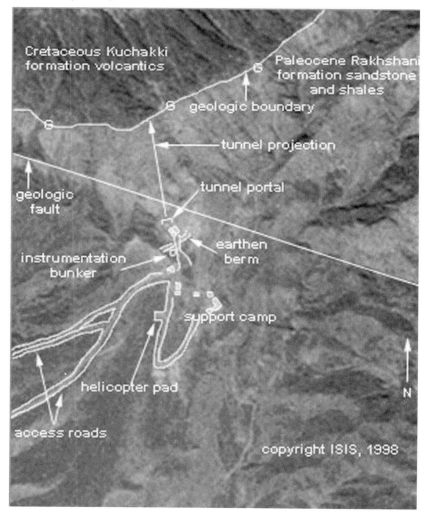

Figure 17.8—Pakistan's First Nuclear Test Site (see color plate)

the forward recording site instrumentation bunker with a single dirt road that leads to the tunnel portal. A loop-shaped road leads from the terrace down to the lower level, where the main support camp is located. Figure 17.9 is an exterior view of the forward recording site's instrumentation bunker, with an interior view showing what appear to be oscilloscopes (including one with a recording camera) connected with extensive input signal cabling. Figure 17.10 shows a frame-capture mosaic of the tunnel portal area, illustrating how the Pakistanis used adobe structures to obscure the portal's presence from overhead observation and monitoring. The exact location of the tunnel entrance was found through an analysis of the Pakistan Television footage, which briefly showed the prime minister inside the tunnel.[8] Similar adobe structures were scattered throughout the site, successfully giving it the general appearance of a simple hamlet (much as the Indians had achieved at their site in Rajasthan). The Pakistanis also used camouflage netting, matting, and canvas to conceal the instrumentation bunker, cabling, and a few other structures, along with various pieces of equipment around the site.

GEOLOGIC SETTING

During the May 28 test, the relative motion in the terrestrial video of the shaking of the mountain suggested that the strongest explosion(s) most likely were directly beneath the light-toned rock-capped false summit located on the southern flank of the Koh Kambaran massif, well within the Kuchakki volcanics, approximately 1 km from the tunnel portal. This is consistent with a media report that the tests occurred 1 km inside the mountain (*News Pakistan*, 1998; Baruah, 1998). Another report indicated that the tests were conducted in an "M-shaped tunnel" (Frontier Post, 1998). Such a tunnel arrangement, when seen in plan view, could be used to support multiple nuclear tests simultaneously, with the zigzag pattern providing assistance with closure to prevent venting of the blasts (see Figure 17.11).

Figure 17.11 compares an April 7, 1998, Indian IRS-1C panchromatic 5-m resolution image (courtesy Space Imaging, Inc., obtained subsequent to purchase and analysis of the Russian image) with a geological map. This image clearly shows the relative location of the peak and permits identification of the likely detonation point on the geologic map. According to the surficial geologic map, rocks located above the shot point consist of hard volcanics,

Figure 17.9—Instrumentation Bunker, Forward Recording Site (see color plate)

Figure 17.10—Concealed Tunnel Portal at Pakistan's First Nuclear Test Site (see color plate)

such as andesites, rhyolites, or welded tuffs, similar to those found in the mesas of the U.S. nuclear test site in Nevada.[9]

Interestingly, the geological map also shows that the tunnel actually crosses the fault line. The fault line can most clearly be seen in the Indian IRS-1C satellite image.

PHYSICAL TEST EFFECTS

Many pieces of distinctively light-toned, hard-capping rock on the peak were broken and dislodged during the test, which resulted in significant rock slides (clearly visible in ravines located on the south side of the mountain in the June 19, 1998, post-test video frame capture of the mountain). None of the rock slides were evident on the April 7, 1998, pretest satellite image of the mountain, the most recent earlier image available (compare Figures 17.8 and 17.11).

Conclusions

The studies described in this chapter demonstrate how new information can be derived through the integration of data from a variety of open sources and highlight the crucial role high-resolution commercial satellite imagery plays in deriving the new information. The results include accurately pinpointing the locations of both India's and Pakistan's nuclear test sites, the detailed characterization of the operational layout of the sites, and the determination of the measures taken to conceal the associated test preparations.

Information of this type would greatly enhance the capabilities already planned for future CTBT monitoring and verification and could prove critical in any future CTBT compliance dispute. However, the process of bringing the CTBT into force has gone slowly. Treaty proponents suffered a major setback when the Republican-controlled U.S. Senate voted against ratification in October 1999. In addition, the potential role of commercial imaging satellites in the verification process is not explicit and continues to be a subject of discussion and evaluation.[10] Pending a fully implemented CTBT monitoring regime, nongovernmental organizations, the public, and governments without national satellite systems will find commercial satellite imagery helpful.

Figure 17.11—IRS-1C 5-m Resolution Image (Courtesy: Spacelmaging.com) (see color plate, which includes a geologic map of the corresponding area)

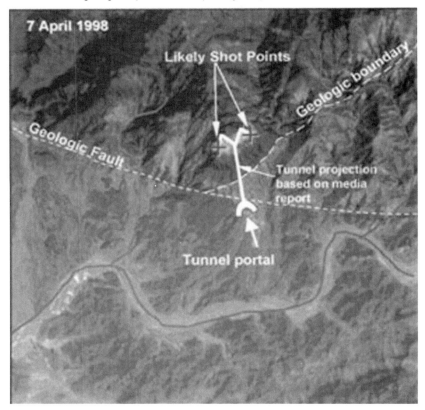

In Pakistan's case, the general geologic setting and the extent of short-term physical effects of the test(s) were also determined. That new information can, in turn, provide a basis for improving the geographic accuracy of processed seismic data for the region, possibly for refining yield estimates, and for verifying previous media reporting. Again, such information provides otherwise unavailable historical and mapping data necessary for the planning of any future on-site inspections that the CTBT Organization may conduct, should Pakistan accede to the CTBT.

These case studies also indicate a need for imagery analysts to differentiate effectively between conventional military activities, nuclear test-site preparations, and innocuous activities (i.e., mining or agricultural activities), while also maintaining vigilance to avoid being deceived by camouflage, concealment, and other such efforts.

The timeliness of available imagery and spatial resolution have, so far, been the only technical factors limiting their use as a CTBT verification tool. With the September 1999 launch of the first 1-m commercial imaging satellite (IKONOS of Space Imaging, Inc.), the problem of spatial resolution is lessened, but timeliness remains a concern.

Notes

1 This section is a condensed version of information contained in Gupta and Pabian (1996).
2 This section is drawn from Albright, Gay [Hinderstein], and Pabian (1999).
3 This report is limited to information generated about the first tests on May 28. Analysis continues of the second test site located about 100 km southwest of the first site, where another nuclear detonation occurred on May 30, 1998. Commercial satellite imagery of a probable location of the May 30 test can be found on the ISIS Web site: http://www.isis-online.org.
4 Personal communication, 1998; the reporter wishes to remain anonymous.
5 Both USGS and the PIDC placed the seismic origin much closer to the Ras Koh range than to the Chagai range. In fact, the uncertainties in both cases covered parts of the Ras Koh range but not the specific mountain face where we finally located the test site. For their final seismic analyses, see http://www.pidc.org; contact USGS directly; or see Barker (1998). See also Wallace (1998).
6 The geology of the mountainous area immediately adjacent to the cluster of seismically derived geocoordinates (and consistent with the televised ground images of the tunnel test location) was determined from a 1:250,000 scale geologic map of West Pakistan. The mountainous area, identified as the Koh Kambaran massif, rises to a maximum elevation of 2,700 m east and north of the Rayo (seasonal) river valley. The geology and the associated steep mountainous terrain of the Ras Koh massif indicate very hard rock of substantial thickness.

7. All images in the Pakistan case study can be viewed at http://www.isis-online.org.
8. To view this video clip, go to http://www.isis-online.org.
9. The specific geology at the emplacement site can vary depending on both the actual length of the tunnel and differences between the geology at the surface and that several hundred meters below. Given the surficial geologic map, a tunnel of roughly 3 km would be required to extend completely through the Kuchakki volcanic group into the Ras Koh formation's diorites and syenites (dark, quartz-poor, granites). If the tunnel were only about 1 km long (as reported in the Pakistani media), the Kuchakki formation would appear to be the more likely geologic setting for the detonation(s). However, given that the subsurface geology can differ considerably from what is mapped on the surface, one has to take quite seriously another May 29, 1998, report in the Pakistani newspaper, *Dawn* (1998a), which stated that the "mountain structure" (used in the May 28 tests) was "composed of black granite rocks." As a result, when considering seismic coupling in yield calculations, one should treat the geologic setting of the test(s) as most likely consisting of diorites and syenites and then, secondarily, hard volcanics such as andesites, ryolites, or welded tuffs. However, yield estimates that assume good coupling in hard rock will not be affected in any case as the rock densities and mechanical strengths of these two different rock types are roughly equivalent (see Government of Canada, 1958).
10. For more information on the use of commercial satellite imagery in a CTBT verification regime, see Gupta and Pabian (1998).

References

Albright, David, Corey Gay [Hinderstein], and Frank Pabian, "New Details Emerge on Pakistan's First Nuclear Test Site," *Earth Observation Magazine*, December 1998–January 1999, pp. 18–21.

Barker, Brian, "Monitoring Nuclear Tests," *Science*, Vol. 281, September 25, 1998, pp. 1,967–1,968.

Baruah, Amit, "Pakistan 'Clears' Mystery Over Nuclear Tests," *The Hindu*, on-line ed., June 30, 1998.

Bedi, Rahul, "Indian Success Casts Doubts Over US Reconnaissance . . . as Nuclear Test Preparations Avoid Detection," *Jane's Defence Weekly*, May 20, 1998, p. 5.

Chengappa, Raj, "The Bomb Makers," *India Today International*, June 22, 1998, pp. 28ff.

Dawn: The Internet Edition, "No Radiation Leakage: Tests Yield up to 40 kt, says PAEC," May 30, 1998a. The Web site is at http://www.dawn.com/ (last accessed June 27, 2000).

—, "Troops Protect Chagai Site," May 30, 1998b. The Web site is at http://www.dawn.com/ (last accessed June 27, 2000).

The Frontier Post—Internet Edition, "PAEC Team to Monitor Radioactivity," June 1, 1998.

Gannon, Kathy, "'No Nuclear-Device Tested yet but It Is Certain,' says Gohar Ayub," *The Times of India*, May 18, 1998.

Government of Canada, Geologic Map No. 19, Kharan Kalat, 34 DH, Reconnaissance

Geology of West Pakistan, published for the government of Pakistan, 1958.

Gupta, Vipin, and Frank Pabian, "Investigating the Allegations of Indian Nuclear Test Preparations in the Rajasthan Desert: A CTB Verification Exercise Using Commercial Satellite Imagery," *Science and Global Security*, Vol. 6, 1996, pp. 101–189. Available at http://www.ca.sandia.gov/casite/gupta/index.html (last accessed June 23, 2000).

—, "Viewpoint: Commercial Satellite Imagery and the CTBT Verification Process," *The Non-proliferation Review*, Spring-Summer 1998, p. 89.

The Hindustan Times, editorial, June 5, 1998.

Hough, Howard, "Special Feature: Asian Nuclear Testing . . . Ground Zero, in the Chagai Hills," *Jane's Intelligence Review*, July 1998, p. 21.

Jasani, Bhupendra, *Civil Observation Satellites and Arms Control Verification*, preliminary report, London: King's College, Department of War Studies, August 1992.

Joshi, Manoj, "Nuclear Shock Wave," *India Today*, on-line ed., May 25, 1998. Available at http://www.india-today.com/itoday/25051998/cover.html (last accessed June 2, 2000).

The News Pakistan, "N-tests yield in accordance with the design calculations: Dr. Ashfaq," on-line ed., June 30, 1998.

Reuters, "Pakistani N-Tests Took Two Hours, Residents Say," May 28, 1998.

Wallace, Terry C. "The May 1998 Indian and Pakistan Nuclear Tests," *Seismological Research Letters*, September 1998.

Weiner, Tim, "U.S. Fears a Weekend Test by Pakistan," *International Herald Tribune*, May 15, 1998, p. 1.

Windrem, Robert, "Pakistan's Nuclear Test Site," MSNBC, May 28, 1998.

18
Nongovernmental Use of Commercial Satellite Imagery for Achieving Nuclear Nonproliferation Goals: Perspectives and Case Studies

David Albright and Corey Gay Hinderstein

In the last few years, high-resolution commercial satellite imagery has become more widely available and more affordable. As a result, we have been examining the value of this imagery for our efforts to develop credible, open information about secret nuclear weapon programs, improve nonproliferation and arms control verification, and bolster work aimed at stopping the spread of nuclear weapons. A key focus of our effort is the integration of images with other information to derive more meaningful products.

We have concluded that the responsible use of high-resolution commercial satellite imagery can make a valuable contribution to the development and achievement of nonproliferation goals and policies. As we evaluated more images, we have learned several lessons and encountered several challenges, including coping with resource constraints on the purchase and interpretation of relatively expensive imagery.

Using case studies that characterize nuclear facilities of proliferation concern, this chapter discusses how satellite imagery can assist nongovernmental organizations (NGOs), such as the Institute for Science and International Security (ISIS), achieve their goals in the field of nuclear nonproliferation. Studies we have conducted are used to illustrate these efforts.[1]

We use panchromatic satellite data in this chapter. This type of imagery has the highest resolution commercially available and thus reveals the greatest spatial detail about buildings and sites. We also found these data to be the easiest to interpret during our initial efforts.

Other types of imagery are also likely to prove useful in the nuclear nonproliferation area. Thermal imagery may be useful for assessing the operational status of nuclear facilities that produce significant quantities of heat. Multispectral imagery, despite having lower resolution than panchromatic imagery, may prove useful in monitoring certain sites. This type of imagery contains more data, such as different color bands, that may be useful in detecting camouflaging or subtle changes at a site. Commercially available hyperspectral imagery may be able to detect chemical effluents, which could be particularly useful in monitoring nuclear facilities. Because synthetic aperture radar (SAR) satellites can provide imagery on cloudy days, RADARSAT-2, with its 3-m resolution, may also be useful for monitoring nuclear sites.

Background on Nuclear Facilities

The case studies in this chapter focus on nuclear sites in proliferant states, which are states seeking or possessing nuclear weapons that are not considered one of the five nuclear weapon states recognized in the Nuclear Non-Proliferation Treaty (NPT).[2] The most significant industrial activity at these sites is the production of plutonium or highly enriched uranium, or the development of the ability to make these materials. The main facilities at these sites are reactors, plutonium separation or "reprocessing" facilities, uranium enrichment plants, and fuel fabrication facilities, or research and development buildings associated with the above activities.

Detecting nuclear facilities in many proliferant states through photo interpretation has proven difficult. Nuclear facilities in such states tend to be smaller than and can be of designs different from similar facilities in the United States and Russia, making them easier to camouflage against overhead detection. This complicates photo interpretation, which often begins by looking for "indicators" or "signatures" that have been identified by studying large facilities in the major nuclear powers.[3]

Nonetheless, nuclear facilities in proliferant states have frequently been identified in imagery, particularly when the analysis includes information from other intelligence sources or on-site inspections by the International Atomic Energy Agency (IAEA). With the advent of commercial high-resolution imagery, members of the public and the media now face the challenge of learning to recognize nuclear sites and combining this imagery with other available information to improve their analyses. Basic to any such endeavor is an understanding of the major nuclear facilities and their characteristics in overhead imagery. The following is intended to provide a short introduction to the main nuclear facilities discussed in this chapter.

In general, many types of nuclear facilities are involved in civil and military applications. The most common nuclear facility is the civilian nuclear power reactor that is optimized to produce electricity, although such reactors also produce plutonium as a by-product. In most proliferation cases, however, reactors of concern tend to be significantly smaller than power reactors, often characterized by proliferant states as "research reactors." Rather than electricity, the main product of these reactors is plutonium.

Power reactors are rather easy to see in overhead imagery—being large, having electrical transmission equipment, and typically possessing distinctive containment vessels surrounding the nuclear core. Research reactors, on the

Figure 18.1—A Small Research Reactor at North Korea's Main Nuclear Site, January 1989

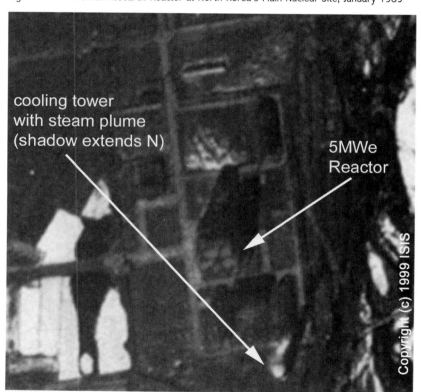

other hand, are relatively small, often lack any electrical transmission equipment, and may be housed inside industrial-looking buildings. However, they may have cooling towers that can be identified in high-resolution satellite imagery. Figure 18.1 is a Russian KVR-1000 satellite image taken in January 1989 of a small research reactor at North Korea's main nuclear site. Although this site will be covered in more detail later, this photo shows a reactor and a cooling tower. Although the reactor is in a rather nondescript building, the cooling tower is plainly visible. Moreover, a plume can be seen coming from the cooling tower south of the reactor, which means that the reactor was operating when the photo was taken.

Some research reactors are housed in easily identified domed buildings. **Figure 18.2** is a 1-m IKONOS image that shows the dome of the Khushab

Figure 18.2—A 1-m IKONOS Image Showing the Dome of the Khushab Reactor, Near Sargodha, Pakistan

Figure 18.3—A Heavy Water Plant Just South of the Khushab Reactor

reactor, located near Sargodha, Pakistan. Near the reactor, cooling towers can be seen, although in this image noticeable steam plumes are not visible, suggesting that the reactor was not operating or was operating at low power when the image was acquired.

The Khushab reactor requires a relatively large amount of heavy water to operate. A heavy water plant with distinctive features, such as distillation towers, can be identified just south of the reactor (see Figure 18.3).

Before plutonium can be used in nuclear weapons, it must be chemically separated from other radioactive material in the irradiated fuel. This is done in special facilities, called plutonium separation plants. Plutonium separation plants often have distinctive characteristics, such as heavy concrete shielding, extensive air filtration systems, or nearby nuclear waste facilities. In addition, many reprocessing plants are long and narrow, reflecting a design that places a series of chemical equipment involving radioactive material in a long, narrow line.

The other main type of nuclear explosive material is highly enriched uranium, which is produced in uranium enrichment plants. Different types of uranium enrichment facilities vary enormously in size and characteristics. Some enrichment plants, such as those based on gaseous diffusion or electromagnetic isotope separation technologies, may have extensive electrical power equipment, large water purification buildings, and cooling towers. Other types, such as gas centrifuge uranium enrichment facilities, can be particularly difficult to identify in overhead imagery. Unlike plutonium production and separation operations, the gas centrifuge process involves little nuclear waste or shielding, requires little electricity, and involves little heat.

Most major nuclear sites have shared characteristics, including a remote location and well-defined high security perimeters. Many sites are also relatively large and may have substantial air defenses.

An example of a nuclear facility with several of these characteristics is the Rocky Flats site near Denver, Colorado, which made nuclear weapon components for the U.S. nuclear arsenal. For most of the site's history, its main product was the plutonium core of a fission explosive, which was often used

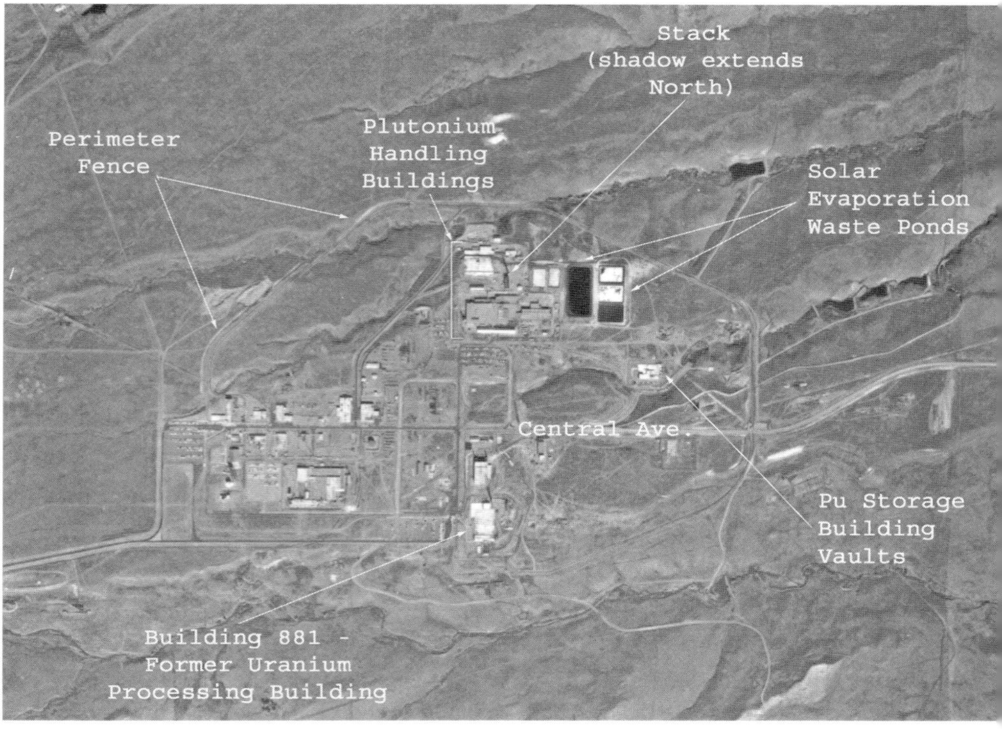

Figure 18.4—Aerial Photograph of the Rocky Flats Site, October 1964

in a thermonuclear weapon to ignite the fusion "secondary." Figure 18.4 is an October 15, 1964, declassified U-2 aerial photograph of the Rocky Flats site that shows several traditional indicators of a nuclear site.[4] In addition to a set of buildings involved in processing plutonium into nuclear weapon components and a uranium processing building, the photo shows the shadow of a high stack that is part of an air filtration system for one of the plutonium handling buildings. Ponds adjacent to the plutonium buildings are for drying liquid radioactive waste. A security perimeter is clearly visible.

Imagery as a Public Education Tool

Properly analyzed commercial satellite imagery can increase the public's awareness of the dangers of nuclear proliferation and the methods through which proliferants pursue nuclear weapon capabilities. As illustrated above, imagery can vividly show a nuclear weapon production facility. Multiple images can indicate progress in construction of a facility and growth in the number of buildings at a site. Annotated images can build appreciation for intrusive verification requirements under the NPT or other arms control treaties. Images can also make important contributions to historical analyses of nuclear programs in proliferant states.

In some cases, the release of commercial satellite imagery could improve and expand the debate on key proliferation issues and concerns. This role may be especially important during crises or when available information is limited.

Becoming familiar with the complexity of nuclear facilities and sites can help the public understand the nature of the process of obtaining nuclear weapons, particularly the production of nuclear explosive materials. In addition, thermal or hyperspectral imagery may provide information about the operational status of nuclear facilities.

NORTH KOREAN YONGBYON SITE

During the early 1990s, outside of public view, high-resolution satellite imagery was critical in proving that North Korea was deceiving the IAEA. This imagery also indirectly indicated that North Korea had produced more plutonium than it had declared to the IAEA. In early 1993, U.S.-supplied high-resolution satellite imagery presented to the IAEA's Board of Governors demonstrated that North Korea had systematically hidden from inspectors two suspected nuclear waste sites that could hold waste left over after separating plutonium from irradiated material. These images were instrumental in convincing the IAEA to call for "special inspections" in North Korea in February 1993.

North Korea refused to allow the special inspections, precipitating a crisis with the United States and its Asian allies. But in 1994, North Korea and the United States signed the Agreed Framework Between the United States of America and the Democratic People's Republic of Korea. North Korea agreed to "freeze" operations and construction at a plutonium separation plant and at its gas-graphite reactors and to allow the IAEA's special inspections at an agreed upon time, currently expected to occur before 2005. In exchange, the United States agreed to organize the supply of two modern power reactors that are significantly more difficult to use to make nuclear weapons than the ones North Korea shut down under the freeze.

This conflict over suspect nuclear sites and additional undeclared plutonium production centered on facilities at the Yongbyon Nuclear Center, about 100 km north of Pyongyang. Although North Korea came under intense scrutiny in the late 1980s and early 1990s, the public did not understand the vast and complex nature of the site. Contributing to the confusion, the United States did not publicly release any of the imagery it provided to the IAEA. During this period, the available commercial imagery was limited to 10-m SPOT imagery. Although SPOT images of Yongbyon were common and educationally useful, this imagery was not sufficient to show much detail about the site. Not surprisingly, annotated images and schematics of the site distributed by the media suffered from mistakes and misinterpretations.

Given the importance of the historical development of the Yongbyon site, we purchased KVR-1000 imagery of the site taken in 1989. We sought other imagery from the early 1990s, but the Russian supplier said it could not provide any other photos for that period.[5]

The 1989 image provides more detailed information than was available from earlier imagery sources. It is detailed enough to provide a time record that shows a suspect nuclear waste site that was later covered up with dirt and vegetation.

The effort to acquire and analyze this imagery is part of an ISIS project to document the history of the conflict between the IAEA and North Korea about how much plutonium North Korea may have produced (Albright and O'Neill, 2000). Despite its obvious inferiority compared to U.S. government imagery, current commercial satellite imagery has allowed us to show far greater detail of the Yongbyon site. Unlike the SPOT imagery, the newer imagery clarifies which nuclear facilities were disguised by North Korea in its attempt to deceive the IAEA.

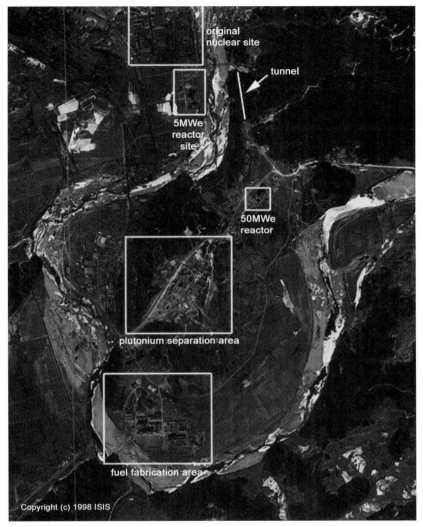

Figure 18.5—The Yongbyon Nuclear Center, 1989

Figure 18.5 is the January 1989 Russian image of the Yongbyon Nuclear Center. Clearly visible is the Kuryong River, which runs through the site and around a portion of it. In the upper portion of the picture is the oldest part of the center, containing a small, Russian-supplied research reactor, the Isotope Production Laboratory, and the Institute of Radiochemistry, which date from the 1960s and early 1970s. The 5 megawatt-electric (MWe) reactor, associated buildings, and a cooling tower can be seen in an area directly below the oldest part of the center (see also Figure 18.1). This reactor is suspected of producing up to 8 kg of undeclared plutonium and is subject to the freeze.

The 50 MWe reactor can be found by crossing a bridge that starts near the 5 MWe reactor, passing through a tunnel, and continuing further down the road. The 50 MWe reactor was still incomplete in 1994 at the time of the Agreed Framework and is another of the frozen reactors. The other reactor covered by the freeze was under construction at Taechon in North Pyongan Province.

Figure 18.6—The Yongbyon Nuclear Center, 2000

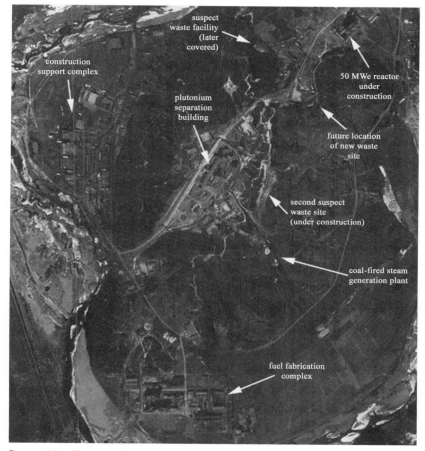

Figure 18.7—Plutonium Separation Building at Yongbyon, 1989

Continuing down the road past the 50 MWe reactor, one of the "suspect" waste sites can be seen on the right of the road and about 200 m from it. In a 2000 IKONOS image, this waste site is no longer visible (see Figure 18.6). Another waste site was built a few years later, apparently as a ruse. Although North Korea claims that the second site has been there since 1977, it is not visible in the 1989 image but is in the 2000 image.

A short distance further down the road is the plutonium separation complex, which contains the distinctive-looking plutonium separation building (see Figure 18.7). Nearby buildings include a spent fuel receiving facility, located across the road from where spent fuel enters the separation building, and interim waste storage and treatment buildings. Further away is evidence of another "suspect" waste site, which we understand was in an early stage of construction in early 1989. This suspect waste building is complete in the 2000 IKONOS image in Figure 18.8.

Turning right at the "T" intersection, one comes into an area where construction support buildings are located. Going left, one soon comes to the fuel fabrication complex in the south end of the loop of the river. Continuing on the road, and looping northward, one passes a coal-fired, steam-generating plant, for use by the plutonium separation complex and other facilities in this part of the site.

IRAQI GAS CENTRIFUGE RESEARCH AND DEVELOPMENT FACILITY

Iraq's pre-Gulf War clandestine gas centrifuge uranium enrichment program presents a case that can sensitize the public about the utility of imagery

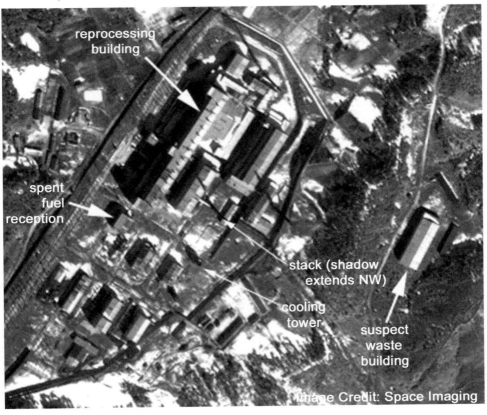

Figure 18.8—Plutonium Separation Building at Yongbyon, 2000

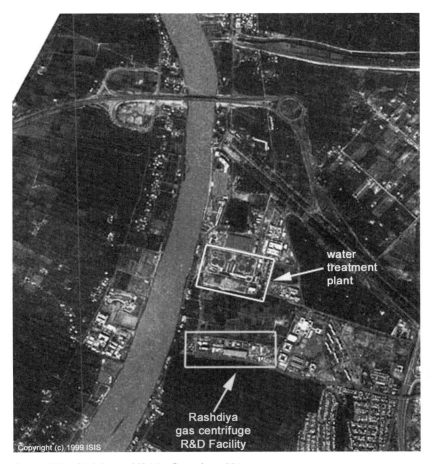

Figure 18.9—Rashdiya and Vicinity, December 1991

analysis when a state undertakes extensive concealment and deception efforts and when the nuclear facility is one with few, if any, clear signatures. This case highlights the importance of multisource information and how satellite imagery can be used to identify a site once other information has revealed its general area.

The Rashdiya site was Iraq's principal gas centrifuge research and development site before the 1991 Persian Gulf War and was slated to hold Iraq's first operational centrifuges for producing enriched uranium. Located in the northern outskirts of Baghdad, this site went undetected by Western intelligence services before the war and was not targeted in the Coalition bombing campaign. The Iraqis went to great lengths to make this site nondescript while maintaining adequate security. In any case, gas centrifuge facilities, particularly those engaged in research and development activities, have few, if any, reliable indicators.

An Iraqi defector revealed Rashdiya's existence and its general location in the summer of 1991. He described this site as near a water-treatment facility just north of Baghdad. According to an inspector involved in the search for this site, inspectors of the IAEA Action Team, with the assistance of member states, used U-2 aerial imagery to identify a likely site. This site was the only unknown facility in the area with the appearance of research and development activities. Figure 18.9 is a KVR-1000 image showing Rashdiya and the surrounding area in December 1991, including the water-treatment facility.[6]

Nonetheless, the subsequent inspection proved ambiguous because of extensive Iraqi concealment activities. Not until 1995, following the defection of General Hussein Kamel, Saddam Hussein's son-in-law, did Iraq finally reveal the truth about Rashdiya. In 1996, one of us visited Rashdiya and can attest to the lack of any clear indicators of gas centrifuge or nuclear activities that would be visible in overhead imagery.

Figure 18.10 shows the layout of the Rashdiya site. We have identified some of the major buildings. As discussed, there are no distinctive indicators of nuclear activity in this image. Security features are apparent; however, the site does not appear to be a high-security military area. Further complicating the interpretation of this image, the centrifuge site is adjacent to apartment buildings.

Figure 18.10—Rashdiya Layout (copyright isis, 1999)

Despite the Iraqi conversion of the Rashdiya site to nonnuclear uses in 1991, a step verified by Action Team inspectors, some in the media identified it as a nuclear site during the U.S.-British bombing campaign of Iraq in December 1998. As far as we know, however, it was not bombed during this campaign.

Analyzing False or Misleading Media Claims

During a crisis involving a country's nuclear activities, high-resolution satellite imagery can be used to sort out exaggerated or untrue claims in the media. For example, following the Indian and Pakistani nuclear tests in May 1998, satellite imagery was important in straightening out many media distortions and answering questions about these tests (see Chapter 17 in this volume).

Another example involves the Yongbyon Nuclear Center in North Korea (see Figure 18.3). Under the 1994 Agreed Framework, key nuclear facilities at the site are subject to a "freeze" monitored by the IAEA. In May 1998, the media reported that North Korea was threatening to open and possibly resume operations at one of the sealed nuclear reactors at the site, calling into question the North Korean commitment to the Agreed Framework.[7] U.S. and IAEA officials responded that North Korea had proposed routine maintenance at the reactor and had not expressed any intention to resume operation. Although commercial satellite imagery could not verify the nature of activity within a building, thermal, multispectral, hyperspectral, and panchromatic satellite images could potentially be used to verify the nonoperational status of the 5 MWe reactor and the main reprocessing plant. In addition, imagery could also be used to combat concurrent rumors of resumed large-scale construction at the unfinished reactor sites. In this way, satellite imagery played a part in sorting out conflicting or ambiguous information that had been used to justify particular policy positions publicly.

Undoubtedly, the frequency of misinterpreted images, "paparazzi images," and other misinformation will increase as high-resolution satellite imagery becomes more prevalent. NGOs and the media are likely to make some serious mistakes. We can only hope that these inevitable mistakes do not undermine credible efforts to effectively use satellite imagery.

Increasing Accountability

Commercial satellite imagery has the potential to make governments more accountable. In the event of suspicious activity in a potential or actual proliferant state, commercial satellite imagery could indicate banned or destabilizing activities. The satellite image, properly interpreted by the media or NGOs, could become a basis for further action by concerned governments or international verification agencies.

In general, commercial satellite imagery can be a resource for analyzing the statements of governments and verification agencies. With commercial satellite imagery available, governments and international organizations will have to take into account that their conclusions and judgments with regard to a country's activities will be examined and questioned in a new way by the public.

Governments could also voluntarily release imagery to NGOs or the media to allow more transparency in their nuclear activities. In some cases, such releases could remove ambiguity about activities that were increasing suspicion in the first place.

LOOKING BACK: UNEXPLAINED GROWTH IN IRAQ'S TUWAITHA NUCLEAR RESEARCH CENTER[8]

What if, in the 1980s, the public and the media had had access to high-resolution imagery of Iraq's Tuwaitha Nuclear Research Center, located about 30 km south of Baghdad? Press speculation at the time was rampant that undeclared nuclear activities were being carried out at Tuwaitha. The IAEA, responsible for verifying that Iraq was complying with the NPT, was confined to inspecting a small set of buildings involved in declared nuclear activities. It was not provided any photographic evidence of the construction activities at Tuwaitha, for reasons known only to the national authorities that had access to imagery at the time. Had high-resolution commercial imagery been available, other actors more aggressive than the IAEA—such as the media, NGOs, or perhaps certain governments—might have used this imagery to identify buildings suspected to be housing prohibited nuclear weapon activities. NGOs armed with commercial satellite images could have prompted official action in the case of ambiguous nuclear facilities at Tuwaitha.

Although the inspection rights of the IAEA were severely limited, the agency had the right to conduct "special inspections" in nonnuclear weapon states party to the NPT. Armed with evidence, the IAEA would have had a legal right to request inspections of additional buildings at Tuwaitha. A refusal by Iraq to allow the inspections would have triggered greater suspicion.

On the other hand, would Iraq have acted differently if it knew the public had access to such imagery? Would it have dispersed its activities at Tuwaitha earlier than it eventually did, fearing that public disclosure would somehow inhibit its progress in making nuclear weapons? We can never know. Such a step, however, could have been costly and would have stretched Iraq's technological resources, particularly in the early and mid-1980s, when it suffered serious personnel and infrastructure shortages.

During the 1980s, the Tuwaitha Nuclear Research Center grew dramatically. Much of this growth involved new buildings and facilities that housed clandestine nuclear weapon-related activities in violation of Iraq's commitments under the NPT. This large-scale construction activity was well removed from known civil facilities.

Figure 18.11 is a Russian KVR-1000 satellite image that shows the Tuwaitha site in December 1990, just one month before it was heavily damaged by the

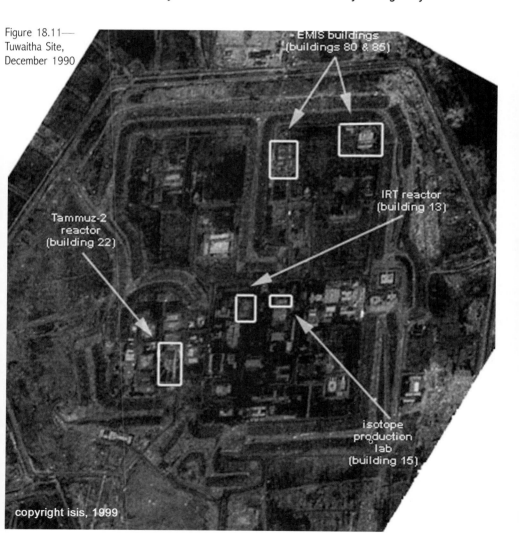

Figure 18.11—Tuwaitha Site, December 1990

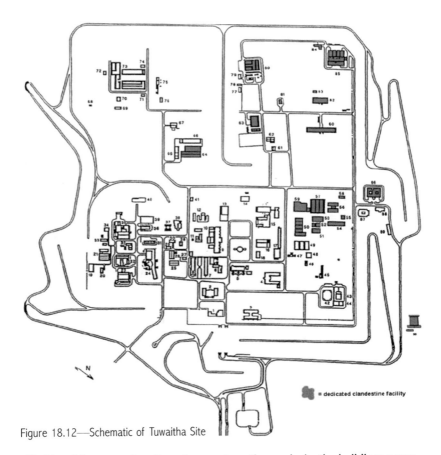

Figure 18.12—Schematic of Tuwaitha Site

allied bombing campaign. Based on a retroactive analysis, the buildings represented in Figure 18.11 were identified. Figure 18.12 shows Tuwaitha schematically, where the dedicated, clandestine facilities are shaded. Weapon-related activities occurred in many other buildings, but the shaded buildings are those in which essentially only undeclared activities occurred.

In the late 1980s, such detail was unavailable; however, it was known publicly which facilities the IAEA inspected or visited. Although satellite images could not have revealed the purpose of the new buildings, they might have at least raised questions about Iraq's intentions. Given the public's and media's intense interest in Iraq's nuclear weapon intentions, many NGOs or members of the media would have sought high-resolution satellite imagery and raised questions about these unexplained buildings.

The IAEA now purchases commercial satellite imagery as part of its strengthened safeguard system, which resulted from an extensive review of safeguards following the Gulf War. Under the new system, the chance of the IAEA missing another Iraq has been greatly reduced.

Armed with satellite imagery, the public also has the opportunity to help prevent another Iraq. At least, NGOs and the media now have the capability to use commercial satellite imagery to look for undeclared nuclear or nuclear-related activities. If done credibly, these actors could stimulate the IAEA or governments to investigate suspicious facilities.

A Final Note of Caution and Hope

Commercial satellite imagery offers enormous advantages to help NGOs achieve their goals. At a minimum, satellite imagery offers another source of data to help sort the truth from conflicting and fragmentary information. Imagery could be a powerful new tool for NGOs to filter plausible from less plausible reports.

The multistep process to arrive at an annotated image requires considerable experience and information about the image's subject. Imagery can thus be only a part of the intensive effort that NGOs, the media, concerned governments, and such international verification agencies as the IAEA will need to make to obtain information about secret nuclear weapon programs.

Beyond the basic need to provide the context of an image, the successful use of imagery confronts a number of challenges. Nonetheless, several developments give cause for optimism, particularly in the long term.

COSTS AND RESOURCES

NGOs' successful use of imagery is currently hindered by a lack of funding from foundations, governments, or international organizations. Purchasing imagery is expensive. Increasingly, one image of a particular date is not sufficient; instead, imagery of a site taken on multiple dates is necessary.

In addition, the equipment and software necessary to properly analyze imagery requires relatively large capital expenditures. For example, adequate software is required to compare images of the same site on different dates to detect changes. The willingness of the company ERDAS to offer discounted rates for its Imagine software to nonprofit research and educational organizations is encouraging.

To reduce costs, we are drawn to share imagery with other organizations, but several problems must be worked out to do so. One in particular is that, in some cases, the commercial license governing the use of an image does not allow such sharing. With the future launching of more satellites, imagery costs should decline. In addition, as firms encounter more competition, they may be more willing to resolve licensing inconsistencies that inhibit NGO co-operation. As NGOs develop their own expertise, cooperation should also be easier to accomplish. One possibility is for NGOs to pool their buying power into a new legal company to obtain a better rate from the imagery sellers.

PHOTO INTERPRETATION

The development of a capability to interpret the image itself is demanding in terms of organizational resources. NGOs need the assistance of experienced photo analysts. Finding such assistance is not easy, given the classification rules that have surrounded satellite imagery of proliferant states and the high cost of hiring the available professional photo analysts. Nonetheless, we remain hopeful that experienced people will help NGOs learn how to interpret and annotate commercial satellite imagery.

As ISIS and others outside the traditional intelligence community climb the learning curve of high-resolution imagery interpretation, they are bound to make mistakes. However, methods of coping with mistaken imagery analysis are emerging as NGOs, independent experts, and the media establish adequate review and oversight in this area.

As NGOs gain experience using commercial satellite imagery in policy debates, the quality of the product will improve. Such experience will further legitimize their role as photo interpreters and possibly lead to offers of assistance from professional or governmental photo analysts.

EFFECTS ON CAMOUFLAGING

With the dawning of the era of high-resolution commercial satellite imagery, much has been discussed about the possibility that some countries may take additional steps to disguise and hide their nuclear activities. Clandestine proliferant programs are taking steps to do this in any case because of the advancements in IAEA safeguards and a broader recognition of national intelligence agency capabilities. But a country may be more concerned about public reaction than about IAEA or intelligence reports. Thus, proliferant states are likely to continue to improve the concealment of their covert activities.

On the other hand, activities that adequately camouflage a nuclear weapon program can also complicate success. Activities may have to be spread out to multiple facilities or moved to keep them secret. This step may be difficult for some programs to achieve economically, delaying the program as a result and possibly exposing it to international inspectors or intelligence agencies in other ways.

If increased monitoring by commercial high-resolution satellites does drive states to increasingly elaborate lengths to hide their clandestine nuclear activities, the public will have joined the long-term cat-and-mouse game that intelligence agencies and international verification agencies have played to defeat those disguising activities and get to the truth.

FUTURE CAPABILITIES

Satellite imaging technologies are expected to improve. Higher resolutions are inevitable. More powerful spectral capabilities—including radar, multi-spectral, hyperspectral, and thermal—may dramatically increase the ability of NGOs, the IAEA, and governments to detect suspicious indicators of undeclared nuclear activities or to combine these data with other available information to strengthen nonproliferation endeavors.

If NGOs can successfully integrate satellite imagery with other sources of information, unique products may result. In some cases, NGOs have access to information that the U.S. government or the IAEA does not. NGOs may be able to use imagery to develop further as international actors in their own right.

It will be exciting to see if commercial satellite imagery can contribute significantly to efforts to stop the spread of nuclear weapons. If the efforts are successful, we will all be more secure.

Notes

1. More information about these particular case studies can be obtained by consulting the References or reading Chapter 17 in this volume. Higher-quality images can be found on the ISIS Web site at http://www.isis-online.org. Other case studies can be found in Gupta (1992) and Handler (1999). In addition, Bhupendra Jasani, Department of War Studies, King's College London, has done numerous studies on the security and arms control role of commercial satellite imagery.
2. The NPT defines a nuclear weapon state as a country that had manufactured and exploded a nuclear weapon or device by January 1, 1967. Under this definition, there are five nuclear weapon states: Britain, China, France, Russia, and the United States. The most important proliferant states currently are India, Iran, Iraq, Israel, North Korea, and Pakistan. Algeria, South Korea, and Taiwan, while not believed to be seeking nuclear weapons, continue to raise concern about their possible future actions.
3. For example, see the U.S. Defense Mapping Agency (1995), Ch. 4 (this work was declassified in April 1996 and obtained through the Freedom of Information Act by the Federation of American Scientists). This source contains an interesting and useful discussion of nuclear facilities indicators, although significant portions of the discussion can be misleading if applied directly to facilities in proliferant states.

4   For declassified U.S. Corona satellite photos of Rocky Flats from the same period, visit the ISIS Web site (http://www.isis-online.org). We decided to use the U-2 image here, because its resolution is better than that of the Corona images and is similar to that of currently available commercial satellite imagery. An extensive collection of declassified Corona photos can be seen on the Federation of American Scientists Web site (http://www.fas.org).
5   Lack of archived imagery is a problem for many of our studies. The declassification of U.S. Corona images from the 1960s and early 1970s has been useful in our work. We remain hopeful that the U.S. government will continue to declassify additional satellite and aerial images.
6   In contrast to many images, which typically cost over $1,000, this one cost only $20. It was available on the Terraserver Web site (http://www.terraserver.com).
7   See, for example, Rosenthal (1998), Sullivan (1998), or Gay (1998).
8   This section is based on Albright, Hamza, and Hinderstein (1999). This study can be viewed or downloaded at the ISIS Web site at http://www.isis-online.org.

References

Albright, D., Khidhir Hamza, and C. Hinderstein,"Development of the Al-Tuwaitha Site: What If the Public or the IAEA had Overhead Imagery?" April 26, 1999. Available at http://www.isis-online.org (last accessed June 12, 2000).

Albright, David, and Kevin O'Neill, eds., *Solving the North Korean Nuclear Puzzle*, Washington, D.C.: ISIS Press, 2000.

Gay (Hinderstein), Corey, "So, Whatever Happened to Checking with Another Source?," *The Bulletin of the Atomic Scientists*, July/August 1998, p. 12.

Gupta, Vipin, "Algeria's Nuclear Ambitions," *International Defense Review*, April 1992, pp. 329–331.

Handler, Joshua, "Lifting the Lid on Russia's Nuclear Weapon Storage," *Jane's Intelligence Review*, August 11, 1999, pp. 19–23.

Rosenthal, Elisabeth, "North Korea Says It Will Unseal Reactor," *The New York Times*, May 13, 1998, p. A10.

Sullivan, Kevin, "N. Korea Threatens Revival of Its Nuclear Program," *The Washington Post*, May 15, 1998, p. A33.

U.S. Defense Mapping Agency, *Photo Interpretation Student Handbook, Vol. 2: Cultural Features*, 1995.

19 Supporting Humanitarian Relief Operations

Einar Bjorgo

Today's disasters, natural or man-made, are often complex and affect a large number of people who are forced to seek refuge in other parts of the country or to cross an international border for aid and protection. Such victims are referred to as *internally displaced persons* (IDPs) if they remain within their country of origin or *refugees* if they cross an internationally acknowledged border. The United Nations High Commissioner for Refugees (UNHCR) currently provides protection and assistance to approximately 21 million refugees and IDPs persons (UNHCR, 1999). This chapter does not distinguish between an IDP and a refugee. Protection and assistance are most often provided when refugees are organized in camps or settlements, which can offer shelter, water, food, medicine, and other basic needs in an emergency situation.

Although we may think of refugee camps as temporary, the average refugee settlement lasts for approximately seven years (Rogge, 1987), much longer than typical international media coverage. This requires refugee camps to be well planned with respect to communication lines, water resources, soil conditions, longer-term self-sustainability and potential environmental effects. In tense situations, camp locations also need to be planned in areas protected from potential gunfire or shelling from hostile forces. Relief agencies require updated maps of the area of interest that include road networks, potential landing sites for airplanes, locations of the refugees and estimates on their numbers, potential refugee camp sites, etc. Furthermore, environmental impacts need to be monitored to ensure sustainable use of natural resources in refugee-hosting areas. All this information is important for the security of the refugees and the relief staff, as well as for the protection of the environment. This chapter illustrates how such humanitarian relief operation information needs can be met using satellite remote sensing technology.

Various satellite imagery sources can be used, depending on the information required. These sources vary from relatively coarse resolutions (such as Landsat) to higher resolutions of 5 m or less (such as IKONOS)[1] and include archived SPIN-2 (KVR-1000 in Europe) imagery. To illustrate the potential level of global transparency satellite sensors offer during refugee relief operations, this chapter will describe four cases in which this technology has been used. The main focus will be on the higher-resolution satellite sensors, but examples from coarser-resolution satellites are also included. The case studies are located in: Asia (Site 2, Thailand), America (Chiapas, Mexico), Africa (Qala en Nahal, the Sudan), and Europe (Albania, Macedonia, and Kosovo). However, this technology has limitations, which are described later in this chapter. It is important to keep these limitations in mind, because they can heavily influence the successful use of satellite technology in relief operations.

Commercial Observation Satellites and Humanitarian Relief Operations

In general, the whole spectrum of commercial observation satellites, from geostationary weather satellites to polar-orbiting higher-resolution satellites, can be used in humanitarian relief operations. Geostationary observation satellites, such as the Geostationary Operational Environmental Satellite (GOES) and the Meteorological Satellite (METEOSAT), are important for weather forecasts (for example, to monitor hurricanes), while polar-orbiting satellites provide more detailed views of the area of interest. However, the level of spatial detail that can be distinguished with, say, METEOSAT, is too coarse for regional or local analyses of areas requiring humanitarian relief. The most commonly used commercial observation satellites up to now have included the U.S. National Oceanic and Atmospheric Administration's Advanced High Resolution Radiometer (AVHRR) and Landsat series, the French Satellite Pour l'Observation de la Terre (SPOT) and Indian Remote Sensing (IRS) -1C and -1D satellites. These satellites provide information at resolutions from 1 km (AVHRR) to 6 m (IRS). In addition, there are the radar satellites, such as the European Remote

Sensing (ERS) -1 and -2 and Canadian RADARSAT-1 satellites. The advantage of radar sensors is that they are independent of cloud cover or sun illumination, which is particularly valuable in equatorial regions, which are often cloud covered.

Relief agencies have not, so far, used commercial satellite imagery routinely. Some offices within large relief agencies, such as the UNHCR Mapping Unit, have experience in satellite imagery (using it, for example, to assess deforestation in refugee hosting areas). However, this technology is, in general, not included in the relief organizations' decisionmaking processes. This may be due to the relatively high level of expertise needed to analyze and interpret the traditional imagery or due to previous "overselling" of what can actually be detected in an image. However, the advent of new higher-resolution commercial satellite sensors may change this. New detailed images in nearly natural colors have the potential to provide important information, in near real time, for relief staff having a low level of expertise in image interpretation. The global coverage (except for cloud-covered regions) and daily updates this constellation of new higher-resolution satellite sensors will offer, along with the existing earth-observation satellites, will fit very well with most relief agencies' needs. Global problems need global tools for efficient management.

The level of detail possible to detect in commercial satellite images has increased by a factor of 100 over the 10-m resolution SPOT satellites (today's most successful commercial satellites). The new 1-m higher-resolution satellite sensors are approaching the level of detail that, in the 1970s, was reserved for the military.[2] And while today's military systems are capable of much finer detail, the relief community does not require images with such extreme resolution.[3]

How, then, can commercial observation satellites be used in humanitarian relief operations? In general, such operations can be divided into three phases: emergency, relief, and repatriation (Bouchardy 1995); see Figure 19.1. In the emergency phase, relief agencies need detailed information in near real time about where the refugees are and how many there are, as well as about the infrastructure, including road networks and terrain features. Many refugee situations occur in developing countries, whose maps, if any, are outdated. This makes the logistics planning difficult and sometimes not as efficient as it could be. Imagery from the new higher-resolution satellite sensors can be used to update maps of the area of interest, and, in some cases, to locate and give rough estimates on the number of people involved.

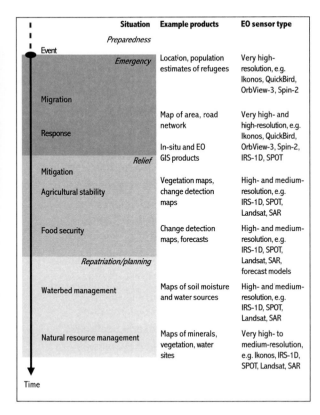

Figure 19.1—Types of Imagery Products Needed at Different Phases of Relief Operations

In the relief phase, the refugee situation is less dynamic, and refugee camps need to be planned and established. This leads to a need for information on potential refugee hosting areas and for environmental analysis of such parameters as soil moisture, terrain slopes, road networks, potential airfields, available water sources, and fuel wood. It is also important to establish baseline information on environmental parameters for later reimbursement of the refugee-hosting authorities. The security aspect of the refugees is, of course, very important when planning refugee camps. The sites must not be constructed so as to be easy targets for artillery or gunmen. Using a combination of satellite image-derived digital terrain models and higher-resolution satellite images can help ensure that these parameters are assessed objectively.

The repatriation phase requires detailed planning of where the refugees should be relocated within their country of origin. For example, refugees may not be able to settle in their preconflict area because of environmental degradation or land mines and thus must settle elsewhere. Updated databases on complex environmental parameters are needed, and commercial satellite images at various levels of details can be help here. Environmental analyses at the village level would need analysis of higher-resolution satellite imagery, while studies at the regional level could use Landsat images. It is important, though, to remember that information from the field will be coming in constantly from various sources, such as local population and relief staff, in all these phases. Information from the field will always be important for verifying and contributing information to relief agencies.

During the last few years, considerable progress has been made in using commercial observation satellites in humanitarian relief operations. However, more work needs to be done to fully prove the technology and to determine how relief organizations can best incorporate the information into their decisionmaking processes. Fortunately, this process has already begun, and the use of commercial satellite imagery in relief operations is an important aspect in several international forums.[4] Commercial observation satellite technology clearly fits well with international relief. But full acceptance and appreciation of this tool may require higher-resolution satellite imagery because nonspecialists can interpret such detailed images more intuitively, as with photographs taken from aircraft.

Satellite-derived information can, in theory, be disseminated to all levels of an organization's structure. However, in practice, people working in the field have limited access to telephones; fax machines; and, especially, Internet-connected computers. Even though this situation should eventually improve, the hard fact is that it will be a very limiting factor in the foreseeable future. Information can, however, be disseminated from headquarters via the organizations' own proven distribution channels, such as courier mail. These are not high-tech, but at least the information will reach the field. Unfortunately, such dissemination channels are not likely to be timely, so near–real-time satellite-derived information may first be disseminated at an agency's headquarters. Incorporated into the decisionmaking process at that the headquarters or regional level, this technology may prove highly efficient. However, for maximum benefit, relief staff in the field must also benefit from satellite-derived information.

Case Studies

To demonstrate the use of commercial observation satellites in humanitarian relief operations, this section presents four case studies. Because one of the most promising tools is the use of higher-resolution satellite sensors, most of the case studies will be related to this technology. The studies will also illustrate the use of various coarser-resolution data from commercial observation satellites and how these can be combined with geographical information system (GIS) data from the field. The subsections below describe studies representing Asia, America, Africa, and Europe that were originally undertaken during the ENVIREF and ReliefSat projects and include both SPIN-2 and IKONOS data. The case studies were selected according to the availability of adequate field data and satellite image coverage.

ASIA—SITE 2, THAILAND, AND BELDANGI, NEPAL

The first case study demonstrates how multispectral higher-resolution satellite images can be used for refugee camp planning and monitoring. The 3.3-m-resolution Russian KVR-1000 sensor image analyzed came from combined panchromatic and near-infrared channels (see Figure 19.2). This image is thus a good example of the kind of images that will be available from combined new higher-resolution 1-m panchromatic and 4-m multispectral imagery. This image of the now-demolished refugee camp Site 2 in Thailand was acquired on February 7, 1989. At the time of image acquisition, approximately 143,000 people lived at Site 2. This camp was established to host refugees previously settled in separate camps closer to the border. For security reasons, following Vietnamese military offensives between 1984 and 1986, the refugees were moved to the area of Site 2 (Lynch, 1990).

A space map of Site 2 (see Figure 19.2) was developed by analyzing the satellite image and comparing the findings to UNHCR data. The image clearly shows the well-organized road network and the housing structures. This map indicates various features important for camp management, such as individual camp administrative centers, water sites, hospitals, and sanitation centers. Figure 19.3 is a blowup of the upper-left-hand section of the camp. As the area marked "scattered vegetation" shows, even individual trees are visible in this satellite image. The analysis also revealed an area likely to have been exposed to deforestation by camp construction (upper left of Figure 19.2).

A camp of this size clearly indicates the need for sustainable camp planning. The satellite image shown here was not used in planning and managing Site 2. For this task, the UN used a special engineering unit, which utilized standard ground measurement tools. However, this image could have contributed toward more efficient planning and management of Site 2. The image is an example of how high-resolution baseline data can be collected and how potential deforestation can be mapped in detail, as can road networks. Water engineers and sanitation staff can use such detailed images to plan and potentially improve the living conditions. The image's nearly natural colors also make interpretation easier for nontechnical relief staff than it would be using the coded coloring the remote sensing community has traditionally used for multispectral images (e.g., imaging vegetation in red).

19 Supporting Humanitarian Relief Operations

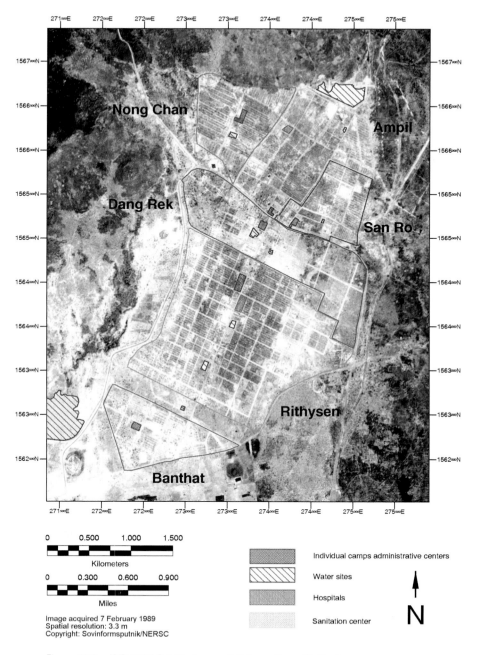

Figure 19.2—KVR-1000 Satellite Image of Refugee Camp, Thailand
(image ©1998 Sovinformsputnik) (see color plate)

Figure 19.3—Detail View of Refugee Camp shown in Figure 19.2 (image ©1998 Sovinformsputnik) (see color plate)

Figure 19.4 is an IKONOS image covering approximately 10 by 10 km. The three Beldangi refugee camps (Beldangi I, Beldangi II, and Beldangi II extension) are in the center of the image, which also shows their surrounding environment, including forest cover, agricultural areas, local villages, road network, and rivers. Figure 19.5 is a close-up of the three camps. The images were produced by merging the 1-m resolution panchromatic (black and white) image with the four 4-m resolution multispectral (color) images. This approach preserves the original details in the highest-resolution image (panchromatic) while improving landcover classification using the multispectral images. The multispectral data have been combined, so the resulting image is displayed in nearly natural colors, which is helpful for image interpretation and utilization by nonspecialist relief staff. Such images are useful for camp planning, monitoring, and environmental assessments.

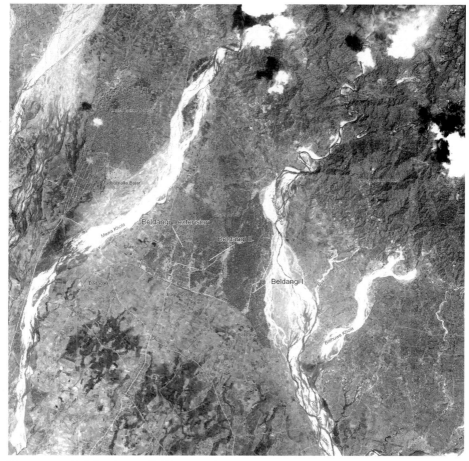

Figure 19.4—High-Resolution IKONOS Image of Beldangi Refugee Camps, Nepal, May 15, 2000 (image ©2000 Space Imaging) (see color plate)

AMERICA—CHIAPAS, MEXICO

Because of military activity in the Ixcan region of Guatemala in 1982, several thousand refugees crossed the border into Chiapas, Mexico. In 1983, there were 46,000 refugees in 90 camps and refugee-hosting villages scattered along the border with Guatemala. Approximately 18,000 refugees were relocated in 1985, while the remaining 28,000 stayed in Chiapas. Although this region was once relatively inaccessible, a road network had been established before the refugee crisis because of increased farming and logging in the late 1970s. The area, also known as Selva Lacondona, features dense rain forest and several major rivers and lakes.

Figure 19.5—Detail View of Refugee Camps shown in Figure 19.4 (image ©2000 Space Imaging) (see color plate)

To cover the entire region of interest, three SPIN-2 panchromatic 2-m higher-resolution scenes were acquired on January 13, 1992. The resulting combined image was used to map road networks, rivers, and lakes (see Figure 19.6), and maps from various other sources were used to verify the result (Manz 1988, UNHCR 1990, O'Brien 1998). Maps and population data from UNHCR (1990) were then used to locate and map individual camps and refugee-hosting villages (see Figure 19.6).

One example of how relief organizations might use such information is to assess deforestation. In this case, the areas deforested as of January 1992 were identified (Figure 19.7). Since satellite-derived information should be compared to field reports whenever possible, this study also compared regions of deforestation between 1993 and 1997, as depicted in radar satellite imagery, with the locations of refugee-hosting villages and camps, as derived from the higher-resolution satellite imagery (see Figure 19.8). The analysis and information from the field together suggest that the refugees were not

19 Supporting Humanitarian Relief Operations

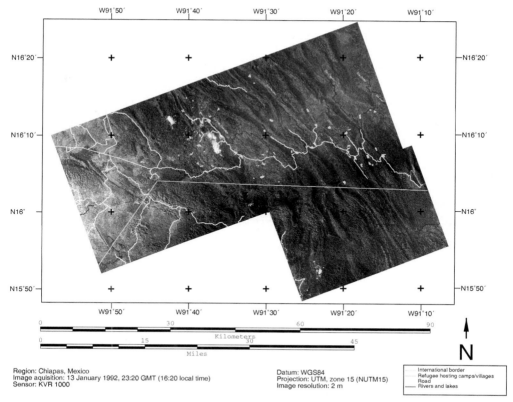

Figure 19.6—Space Map of Refugee Hosting Area in Chiapas, Mexico, January 13, 1992 (image ©1998 Sovinformsputnik) (see color plate)

solely to blame for the deforestation, because the deforested areas do not seem to be spatially coordinated with the locations of refugee camps. It is more likely that logging, to allow the expansion of such things as cattle farms, caused the observed deforestation.[5]

Such images and maps are valuable for such regions as Chiapas, for which detailed and up-to-date maps are difficult to obtain. An updated road network from the rapidly expanding construction in this region was relatively easy to detect in the higher-resolution images, as were the various rivers. Images such as those shown here could have been of great benefit in planning the logistics of relief efforts for this area because they enable newly constructed individual camps and roads to be located quickly. Planning new camps, further away from a border, could be another application of such imagery.

Figure 19.7—Deforested Regions in Refugee Hosting Areas in Chiapas, Mexico (image ©1998 Sovinformsputnik)

Figure 19.8—Deforestation over time (1993–1997) in Chiapas, Mexico, as recorded in satellite imagery (image ©1998 European Space Agency) (see color plate)

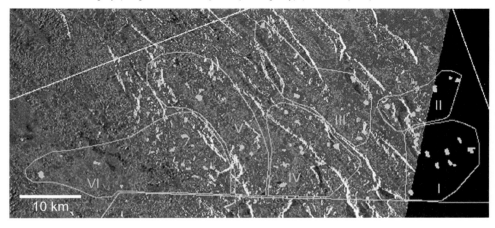

AFRICA—QALA EN NAHAL, THE SUDAN, AND DADAAB, KENYA

Conflicts and poor harvests in Ethiopia forced people to cross the border into Sudan in the late 1960s and early 1970s. The area allocated for these refugees, Qala en Nahal, was exceptionally dry and had relatively poor soil. A water-distribution system was constructed; although it did not function according to design, it did supply several of the refugee camps with water. The 2-m higher-resolution SPIN-2 image of the Qala en Nahal refugee settlement plan was acquired on January 20, 1992. Figure 19.9 is a space map derived from the satellite image and from GIS data on refugee camp population (purple circles). A total of 22,000 people lived in Qala en Nahal, including 16,000 refugees, at the time of image acquisition. Individual refugee camps, local villages, and road networks were detected and mapped from a visual (nonautomated) inspection of the detailed satellite image (Bjorgo, 2000b).

Man-made water sites (*hafirs*) were identified and verified from sketch maps. No good-quality maps of this area could be obtained from external sources during this study. The details of camps and road networks between camps and in mountainous areas on the map clearly illustrate the benefits of higher resolution in producing updated space maps. An agricultural area can also be seen in the lower right of Figure 19.9. This area was created partly to facilitate self-sustainability for the refugees.

Figure 19.10 shows the Um Burush refugee camp, which hosted 2,910 refugees at image acquisition. The left (west) part of the camp has a relatively organized street network, while the right (east) part of the camp does not have such a rigid structure. In the upper middle of the image, one can see the road connecting Um Burush to neighboring camps. The white and dark dots seen in this image are houses and fencing, as well as bushes and small trees.

UNHCR has numerous environmental activities in refugee hosting areas in Africa. One of these focuses on the management and sustainable use of natural and forest resources around the Dadaab camps in Kenya. These resources are used for grazing, firewood collection, and construction materials, particularly in the immediate vicinity of the camps. As part of an integrated package of environmental activities that include awareness raising and energy conservation, regeneration schemes were implemented in the areas close to and inside the camps. The IKONOS image of Ifo refugee camp (Figure 19.11) shows "greenbelts" of trees planted as rectangles on the outskirts of the camp. In addition, trees have been planted inside the camp to regenerate the relatively dense tree cover and to provide shade for the refugees. The results of this

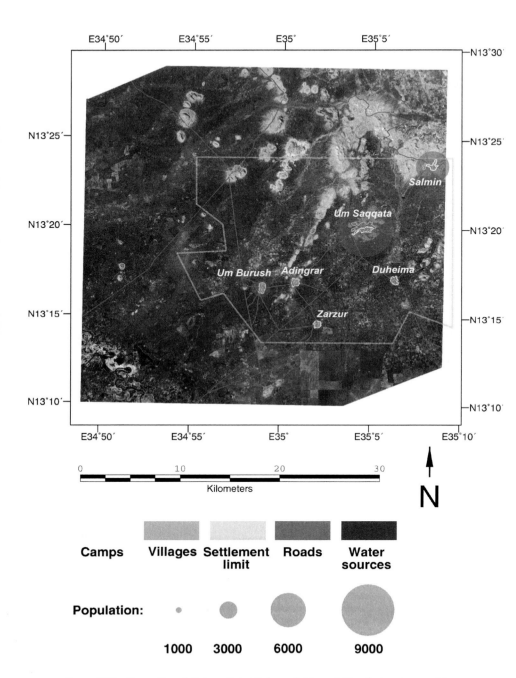

Figure 19.9—Space Map of Qala en Nahal Refugee Settlement Plan, Sudan, January 20, 1992 (image ©1996 Sovinformsputnik) (see color plate)

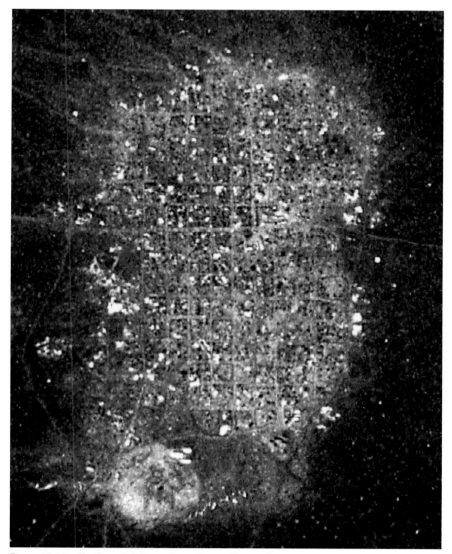

Figure 19.10—Um Burush Refugee Camp (image ©1996 Sovinformsputnik)

activity can be seen in Figure 19.12, where individual trees can be seen between the housing structures.

EUROPE—KOSOVO, MACEDONIA, AND ALBANIA

In the spring of 1999, several hundred thousand Kosovars were forced to flee their country because of the military and paramilitary actions taken against them.

Figure 19.11—IKONOS image of the Ifo Refugee Camp, Dadaab Settlements, Kenya, May 25, 2000 (image ©2000 Space Imaging) (see color plate)

These refugees were settled in camps in Albania and Macedonia, close to their borders with Kosovo. Most relief agencies found the magnitude and suddenness of this exodus unexpected. Consequently, the relief community was not prepared to handle the huge number of refugees crossing the borders. Updated information on the number waiting to cross the border was one of the key issues during the emergency phase of this crisis. However, no commercial higher-resolution satellite images were available in near real time. A "historic" Landsat image was combined with GPS data on refugee camp locations from UNHCR and on road networks and hydrology from U.S. Agency for International Development (USAID) to a simple map, which was disseminated to several relief agencies (see Figure 19.13).

The heavy cloud cover during the crisis made cloudless satellite imagery difficult to obtain. Fortunately, a 5.8-m resolution panchromatic image from

19 Supporting Humanitarian Relief Operations

Figure 19.12—Detail View of Refugee Camp Shown in Figure 19.11 (image ©2000 Space Imaging) (see color plate)

IRS-1D, acquired on May 27, 1999, did cover several of the refugee camps. Space maps of these camps were produced but, because of the relatively rapid end of this situation, did not reach the relief community in time to be used operationally. However, the space maps did show exact placements of the camps in relation to road networks, railways, rivers, and lakes. Combining the Landsat image with the 5.8-m resolution IRS-1D image yielded an image of the camps and their environment in nearly natural colors. A portion of this, in combination with GIS data, is shown in the space map of Figure 19.14. The original GIS data were improved using details retrieved from the IRS-1D image. Note the agricultural areas and villages located close to the Neprosteno refugee camp (outlined in green). The bright green-yellow areas in the upper middle and lower right of the image correspond to clouds in the IRS-1D image but to cloud-free areas with forest or agriculture signatures in the Landsat image. Combining two such satellite images is relatively straightforward as long as common geographic reference points can be identified in the two satellite images.

Digital terrain models, produced from satellite imagery using off-the-shelf image processing software can help in planning refugee camp locations and can provide a realistic view of the refugee-hosting area. For this study, two

Figure 19.13—Satellite image of Kosovar Refugee Hosting Area, Albania and Macedonia (image ©1999 Eurimage) (see color plate)

19 Supporting Humanitarian Relief Operations

Figure 19.14—Satellite Image of Neprosteno Refugee Camp (image ©1999 Eurimage) (see color plate)

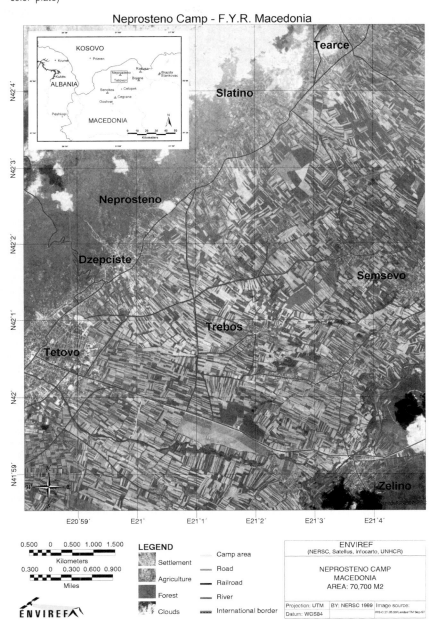

tandem-phase[6] ERS radar satellite images were used to create a digital terrain model of the area of interest. This model can be used to compute terrain slope, which, in combination with soil data (from a Landsat image, for example) can be used to pinpoint areas suitable for refugee camps. Here, the IRS-1D image was draped over the digital terrain model, creating a realistic view of the area. Figure 19.15 illustrates what the view of the Kukës refugee camps in Albania looked like from the southeast. The lake is shown in black, while the airport, refugee camps, and the city of Kukës are all clearly visible. The border with Kosovo is in the mountainous areas near the horizon. This image was not used for practical purposes in the Kosovo crisis, since it was produced after the return of refugees.

Limitations

The Kosovo case study clearly illustrated some of the limitations of commercial satellite imagery in supporting humanitarian relief operations. The persistent cloud cover and the relatively few commercial observation satellites capable of providing detailed images (SPOT and IRS were really the only sources for somewhat detailed views of the region) resulted in several weeks of waiting and searching for potential imagery. The launching of new satellites will improve this situation. Within the next few years, there should be enough

Figure 19.15—Three-dimensional perspective of Kukës refugee camps, Albania (image ©1999 Eurimage and European Space Agency) (see color plate)

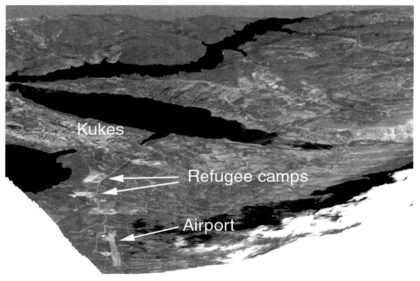

satellites to provide higher-resolution imagery for any region in the world daily. However, the fact that the area of interest must be cloud free severely limits the ability to use such images in humanitarian relief operations. Cloud-free skies cannot be counted on, especially, for example, in equatorial regions. While radar satellites are not limited by cloud cover, they provide coarser images, and image analysis requires a higher degree of specialization.

Another limiting factor is the cost of satellite imagery. This is a new technique, and relief organizations do not generally have funding for testing technology. It is therefore of paramount importance to demonstrate the cost-benefit of this tool. For international coverage, a single higher-resolution 11-by-11 km panchromatic *or* multispectral satellite image starts at approximately $3,000 U.S., and twice that for *both* panchromatic and multispectral images of the scene. If near–real-time data are needed, this price increases by approximately the same amount because the satellites must be programmed to acquire previously unplanned images, for a total of $12,000 U.S. for a single set of near–real-time panchromatic and multispectral images.[7] The trade-off is that a larger logistical effort could easily "waste" this much money if planners lack information on road networks during the emergency phase of a refugee crisis. Also, a poorly chosen camp location and camp design could increase the costs of running the camp by much more than the cost and analysis of higher-resolution images. Using commercial higher-resolution satellite imagery is likely to reduce such risks. Near–real-time data are not generally needed for camp planning, so the cost will be approximately $3,000 U.S. for a single panchromatic or multispectral scene. Furthermore, environmental assessments of relatively large areas can be carried out using commercial satellite images, such as those from SPOT or Landsat. Compared to traditional field surveys, using satellite images is a good investment for relief agencies because these images provide highly aggregated data at relatively low cost.

Even if technical image acquisition is streamlined, it is very important that the data providers have an efficient order and distribution system. Image orders for emergency situations must receive priority, even if more-substantial customers are already waiting to have their data processed. If it takes too long to receive the data and derive the required information, the images will be useless to the relief community. For this technology to be of use in humanitarian emergencies, the data providers, especially those of the new higher-resolution sensors, must prove that they are able to deliver near–real-time data at short notice. This will, in some cases, require flexibility with respect to internal procedures. The users should be able to access the data via the Internet or using file transfer protocol to minimize the time between image processing and analysis—two days by courier is often too slow for emergency applications.

The international relief community should develop a policy for using commercial higher-resolution satellite imagery during relief operations. The details possible to detect in these images can easily increase tension between relief agencies, local authorities, potential guerrilla groups, etc. For example, if such military objects as tanks or command posts are identified, that should be acknowledged; such information might be masked out if it does not serve a purpose for the relief agencies. Relief workers can be, have been, arrested in the field for carrying information much less sensitive than a highly detailed satellite image of a military command post. In the Sudan, for example, military restrictions currently prevent relief agencies from legally using GPS receivers.

If U.S. interests are at stake, the commercial higher-resolution satellite image distributors may be forbidden to collect imagery for or disseminate it to relief organizations. Such "shutter control" may seriously limit the efficient use of higher-resolution imagery in refugee relief. So far, shutter control has only been applied in a few cases. During the Gulf war, use of the SPOT satellite over Kuwait and Iraq was limited to the military. Also, U.S. commercial companies currently are not allowed to disseminate images of Israel of better resolution than 2 m. Shutter control is discussed in more detail in Section IV of this volume.

Conclusions and Future Aspects

From an academic viewpoint, the use of commercial observation satellites to support humanitarian relief operations can be viewed as bridging the gap between the "hard" science of remote sensing and the "soft" social sciences. However, the methods and tools discussed in this chapter have a significant practical angle. More-detailed academic analysis and discussion on bridging this gap can be found in other sources, such as Livermann et al. (1998).

The case studies described above (which are discussed in more detail in Bjorgo, 1999, 2000a, 2000b, except for the Kenya and Nepal cases) point toward ways of utilizing commercial higher-resolution imagery. This kind of information is highly useful for creating up-to-date maps of road networks, water sources, camp locations, and terrain features; for assessing environmental degradation; and for planning refugee camps, especially by combining the 1-m panchromatic and 4-m multispectral channels. Even data days or weeks old can still be highly applicable—depending on the time frame in which the information must be analyzed and disseminated. Satellite images may be the only source of relatively up-to-date information on remote or inaccessible areas.

The ability to estimate population from higher-resolution sensors should be investigated further. The inaccuracies are currently too large for high-confidence population estimates. However, when no other information is available, analyses of higher-resolution satellite images can provide rough estimates of the number of people affected. To be most valuable, information from commercial observation satellites—Landsat, SPOT, or higher resolution—should be combined with GPS field data and other on-site sources. Higher-resolution satellite images should be considered complementary to, not a substitute for, field data collected by relief staff.

Even if we have so far only had a sneak preview of what will be available in the short term, the longer-term future is likely to bring even more detailed imagery (Ferster 1999), including radar images independent of cloud cover and hyperspectral images for in-depth environmental analyses. The bandwidth will improve, and near–real-time commercial satellite imagery, or information derived therefrom, may become available to the relief worker in the field, although this will take time. Limitations and potential risks should be openly discussed to avoid "overselling" commercial satellite imagery to the humanitarian relief community.

Notes

1. IKONOS was successfully launched September 24, 1999. The latest higher-resolution satellite sensors, such as IKONOS, have a resolution of 1 m in panchromatic (black and white) mode and 4 m in multispectral (color) mode. The definition of higher resolution, as provided here, is sometimes also referred to as very high resolution (VHR), or very high spatial resolution (VHSR).
2. The Federation of American Scientists (FAS) estimates that today's most advanced military satellites provide imagery with a spatial resolution of 10 cm (http://www.fas.org/spp/military/program/imint/).
3. Two recent projects mapped the users' information requirements for the various phases of a refugee relief operation: Environmental Monitoring of Refugee Camps Using High Resolution Satellite Images (ENVIREF), a project supported by the European Commission through DGXII. Data are available at http://www.enviref.org/documents.htm, and Relief Management Using Very High Resolution Satellite

Images (ReliefSat), a project sponsored by the U.S. Institute of Peace. A briefing on this work is available at http://www.nrsc.no/reliefsat.
4 The Global Disaster Information Network (Disaster Information Task Force, 1997), the operational UN Office for the Coordination of Humanitarian Affairs (UN OCHA) ReliefWeb server (http://www.reliefweb.int), UNHCR, OpenEyes e-mail list, ENVIREF, ReliefSat and others currently demonstrate and discuss how to take full advantage of this technology during relief operations.
5 Industrial logging activities were not permitted in this region between 1993 and 1997.
6 Tandem-phase here means that the two ERS satellites flew in an identical orbit and that the time-lag between image acquisitions was 24 hours. The resulting images are therefore acquired under similar surface conditions as long as the meteorological conditions are stable, and so the radar signals are highly comparable, and useful for producing digital terrain models.
7 The cost of higher-resolution satellite images may change, but these are the figures listed by the data providers as of the time of this writing.

References

Bjorgo, E., *Use of Very High Spatial Resolution Satellite Sensor Imagery in Refugee Relief Operations*, Ph.D. thesis, Bergen, Norway: University of Bergen, Nansen Environmental and Remote Sensing Center, 1999.

—, "Using Very High Spatial Resolution Multispectral Satellite Sensor Imagery to Monitor Refugee Camps," *International Journal of Remote Sensing*, Vol. 21, No. 3, 2000a.

—, *Refugee Camp Mapping Using Very High Spatial Resolution Satellite Sensor Images*, Hong Kong: Geocarto International, 2000b.

Bouchardy, J. Y., *Development of a GIS System in UNHCR for Environmental, Emergency, Logistic and Planning Purposes*, Environment Unit report 1-95, Division of Programs and Operational Support, United Nations High Commissioner for Refugees, Switzerland, 1995.

Disaster Information Task Force, *Harnessing Information and Technology for Disaster Management*, Washington, D.C.: The Global Disaster Information Network, November 1997.

Environmental Monitoring of Refugee Camps Using High Resolution Satellite Images (ENVIREF), data, no date. Available at http://www.enviref.org (last accessed September 28, 2000).

Ferster, W., "After IKONOS, Space Imaging Plans Even Better Satellite," *Space News*, Vol. 10, No. 36, 1999.

Livermann, D., E. F. Moran, R. R. Rindfuss, and P. C. Stern, *People and Pixels: Linking Remote Sensing and Social Science*, Washington D.C.: National Academy Press, 1998.

Lynch, J. F., *Border Khmer: A Demographic Study of the Residents of Site 2, Site B, and Site 8*, Thailand: U.S. Embassy, Joint Voluntary Agency Program, report no. 1/90, 1990.

Manz, B., *Refugees of a Hidden War*, Albany: State University of New York Press, 1988.
O'Brien, K., S*acrificing the Forest: Environmental and Social Struggles in Chiapas*, Boulder: Westview Press, 1998.
Rogge, J. R., *Refugees: A Third World Dilemma*, Totowa: Rowman & Littlefield, 1987.
United Nations High Commissioner for Refugees (UNHCR), *Review of Health and Nutrition Programmes for Guatemalan Refugees in Three States, 29 March–13 April 1990*, UNHCR Technical Support Section Mission Report 90/14, Geneva, Switzerland, 1990.
—, *Global Appeal 2000*, Geneva, Switzerland, 1999.

Section IV: Emerging International Policy Issues

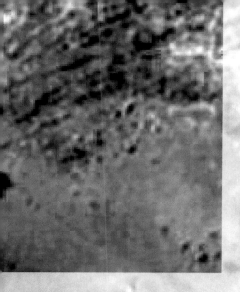

Access to satellite images, particularly high-resolution images, was once confined to a relatively small number of countries. Within these countries, the use of satellite imagery was largely dominated by highly trained experts working at defense and intelligence organizations, scientific projects, natural resource agencies, or in narrow commercial sectors, such as industries focused on exploring for energy resources. The advent of dedicated commercial observation satellites in 1999, along with a new generation of civilian imaging satellites and information systems, signals a fundamental expansion in international and public access to satellite imagery data. Section IV presents chapters that explore various aspects of this basic change in the degree of international and domestic access to satellite imagery. Some chapters focus on the international dimensions of proliferating imaging satellite technologies and expertise. Others assess the policy implications of having new types of imagery users, such as the news media and nongovernmental organizations (NGOs), with unprecedented interest in making use of higher resolution satellite imagery to support their activities.

In Chapter 20, Ann Florini and Yahya Dehqanzada conclude that greater transparency is inevitable as the barriers for countries to acquire their own earth-observation satellites diminish with the growing availability and affordability of satellite imaging technologies. While recognizing that some countries can exploit this new source of information to support their military operations, the authors contend that—on balance—commercial observation satellites will enhance international security by contributing to growing global transparency. Broader access to data from higher-resolution imagery satellites will help the international community anticipate and document aggressive international behavior or even large-scale internal atrocities.

Taking a somewhat different perspective on the potential influence of commercial satellite imagery, Mark Gabriele offers a conceptual analysis of the international and domestic implications of transparency in Chapter 21. He accepts the notion the transparency—or open skies—is both inevitable and desirable in the long run as high-resolution imagery becomes readily available to all nations. Yet, Gabriele cautions, a destabilizing asymmetric international situation could arise between rival countries, such as India and Pakistan, if either side enjoys "unilateral transparency" by having an unequal advantage in access to high-resolution imagery of the other country. The author extends his exploration of the risks that asymmetrical access to information creates by considering the domestic implications of the effects of commercial satellite imagery on privacy rights.

Two chapters in Section IV assess how nonstate actors are likely to take advantage of their unprecedented access to commercial satellite imagery. In Chapter 22, Karen Litfin explores the role that arms control and nonproliferation activists, environmental groups, and NGOs concerned with human rights or humanitarian relief activities have in using satellite imagery to press for their agendas. She distinguishes among the potential capabilities of different types of NGOs to make effective use of imagery data in their work.

In comparison, Steven Livingston in Chapter 23 assesses the news media's interest in taking advantage of high-resolution commercial satellite imagery within the context of other cutting edge information technologies, such as portable satellite uplinks and the increasingly ubiquitous camcorders, that have already greatly expanded the global reach of news organizations to remote areas and conflict zones. Livingston also discusses the challenges that news organizations, which are driven by relatively short turnaround times in acquiring and presenting information, face in acquiring timely satellite imagery and taking time for accurate imagery interpretation and sound analysis.

In Chapter 24, Bob Preston analyzes the regulatory landscape that shapes the legal and policy environment for the operations of U.S. commercial observation satellites. This chapter offers a detailed review of the legal basis for imposing restrictions on the operating practices of U.S. commercial remote sensing firms. Preston gives particular attention to assessing the controversy concerning the government's authority to impose the so-called "shutter controls" on the collection and dissemination of commercial satellite imagery data for a particular region in the interests of national security, international obligations, or foreign-policy concerns, as defined by the Executive branch of the U.S. government.

Finally, John Baker explores in Chapter 25 the issue of imagery credibility that could arise as new users, such as NGOs and news media organizations, seek to take advantage of high-resolution commercial satellite imagery. Although these new "imagery activists" possess relatively limited expertise and experience in image interpretation and analysis, their use of overhead imagery is likely to attract much public attention. Baker analyzes the types of deliberate and inadvertent errors that could arise with greatly expanded access to commercial satellite imagery. He proposes various ways to improve the imagery-handling expertise of new users, including greater cooperation with the more-experienced imagery analysts found in government, industry, academia, and professional organizations concerned with remote sensing.

20
The Global Politics of Commercial Observation Satellites

Ann M. Florini and Yahya A. Dehqanzada

Information is power, says the well-worn cliché. The advent of commercial high-resolution remote sensing provides a particularly comprehensive and vivid type of information to many actors who previously lacked access to it. As the number of state and non-state actors with access to satellite imagery increases, governments worry that they may lose a significant degree of control over information about their territories. And governments that formerly held a monopoly over such information worry about the consequences of the loss of that monopoly. How valid are these concerns? Will the newly informed actors, be they other nation-states, international organizations, or nongovernmental groups, gain meaningful power at the expense of those formerly privileged with unique access to satellite imagery? Or will the status quo be able to fight back effectively?

This chapter explores the broad political implications of commercial high-resolution observation satellites. It does so by considering two issues: first, the degree to which satellite imagery constitutes the kind of information that translates into power, and second, the degree to which status quo powers, and particularly the United States, might realistically hope to maintain control over the acquisition and dissemination of high-resolution imagery.

The Power of Imagery

Power—the ability to induce others to do what they would not have done without such intervention—comes in two forms: compellence and persuasion. It is important, therefore, to begin our discussion with an analysis of the possible effects of high-resolution satellite imagery on these two different types of power.

COMPELLENCE

Compellence refers to the capacity to raise the costs of noncompliance high enough to deny others the option of acting freely and in their own best interests. While there are many different instruments of compellence, the most dramatic—and the one that is likely to be most directly affected by the widespread availability of high-resolution satellite imagery—is the military. Satellite imagery can be a significant multiplier of military power. As Table 20.1 illustrates, 1-m imagery is quite good enough to provide detailed information about a wide variety of military equipment and activities. Chapter 22 in this volume and a study by the Cooperative Monitoring Center at the Sandia National Laboratories (Gupta and Harris, 1999) further highlight the range of military information that civilian analysts have been able to glean even from much coarser satellite imagery. The question, however, remains: Whose power will the new availability of imagery multiply the most?

The new commercial imagery could reinforce the existing distribution of military power, benefiting primarily states with advanced military capabilities and a well-developed capacity for interpreting and utilizing imagery. In the United States, one of the factors that led to renewed support for the continuation of the Landsat program was the U.S. government's realization following the 1990–1991 Persian Gulf War that even lower-resolution civilian satellite imagery can effectively supplement existing national reconnaissance capabilities. During the planning and execution of operations Desert Shield and Desert Storm, the U.S. Department of Defense (DoD) allegedly spent $5 to 6 million on Landsat imagery alone (U.S. House, 1992, p. 26). In addition, the DoD reportedly purchased more than 200 SPOT scenes of the Kuwaiti theater of operations (Zimmerman, 1992, p. 233). Given the experience and analytical skills of U.S. imagery interpreters, the U.S. military was able to use commercial satellite data effectively for terrain analysis, operational planning, and concealment detection. (Kutyna, 1991, p. 733; see also Gordon, 1991, p. 29).

It is just as likely, however, that the widespread availability of satellite imagery will redistribute military power. Even if the United States and the other advanced states gain something, countries that have previously lacked access

Table 20.1—Approximate Ground Resolution for Target Detection, Identification, Description and Analysis (m)

Target[a]	Detection[b]	General ID[c]	Precise ID[d]	Description[e]	Technical Analysis[f]
Troop units	6	2	1.2	0.3	0.15
Vehicles	1.5	0.6	0.3	0.06	0.045
Aircraft	4.5	1.5	1	0.15	0.045
Airfield facilities	6	4.5	3	0.3	0.15
Nuclear weapon components	2.5	1.5	0.3	0.03	0.015
Missile sites[g]	3	1.5	0.6	0.3	0.045
Rockets and artillery	1	0.6	0.15	0.05	0.045
Surface ships	7.5–15	4.5	0.6	0.3	0.045
Surfaced submarines	7.5–30	4.5–6	1.5	1	0.03
Roads	6–9	6	1.8	0.6	0.4
Bridges	6	4.5	1.5	1	0.3
Communications					
Radar	3	1	0.3	0.15	0.015
Radio	3	1.5	0.3	0.15	0.015
Command and control headquarters	3	1.5	1	0.15	0.09
Supply dumps	1.5–3	0.6	0.3	0.03	0.03
Land minefields	3–9	6	1	0.03	—
Urban areas	60	30	3	3	0.75
Coasts, landing beaches	15–30	4.5	3	1.5	0.15
Ports and harbors	30	15	6	3	0.3
Railroad yards and shops	15–30	15	6	1.5	0.4
Terrain	—	90	4.5	1.5	0.75

SOURCES: Table modified from Florini (1988), p. 98. The original sources were U.S. Senate (1978), pp. 1,642–1,643, and McDonnell-Douglas Corporation, 1982, p. 125.
a The chart indicates the minimum resolution at which the target can be detected, identified, described, or analyzed. No source specifies which definition of resolution (pixel size or white dot) is used, but the chart is internally consistent.
b Location of a class of units, object, or activity of military interest.
c Determination of general target type.
d Discrimination within target type of known types.
e Size or dimension, configuration or layout, components construction, equipment count, etc.
f Detailed analysis of specific equipment.
g Surface-to-surface and surface-to-air missiles.

to imagery may gain relatively more. Countries with relatively weak military forces, such as Iran, Iraq, and North Korea, can use commercial satellite imagery to monitor and collect intelligence on their more potent rivals and, sometimes, threaten their personnel, assets, and ultimately, interests. The results of two studies sponsored by the U.S. Air Force's Philips Laboratory (operations Seek Gunfighter and Seek Foe) clearly illustrate this point. In the first exercise, using only publicly available 10- to 30-m resolution satellite imagery, a single relatively low-tech personal computer, and commercial off-the-shelf imagery processing software, the Space Aggressor Team of the Space Warfare Center at Schriever Air Force Base, Colorado, was able to piece together an alarmingly complete picture of the deployment of the 366th Air Expeditionary Wing from Mountain Home Air Force Base, Idaho, to Sheik Isa Air Base in Bahrain during September and October 1997 (U.S. Air Force, 1999, p. 1-4). In addition, the Space Aggressor Team mapped a complete layout of Sheik Isa Air Base, correctly identifying its security perimeter and some potential targets, such as "tent city" areas, aircraft shelters, refueling stations, and a variety of parked aircraft (U.S. Air Force, 1999, p. 15-9). During the second exercise, which took place in January 1998, the Space Aggressor Team used the same sources and methods to locate the Prince Sultan (al Kharj) Air Base in Saudi Arabia geographically and to locate militarily significant sites, including the entry control points, air crew barracks, aircraft shelters, weapon storage areas, and Patriot missile defense batteries (Hays and Houchin, 1999).

The Space Aggressor Team's experiences demonstrate that even relatively poorly financed and technologically unsophisticated states can gain military benefits from commercially available medium- and high-resolution satellite imagery. Indeed, there are some indications that the Iraqi military used 10-m resolution SPOT satellite imagery for attack planning and post-attack assessments, both during the eight-year Iran-Iraq war and immediately before its invasion of Kuwait (Zimmerman, 1999; see also Wright, 1999). Following the invasion, had the United Nations not imposed the economic embargo against Iraq, Saddam Hussein's military might have been able to acquire imagery damaging to the allied forces (Gupta and Harris, 1999). Although it is unlikely that Iraqi access to satellite imagery would have altered the final outcome of the war, it is possible that U.S. casualties would not have remained so low had the Iraqi military detected the massing of the U.S. Army's VII Corps and XVIII Airborne Corps along Iraq's Western border with Saudi Arabia, preparing for the famous "left hook."

Although the United States is unlikely to face a military equal in the foreseeable future, it may well face a better-equipped or more determined foe than it found in Iraq. And countries that undertake humanitarian interventions on behalf of the international community may face violent organized opposition.

In such cases, ready access to militarily significant satellite imagery could be used against the countries acting on behalf of the international community. Unless vital national interests are at stake, the major powers—especially those with democratic systems of governance—may find themselves deterred by much weaker opponents.

In addition to its direct application in military conflicts, satellite imagery has also provided the United States with other kinds of leverage and influence in security-related situations. During the Cold War, as long as the United States was the only North Atlantic Treaty Organization member with a satellite reconnaissance program, it occupied a privileged position in Alliance debates over Soviet military capabilities and compliance with arms control accords. With the advent of commercial high-resolution satellites, other countries no longer have to depend solely on the United States for imagery and can instead have their needs fulfilled by commercial operators. This recent development diminishes a traditional source of U.S. influence in global security affairs.

PERSUASION

Power is not just about compelling others to act. It can also take the form of persuasion. Whereas compellence elicits compliance through coercion, persuasion seeks to influence behavior by convincing others that altering their present course of action is either the right thing to do or, more often, is in their own best interests. High-resolution satellite imagery can be a powerful tool of persuasion. During the four-year-long Bosnian civil war, dozens of reports emerged alleging that Bosnian Serbs had committed acts of genocide against the Bosnian Muslims. Despite ample evidence to support such allegations, the international community for the most part remained silent. Human rights took center stage, however, when on August 10, 1995, Madeleine K. Albright, then U.S. ambassador to the United Nations, called attention to the atrocities by presenting the United Nations Security Council with U.S. spy satellite images of mass graves in which hundreds of Muslim civilians were thought to have been buried. These images prompted the U.N. Security Council to pass Resolution 1010 on the same day (U.N., 1995), demanding that the Bosnian Serbs "respect fully the rights of all [Bosnian Muslims] and ensure their safety." The resolution further warned the Bosnian Serb military "that all those who commit violations of international humanitarian law will be held individually responsible in respect to such acts." These images were then forwarded to the International War Crimes Tribunal in the Hague for possible use against indicted Serbian war criminals, including the Bosnian Serb leader Radovan Karadzic and his top general Ratko Mladic (Dobbs and Smith, 1995).

Once high-resolution satellite imagery becomes available to everyone, any state or non-state actor will be able to publicize mass violations of human rights, regardless of whether such violations occur in the heart of Europe or in remote villages of sub-Saharan Africa, as long as such violations leave evidence visible from above. Commercial satellite operators are generally not restrained by the political constraints that often muzzle state governments. Whereas states may refrain from publicizing acts of humanitarian violence by allies or in regions in which they do not wish to get involved, commercial satellite operators will readily market such imagery to a host of media and human rights groups. Such widely available high-resolution satellite imagery may make it harder for the major powers, especially those that claim to adhere to a higher moral standard, to stand idly by and allow acts of genocide to continue unopposed.

Human-rights groups are not the only ones empowered by the emergence of commercial high-resolution satellites. States and non-state actors can use satellite imagery to monitor the activities of other countries, verify compliance with arms control agreements, and document cases of environmental pollution by corporations. In a majority of these cases, satellite imagery can be used to publicize and exert pressure on states and corporations to modify their behavior. All in all, greater access to satellite imagery promises to increase significantly the number of whistle-blowers in the international system. Unless states can devise policies to maintain the status quo, they will have to get used to having many of their affairs scrutinized by their peers and their citizens.

Maintaining the Status Quo?

Given the significant implications of commercial imagery for power, it seems reasonable to assume that countries that benefit from the existing distribution of power will attempt to maintain control. The United States, in particular, as the world's dominant power and as the possessor of the world's best space reconnaissance capabilities, stands to see some of its power shift to other actors if commercial high-resolution imagery becomes easily available to all. What, if anything, can and should governments do to preserve some control over the flow of information from satellite imagery? The options include unilateral measures, bilateral or multilateral accords, or arrangements that build on existing technology-control regimes.

UNILATERAL MEASURES

Most remote sensing satellites are operated more or less directly by their governments, making it easy to maintain state control over access to imagery from those satellites. As other chapters in this volume make clear, just about every government that operates civilian satellites maintains some control over the imagery that can be sold. Russian marketers cannot sell any imagery of sites of current armed conflict (see Chapter 8 in this volume). The Indian government restricts the sale of IRS imagery of India's border and coastal zones, as well as of the troubled state of Jammu and Kashmir (see Chapter 12 in this volume). The United States, which has licensed more than a dozen private businesses to operate remote sensing satellites, includes in the licensing agreement a provision enabling the Secretary of Commerce to limit commercial satellite operations during a period when "national security, international obligations, or foreign policy interests [of the United States] may be compromised" (U.S. Department of Commerce, 1997, pp. 59, 318). In 1999, Canada announced that it would pursue a similarly stringent policy of controlling access to commercial imagery (Ferster, 1999).

This policy, known as "shutter control," tries to balance competing governmental interests with respect to commercial satellite imagery. Satellite imagery represents a classic case of the difficulty of regulating the "export" of dual-use goods (i.e., those with both civilian and military applications). There are powerful incentives working at cross purposes. Governments have economic interests in maintaining a national presence in what could be a large and highly profitable industry. But they also have national security interests in preventing potential adversaries from using the imagery against them and foreign policy interests in not having certain situations publicized. Shutter control attempts to give governments the best of both worlds.

It will not work. Attempts to deny imagery to potential enemies through unilateral action are just likely to leave the field to competitors. As one former U.S. government official has observed, "the surest way to lose to increasing international competition is to adopt a restrictive regulatory environment at home which encourages customers to seek out foreign sources" (Pace, 1994). Imagery consumers who know that their access to imagery may be cut off at any time can find more reliable providers. Furthermore, Canada, France, India, and Russia are already providing high-resolution optical and radar imagery to consumers throughout the world. Over the next five to ten years, at least eight more countries—Argentina, Brazil, China, Israel, Japan, Pakistan, South Korea, and Taiwan—are planning to enter the commercial remote sensing market. Given the large number of alternative sources of imagery, unilateral shutter control by itself cannot afford any single country a meaningful level of protection.

In the United States, shutter control faces an additional challenge: It may be unconstitutional. The media have already made extensive use of satellite imagery, and some news producers are eagerly anticipating the emergence of the new high-resolution systems (see Chapter 23 in this volume). The Radio-Television News Directors Association argues vehemently that the existing standard violates the first amendment by allowing the government to impose "prior restraint" on the flow of information, with no need to prove clear and present danger or imminent national harm to an impartial judge (Cochran, 1999). If shutter control is exercised in any but the most compelling circumstances, a court challenge is inevitable.

BILATERAL AND MULTILATERAL APPROACHES

Given the impossibility of any single government exercising meaningful unilateral control over access to high-resolution imagery in a world of so many providers, some governments are seeking to negotiate agreements to restrict the availability of satellite imagery. Israel, for example, lobbied hard and successfully, for a U.S. law prohibiting U.S. commercial satellite operators from selling imagery of Israeli territory better than that available from other sources. This is only a stopgap measure, since the U.S. commercial operators will only have the 1-m-resolution field to themselves for a short while. (And since one of the non-U.S. competitors, ImageSat International, is actually a consortium of two Israeli and one American companies using Israeli spy-satellite technology, Israel's lobbying for the U.S. law may have been motivated as much by competitive concerns as by worries about national security.) Although the law would seem to set a precedent for bilateral controls on access to satellite imagery, it is unlikely to be replicated. Few countries have the close relationship that Israel and the United States enjoy.

The United States and Israel are not alone in their discomfort with the prospect of increased global scrutiny. France, which has been selling 10-m SPOT imagery since the mid-1980s, objected to the U.S. decision to allow 1-m imagery to be sold. The French objections seem to be motivated by security concerns and by a reluctance to see more competition for SPOT (see Chapter 9 in this volume). As governments around the world start to take notice of the high-resolution satellite imagery appearing on the front pages, it would seem plausible that other governments might begin to consider the potential merits of agreed multilateral restrictions on access to imagery. Developing countries might fear that high-resolution imagery could be used to exploit them economically or contribute to their military vulnerability.

So far, developing countries seem to have mixed feelings. Some are arguing for increased access to imagery. At the Third United Nations Conference on the Exploration and Peaceful Uses of Outer Space (UNISPACE III) in Vienna in July 1999, most representatives of developing countries seemed more concerned with reducing the cost of imagery to levels they considered affordable than with raising barriers to its availability.[1] But governments are also concerned with who will have access to imagery of their territory. In addition to obvious concerns about regional rivals, some governments are dubious about allowing their own citizens to have unfettered access. India is drafting a policy that will grant Antrix Corp., Ltd., of Bangalore (the commercial arm of the Indian Space Research Organization) the sole right to purchase high-resolution imagery from commercial providers, and India's National Remote Sensing Agency will have sole authority to distribute that imagery (*Space News*, 2000b). That policy will enable the government to control who within India would have access to the imagery.

The prospect of controlling access to sensitive high-resolution satellite imagery through bilateral or multilateral agreements becomes ever more precarious as remote sensing technology proliferates further. As was mentioned above, several countries are already providing high-resolution optical and radar imagery to consumers worldwide, and more are following. There is no reason to believe that all of these current and emerging satellite operators will respect one another's wishes on what imagery should be disseminated or to whom.

TECHNOLOGY CONTROL REGIMES

Although multilateral agreement on controlling access to imagery seems unlikely to be achieved, one effort at multilateral control over the proliferation of remote sensing instruments already exists. The 33-nation Wassenaar Arrangement on Export Controls for Conventional Arms and Dual-Use Goods and Technologies, which was signed in July 1996, aims to "contribute to regional and international security and stability by promoting transparency and greater responsibility in transfers of conventional arms and dual-use goods and technologies."[2] The agreement, among other things, requires member states to maintain effective controls over the export of monospectral and multispectral imaging sensors designed for "remote sensing applications," and to exchange information regarding transfers of such items (Wassenaar Arrangement, 1996).

Despite its good intentions, however, the Wassenaar Arrangement is unlikely to have a significant influence on the proliferation of remote sensing technologies. The Wassenaar regime, like other control regimes before it, rests on two implicit assumptions: that the number of actors able to supply the requisite technologies remains limited, and that all relevant actors are willing to cooperate in pursuit of a common goal. In the case of high-resolution remote sensing, neither of these assumptions seems valid. As one long-time observer recently noted (Stoney, 1999), when it comes to high-resolution land-observation satellites "the technology genie is out of the bottle." An example illustrates the point. In March 1998, the Canadian Space Agency selected Macdonald Dettwiler and Associates Ltd., to develop and operate the second generation of Canadian radar satellites, called RADARSAT-2. Macdonald Dettwiler in turn contracted its parent company, the U.S. company Orbital Sciences Corporation, to provide the bus—the shell that houses the satellite sensor and related subsystems—for the satellite. In early 1998, Orbital filed its application with the U.S. Department of State for a license to provide the necessary hardware. The Department of State, while not rejecting the license, postponed a final decision on the delivery of the satellite spacecraft pending a resolution of U.S.-Canadian dispute over an acceptable Canadian shutter control policy. U.S. officials feared that data from RADARSAT-2, which can provide synthetic aperture radar images as small as 3 m through darkness and cloud, could harm U.S. national security interests. Frustrated by the U.S. government's recalcitrance, the Canadian Industry Minister, John Manley, warned in August 1999 that unless the U.S. Department of State issued Orbital Sciences the license to deliver the RADARSAT-2 bus, Canada would "cut [the] crucial American connection with its RADARSAT-2 satellite project" and would instead seek a European solution (Mertl, 1999). And Canada did just that, awarding the $50-million contract to provide the RADARSAT-2 platform to Alenia (Aerospazio) in January 2000 (*Space News*, 2000a).

The Radarsat controversy demonstrates that U.S. efforts to impose unilateral rules will not work, even with its closest allies. Remote sensing technology is easily accessible from a number of different suppliers, some of whom, such as China and India, are not party to the Wassenaar Arrangement. Moreover, the Wassenaar Arrangement is not a legally binding treaty and has no enforcement mechanisms. The decision to transfer or deny transfer of any items remains entirely within the purview of each member state. This raises serious questions about the effectiveness of the agreement, particularly since there is no consensus among member states as to what kind of transfers are destabilizing and ought to be avoided. While the United States has consistently rejected sales of remote sensing technology to the Arab world, in part because of intense lobbying by pro-Israeli groups, France has been actively marketing its high-resolution satellites among Middle Eastern countries (de Selding, 1999).

In short, the prospects for forging a durable and effective remote sensing technology control regime seem quite unfavorable. High-resolution satellite technology has already reached virtually every continent in the world. Unlike the United States, which has security interests around the globe and, therefore, a strong interest in restricting the proliferation of power-multiplier technologies, other satellite manufacturers are likely to find profit motives more compelling.

COST BARRIERS

To some extent, hopes for containing the diffusion of imagery rest on the assumption that only a few companies will be able to make a profit and thus stay in business, given the high costs of entry into the space-based remote sensing industry. Developing a single high-resolution commercial satellite costs anywhere between $45 and 300 million.[3] Building the associated ground segment would add another $33 to 65 million to the total cost (Martin, 1999). A basic insurance package and the launch vehicle further raise the price tag by an estimated $19 to 132 million.[4] That is, commercial companies need to invest between $97 and 497 million before they collect any revenues (see Table 20.2). For many companies and even some nation-states, the cost of developing and deploying such a system may at present be insurmountable.

But while costs may deter most potential commercial satellite operators from entering the fray, governments may be less sensitive to cost constraints. Many governments that initiate a space reconnaissance program do so at least partly for national security purposes. For some, commercial viability is an issue only to the extent that it helps alleviate the financial burden of these programs. Thus, regardless of the cost barriers, states that are threatened by neighbors or regional rivals will continue to invest whatever is necessary to build, launch, and operate their own high-resolution systems.

In any case, technological progress and the ability to miniaturize satellites and the associated subsystems may dramatically reduce the costs of developing and deploying commercial satellites. Several companies worldwide are currently producing microsatellites with capabilities approaching those of high-quality commercial satellites but at a fraction of the cost (see Table 20.3). The Satellite Technology Research Center (SaTReC) of the Korea Advanced Institute of Science

Table 20.2—Estimated Costs of Entering the Market for High-Resolution Satellite Imagery

Item	Estimated Item Cost ($M U.S.) Unit	Estimated Item Cost ($M U.S.) Total	Payload to Low Earth Orbit (kg)
Satellite			
Sensor and spacecraft		45–300[a]	
Ground segment		33–65	
Primary ground station[b]	20–35		
Backup ground station[c]	10–15		
Remote tracking sites[d]	3–5 each		
Launch		12–60[e]	
Ariane 42P (French)	65–85[f]		6,100
Long March 2C (Chinese)	20–25[f]		3,200
PSLV (Indian)	15–25[f]		2,900
Athena 2 (United States)	22–26		1,990
Delta 2 (United States)	45–60		1,982
Cosmos (Russian)	12–14		1,400
Taurus 1 (United States)	18–20		1,400
Pegasus XL (United States)	12–15		460
Insurance		7–72[g]	
Total		97–497	

SOURCES: U.S. Department of Transportation (1999), Martin (1999), Mowry (1999).
a Space segment costs are directly related to key system characteristics such as the number and type of sensors, spatial and spectral resolutions, swath width, platform stability, and satellite agility. These costs also reflect whether existing technologies are used or whether new technologies are developed specifically for the satellite.
b Includes uplink and downlink terminals and image processing and storage facilities.
c Same as above, but more austere.
d To relay satellite commands and telemetry; 1–3 nominally.
e Launch costs depend on the weight of the satellite and the type of rocket used to deploy the system.
f The Ariane 42P, the Long March 2C, and the PSLV launch vehicles are designed for much bigger payloads than the 350–2,000 kg commercial satellites. When used to loft commercial systems, these launchers always carry more than one payload.
g Insurance costs range between 13 and 20 percent of the total value of the satellite plus the launch vehicle.

and Technology (KAIST) has developed and launched several microsatellites of varying capabilities. The Center's most recent microsatellite, KITSAT-3, which was launched on May 26, 1999, weighs approximately 110 kg and cost merely $8 million (Lee, 1999). This low-cost system, however, can provide 13.8-m multispectral images of objects on earth's surface (Lee, 1999).

In addition, the British company Surrey Satellite Technology LTD. (SSTL), a wholly owned subsidiary of the University of Surrey, has been heavily involved in the microsatellite industry. SSTL's 325-kg UoSat-12, for example, with a 10-m panchromatic sensor and a 32-m multispectral sensor, cost only $10 million

Table 20.3—Overview of the Land Observation Microsatellite Industry

Satellite	Launch (Year)	Resolution (m)	Type	Swath Width (km)	Weight (kg)	Revisit Time (Days)	Cost ($M U.S.)	Status
KITSAT-1[a]	1992	400	Panchromatic	136 x 360	48.6	10	5	Failed
		4,000	Panchromatic	2,170 x 1,450				
KITSAT-2[a]	1993	200	Panchromatic	68 x 180	47.5	20	5	Operational
		2,000	Panchromatic	1,085 x 725				
KITSAT-3[a]	1999	13.8	Multispectral	50	110	34	8	Operational
DLR-TUBSAT[b]	1999	6	—[c]	45	—[c]	—[c]	Operational
		120		—[c]				
		370		—[c]				
UoSat-12[d]	1999	10	Panchromatic	10	325	7	10	Operational
		32	Multispectral	64				
Tsinghua-1[d,e]	2000	35	Multispectral	70	50	—[c]	5	Operational
TiungSat-1[d,f]	2000	120	Multispectral	120	50	—[c]	5	Operational
DMC[d,g]	2002	35	Multispectral	600	70	1	18	Planned
RapidEye[d,h]	2002	6.5	Multispectral	150	380	2	30	Planned
TSSCD[d,i]	N/A	2.5	Panchromatic	N/A	150	—[c]	—[c]	Planned

[a] Produced in South Korea by SaTReC.
[b] Produced in Germany by the German Center for Air and Space Travel (DLR) and the Institute of Aeronautics and Astronautics at the Technical University of Berlin. The satellite transmits a television signal to ground stations within a 1,000-km radius
[c] Not available.
[d] Produced in the United Kingdom by Surrey Satellite Technology LTD.
[e] Developed for the Tsinghua University of China—a constellation of seven microsatellites is planned.
[f] Developed for the Astonautic Technology (M) SDN, BHD, of Malaysia.
[g] Disaster Monitoring Constellation (DMC)—a constellation of five to seven microsatellites is planned.
[h] This microsatellite is being developed for RapidEye Industries in Germany—a constellation of four microsatellites is planned.
[i] Tactical Small Satellite Concept Demonstrator (TSSCD).

to develop (Sweeting, 1999). The company is presently developing a 6.5-m multispectral system called RapidEye, which weighs about 380 kg, at a cost of $30 million (Sweeting, 1999). In addition, SSTL has reached an agreement with the British government to construct a 150-kg microsatellite called Tactical Small Satellite Concept Demonstrator (TSSCD) for the British Ministry of Defence that will be capable of detecting objects as small as 2.5-m in black and white (Sweeting, 1999). Although the system's development costs have not been publicly disclosed, SSTL maintains that the price tag will be considerably less than any comparable government- and privately owned land-observation satellites (Sweeting, 1999).

It is not clear whether the emerging microsatellites can match either the performance or capabilities of their much larger and more expensive counterparts. Some have argued that the traditional systems possess more-stable platforms and can produce better-quality images. While there maybe some merit to this argument, it is also true that even these less-capable systems can fulfill many imagery needs. And more importantly, the growth of the microsatellite industry will further limit the ability of a handful of nation-states to control access not only to satellite imagery but also to remote sensing technology.

Conclusion

Within a decade or two, quite a number of countries and several private companies will have high-resolution remote sensing satellites in orbit, and most of these will be selling an appreciable share of their imagery on the open market. As technological and cost barriers continue to fall, the likelihood grows dimmer by the day that any single government, or likely consortium of governments, will be able to put meaningful constraints on access to imagery.

The consequences of this shift to greater transparency are not fully predictable, but the good will likely outweigh the bad. Cold War experience teaches that imaging satellites can stabilize even virulently hostile relationships by calming unwonted fears. All sorts of actors, from environmentally despoiling corporations to human-rights-violating despots, will face new pressures to clean up their acts or be shamed in the global spotlight. Although adapting to increased scrutiny is always uncomfortable, the increased scrutiny is inevitable in this case, and the potential benefits are enormous.

Notes

1. For the full report on the UNISPACE III conference, see http://www.un.org/events/unispace3/.
2. Member countries include Argentina, Australia, Austria, Belgium, Bulgaria, Canada, the Czech Republic, Denmark, Finland, France, Germany, Greece, Hungary, Ireland, Italy, Japan, Luxembourg, Netherlands, New Zealand, Norway, Poland, Portugal, the Republic of Korea, Romania, the Russian Federation, the Slovak Republic, Spain, Sweden, Switzerland, Turkey, Ukraine, the United Kingdom, and the United States.
3. Satellite costs are directly related to key system characteristics, such as the number and type of sensors, spatial and spectral resolutions, swath width, sensor stability, and satellite agility. Moreover, satellite costs also reflect whether a planned system makes use of existing technologies or whether substantial funds are needed for additional research and development activities.

4 In general, insurance costs can be expected to range between 13 percent and 20 percent of the combined value of the satellite and the launch vehicle. Launch costs depend on the weight of the satellite and the type of launch vehicle used to deploy the system. Clayton Mowry of Satellite Industries Association, Alexandria, Virginia, September 28, 1999, provided insurance costs. The launch vehicle costs can be found in U.S. Department of Transportation (1999).

References

Cochran, Barbara, comments during the Carnegie Endowment for International Peace conference titled "No More Secrets? Policy Implications of Commercial High-Resolution Satellites," Washington, D.C., May 26, 1999.

de Selding, Peter B., "France Weighs Purchase of Commercial Spy Data," *Space News*, Vol. 10, No. 38, October 11, 1999.

Dobbs, Michael, and R. Jeffrey Smith, "New Proof Offered of Serb Atrocities—U.S. Analysts Identify more Mass Graves," *The Washington Post*, October 29, 1995, p. A1.

Ferster, Warren, "U.S., Canada Reach Radarsat 2 Accord," *Space News*, Vol. 10, No. 39, October 18, 1999.

Florini, Ann M., "The Opening Skies: Third Party Imaging Satellites and U.S. Security," *International Security*, Vol. 13, No. 2, Fall 1988.

Gordon, D. Brian, testimony before the U.S. Congress Committee on Science, Space, and Technology and the Permanent Select Committee on Intelligence, June 26, 1991, p. 29.

Gupta, Vipin, and LTC George Harris, *Detecting Massed Troops With the French SPOT Satellites: A Feasibility Study for Cooperative Monitoring*, Sandia National Laboratories: Cooperative Monitoring Center, January 1999.

Hays, Lt. Col. Peter L., and Lt. Col. Roy F. Houchin, II, "Commercial Spysats and Shutter Control: The Military Applications of U.S. policy on Selling and Restricting Commercial Remote Sensing Data," Maxwell Air Force Base, Ala.: School of Advanced Airpower Studies, unpublished paper, 1999.

Kutyna, Donald J., prepared statement, Hearings before the U.S. Senate Armed Services Committee, April 23, 1991.

Lee, Seorim, Satellite Technology Research Center, Korean Advanced Institute of Science and Technology (KAIST), response to questionnaire, October 29, 1999.

McDonnell-Douglas Corporation, *Reconnaissance Hand Book*, 1982.

Martin, Jim, Raytheon Systems Company, Arlington, Va., personal communication, September 29, 1999.

Mertl, Steve, "Industry Minister Warns U.S. on Stalled Satellite Project," *The London Free Press*, August 10, 1999, p. D3.

Mowry, Clayton, Satellite Industries Association, Alexandria, Va., personal communication, September 28, 1999.

Pace, Scott, testimony before the U.S. House of Representatives, Committee on Science, Space, and Technology and the Permanent Select Committee on Intelligence, February 9, 1994.

Space News, "Alenia Wins Radarsat 2 Contract," January 10, 2000a, pp. 3, 20.

——, "New Delhi Considers Strict Policy on Imagery," May 1, 2000b, pp. 3, 19.

Stoney, William, comments during the Carnegie Endowment for International Peace conference titled "No More Secrets? Policy Implications of Commercial High-Resolution Satellites," Washington, D.C., May 26, 1999.

Sweeting, Martin, Director Surrey Space Centre, Surrey Satellite Technology LTD., response to questionnaire, December 15, 1999.

The Wassenaar Arrangement on Export Controls for Conventional Arms and Dual-Use Goods and Technologies, Dual-Use List, Category 6, Section A(2b), The Hague, July 1996. Available at http://www.wassenaar.org/ (last accessed June 12, 2000).

United Nations Committee on the Peaceful Uses of Outer Space, Third United Nations Conference on the Exploration and Peaceful Uses of Outer Space, July 19–30, 1999. See http://www.un.org/events/unispace3/ (last accessed June 12, 2000).

United Nations Security Council, Resolution 1010, August 10, 1995.

U.S. Air Force, Philips Laboratory Aggressor Space Applications Project, Operation Seek Gunfighter, Final Report, January 23, 1999.

U.S. Department of Commerce, National Oceanic and Atmospheric Administration, 15 CFR Part 960, "Licensing of Private Land Remote-Sensing Space Systems, Notice of Proposed Rulemaking," *Federal Register*, Vol. 62, No. 212, November 3, 1997.

U.S. Department of Transportation, Federal Aviation Administration, Commercial Space Transportation, Washington, D.C., 1999.

U.S. House of Representatives, Technology, National Landsat Policy Act of 1992, Report 102–539, Washington, D.C.: Government Printing Office, May 28, 1992.

U.S. Senate Committee on Commerce, Science, and Transportation, NASA Authorization for Fiscal Year 1978, pp. 1642–1643

Wright, Robert, "Private Eyes," *New York Times Magazine*, September 5, 1999.

Zimmerman, Peter D., "The Use of Civil Remote Sensing Satellites During and After the 1990–91 Gulf War," *Verification Report*, London: VERTIC, 1992.

——, "From the SPOT Files: Evidence of Spying," *The Bulletin of the Atomic Scientists*, Vol. 45, No. 7, September 1999, p. 24.

21
How Open Will the Skies Really Be?

Mark David Gabriele

Let us begin by conceding that in the long term, open skies will prevail. While commercial imagery with a resolution of approximately 10 m has existed for many years, 1-m satellite imagery is finally commercially available. This new technology, with its dramatically improved resolution, provides the ability to perform tasks that were not possible in the past. While there is now only a limited supply of 1-m resolution imagery, there is cause for optimism that such high-quality satellite imagery will become ubiquitously available to all interested parties in the long-term future. When this comes to pass, the global community will enjoy the security benefits of multilateral transparency as a stabilizing force, and governments will be able to use high-resolution imagery of their nations for urban planning, natural resource exploration and exploitation, and a host of other benefits.

Unfortunately, there are clouds to accompany even the brightest of silver linings; or, as John Maynard Keynes put it in *A Tract on Monetary Reform* (1923, Ch. 3; emphasis in the original), "*Long run* is a misleading guide to current affairs. *In the long run* we are all dead." Openness may not always provide the security benefits claimed, and it is not obvious that all will benefit from the promise of efficient resource planning and allocation—at least not initially. While the availability of commercial satellite imagery might ideally prompt aspirations to egalitarian advancement and cooperation among nations, there is some reason to believe that events will not play out in the most benign fashion. Once information from satellites—preferably from multiple providers—becomes universally available, we may expect the promised benefits to appear. However, in the near future, the access to information that the few high-resolution commercial remote sensing systems provide may serve to exacerbate sensitivities and difficulties. This is because 1-m resolution imagery provides dramatically different capabilities than those available in previous-generation systems. For example, 1-m imagery may be used to locate and identify military targets and installations down to the level of individual tanks and artillery pieces, with sufficient precision to create an advantage for one side over another. This chapter will consider how the unilateral introduction of such capabilities to the international and domestic markets on a commercial basis may present certain risks that should be accounted for in the deliberations of policymakers.

Open Skies, Commercial Imaging, and the International Community

Open skies refers to an arrangement between nations to allow overflights for the purpose of collecting imagery. Volumes have been written about the origins of the concept of open skies, and about the original and current incarnations of the Treaty on Open Skies.[1] A substantial literature in the fields of political science and international relations is devoted to confidence-building measures writ large; open skies, and more generally transparency, figure prominently in this work.[2] For the purpose of the present discussion, transparency between nations refers to a state having visibility into the interior of a counterpart; this visibility could be photographic in nature (as provided by a satellite) or could be provided by a free flow of goods, services, and information across the border. The intention of an open skies agreement is to provide a well-defined mechanism through which some degree of cross-border transparency may be implemented. To further oversimplify, there are basically two schools of thought on the issue of transparency: One side argues that transparency lends a stabilizing influence and is good for international security, the other that it is potentially destabilizing and a detriment to security. Both sides are probably right.

Transparency across national borders can certainly be a destabilizing factor in regions of tension, where high-resolution imagery can be used to ascertain information on an adversary's order of battle and disposition of forces. In such cases, the side having a clear military advantage will find its superiority confirmed by the imagery collected and may choose to exploit their opponent's weaknesses. Alternatively, the disadvantaged party may come to the conclusion that, in view of its evident handicap, the best opportunity for survival against a superior enemy force would be to execute a preemptive strike in the hope of mitigating a probable future loss. How either side views the issue depends on a variety of factors, among which are the exclusivity of the information and the perception of the accuracy of the information. If the recipients of the information accept that it is accurate, exclusivity becomes a more important factor in the equation.

Consider two nations that are inclined to be hostile to each other; call them Red and Blue. Red is aggressively inclined toward Blue and may seek to prosecute a military advantage if one is known to exist. If there is high-quality information of which only Red is aware relating to Blue's military readiness, information asymmetry exists and an aggressive Red force has a very real advantage. If Blue can obtain similar information relating to Red's readiness, no information asymmetry exists, and the two nations operate under conditions of bilateral transparency. The idea behind open skies is that, if both Red and Blue have high-quality information about each other's readiness—and if each knows what information the other has—both sides can work to mitigate their apparent weaknesses and improved stability should result.

There are, however, different cases that may arise if the transparency is not bilateral. Say that Blue is unable to obtain information on the state of readiness of Red's forces, but that Blue does have knowledge of exactly what information Red has access to. In this case, Blue forces can take appropriate measures to mitigate any weaknesses that Red is known to have observed, and Red's information asymmetry advantage is somewhat diminished. The worst case for Blue would occur if Red were able to obtain information on the disposition of Blue forces surreptitiously, so that Blue forces could be taken entirely by surprise. Where there is a great disparity between the actors in military strength, the benefits from information incongruity can be further magnified. There are similar cases in which political, rather than military, objectives are at stake; transparency can play a similar role in these circumstances.[3]

Consider as an example the case of India and Pakistan. In this case, there is a disputed territory in Kashmir that has been the scene of border skirmishes for many years. Hypothetically, assume that Pakistan becomes able to get access to very accurate information about the disposition of Indian forces in the area and the information showed significant and exploitable weaknesses in those forces. Pakistan may be tempted to launch an attack in the hope of exploiting its perceived advantage and setting up a new frontier. Exactly how information asymmetry would contribute to a situation like this is of course impossible to predict, and whether the asymmetric availability of high-resolution imagery would tip the scale in favor of attack is unknowable. In general, the decision of one nation deciding to attack another based upon the detection of an exploitable weakness will of course depend on the wide variety of other factors that also influence a strategic balance. Among these are underlying asymmetries in force structure, domestic and international political considerations, and a host of others. Accurate information about enemy disposition is appropriately described in U.S. doctrine as a "force enhancement" or "force multiplier"; consequently, the ability to provide "force multiplication" to potential aggressors could be regarded as similar to providing any other dual-use materiel.

While information regarding the disposition of enemy forces has always been the stock-in-trade of spies and intelligence officers, the nature of photographic (or electro-optical) reconnaissance imagery leads to a difference in the acceptance of this information. Imagery intelligence—particularly high-resolution, visible-light imagery, which looks like a snapshot—is perceived as more readily interpretable (or perhaps believable) by line officers and political leaders. Imagery also benefits in that it is a mechanical system and has no points of view or interests to further, so it does not face the credibility problems of human intelligence agents. Of course, an image must be interpreted, and the interpreter will provide an assessment that incorporates his or her opinion; however, this is a more objective and controllable process than many other technical and nontechnical sources of intelligence.

Clearly, there are advantages for an actor who has more, or better, information than his or her opponent. Where transparency-enhancing activities serve to alleviate such disparities and provide a stabilizing influence, they aid in the achievement of a laudable goal. It is important to consider that, to manifest these desirable stabilizing effects, the implementation and practice of transparency must be carefully undertaken, in a way that avoids creating potentially destabilizing asymmetric distributions of information. Indeed, the asymmetry of information that existed during the Cold War between the open societies of the West and the Eastern Bloc led the West to suggest introducing measures that would create greater multilateral transparency across the Iron Curtain. Thus evolved the concepts that were eventually enshrined in the

Treaty on Open Skies: All participants should have equal access opportunities; all participants should use standardized equipment capable of delivering the same types and quality of information; and all information thus collected should be made available immediately to the nation collecting the data and the nation from which the data was collected (and to other signatories, as requested). The successful practice of transparency as envisioned by this policy could be thought of as providing a window that allows a clear view back and forth between the interested parties. Successful examples of this type of arrangement are the U.S. provision of imagery to interested parties during the course of peace negotiations (such as those at the conclusion of the Bosnian conflict, which ended in the so-called "Dayton Accord"; see Chapter 14 in this volume) or as a routine practice in sensitive border areas (as part of the Middle East peace process) (Smithson, 1992).[4]

Unfortunately, the United States commercial imagery licensing policies (and indeed the United Nations policies as well) as they are currently structured bear a greater resemblance to a one-way mirror than to a window. Section 960.9 of the regulation implementing the Land Remote Sensing Policy Act (Federal Register, 1997, p. 59, 327), which derives from Principle XII of the United Nations' *Principles Relating to the Remote Sensing of the Earth from Space* (1986), notes that any sensed state will have access to data collected "concerning its territory . . . as soon as the data are available and under reasonable terms and conditions." This appears to be a laudable goal, and it is most readily understandable as such when considered with respect to imagery that might be collected for the purposes of natural resource mapping or exploitation. In that case, a developing nation would (properly) be informed of imagery collections over its territory and could choose to acquire the same images for presumably similar purposes. But the concept of open skies—with its implication of equal access on equal terms—is not part of the UN codification. In practice, if the party requesting the image collection is malevolent and if resource mapping is not the goal, the UN principles fail, because adherence to the letter of the principles but not to their spirit can create the same type of asymmetry described above.

As an example, if an aggressor nation wishes to obtain high-resolution remotely sensed data about a neighbor, all it needs to do is order imagery of the target from a U.S. commercial imagery provider. Once the imagery is collected, the sensed nation will (eventually) be informed of the collection and the collected data will be made available to it under the "reasonable terms and conditions" noted above. By that time, the would-be aggressor will presumably have had its order filled and may have had time to use that information to achieve some advantage over its neighbor.[5] The target nation will theoretically have the opportunity to obtain the data for itself and may be able to see what the aggressor has seen, but that is a far cry from true bilateral transparency. Even this opportunity for a "look in the mirror" may not occur for significant lengths of time because of latency in the notification of the imagery collection or any number of delays that could happen during the process of production or distribution of the image data. These difficulties may be exacerbated through the establishment of exclusive "regional distributorships" that could offer preferred treatment to one nation at the expense of another.

Consider a nation that obtains a regional distributorship for data from a high-resolution satellite. As a regional distributor, it would have immediate access to imagery of neighboring states but would also have the ability to decline to sell imagery of any particular site in its own nation to anyone within the bounds of its regional distributorship. Indeed, since there is no requirement that a regional distributor must even sell data to any given client, a nation could find itself practically shut out of the high-resolution remote sensing market entirely. As a historical precedent, India has been widely accused of using its Landsat downlink site in this manner against other regional actors (Spector, 1990); in that case, it was reported that India restricted the distribution of images of itself and of surrounding nations it receives via its Landsat ground station. In a recent and interesting continuation of this behavior, India has recently tried to prevent the sale of high-resolution imagery of portions of its own terrain from distribution to its own citizens (Jayaraman, 2000).

This is what might politely be called "unilateral transparency," by which one nation is capable of obtaining access to data about the other without the need to reveal anything of itself. In such cases, the ideal of multilateral transparency via open skies and remote sensing falls by the wayside. From this type of "exclusive distributorship," arrangement, it is not a great conceptual leap to the case in which the observed nation knows nothing of the fact that it has been observed or of the specific locations targeted; this might simply be called "spying." Even the imperfect ideal of unilateral transparency with the stipulation of required notification and access for the observed state is superior to this. If unilateral transparency can be thought of as transparency through a one-way mirror, the no-notification case is more similar to transparency through a hidden camera. Unfortunately, the only substantive difference

Table 21.1—The Effects of Image Delivery Delays on the Strategic Balance Between Two Nations

Image Delivery Time	Military Power of Observer Versus Observed Nation	
	Relatively Strong	Relatively Weak
Images available to both observer and observed nations at the same time	Unilateral transparency	Could expose vulnerabilities that require immediate rectification to maintain balance
Observed nation experiences significant latency in image delivery	Potentially destabilizing if observer intends a preemptive strike	Without ability to find and fix vulnerabilities, could result in catastrophe for observed nation

between these two cases is the time that passes before the observed party receives notification that it is, in fact, being observed. Some potential effects of unilateral imagery collections on the strategic balance between an observer and an observed nation are described in Table 21.1.

Given the existence of a healthy commercial remote sensing market, this problem should ameliorate itself. Over time, remote sensing companies will arrive in the market to supply imagery to nations that may initially be more or less shut out of the market by "regional distributorships" or similar arrangements; however, that will not happen until and unless the market becomes competitive. Therefore, at least initially, before market forces have the opportunity to act, a set of information "haves" and "have-nots" may develop. In the bipolar days of the Cold War, there were two clear "haves"—the United States and the USSR—and the remainder of the world was without access to high-resolution satellite imagery, except as it may have been provided under intelligence-sharing agreements or to military and political allies. Before the uneasy acceptance of a sort of information parity had set in, once the United States and Soviet Union had both developed their national technical means of reconnaissance, the information asymmetry that existed between them contributed to the development of an unprecedented arms race. It is worth noting that both proposals for a Treaty on Open Skies (the original Eisenhower proposal of 1955 and the Bush proposal of 1989) provided for truly multilateral transparency, presumably to avoid the creation or propagation of the information asymmetries that ultimately led to the arms race. Indeed, the current version of the treaty additionally stipulates that copies of all data collected are to be provided to the observed nation at the conclusion of the mission and are to be made available to all member nations thereafter. Now that the commercial world is collecting and disseminating data, sales of imagery will eventually put the majority of nations onto the same, level playing field, but only if and when a competitive market develops. Because there are significant barriers to entering this marketplace, this is unlikely to occur very soon.

Of the many barriers to entering the commercial remote sensing marketplace, the most formidable may be technical risk, cost, commercial risk, and licensure. Among these, technical risk may present the greatest challenge. *Technical risk* refers to the engineering challenge involved in manufacturing, launching, and controlling the spacecraft with sufficient precision to allow the collection of very high-resolution imagery from space. To accomplish this, mechanical tolerances must be very fine, and a wide assortment of high-performance materials must be fabricated into the myriad of components that compose a launch vehicle and spacecraft. Once the engineering and mechanical tasks have been mastered, the barriers to entry presented by cost and commercial risk present themselves. These two go hand in hand; estimates of the cost of a commercial remote sensing satellite with resolution in the 1-m range vary widely, but by all accounts it is an expensive undertaking ranging well into the tens of millions of dollars. To recoup those costs, imagery is sold on the commercial market. However, there is as yet great uncertainty about the commercial market for high-resolution imagery and thus about the ability to recoup the costs of design, construction, and launch of the spacecraft. Finally, licensure may act as a barrier to entry regardless of mastery of all the others. This is because operating any system with significant U.S. involvement requires a license from the Department of Commerce. To obtain such a license, the prospective system operator must demonstrate to the satisfaction of the Secretary of Commerce that the system will comply with any and all relevant U.S. government regulations. Given the number and magnitude of these barriers to entry, it appears unlikely that a large number of competitors will succeed in bringing their systems to market anytime soon. Until that does happen and the market becomes competitive, these systems may cause more information-based asymmetries than they cure.

"Open Skies" in the Domestic Context

The international issues arising from open skies and from cross-border transparency in general have been argued across many venues and are well understood. They may in some sense be more easily dealt with than the issues that arise from the domestic applications of what might be called "transparency for hire." Although it has been a staple of the military since World War II, overhead imagery has generally been perceived as a tool that is best suited to filling a few specific niches in the commercial and civil sectors. Because of its relegation to these niche markets, overhead imagery has existed largely in relative obscurity. However, the availability of high-resolution commercial remote sensing data that can be easily ordered and obtained by anyone may, in the near future, necessitate consideration of some basic issues relating to information and privacy here in the United States.

There is some case law regarding the use of imagery collected from aircraft by law enforcement agencies as evidence in criminal proceedings. The U.S. Supreme Court has ruled on at least three separate occasions that observations made from aircraft do not violate fourth amendment privacy rights. The fourth amendment stipulates that

> The right of the people to be secure in their persons, houses, papers, and effects, against unreasonable searches and seizures, shall not be violated, and no Warrants shall issue, but upon probable cause, supported by Oath or affirmation, and particularly describing the place to be searched, and the persons or things to be seized.

The court applied a two-part test to determine whether the action taken by the enforcement agency constituted a "search" in the legal sense, which would have required the issuance of a warrant. The test required that (1) the party possess a desire to maintain privacy and (2) that the expectation of privacy be reasonable. In these cases, the court found that the expectation of privacy was unreasonable because the objects seen were visible from freely navigable airspace and could have been observed by a passenger in any airplane overflying the area (Kelly, 1995).

By this reasoning, a spacecraft flying through freely navigable space must be expected to be able to collect imagery of any item within its field of regard. Further, because this technology, and the ability to employ it, is not limited to law enforcement agencies but is instead available to the general public, it is even less "reasonable" to believe that an object left out in the open will not be observed. The skies are apparently wide open over the United States, insofar as the Supreme Court is concerned.

In a very real sense, nothing regarding domestic collection of imagery has changed with the introduction of the current generation of high-resolution commercial remote sensing satellites. Commercial remote sensing satellites are not the only source of overhead imagery of the United States—a light aircraft and a camera are all that are needed to collect imagery of most domestic areas. Because the vast majority of U.S. airspace is freely navigable, aerial overflight for the purpose of imaging is basically unrestricted; for areas over which airspace is restricted, remote sensing spacecraft can provide coverage. This is the inverse of the international case. In international law, the airspace of any given nation is the sovereign territory of that nation. It is therefore possible for one nation to deny any other nation the ability to collect imagery of its territory (at least the portion of its territory not visible from international airspace) via airborne platforms simply by denying the use of its airspace. Because spacecraft in their orbits do not respect international boundaries, there is no practical and enforceable prohibition on the conduct of imaging operations from spaceborne platforms. These spacecraft may be commercially owned and operated either by U.S.-based or by foreign entities; U.S. firms will be answerable to the U.S. government, while those operating from abroad will not.

Regardless of whether data are collected by commercial remote sensing firms or by a government, the issue is the same on the domestic scale as it was on the international scale: the asymmetry of information. One party—a government, a plaintiff in a lawsuit, a law enforcement agency—may have the ability to acquire information that the other cannot. When it happens in the international arena, the consequences are felt between nations. In the domestic arena, it may affect the expectations and the basic rights of the populace to privacy (notwithstanding the constitutional arguments regarding the nature and scope of "privacy rights"). In the United States, the fourth amendment holds that the government must be held to a rigorous standard when it comes to law enforcement's capabilities to obtain private information surreptitiously.

Because of the long-standing separation between law enforcement and military–national security function in the United States, it appears that the concept of high-resolution commercial remote sensing may be broken into two distinct regimes: imagery collected over the United States (especially as collected by U.S. commercial platforms; open skies treaty flights; and, potentially, by foreign high-resolution platforms) and data collected over other nations. This leads to questions about whether imagery collected over the United States can or should be handled any differently from imagery collected over any other nation. Would the use of data collected by these sensors be regarded as constituting an invasion of privacy? How does the issue of personal privacy for U.S. citizens relate to the concept of international transparency?[6]

Reconciling Transparency and Privacy

Fortunately, the difference is rather straightforward. Transparency and privacy are two sides of the same coin. However, in this case, transparency refers to the ability of a nation to understand the politics and policies of its neighbor, while privacy refers to the desire of individuals to shield their actions and beliefs from public scrutiny. Transparency between nations, when it occurs on an equitable basis, can lay a foundation for a peaceful and trusting relationship.

The current generation of high-resolution commercial remote sensing satellites is limited in terms of the highest resolution that a system can collect. Further, this generation of systems collects data only in the visible and near-infrared portions of the electromagnetic spectrum. Future systems may be capable of collecting imagery using novel technologies or using other portions of the electromagnetic spectrum, notably synthetic aperture radar (SAR) imagery or so-called "hyperspectral" imagery; presumably they, too, will face limitations on the best allowable resolution. Such systems may collect information that is both qualitatively and quantitatively different from what more traditional systems can collect. Such systems collect information that cannot be sensed or processed by any human who does not have access to highly sophisticated and specialized technology.

As technology and systems continue to improve, such systems will offer the potential to gather data that are more closely associated with a gathering of technical intelligence than with any concept of openness or transparency. A very high-resolution (e.g., 1-m or better) SAR or hyperspectral sensor may more be likely to be a tool for hire by intelligence services than a tool for exercising more traditional implementations of transparency in the political-military context. Likewise in the domestic context: While reasonable people may expect themselves or their property to be observed from any public vantage point, the prospect of being subjected to analysis by an imaging spectrometer may be more than a bit surprising.

The domestic privacy issues and the international transparency issues relate to information disparities that will occur as society comes to grips with the existence of such data. In the international case, a sensed nation may be aware of a potential information asymmetry but may be unable to obtain a solution until a more competitive high-resolution remote sensing market has developed. In the domestic case, the information asymmetry that exists lies in the knowledge and expectation of a satellite's imaging operations. For the present, the singular difference between the two is that domestic imaging merely provides an additional source of data that can be obtained by a variety of other means (e.g., aerial overflight, observation from public places). Imagery of international targets differs because the data may be of a "denied area"—a place exclusively available via a particular source—and that source may choose not to provide the data on equitable terms, if at all. Thus the question arises: How open will the skies really be, and is there some way to ensure that they are open for the development of stability and transparency, rather than for the development of destabilizing and asymmetric intelligence gathering?

As technology improves, a time may come when the resolution and capabilities of the available sensors becomes too good; that is, the information that can be gathered by the sensors does not fit some legal definition of "reasonable." When that happens, many people will become uncomfortable with the concept that an unseen and somehow omniscient "eye in the sky" really might be watching them. Realistically, mitigation of the risks to privacy and equity that stem from information disparities in this context may not be possible, except by the rather draconian measure of prohibiting the release of imagery that has a resolution beyond some limit, as the Department of Commerce guidelines now provide. What that limit ought to be and how that determination is made should, however, be matters of public concern and debate.

In the international arena, it is easier to conceive a solution to the problem of inequitable access to imagery; such a solution could help to alleviate the threat of information asymmetry. This is readily accomplished because there are a fairly small number of independent states, and almost all of them have dedicated significant resources to their national security. The commercial remote sensing providers therefore have available a reasonable set of contacts through which information regarding ongoing collections can be provided. The exact coordinates of each image collected can be compared to international boundary information by means of a geographic information system; a notification can then be made when any collection either crosses an international boundary or takes place within the boundary of any sovereign nation that has requested immediate notification of such collections. This information should be provided immediately, as the collections are made or as the data are processed, before image distribution.

Depending on the diplomatic climate in any given area, a requirement could be levied for strict enforcement of the UN "nondiscriminatory access" provision of imagery to a sensed state under certain conditions, such as if two states are known to be involved in a diplomatic conflict over particular issues. In this case, sales of sensitive imagery to one party (e.g., an image of a disputed border area) should be countered by provision of that image to the other party, with delivery occurring simultaneously. Additionally, the "reasonable cost" fee might be paid not by the sensed state but by whoever ordered collection of the image. If the states that license the providers of these remote sensing services reach the conclusion that transparency in international relations—particularly bilateral transparency—is beneficial, the contractual arrangements for the provision of data can be structured to emphasize characteristics that favor transparency rather than the secretive collection and analysis of military data. This is not an implementation of true bilateral transparency, but it is at least a mitigation of dramatic and potentially destabilizing information asymmetry.

To address the title question of this chapter directly: In either the international or the domestic case, as we enter the era of high-resolution commercial remote sensing satellites, the skies will be open. Specifically, they will be open for business; until a competitive market arises to ensure all players equal opportunity and access to high resolution remote sensing data, the side with the advantage will be the side that is able to purchase an advantage. While mitigative techniques are both reasonable and practical to implement, they will at best limit the damage done by the "unilateral transparency" that the commercial remote sensing market might initially enable. The final question, as we consider the remarkable potential that lies in the data returned by these systems, is whether there is a way to resolve the inherent issues of data access, equitability, and privacy fully or whether we should simply allow the market to dictate its own answer.

Notes

1. For example, Chapter 2 in this volume, Krepon and Smithson (1992), Rostow (1982), and Gabriele (1998).
2. Krepon (1995) provides an excellent bibliography for this literature.
3. This brief discussion obviously cannot do justice to an inevitably complex and highly nuanced topic. Chodakewitz and Levy (1990) provide a good short treatment.
4. An official mention of this program can also be found in Pedlow and Welzenbach (1998, p. 257).
5. The advantage obtained might be military, but may also be diplomatic or political. In any case, it may be a strategic or tactical advantage.
6. There are other, more immediate threats to personal privacy in the United States (sharing of medical records, tracking of credit purchases, etc.). As such, the

issue of domestic privacy as it relates to current high-resolution remote sensing systems may become a greater concern in relation to future systems that could offer higher resolution imagery.

References

Chodakewitz, S. B., and J. L. Levy, "Implications for Cross-Border Conflict," in Michael Krepon, Peter D. Zimmerman, Leonard S. Spector, and Mary Umberger, eds., *Commercial Observation Satellites and International Security*, New York: St. Martin's Press, 1990, pp. 90–103.

Department of Commerce, National Oceanic and Atmospheric Administration, "Licensing of Private Land Remote-Sensing Space Systems," draft rules, *Federal Register*, Vol. 62, No. 212, 15 CFR Part 960, November 3, 1997, pp. 59, 317–59, 331.

Gabriele, M. D., *The Treaty on Open Skies and its Practical Applications and Implications*, Santa Monica, Calif.: The RAND Graduate School, 1998.

Jayaraman, K. S., "Imagery Deal Spurs Indian Policy Review," *Space News*, January 3, 2000.

Kelly, K. C., "Warrantless Satellite Surveillance: Will Our 4th Amendment Privacy Rights Be Lost In Space?" *The John Marshall Journal of Computer and Information Law*, Vol. XIII, No. 4, 1995, pp. 729–762.

Keynes, John Maynard, from *A Tract on Monetary Reform*, as quoted in *Bartlett's Familiar Quotations*, 16th ed., New York: Little, Brown, & Co., 1992, p. 652.

Krepon, M., ed., *A Handbook of Confidence-Building Measures for Regional Security*, Washington, D.C.: The Henry L. Stimson Center, 1995.

Krepon, Michael, and Amy E. Smithson, eds., *Open Skies, Arms Control, and Cooperative Security*, New York: St. Martin's Press, 1992.

Pedlow, Gregory W., and Welzenbach, Donald E., *The CIA and the U-2 Program, 1954–1974*. Washington, D.C.: Central Intelligence Agency, 1998, p. 257.

Rostow, W. W., *Open Skies*, Austin: University of Texas Press, 1982.

Smithson, A. E., "Multilateral Aerial Inspections: An Abbreviated History," in Krepon and Smithson (1992), p. 116.

Spector, L. S., "The Not-So-Open Skies," in Krepon, et. al., eds., *Commercial Observation Satellites and International Security*, pp. 165–166.

United Nations, General Assembly, *Principles Relating to Remote Sensing of the Earth from Outer Space*, resolution 41/65, 1986. Available at http://www.un.or.at/OOSA/treat/rs/rstxt.html (last accessed May 31, 2000).

22 The Globalization of Transparency: The Use of Commercial Satellite Imagery by Nongovernmental Organizations

Karen T. Litfin

While the primary users of commercial high-resolution satellite imagery will be government agencies and private firms, a smaller but politically significant market consists of nongovernmental organizations (NGOs), whose numbers and influence have skyrocketed in recent decades. NGOs have been alternatively analyzed as "the conscience of the world," as "sovereignty-free actors" capable of acting in the porous space beyond the nation-state, and as critical nodes in an emerging "global civil society." (Willetts, 1996; Rosenau, 1990; Lipschutz, 1996.) The political clout of transnational NGOs—such groups as Amnesty International, Friends of the Earth, and Doctors Without Borders—has increased to the point that, on many issues, especially environmental and human-rights ones, they are central actors in agenda setting, policy negotiation, implementation, and monitoring (Keck and Sikkink, 1998). NGOs work closely with United Nations (UN) agencies on every contemporary global issue imaginable. While global democratic institutions are relatively undeveloped, the globalization of social movements and pressure-group politics should be understood as a corollary to economic and technological globalization. The globalization of transparency through the diffusion of commercial satellite imagery could make a meaningful contribution to the globalization of democracy.

While some NGOs, particularly environmental NGOs, have been using satellite data for many years, others, such as humanitarian and human-rights

NGOs, are only beginning to explore its utility. Although NGOs generally have small budgets compared with their counterparts in government and business, they have demonstrated substantial ingenuity in gaining access to satellite imagery. Environmental NGOs have a long history of exploiting free or inexpensive imagery and networking with other groups to further ecological objectives. Although the use of satellite data by other NGOs is in the embryonic stages, these groups have been remarkably creative in exploiting public-domain images; strategically purchasing inexpensive images from such sources as TerraServer; and forming partnerships with university researchers, the media, and corporate sponsors. At this point, the utility of commercial satellite images for most NGOs is more in the realm of potential than proven success, especially for humanitarian and human-rights applications. Yet, given that the industry itself is in its early stages, this should be neither surprising nor discouraging.

NGOs can be important players in the dissemination of commercial satellite data, not just as direct users of images but also by being able to pressure governments to procure, release, or transfer images relevant to politically sensitive events. The rise of public diplomacy means that governments must increasingly disseminate convincing evidence not only to other states to persuade the states of the rightness of official policies, but also to relevant NGOs and the public at large. The visual evidence satellite images provide can be a powerful tool of legitimation for a wide range of nonstate actors on a host of issues. Particularly as the sources of government data become increasingly commercial, they are likely to come under pressure from NGOs to make imagery public. The protection of "sources and methods" that was the sine qua non of "national technical means" during the era of Cold War reconnaissance satellites is no longer a top priority in the new era of commercial observation satellites.

In the following pages, I examine a number of different ways in which NGOs have used or could use satellite imagery in their work. Because environmental NGOs in some ways have set the stage, I look first at these and then turn my attention to arms control NGOs, humanitarian NGOs, and human-rights NGOs. In each case, I look at how satellite imagery can be useful to NGOs, what sorts of things can be detected with commercially available images, and what factors both facilitate and impede the application of satellite data to the work of NGOs.

Environmental NGOs

Since the 1980s, environmental NGOs have been using satellite data extensively in their work, and then continue to do so more than any other kind of

NGO. Five factors account for this discrepancy. First, the lower spatial resolution of previously available satellite imagery was better suited to the large-scale problems that typically concern environmental NGOs. Second, much of the data (especially from Landsat) was available at minimal cost. Third, while environmental NGOs often wish to track degradation over months, years, or even decades, NGOs addressing arms control or international crises typically require more timely data. Delivery of Landsat and Satellite Pour l'Observation de la Terre (SPOT) products, until recently the only satellite data commercially available, can take weeks or months from time of observation. Additionally, archives of imagery that might be a gold mine for an environmental NGO would be relatively useless to other NGOs. Fourth, governments have worked closely with environmental NGOs in some countries, sometimes financing the purchase of satellite imagery. Fifth, environmental NGOs generally have a stronger scientific subculture than other NGOs. While NGOs devoted to issues of peace, sustainability, and human security are all imbued with an ethic of care, arguments based upon scientific causality have an especially prominent place in environmental issues. These five reasons—spatial requirements, cost, temporal requirements, governmental collaboration, and cultural considerations—have made environmental NGOs the most common NGO users of satellite imagery.

Although environmental NGOs are unlikely to become major users of commercial high-resolution satellite data in the near future, a review of their experience with lower resolution data reveals some of the basic issues confronting NGOs in their use of satellite imagery. The factors facilitating environmental NGO applications include the visual appeal of images; the sense of legitimacy associated with information gleaned from technologically sophisticated sources; the ability of environmental NGOs to forge helpful relationships with researchers, the media, government agencies, and the value-added sector; and the accessibility and low cost of Landsat data (before and after the privatization debacle during the mid-1980s). Yet satellite imagery is not a panacea. It can displace other forms of knowledge, accentuate the yawning disparity in wealth and technology between industrialized countries and developing countries; mask the deeper socioeconomic causes of physical phenomena; and obscure human agency. Thus, other NGOs are likely to find the experiences of environmental NGOs instructive.

Satellite data can serve as a powerful tool of legitimation. The Nature Conservancy and the Natural Heritage Network, for instance, have used satellite images to evaluate biodiversity and assess the health of ecosystems in their efforts to monitor enforcement of the U.S. Endangered Species Act (Stein, 1996). The Britain-based Coral Cay Conservation uses Landsat thematic mapper data along with aerial photography not only to document reef destruction around the world but also to generate geographical information system (GIS) maps of coral reef habitats, which are used in ecosystem management initiatives (Harborne, 1998). And the widely read World Resources Institute (WRI) annual reports base their estimates of global and regional deforestation primarily on satellite data (WRI, 2000). Similarly, a recent WRI report used a combination of satellite data and fossil records to conclude that only 20 percent of the planet's original frontier forests—that is, large, ecologically intact forests—remains today (Bryant, Nielsen, and Tangley, 1997). In all of these instances, satellites helped to authenticate findings for an audience that could include policymakers, the public, and/or potential donors.

Perhaps the greatest asset of satellite imagery is its visual character. As one individual working to stop deforestation in British Columbia's Clayoqot Sound —the world's last remaining expanse of temperate rain forest—declared, "Satellite images are totally convincing. You show people a map, and they can see clearly what's left." (Quoted in Clayton, 1996; p. 5.) Environmental NGOs have found that satellite imagery can make a distant, abstract ecological crisis seem present and tangible. Consider, for instance, the widely circulated video representation of the Antarctic ozone hole during the negotiations for the 1987 Montreal Protocol. That composite of satellite imagery, which was broadcast to treaty negotiators, the U.S. Congress, and television audiences worldwide, transformed an invisible phenomenon into a dramatic global ecological crisis (Litfin, 1994).

While satellite images have been deployed for many global environmental issues, they can also facilitate the localization of control in some surprising ways. Perhaps most interesting is the use of satellite data by indigenous peoples for mapping their customary land rights and documenting the role of governments and industry in environmental destruction. Environmental advocates have teamed up with indigenous rights groups and university researchers in Indonesia, Nepal, Thailand, Canada, the Caribbean, and the Amazon to integrate traditional indigenous knowledge claims into conventional scientific methodologies through the use of satellite images and GISs in order legitimate territorial claims (see Sirait et al., 1994; *Cultural Survival,* 1995). Yet while satellite imagery serves as a powerful advocacy tool, it can also displace other forms of information. The use of satellite data by indigenous groups, for instance, requires these groups to legitimate their positions on the basis of high technology rather than their own traditional ways of knowing. Similarly, NGOs based in Europe and North America have found that using satellite data for conservation purposes in the tropics can have the unintended consequence of undercutting local aerial photography companies, an ironic consequence if economic development is one of their objectives.

Moreover, satellite data are historical in the sense that they show current, but not past, levels of ecological degradation. Landsat and SPOT images of tropical deforestation in developing countries, for instance, not only render human agency invisible, they also render invisible previous centuries' deforestation in industrialized countries, thereby reinforcing the perception that deforestation is ultimately a problem caused by developing countries. And satellite images cannot reveal that most tropical timber exports are consumed by industrialized countries. Not surprisingly, groups such as Conservation International and the World Wildlife Fund that use satellite images for their work in developing countries almost inevitably meet some initial resistance from local environmentalists and development workers.

Despite the relative familiarity of the environmental NGO community with satellite data, very few environmental NGOs have the technically trained staff needed to acquire, interpret, and manipulate satellite images. For instance, the U.S.-based Wilderness Society acquired its in-house remote sensing facility as a donation, but the group was able to hire the technical staff needed to operate it only after a substantial fundraising effort (Thomson, 1998). Because of the costliness and technical expertise involved in using satellite imagery, environmental NGOs have engaged in partnerships with universities and research institutions, state agencies and international organizations, businesses, and even the military and intelligence communities.

For a combination of philanthropic and commercial reasons, a number of value-added companies working with earth-observation data, particularly in the GIS sector, have made a concerted effort to disseminate their products to environmental NGOs over the past 20 years. For instance, two well-known GIS companies, Environmental Systems Research Institute, Inc. (ESRI) and ERDAS, have implemented grant programs involving thousands of NGOs all over the world. Through this program, environmental NGOs have received software packages, technical training, international mentoring, and a range of technical assistance. While desktop software capable of handling large databases associated with satellite imagery did not exist even five years ago, ESRI's new desktop ArcView program, released in 1994, can be learned in a matter of days.

In the near term, the advent of commercial high-resolution panchromatic satellite imagery is unlikely to have a significant effect on the work of environmental NGOs, particularly those concerned with large-scale and long-term ecological degradation. Detecting clear-cut forests, major oil spills, and biodiversity loss simply does not require 1-m resolution imagery. Some environmental NGOs, however, would have an interest in data from high-resolution space-based hyperspectral sensors because of these instruments' potential for finer discrimination of specific land and ocean features. To the extent that environmental NGOs are concerned with smaller-scale problems, they may rely upon existing relationships with aerial photography firms. Unlike NGOs active in crisis situations, environmental NGOs generally do not need to worry about aerial photographers being shot down.

Arms Control NGOs

The profusion of civilian and commercial satellite imagery opens up new opportunities for nonstate actors to be involved in traditional national security issues. A handful of NGOs have used open-source satellite images to advocate peace and disarmament, to involve themselves in verifying arms control agreements, or to legitimize themselves as alternative sources of information to governments. Given that governments are especially secretive about military matters, this development has important implications for the growth of open and democratic societies. On the other hand, it carries with it certain risks, because what is knowable to arms control NGOs can also be known by more nefarious actors.

Even before the advent of commercial high-resolution satellites, some groups were creatively deploying open-source images.[1] The United Kingdom-based Verification, Research, Technology, and Information Centre (VERTIC) uses imagery to compile its yearly *Verification Report* on arms control and environmental treaty compliance.[2] Vipin Gupta, who worked with VERTIC on some of his early photointerpretation projects, has demonstrated the utility of publicly accessible satellite imagery in monitoring the nuclear activities of China, India, and Algeria (see Gupta, 1998; McNab, 1993). While Landsat and SPOT images have been available for some years, their low spatial resolution makes it impossible to resolve many objects of interest to arms control NGOs, such as weapon bunkers, tanks, and aircraft. With new high-resolution images commercially available, the public interest community "will have direct access to global imagery at resolutions previously only available to the national intelligence communities of a few nuclear weapons states." (Federation of American Scientists [FAS], 1999.) However, if Space Imaging's prices for IKONOS data—upwards of $2,000 per scene—are any indication of future trends, the cost of timely data will be prohibitive for most NGOs. Yet arms control NGOs have shown that even archived images can be valuable sources of information.

NGO use of Russian KVR-1000 data, available from archives for as little as $40 for a 25 km^2 scene, offers a glimpse into the potential. The Institute for Science and International Security (ISIS), a Washington, D.C.–based NGO, has used imagery to help generate evidence of nuclear activities in Iraq, Pakistan, and North Korea. One of the organization's most successful projects supplemented KVR-1000 scenes with SPOT imagery, seismic data, and television news coverage to precisely locate Pakistan's first nuclear test site in May 1998. Such information could provide historical and mapping information to help the CTBT Organization's on-site inspections if Pakistan were to become party to the treaty (Albright, Gay, and Pabian, 1999).[3]

John Pike, who directed FAS's Public Eye Project, hopes to obtain funding to purchase new commercial high-resolution images that will enable citizens to monitor military developments around the world. The Public Eye Project used the Internet to post hundreds of declassified high-resolution images from the Central Intelligence Agency's Cold War–era Corona satellites, enabling citizens around the world to gain a clearer view of Cold War history (see http://www.fas.org). The value of these images, however, is not limited to historical curiosity. Pike and his colleagues have used Corona imagery, for instance, to work with NBC News and *The New York Times* to investigate the ecological and human health effects of a former Soviet biological weapons facility on an island in the Aral Sea (Pike, 1999).

These developments demonstrate the potential for commercial high-resolution satellites to open to ordinary citizens the relatively closed world of national security politics. One of the stated goals of the Public Eye Project is "evangelize," i.e., to "actively promote public interest applications of imagery intelligence" (FAS, 1999, p. 4). Of course, few NGOs presently possess the technical and financial resources necessary to use satellite imagery effectively, but the combination of a plethora of competing commercial satellites and the availability of cheap personal computer–based software for manipulating images could bring down the cost sufficiently to enable them do so. Arms control NGOs, like their environmental counterparts, are characterized by a strong scientific culture; many prominent arms control advocates have scientific and technical degrees. This orientation appears to make arms control NGOs more predisposed than humanitarian and human-rights NGOs to apply satellite imagery. Outside the environmental community, there has been little cross-fertilization among NGOs in terms of skill-sharing or resource-pooling to purchase imagery.

Because the news media are less likely than NGOs to be deterred by cost, issue networks comprising NGOs and news groups could emerge for using satellite imagery. News organizations have used satellite imagery since 1986 when ABC, followed by NBC, CBS, and *The New York Times*, used SPOT photos to report on the Chernobyl nuclear disaster (Blumberg, 1991). The potentially dramatic impact of the media's use of commercial high-resolution images on issues of war and peace was demonstrated just prior to Operation Desert Storm, when ABC purchased two 2-m resolution images from the Soviet commercial satellite company, Soyuz-Karta, for $1,560 apiece. Photoanalysts working for ABC News could find no trace of the 265,000 Iraqi troops that U.S. officials claimed were massed at the Saudi-Kuwait border in the fall of 1990. Nonetheless, ABC decided not to run the story because the photos did not include a section of southern Kuwait where the troops may have been. The *St. Petersburg Times*, however, purchased the third photo and contracted with photointerpreters who agreed that there was no evidence of a massive Iraqi presence. The Pentagon continued to insist that the troops were there, but would provide no visual evidence (Heller, 1991). This case suggests that the commercial availability of spy-quality data could partially lift the veil of secrecy behind which states have traditionally conducted their military operations.[4]

NGOs working on military-related issues are especially likely to encounter institutional obstacles to using high-resolution satellite imagery, especially during times of crisis. Even among democracies, governments traditionally have claimed a right to withhold information deemed vital to national security. The United States, for instance, might exercise "shutter control" around militarily sensitive areas during a crisis, thereby preventing both NGOs and the media from obtaining valuable information. While such a move would no doubt raise important legal issues of free speech, current satellite licensing regulations include provisions for such an eventuality. Moreover, because of their work on sensitive issues, arms control NGOs may find it more difficult than other NGOs to gain corporate support from imagery providers and processors. Because the primary early markets for high-resolution images will be military and intelligence agencies, arms control NGOs are unlikely to enjoy the level of corporate support (e.g., training programs and product discounts) extended to environmental and humanitarian NGOs. The work of "imagery activists" on military issues is more likely to be supported by private foundations.

Humanitarian NGOs

While humanitarian NGOs have little experience using satellite data, this situation may soon change. With the advent of commercial high-resolution imaging satellites, new information technologies that make satellite data accessible in a timely manner and manipulable for a multitude of purposes, and a heightened involvement of the military and intelligence establishment in humanitarian crises, the stage has been set for a shift. Humanitarian NGOs have extensive experience in bringing seemingly remote crises into the living rooms of television audiences around the world but often need a visual source of objective information. As Einar Bjorgo's research demonstrates, satellite imagery can help answer such questions as how many people are in need of assistance, precisely where they are and which way they are going, and what local geographic and resource factors are likely to affect assistance efforts (see Chapter 19 in this volume). Not only can satellite imagery help humanitarian NGOs formulate their own relief operations, it can help them to communicate the gravity and scope of a problem to policymakers, the public, and donors.

Normative understandings about which human beings merit international intervention have changed dramatically in this century (Finnemore, 1996). In past centuries, Western norms of humanitarian intervention applied only to whites and Christians. By the late 20th century, a combination of decolonization and a globalized telecommunications system has facilitated what we might call the universalization of "humanity," so that famines, natural disasters, and genocide anywhere on the planet have sparked international concern and action. The global reach of earth observations from space is not only *compatible* with this normative trend; it can also help *reinforce* that trend in some practical ways.

Terrain mapping can help locate areas with appropriate topography, land cover, and access to water for the settlement and repatriation of refugees. Likewise, for purposes of reforestation and habitat conservation, satellite data can help determine the environmental effects of large-scale movements of refugees. In areas with extensive cloud cover, especially in the tropical "cloud belt," radar images can help locate and track refugees; satellite-based observations of campfires can aid in estimating total numbers of refugees. For rapid-onset disasters, such as floods, cyclones, earthquakes, and volcanic eruptions, satellite imagery can be coupled with on-site GIS and/or Global Positioning System capacity to produce maps with key hazards and infrastructural damage pinpointed accurately. In programs using food aid, imagery can help track the volume and movement of food commodities along supply chains. Satellite-based measurements of rainfall data, crop acreage, and soil aridity can provide crucial information about the likelihood and extent of famine, even in closed societies, such as North Korea.

With the avalanche of regional crises in the post–Cold War period, the 1990s could be labeled the decade of refugees. Crises in Somalia, the former Yugoslavia, Rwanda, Afghanistan, and Sudan have produced an unprecedented number of internally and externally displaced refugees. Humanitarian NGOs have worked alongside UN and national relief agencies in each case and have sometimes been the primary source of relief assistance. Each crisis presents its own logistical difficulties: lack of water, great distances to supply roads, inclement weather, mountainous terrain, etc. Before the advent of commercial high-resolution satellites, data from Landsat and SPOT satellites gave broad hydrological and topographical information for refugee camp planning, as well as providing subsequent information on the environmental effects of refugee settlements (Lodhi, Echarria, and Keithley, 1998). Relief agencies are beginning to realize the potential utility of high-resolution satellite imagery (Dowd, 1996). Using KVR-1000 images with a resolution of 2 m, Einar Bjorgo has shown that such data can be useful in estimating numbers of refugees, locating supply roads for settlements, and locating water sites (Bjorgo, 1999).

Nonetheless, the experiences of environmental NGOs advise an element of caution. First, even at 1-m resolution, imaging satellites cannot discern the social causes of crises. Because human agency all but vanishes from the vantage point of outer space, crises may be mistakenly reduced to physical processes, and the root social causes may thereby be ignored. For humanitarian relief efforts concerned with short-term amelioration, this may not be a huge problem. But if either prevention or a deeper understanding of the roots of the problem is the objective, satellite imagery is of limited utility. In any case, satellite data will need to be supplemented with substantial "ground truthing." Satellite images, for instance, cannot reveal that families in a drought-ridden part of Zambia are not starving because the men from those villages work in copper mines elsewhere and send their wages home

(Hutchinson, 1998). Neither can they discern whether refugee camps are populated with terrorists or their innocent victims.[5] Thus, there is a strong need to pair satellite data with sociological and anthropological appraisal tools on the ground.

The fact that satellite data renders human agency invisible is related to another caveat. Although having good information is preferable to having poor information, it does not necessarily generate good policy. Environmental NGOs can cite numerous examples in which further research has been more a substitute, rather than the basis, for sound policy. Although the short-term nature and the high visibility of humanitarian emergencies make this less of a risk for these issues, it is a risk that should be considered. One can imagine a crisis in which precise numbers and geolocations of starving refugees are known, along with digital elevation maps of their camps, but little is done to alleviate their suffering. There is no simple correlation between having detailed information about a problem and knowing how to respond. If the habitat of an endangered species is being destroyed on private land, should the federal government establish a protected wilderness area, or should it offer tax incentives and zoning provisions to induce the owners to preserve the ecosystem? If detailed imagery shows the existence of half a million refugees in a conflict-ridden zone, should the United States or the UN send troops (and, if so, on what sort of mission)? Or should relief groups assume full responsibility?

This, in fact, was the situation during the massive refugee migration in Central Africa's Great Lakes region from 1996 to 1997. While the U.S. intelligence community had access to high-resolution images, it refused to share them with NGOs, most of which were pressing for military intervention at the time. Had commercial high-resolution satellite data been available at the time, NGOs might have used the data to strip away the government's veneer of deniability but may also have used it to press for U.S. military engagement, a policy whose success would have been highly questionable. Accurate information may be an important factor in formulating effective policy responses, but it is by no means a panacea.

For humanitarian NGOs, the effective use of earth observations from space would also be costly.[6] A humanitarian crisis by definition demands timely high-resolution data; archival data sets, such as TerraServer, will be of limited help. A single image, by itself relatively useless without interpretive techniques, can cost anywhere from $500 to $5,000. At present, no humanitarian NGOs have the technically trained staff needed to acquire, interpret, and manipulate satellite images, although there has been some recent discussion about developing a pooled base of expertise. While humanitarian NGO ties to the scientific community are not as strong as those of their environmental counterparts, there is some evidence that those ties are increasing. Einar Bjorgo has begun to develop near–real-time image processing and interpretation techniques for using high-resolution satellite images to support humanitarian relief operations (Bjorgo, 1996). Both ESRI and ERDAS, major GIS and imagery processing companies that have worked closely with environmental NGOs, have begun to expand their grants and training programs to include humanitarian NGOs.

Interestingly, the end of the Cold War has led the U.S. and European defense and intelligence community to become more deeply involved in humanitarian emergencies. Because of their early and continuing presence on the ground, relief groups are often key players in these situations. Thus, military agencies, NGOs, and international organizations are finding that they need to work together. But relations among military and intelligence agencies and humanitarian NGOs, which have very different missions, are likely to be tense —particularly regarding access to overhead imagery. For instance, following the Great Lakes crisis in Central Africa (1996–1997), many in the humanitarian community believed that the reconnaissance operations were set up as a fig leaf to justify an eventual high-level, international political decision not to go ahead with any kind of significant military intervention by a Multi-National Force (Gowing, 1998, p. 47).

While UN and NGO aid workers estimated that 600,000 refugees remained in the eastern Zaire forests, U.S. and British intelligence experts claimed that imagery revealed no significant refugee presence. More ominous is the belief among aid workers that overhead imagery was funneled to the Rwandan military forces, possibly enabling them to achieve more swiftly and completely the systematic slaughter of Hutus in eastern Zaire (Gowing, 1998, pp. 50–51). The full story of how imagery was deployed in the Great Lakes crisis may never be known, but these events bring into sharp relief the potential for distrust between NGOs and the military and intelligence communities.

While imagery from the U.S. intelligence community's state-of-the-art satellites would be of great use to humanitarian NGOs, the key obstacle is secrecy. In its support of disaster and humanitarian relief operations, however, the intelligence community may have a greater incentive to share imagery when U.S. troops are involved and the lives of Americans are at stake.

Military logistics in such situations are likely to involve coordinating activities with humanitarian NGOs and even protecting their supply lines. In many cases (as in Somalia, Bosnia, and Rwanda), humanitarian NGOs operate on the ground long before military personnel arrive on the scene. Consequently, the intelligence community is under increasing pressure to share information, including satellite imagery, with NGOs. Indeed, one of the key findings of a 1995 CIA-sponsored study is that intelligence for humanitarian relief purposes too often "excludes important players, and thereby limit[s] the value of the information provided." (Constantine, 1995, p. 1.) The defense and intelligence communities may prefer to share commercial images with NGOs rather than risk divulging its "sources and methods."

At this point, the utility of commercial high-resolution satellite imagery is more in the realm of promise than practical reality. A recent conference, OpenEyes 99: Remote-Sensing Operations in Humanitarian Emergencies, sponsored by the U.S. Institute of Peace, concluded that "the promise of emerging remote-sensing and GIS tools and platforms has not yet been fulfilled as a result of operational and cultural challenges." (U.S. Institute of Peace, 1999.) Participants identified numerous obstacles to successfully deploying satellite imagery in humanitarian operations: the lack of standardization of the content of humanitarian data; response delays on the part of vendors; and poor lines of communication among governmental, intergovernmental, and nongovernmental relief workers. Given that the use of satellite imagery in the humanitarian NGO community is in its infancy, we should see improvement in all of these areas in the coming years.

Human-Rights NGOs

Because large-scale violations of human rights can precipitate a humanitarian crisis, there is some overlap between humanitarian and human-rights issues. But they are not the same; indeed, there is sometimes tension on the ground between these two types of NGOs. While humanitarian NGOs are principally concerned with helping communities devastated by either natural or man-made disasters, human-rights NGOs generally have a broader and more overtly political agenda. They are rarely concerned with natural disasters and have a far greater interest in preventing or mitigating political violence than in treating its victims. Because a human-rights NGO's agenda is more focused on questions of agency and culpability, their work tends to be more politically controversial.

With a few exceptions, human-rights NGOs have not expressed any strong interest in using satellite images. This is not surprising, since only certain types of human-rights abuses can be detected with even the highest resolution satellite imagery. The arrest, detainment, and torture of political prisoners, for instance, cannot be detected with 1-m images. Neither can the use of child labor, the systematic violation of women's rights, or racial discrimination. In fact, satellites can detect only a relatively small portion of possible human-rights violations: most obviously, the large-scale violation of human rights that accompanies campaigns of genocide. Even here, questions of agency and culpability typically require validation from observers on the ground.

It is important to distinguish between two ways in which human-rights NGOs might become involved in using satellite images. First, these groups might wish to acquire and use the images themselves, although I am unaware of any instances in which they have done so. Second, they might pressure governments to procure, release, or otherwise transfer images to third parties. There are many conceivable reasons for the second type of involvement, including the desire to bring guilty parties to justice. Human-rights NGOs might use satellite images to prevent or mitigate a human-rights crisis. Noxious political leaders may be less willing to violate human rights if such actions can be observed. Satellites have an important advantage over other sources of information in a human-rights crisis: They can operate even when aerial or ground access is denied. Not only might satellites have some deterrent value in a human-rights crisis, images can also supply evidence to support international sanctions, military intervention, or peacekeeping operations. While governments may hold up their own satellite data to support their policies, human-rights NGOs may prefer to rely on commercially available sources of information to validate or contest official reports. And, as in humanitarian emergencies, governments might be more willing to share open-source imagery with NGOs rather than risk divulging their intelligence sources.

During the Russian bombardment of Chechnya and shortly after the launch of Space Imaging's IKONOS satellite, NGOs recognized that commercial imagery could help verify refugee reports of human-rights abuses. A spokesperson from Human Rights Watch, a U.S.-based NGO, noted that because the Russian government denied access to journalists and foreign observers, satellite imagery would have been especially useful there. At $2,000 per IKONOS scene, the cost would be prohibitive for an NGO. To reliably establish human-rights abuses, several images of the same area would need to be acquired over time. The possibility that NGOs could obtain images from open sources, however, could make governments more willing to release data from spy satellites (Willum, 1999a).

As we have seen, satellite sensors cannot detect intention, historical background, individual persons, or specific agency. Each of these limitations carries special significance in the arena of human rights. Nonetheless, 1-m images can detect the following: tire tracks that might indicate troop movements or convoys of displaced people; burned or burning villages; smoke from campfires, revealing possible refugee camps (or, with synthetic aperture radar data, the heat from campfires); and recently overturned soil, which could indicate the location of mass graves. Landsat multispectral imagery could help detect bodies decomposing underground and vegetation changes that could indicate possible mass graves (O'Connor, 1998).

The utility and, to some extent, the limitations of overhead imagery in revealing massive human-rights violations have been demonstrated clearly in the crises in Bosnia (1993–1995) and Kosovo (1999). For instance, a 1993 CIA study combined refugee testimony with aerial and satellite imagery to document the destruction of 3,600 Bosnian villages and the murder and expulsion of tens of thousands of people, mostly Muslims (Lane and Shanker, 1996). Satellite and aerial U-2 imagery helped detect hundreds of concentration camps and mass graves. Yet, the findings of that report were not made public until 1995. Even in the face of clear evidence of massive human-rights abuses, Western governments were reluctant to act decisively. If NGOs had had access to commercial high-resolution imagery during the Bosnian crisis, they might have been able to supplement it with eyewitness accounts to press governments into action. During the Kosovo crisis, a steady stream of refugees into Albania and Macedonia brought continual reports of atrocities, diminishing the need for validation through imagery. Nonetheless, during its air attacks on Serbia, the North Atlantic Treaty Organization posted daily images of its strategic bombardment on the Internet. (An image of the bombing of the Chinese embassy was conspicuously missing.) While this new openness was not directly intended to promote human rights, it does show the perceived political value of releasing imagery to the public.

While human-rights NGOs have not actively pursued the use of satellite imagery in crisis prevention and mitigation, they have begun to press for its use after the fact—in investigations of war crimes. Two NGOs primarily consisting of international lawyers, the Italy-based No Peace without Justice and the U.S.-based International Coalition for Justice, have been actively pursuing the declassification of intelligence for use as evidence in prosecuting war criminals. The use of imagery to document war crimes is a new and fascinating endeavor, offering great promise but also raising thorny issues. The rules of evidence in traditional legal proceedings would have a hard enough time incorporating the use of satellite imagery. In an international war crimes tribunal, the problems are complicated by the fact that, at least until recently, satellite imagery of sufficiently high resolution to detect evidence of war crimes would probably originate from U.S. intelligence sources. The old Cold War problem of protecting "sources and methods" returns in a post–Cold War setting. An uneasy compromise has been reached in the Bosnian War Crimes Tribunal. According to Rule 70, national intelligence sources can be transferred to the tribunal for analysis, but cannot be used in an actual trial. This compromise, while protecting U.S. sources, is not particularly helpful from the standpoint of war crime prosecutors.

The advent of commercial high-resolution images with near–real-time availability offers a potentially promising way to circumvent this problem. Humanitarian and human-rights NGOs, both of which are becoming more technologically sophisticated, could become interested in acquiring 1-m images during a human-rights crisis. Although human-rights NGOs have yet to express a strong interest in using satellite imagery, a number of factors could facilitate such an interest. Like most NGOs, human-rights NGOs do much of their work through the Internet. Cost will be a concern, but prices will decrease as the number of commercial satellites increases. Image providers may also discover that offering steep discounts to high-profile NGOs in exchange for public acknowledgment is good for public relations. Although the human-rights community, like the humanitarian relief community, lacks the technical expertise to acquire, process, and interpret satellite images, such companies as ESRI and ERDAS, which have historically assisted environmental NGOs, may make their services available. Moreover, like arms control NGOs, they are likely to find helpful allies in the media.

There are some significant obstacles, however, to using satellite imagery for human-rights advocacy. First, without ground-based information from witnesses or victims, satellite imagery on its own is of little value.[7] In every case in which satellite images have been used to document massive human-rights violations, collateral information was needed to guide collection of imagery and aid in its analysis. Second, if human-rights violations are occurring in a militarily sensitive environment, the licensing government for a commercial satellite may exercise shutter control, thereby making it impossible for NGOs, the media, or outside governments to acquire images. Already, the Russian government has refused to sell KVR-1000 high-resolution images of North Korea, China, Serbia, Bosnia, and Russia. A combination of factors could precipitate a policy dilemma for the United States in the near future. Most commercial high-resolution satellites will be licensed by the United States; the most visible human-rights NGOs are based in the United States; the United States (unlike Russia) has a long legal and cultural tradition of free speech; and the United States often finds itself in the position of considering military intervention in hot spots around the world. Third, a host of technical issues present themselves: the inability of even very-high-resolution images to reveal many human-rights violations; the inability of optical systems to image through cloud cover and dense forest or under night skies; and the possibility that perpetrators, knowing the sun-synchronous orbit of imaging satellites, could camouflage their actions. Fourth, NGO imagery analysts, who will be less well-trained than their counterparts in the intelligence community, may make mistakes, thereby diminishing their own credibility. This problem may be linked to the problem of ascribing agency cited earlier; either military forces or a forest fire, for instance, could cause a burned village.

Nonetheless, despite the obstacles and the potential for error, commercial satellite images may be useful in the work of human-rights NGOs in the near future. To a far greater extent than national governments, which often find themselves hamstrung by interagency or geopolitical pressures, such groups as Amnesty International and Human Rights Watch may be in a position to address the demand for more-objective sources of information during international crises. As both symptom and cause in the globalization of democratic sentiments, human-rights NGOs have an important role to play in the globalization of transparency. Thus, we can expect that, as commercial high-resolution satellite images become increasingly available and affordable, these groups will find innovative ways of deploying the new technology.

Conclusions

As a multitude of social, environmental, and economic problems have proven too intractable for states alone to address, NGOs have proliferated to fill the void. Simultaneously, political developments and the "information revolution" have enabled NGOs to use observation technologies previously confined to the military and intelligence agencies of the superpowers. The users of satellite data are increasingly nonstate actors, including a wide variety of industries, scientists, the media, and NGOs. User communities, primarily scientific and government, were initially stimulated by Landsat and later by SPOT. The multi–billion-dollar GIS industry and new Internet accessibility have helped spread satellite imagery far and wide, with the commercialization of high-resolution satellite data contributing to that momentum. High-resolution satellite imagery, until recently monopolized by the national security agencies of the superpowers, is now freely available to anyone with access to a credit card and the Internet. With the recent launch of Space Imaging's IKONOS satellite, such images may be available on a near–real-time basis in crisis situations. The timely availability of data will only increase as more commercial satellites go into orbit.

With civilian and commercial imaging satellites, new surveillance networks have sprung up around a host of issues and have drawn in an increasingly diverse array of state and nonstate actors. This general trend supports the claim that the diffusion of information technologies is undercutting the ability of states to exercise control and authority. Yet governments still enjoy a significant margin of control, even in the face of commercialization, because of their ability to license commercial satellites, to outlaw the export of turnkey systems, to exercise shutter control, to compete in markets, to mandate data purchasing policies for their agencies, and (in the most farfetched scenario) to shoot down the satellites. Thus, while the general trend is clearly in the direction of the diffusion of transparency beyond the state, the case should not be overstated. New surveillance networks may decenter the state but do not render it obsolete.

The diffusion of civilian and commercial satellite imagery, coupled with the advent of desktop computer programs capable of manipulating the large databases associated with satellite imagery and transmitting them over the Internet, is creating new information networks involving NGOs working on environmental, arms control, humanitarian relief, and human-rights issues. Satellites can enable citizens, mostly in industrialized countries but to a growing extent in developing countries as well, to monitor such diverse phenomena as crop and weather conditions, deforestation, marine habitats, and the movement of large numbers of refugees. By serving as a powerful tool of legitimation, earth observations from space can offer visual evidence to support NGOs' positions on a multitude of issues. Yet the slipping of information beyond the state does not mean that NGOs and states necessarily stand in an adversarial relationship to one another. In some cases they do; in others they do not. But it does mean that states' ability to control the flow of information about their own activities and within their own territorial borders is being eroded.

Transparency is increasing as a corollary to two other trends: globalization and democratization. Globalization has many facets: technological, economic, and cultural. The global spread of technology means that governments often find it futile to attempt to regulate technologies within their borders. A law prohibiting the dissemination of high-resolution images in one nation is drastically weakened by the availability of those same images on international markets. A global media system now means that events in one place—oil spills, nuclear accidents, genocide, earthquakes, etc.—can be "viewed" almost instantaneously on the other side of the planet. Moreover, the globalization of democracy, spurred on by global communication links, is the political expression of globalization. As democratic norms take root around the world, those in positions of power—whether in governments, corporations, or international organizations—are increasingly held accountable for their actions. That demand for accountability comes most often from NGOs. Despite the barriers of cost and the lack of technical expertise, the availability of commercial satellite imagery is likely to be a powerful tool for NGOs in their pursuit of accountability and openness.

Acknowledgments

The author would like to thank Einar Bjorgo, William Harborne, Ted Okada, and Jon Western, former State Department Officer for East European Affairs, Bureau of Intelligence and Research, for their contributions through various conversations to this chapter.

Notes

1 For an early work that demonstrates the applicability of commercial satellite imagery to arms control and military problems, see Krepon, et al. (1990).
2 These reports can be ordered from http://www.vertic.org (last accessed June 26, 2000).
3 For more detail on ISIS's work, see Chapters 17 and 18 in this volume.
4 We should note the danger, however, that news organizations, in their rush to break an important story or because of their lack of familiarity with photointerpretation, could be wrong. For instance, following India's nuclear weapon tests in May 1998, *Newsweek* included a satellite image that mistakenly

identified the Indian nuclear test site (see Chapter 23 in this volume). NGOs must be wary of committing the same error.
5 If detectable signatures of weapons and emplacements used by terrorists were available, it might be possible to distinguish terrorist encampments from refugee encampments. I am grateful to an anonymous reviewer for making this point.
6 In humanitarian crises that do not involve violent conflict, and therefore do not entail a high risk of being shot down, the use of aerial remote sensing may be a cheaper and easier way for humanitarian NGOs to get synoptic information from the ground.
7 This is a primary conclusion of Willum (1999b).

References

Albright, David, Corey Gay, and Frank Pabian, "New Details Emerge on Pakistan's First Nuclear Test Site," *Earth Observation Magazine*, December 1998–January 1999, pp. 18–21.

American Council for Voluntary International Action, Web site at http://www.interaction.org (last accessed June 26, 2000).

Bjorgo, Einar, "RefMon: Refugee Monitoring Using High-Resolution Satellite Images," Bergen, Norway: Nansen Environmental and Remote Sensing Center, 1996. Available at http://www.nrsc.no/~einar/UN/refmon_abstract.html (last accessed June 26, 2000).

---, ReliefSat Web page, 1999; see http://www.nrsc.no/reliefsat/ (last accessed June 26, 2000).

Blumberg, Peter, "Satellite Picture Puzzle: No Iraqis," *Washington Journalism Review*, Vol. 13, No.4, May 1991, p. 14.

Bryant, Dirk, Daniel Nielsen, and Laura Tangley, *The Last Frontier Forests: Ecosystems and Economies on the Edge*, Washington, D.C.: World Resources Institute, 1997.

Clayton, Mark, "Got an Earthly Cause?" *Christian Science Monitor*, May 8, 1996, p. 1.

Constantine, G. Ted, *Intelligence Support to Humanitarian-Disaster Relief Operations*, Langley, Va.: Central Intelligence Agency, Center for the Study of Intelligence, 1995.

Coral Cay Conservation (CCC), Web site at http://www.coralcay.org (last accessed June 26, 2000).

Cultural Survival, "Geomatics: Who Needs It?" Vol. 18, No. 4, Spring 1995.

Dowd, S., "Relief Groups Need Satellite Data," *Space News*, January 1996, pp. 15–21.

Federation of American Scientists, Public Eye Project Description, 1999. Available at http://www.fas.org/eye/project.htm (last accessed June 26, 2000).

—, Web site at http://www.fas.org (last accessed June 26, 2000).

Finnemore, Martha, "Constructing Norms of Humanitarian Intervention," in Peter J. Katzenstein, ed., *The Culture of National Security: Norms and Identity in World Politics*, New York: Columbia University Press, 1996, pp. 153–185.

Florini, Ann M., and Yahya A. Dehqanzada, *No More Secrets: Policy Implications of Commercial Remote Sensing Satellites*, Washington, D.C.: Carnegie Endowment for

International Peace, Global Policy Program Project on Transparency, 1999.

Gowing, Nik, "New Challenges and Problems for Information Management in Complex Emergencies," paper prepared for "Dispatches from Disaster Zones," Conference in London, May 27–28, 1998.

Gupta, Vipin, "Detecting Troop Concentrations Using SPOT Data: The 1991 Persian Gulf War," presentation for "Secret No More: The Security Implications of Global Transparency," workshop sponsored by the National Air and Space Museum, Washington, D.C., May 21–22, 1998.

Harborne, William, personal communication, May 6, 1998.

Heller, Jean, "Photos Don't Show Buildup," *St. Petersburg Times*, January 6, 1991, p. A1.

Hutchinson, Charles, personal interview, May 1, 1998.

Institute for Science and International Security (ISIS), Web site at http://www.isis-online.org (last accessed June 26, 2000).

Keck, Margaret, and Kathryn Sikkink, *Activists Beyond Borders*, Ithaca: Cornell University Press, 1998.

Krepon, Michael, et al., eds., *Commercial Observation Satellites and International Security*. New York: St. Martin's Press, 1990.

Lane, Charles, and Tom Shanker, "What the CIA Didn't Tell Us," *The New York Review of Books*, May 9, 1996, p. 3.

Lipschutz, Ronnie D., with Judith Mayer, *Global Civil Society and Global Environmental Governance: The Politics of Nature from Place to Planet*, Albany: State University of New York (SUNY) Press, 1996.

Litfin, Karen T., *Ozone Discourses: Science and Politics in Global Environmental Cooperation*, New York: Columbia University Press, 1994.

—, "The Status of the Statistical State: Satellites and the Diffusion of Epistemic Sovereignty," *Global Society*, May 1999, pp. 95–116.

Lodhi, M. A., F. R. Echarria, and C. Keithley, "Using Remote Sensing Data to Monitor Land Cover Changes Near Afghan Refugee Camps in Northern Pakistan," *GEOCARTA International*, Vol. 13, No. 1, 1998, pp. 33–39.

McNab, Philip, "Sleuthing From Home," *The Bulletin of the Atomic Scientists*, Vol. 49, No. 10, 1993, pp. 44–47.

O'Connor, Mike, "Bosnia War Tribunal Finds Hidden Bodies of Slain Muslims," *New York Times*, May 13, 1998, p. A3.

Pike, John, interview with the author, August 23, 1999.

Rosenau, James N., *Turbulence in World Politics: A Theory of Change and Continuity*, Princeton: Princeton University Press, 1990.

Sirait, Martua, et al., "Mapping Customary Land in East Kalimantan, Indonesia: A Tool for Forest Management," *Ambio*, Vol. 23, No. 7, November 1994, pp. 411–417.

Stein, Bruce, "Putting Nature on the Map," *Nature Conservancy*, January–February 1996, pp. 24–27.

Thomson, Janice, Remote Sensing Coordinator for the Wilderness Society, interview by the author, March 21, 1998.

U.S. Institute of Peace, "OpenEyes 99: Remote-Sensing Operations in Humanitarian

Emergencies," 1999. Available at http://www.usip.org/oc/vd/oe99/oe99recommendations.html (last accessed June 26, 2000).
——, Web site at http://www.usip.org (last accessed June 26, 2000).
Verification Research, Training & Information Centre (VERTIC), http://www.vertic.org (last accessed June 26, 2000).
Willetts, Peter, *The Conscience of the World: The Influence of Nongovernmental Organizations in the UN System*, Washington, D.C.: The Brookings Institution, 1996.
Willum, Bjorn, "From Kosovo to Chechnya, Selling Images from Above," *Christian Science Monitor*, November 30, 1999a.
——, "Human Rights Abuses Monitored with Satellite Imagery: Myth or Reality?" Unpublished M.A. Thesis, Department of War Studies, King's College London, 1999b.
World Resources Institute, "Global Forest Watch," February 2000. Available at http://www.wri.org/gfw/index.html (last accessed June 26, 2000).
——, Web site at http://www.wri.org (last accessed June 26, 2000).
WRI—*see* World Resources Institute.

23
Remote Sensing Technology and the News Media

Steven Livingston

The September 1999 launch of the IKONOS remote sensing satellite brought news organizations unprecedented access to high-resolution panchromatic and multispectral imagery, adding to an existing array of remote sensing platforms. As Daniel Dubno (1999), a CBS News producer and technologist wrote at the time, "In the past twelve months, powerful new computer technology coupled with greater access to foreign and commercial satellite imagery has begun to revolutionize broadcast journalism." The availability and use of this technology, said Dubno, represent "a sea-change in news presentation."[1]

The point of this chapter is to suggest that the same can be said of other emerging information technologies and the news business. The objective is to examine the place of remote sensing in the larger array of new information technologies available or soon to be available to news organizations and to assess the logic of shutter control in this light. Focusing on commercial remote sensing as a tool of news gathering and as a potential threat to U.S. national security without comprehensively considering other complementary technologies would hamper accurate assessment of the effectiveness of shutter control.[2] First amendment considerations aside, it *may* be possible to control cameras in space, but controlling the many thousands of cameras, satellite and cellular telephones, and other similar technologies found at ground level worldwide is much more problematic. To the degree that information technologies other than remote sensing satellites are able to penetrate denied-access areas, efforts at shutter control are futile: Plugging one hole in the dike will not stem the tide of global transparency.

The chapter closes with some speculations about the likely means security officials will use to cloud the transparency that commercial remote sensing satellites and other image and information sources create.

No single aspect of the contemporary information environment works in isolation from the rest. Together, they constitute a civilian "system of systems." Each system works separately but often with others as well, filling in where and when some other system falters or fails. Efforts to regulate one part of this system of systems will work for only a limited time and will ultimately fail in the absence of more extreme measures, such as those discussed below. The objective here is to explain why this is so.

Operational Security and News

In Chapter 6 of this volume, John Baker and Dana Johnson review the security implications of commercial remote sensing satellites. Suffice it to say here that IKONOS and other similar satellites create various costs and benefits, among them the potential costs associated with a loss of operational security. For instance, one study found that open systems could create a composite picture that was "alarmingly complete and potentially very useful to adversaries" and concluded that commercial satellites will "create a seemingly insurmountable challenge for keeping almost any type of militarily significant activity successfully hidden." (Hays and Houchin, 1999, p. 16.)

In limited measure, the debate is not about the absolute availability of remote sensing data but rather about the *timing* of their availability.[3] For news organizations, the principal concern is whether images will be available quickly enough to use in telling a normal news story or covering breaking news. For defense planners, part of the concern is that the images will be available too quickly and will therefore threaten operational security.

At least two factors affect the timing of satellite data release. First, the number of high-resolution satellites is currently limited and, therefore, so is the revisit frequency.[4] According to Space Imaging's Web site, IKONOS has a revisit frequency of 2.9 days at 1-m resolution and 1.5 days at 1.5-m resolution. Mark Brender, the company's Washington representative, indicates that Space Imaging has demonstrated an ability to fulfill orders for imagery in much less than two days (Brender, 2000). Cloud cover has been the biggest obstacle to faster turnaround times. Of course, if everything goes according to plan, other 1-m imaging satellite systems will join IKONOS by the end of 2000. Furthermore, radar and higher-resolution multispectral and hyperspectral systems will become an important element of the mix.[5]

For news organizations, "turnaround" can be a critical factor. Bob Windrem, an investigative producer for NBC News, said his main concern regarding commercial satellites is how quickly "they can turn stuff around." It is not only a matter of image acquisition and revisit frequency; images must also be interpreted. The time required for professional analysis must be factored into the turnaround time. According to Windrem (1999),

> The total turn-around package (picture acquisition and analysis) has an impact on how it will be used by NBC. If the pictures come in a day or so, they will become an element of the story. They may show the damage in a hurricane, for example. If (the analyzed pictures are available) in hours, they can be used to help make reporting decisions. How do we cover this story? Where do we send the resources?

The availability and speed of image analysis may be an Achilles' heel for news media attempting to use satellite images to report breaking news. Of course, journalists pursuing feature stories or investigative pieces do not face the same time pressures, and here, satellite imagery may play an important role. Even in cases of extended military buildups similar to those seen in the Persian Gulf in 1990 and 1991, commercial remote sensing data may be valuable to news organizations and, most troubling to national security officials, a threat to operational security. But in general, commercial remote sensing will be used more for context and analysis than for covering breaking news.

How do the news media analyze images? According to Brett Holey (1999), director of *NBC Evening News,* NBC has no clear procedure for analyzing satellite images. The network often relies on retired military or intelligence analysts. The same is true at CNN. At CBS, on the other hand, more thought has gone into the problem of image analysis, and clearer procedures and relationships have been forged with professional value-added firms. In general, news organizations now either attempt in-house, nonexpert analysis or rely on professional value-added firms or contracts with individual analysts.

Although relying on professional value-added firms is the safest route for news organizations concerned with accuracy, it is also the most time-consuming and expensive alternative. With a staff of about 100 image analysts,

Autometric is one of the larger value-added firms. It has professional ties with CBS News, Newschannel 8 in Washington, D.C., *Time*, *US News and World Report*, and *Newsweek*.[6]

News organizations that fail to use professional value-added services are more likely to make mistakes. For example, on May 25, 1998, *Newsweek* magazine ran what it said was a satellite image of the site of India's most recent nuclear test detonations (Figure 23.1). According to the caption, the image was taken a week before the detonations. The two-page photo carried annotations identifying what were said to be the detonation site, an abandoned village, and buildings thought to be a part of the nuclear test facility. Unfortunately for *Newsweek*, the image turned out to be something else entirely.

The nuclear detonation site was, in fact, an animal holding pen. How did the mistake happen? According to one of the *Newsweek* reporters involved with the story, "I should have gone to . . . Vipin (Gupta), but it was a Friday and there was pressure to publish."[7] Deadlines create an incentive to take shortcuts. Newsmagazines, like *Newsweek,* are the media least subject to deadline pressures. But media with rolling 24-hour deadlines, which today in-

Figure 23.1—Russian Satellite Image Purporting to Show the May 1998 Indian Nuclear Test Site That Appeared in Newsweek. Copyright 1998 by Aerial Images, Inc. and SOVINFORMSPUTNIK

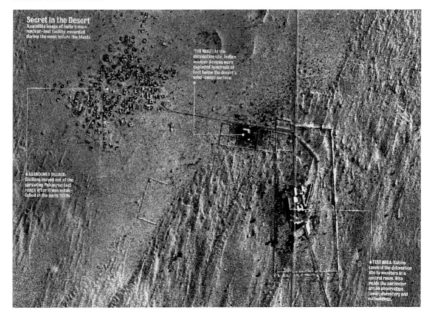

clude newspapers with Web editions, face constant, unremitting pressure to publish something new, something the competition does not have (Kurtz, 1999). *Newsweek* is not alone among media organizations in making this sort of mistake.

What conclusions might be drawn from this? It seems reasonable to anticipate two initial trends. First, given the news media's need for timeliness in reporting breaking news and the challenges associated with image analysis, news organizations will limit their use of satellite imagery for covering breaking news.[8]

Second, satellite imagery will be used in less time-sensitive investigative news stories and features. Cooperative relationships among news organizations and nongovernmental organizations interested in documenting developments of concern may facilitate this use of images in less time-sensitive stories. This has the added value of spreading out the costs of acquiring and properly analyzing the images. Third-party involvement in developing news content is a widely recognized and much-studied phenomenon. Environmental groups, urban-planning activists, and arms-control advocacy groups, among others, may acquire images as a part of their evidentiary and advocacy efforts and, in turn, provide the images to news organizations.[9] For example, in early 2000, the Federation of American Scientists (FAS) purchased a Space Imaging satellite photograph of the No-Dong missile facility in North Korea. The IKONOS image was taken on November 1, 1999; appeared on CNN on January 3, 2000; and was subsequently posted on the FAS Web site and picked up by other media (Brender, 2000).[10]

At least over the short run, satellite imagery will not be as common as one might expect. As suggested above, part of the reason is cost. As one industry executive recently stated on background, "Media seem to think $500 is a king's ransom." A more important reason, however, is that satellite imagery, even high-resolution imagery, is a second-order alternative used only when other options fail. The point is that the other options—other means of acquiring images of remote *but increasingly accessible* locations—are developing as rapidly as commercial remote sensing technology. In most cases, producers, reporters, and editors prefer pursuing the other options.

In breaking news of international political crisis—the circumstance of greatest concern to advocates of shutter control—satellite imagery is not the image product of first choice for journalists. NBC's Windrem has remarked, "It is not so much that [such imagery] will ever replace 'chopper 5' phenomenon," referring to television news's use of video shot from helicopters. CBS's Dubno echoed the same sentiment in remarking that satellite images are most useful in presenting images of denied-access areas into which other kinds of cameras cannot go. Otherwise, satellite images are useful in "augmenting or situating a story, but *they are sorry substitutes for real video*." (Dubno, 1999b; emphasis added.) The news media's preference for vivid and immediate inputs is important and serves as the foundation for the core supposition of this chapter.

Still, some stories are told best with overhead imagery. One example is the ability—literally—to see the forest for the trees. For example, environmental changes, such as deforestation, are best illustrated with overhead images, as are the depletion of the ozone layer over the poles and the enormity of the flood damage in North Carolina in 1999. But these examples are less important for the central question of this chapter, whether shutter control makes sense in a world awash with other means of transparency. The point is that, although denied-access areas certainly still exist, they are now often penetrated by an array of information technologies.

Information Technologies and Denied Access Areas

The key motivation in the creation of remote sensing satellites 40 years ago was fear of surprise attack engendered by the relative inaccessibility of Soviet-controlled territory. As Leghorn and Herken discuss in Chapter 2 of this volume, the Iron Curtain substantially blocked the flow of information and people, creating a vast territory from which a surprise attack might be launched on the West.

This territory was pried open through "national technical means." Obviously, much has changed, and although the once vast denied-access region is not entirely gone, the original motivations behind the development of spy satellites no longer apply. Not only have the Eastern Bloc and the Soviet Union collapsed, but the technical means of unsealing denied-access areas have changes.

A civilian system of systems is developing, a counterpart to U.S. strategic planner's notion of an integrated information battlespace-monitoring system. The civilian counterpart to the military system of systems includes newspapers, television networks, and their Internet manifestations; wire services; Internet news services, such as foreignnews.com; Internet gossip mongers, such as Matt Drudge; nongovernmental organizations that maintain Web sites; and visual information companies, such as Space Imaging. At its greatest extent, the civilian system includes anyone with a cellular or satellite telephone, an Internet connection, a fax machine, an audiocassette recorder, a camcorder, or a videocam, and all can be tied together by means of wide-bandwidth communication links. This is a radical expansion of transparency. Cellular and satellite phones, portable satellite uplinks, fiber optic cable, and wide-bandwidth communication satellites mean we are moving in the direction of a total information environment (Livingston, 2000 b).[11] This emerging information order can be called *tectum* transparency drawing on the Latin word for "covering" or "to put a roof over."[12] What follows are several examples intended to illustrate *tectum* transparency.

Video cameras have become smaller, require less power, are more versatile, and are now extraordinarily common. Table 23.1 illustrates the growth in camcorder sales in the United States alone since 1985. The global market is equally robust (Brockhouse, 1998; *The Daily Yomiuri*, 1998). As a result, amateur video is fairly common in entertainment and news programming.

Digital cameras and camcorders are also affecting professional practices. In the estimation of one trade publication, "The increasing use of hand-sized digital video cameras in news gathering may ultimately have much the same impact on television news as did the advent of videotape and communications

Table 23.1—Camcorders: U.S. Sales to Dealers

Year	Units (000)	Sales ($M)	Household Penetration (%)	Average Unit Price
1985	517	793	1	1,534
1986	1,169	1,280	2	1,095
1987	1,604	1,651	4	1,029
1988	2,044	1,972	5	965
1989	2,286	2,007	8	878
1990	2,962	2,260	11	763
1991	2,864	2,013	15	703
1992	2,815	1,841	18	654
1993	3,088	1,958	19	634
1994	3,209	1,985	21	625
1995	3,560	2,130	22	598
1996	3,634	2,084	25	598
1997	3,650	1,894	28	519
1998	3,829	1,828	32	477

SOURCE: Consumer Electronics Manufacturing Association Market Intelligence Center.

satellites." (Hall, 1998.) "What you will see," remarked Ken Tiven, vice president of television systems at CNN, "is the addition of hundreds of these small cameras all over the world in the next couple of years." (Quoted in Hall, 1998.)

In 1999, a Kosovar Albanian physician named Liri Losci illustrated the potential of amateur video. Losci's videotape showed the bodies of 127 ethnic Albanians said to have been murdered on March 28 by Serb forces in the village of Izbica. Beginning on May 14th, for two weeks CNN repeatedly aired portions of Losci's video (Woodruff, Sadler, Kibel, 1999).[13] On May 19th, Department of State spokesperson James Rubin used Losci's video to verify overhead imagery the U.S. government had released on April 17th. Figure 23.2 presents the images used at the April 17th briefing to which Rubin referred.[14]

At his May press conference, Rubin matched prominent features seen in the overhead imagery with what appeared to be the same features in Losci's video. Said Rubin,

> But I will take this opportunity to walk you through the combination of the videotape and the overhead imagery on the same screen, and what that combination will show is, conclusively, that the videotape that was released earlier today of a massacre of over 100 Albanians in

Figure 23.2—Mass Burial Site near Izbica, Kosovo (courtesy of Reuters/Handout/Archive Photos)

Izbica, Kosovar Albanians, is the very same location that we were able to release from overhead imagery earlier in the year on April the 17th. (Woodruff, Rodgers, and Lewis, 1999.)

This clearly shows the power of information systems working in tandem, here, overhead imagery and video. But it is important to keep in mind that everyone, from the news media to the Department of State, recognized that Losci's videotape offered the most powerful illustration of what had happened at Izbica. Satellite imagery was the best available evidence until Losci's videotape. Together, the videotape and imagery, in addition to independent eyewitness testimony, told the most complete story.

Camcorders are not alone in expanding access to sealed areas. The portable equipment used to transmit video images from remote locations—known as flyaway units—has diminished in size and weight (*Electronic Media*, 1993).[15] Likewise, field editing equipment is smaller and more sophisticated (*Business Wire*, 1998). Together, smaller cameras, flyaway units and editing equipment allow journalists greater mobility and flexibility in covering international affairs.

Cellular telephones are another part of the civilian system of systems. According to industry estimates, there would be 200 million wireless subscribers worldwide. Table 23.2 traces the rapid expansion of cellular technology in the United States. Worldwide, wireless service is growing 15 times faster than the subscriber base for wired telephone service (NAS, 1997).

Cellular telephones have become an indispensable tool for journalists. The 1999 war in Kosovo made this clear:

> The ubiquitous mobile phone has made a major difference in the flow of information out of the Balkans. The technology is affordable and widely available—cell phones can be rented at Zurich and other airports on the way into the war zone. They provide 24-hour access to newsrooms, to NATO officials and other sources, and to fellow journalists on the hunt for news. (Ricchiardi, 1999.)

The Washington Post's Michael Dobbs has remarked that "News bounces around almost instantaneously through a network of cell phones, satellite phones and the Internet." Satellite telephones rely on satellites passing overhead to relay a signal. The availability of data and communication satellites is expected to expand considerably over the next few years.

With the success of the Internet and other broadband services, there is greater need for wide-bandwidth capacity. Conventional (wire and fiber optic)

Table 23.2—Cellular Telephone Expansion: 1985–1998

Year	Estimated Total Subscribers	Cell Sites	Cumulative Capital Investment
1985	340,213	913	911,167
1986	681,825	1,531	1,436,753
1987	1,230,855	2,305	2,234,635
1988	2,069,441	3,209	3,274,105
1989	3,508,944	4,169	4,480,142
1990	5,283,055	5,616	6,281,596
1991	7,557,148	7,847	8,671,544
1992	11,032,753	10,307	11,262,070
1993	16,009,461	12,824	13,956,366
1994	24,134,421	17,920	18,938,678
1995	33,785,661	22,663	24,080,467
1996	44,042,992	30,045	32,573,522
1997	55,312,293	51,600	46,057,910
1998	69,209,321	65,887	60,542,774

SOURCE: The Cellular Telecommunications Industry Association (1998) (Partially reproduced from http://www.wow-com.com/images/1298datasurvey2.gif).

landline connections will continue to be the primary means for data transmission in advanced postindustrial societies but are inadequate for reaching remote regions. Therefore, next-generation communication satellites will serve as an "Internet in the sky," providing wide-bandwidth service to remote locations. The practical effect will be to increase the ease of sending digital data (including streamed video) from remote locations. These are just a few of the components of the larger civilian system of systems, of which commercial remote sensing is a part, but only a part.

What Is the Meaning of All This?

Overlapping and redundant information collection and distribution systems mean that the ability to regulate data at the point of collection has now weakened to the point of impracticality. If shutter control is impracticable, how else might national security officials defend against the effects of *tectum* transparency? There are at two general methods.

First, more extreme means may be used. In 1999, COL Michael McKeeman, head of Army Space Command (Forward), was quoted in *Defense News* as saying that increased U.S. reliance on commercial communication satellites creates problems with the "open access" that exists beyond military control. Although his remarks focused on communication satellites, the principle applies across satellite platform functions. "We are going to be vulnerable because anybody can gain access to the same satellites the military uses for battlefield communications and data links." VADM Herbert Browne, deputy commander in chief, U.S. Space Command, added that the Department of Defense calls for a

> limited offensive capability to counter the problem of uncontrollable space access, *without doing permanent harm to assets that the United States or its allies may need later on*. "The nation needs to look at a localized, reversible way to deny enemies access to space technology." (Keeter, 1999; emphasis added.)

One might conclude that commercial remote sensing satellites fall into this category.

The initial design of the Army's Kinetic Energy Antisatellite system envisioned a Kevlar sail that would function as a satellite swatter, breaking the antennas off a targeted satellite or simply knocking it out of orbit. More recently, an alternative "soft-kill mechanism" has been proposed. Such a system would drench the target satellite in a cloud of chemicals designed to degrade components and obstruct optics. Most interestingly, chemicals could be used that would degrade over time, therefore only temporarily disabling a satellite (Isby, 1999). Such a system could be used to disable but not destroy commercial satellite systems.

Second, deception may be used to undermine the transparency that commercial remote sensing satellites and other technologies create. After all, the elimination of remote sensing satellites will not prevent the effects of other sources of transparency. Deception, particularly in the digital age, can at least introduce an element of uncertainty into an adversary's decisionmaking cycle, causing delay and hesitation.

In his classic study of deception in warfare, Michael Dewar (1989), suggests the use of a mirrorlike form of deception to prevent the enemy from discovering key information (see Table 23.3). The objective is "to fabricate a pattern, albeit a bogus one, which will result in your adversary building up a false picture of reality." (Dewar, 1989, p. 19.) Similarly, Donald C. Daniel and Katherine L. Herbig (1981, p. 5) observe that strategic military deception

> confuses a target so that the target is unsure as to what to believe. It (deception) seeks to compound the uncertainties confronting any state's attempt to determine its adversary's wartime intentions.

Table 23.3—Measures to Distort Information for an Enemy

To Distort—	Convince the Enemy That—
Your location	You are somewhere else
The enemy's location	He is somewhere else
The types and numbers of your weapons and forces	Your weapons and forces are different from what they actually are
What you intend to do	You intend to do something else
Where you intend to do it	You intend to do it elsewhere
When you intend to do it	You intend to do it at a different time
How you intend to do it	You intend to do it differently
What you know about the enemy's intentions and techniques	Your knowledge of the enemy is greater/lesser than it actually is
How successful the enemy's operations were	His operations were more or less successful than they actually were

SOURCE: Dewar (1989).

The digital information age makes deception easier. Elsewhere, we have referred to "pixel plasticity" to denote the ability to manipulate, often in real time, digital images, including video streaming and satellite images (Livingston, 1999). The basic ones and zeros of digital information can be rearranged seamlessly to create knew images: Faux Web sites can be created (Neal, 1999), and live video can be altered, as CBS has done on several occasions since 1999 (Herring, 2000; Carter, 2000; Kuczynski, 2000). During hostilities, military and intelligence agencies could flood an adversary's information environment with a multitude of false signals, decoy images, and altered data streams. In doing so, any advantage gained by the availability of open access data is neutralized.

In sum, while advocates of shutter control regulatory mechanisms underestimate the effects of other sources of transparency, and therefore overstate the viability of shutter control, other mechanisms exist to undermine confidence in and ability to use remote sensing data. The same applies to digital information from other sources. In a hall of digital mirrors, an adversary may be immobilized by distortion.

Notes

1. For an example of a 1-m image, see Space Imaging's Web site: www.spaceimaging.com.
2. This chapter does not review relevant case law or speculate on the constitutionality of shutter control. See instead Hays and Houchin (1999), p. 16; Kirby (1998); and Chapter 24 in this volume.
3. It is important to bracket this claim. Timing indeed is important, but the mere

availability of remote sensing data will have consequences for militaries around the world. Countries without remote sensing satellites of their own may now, for example, build detailed geospatial information archives on potential adversaries. Images may be used for mapping, target analysis, and simulations.

4 Our comments here are limited to satellites with a resolution of 1 m or less.
5 See Chapter 5 in this volume.
6 Autometric is but one of many value-added firms. For a list, see www.asprs.org.
7 Anonymous telephone interview, May 3, 1999. Several reporters were associated with the *Newsweek* article. The reporter interviewed for this chapter, as well as one of the two experts the reporter identified, asked to remain anonymous. Dr. Vivin Gupta is a recognized U.S. expert on commercial satellite imagery analysis. See also his imagery analysis presented in Chapter 16 in this volume.
8 This is likely to change, of course, as satellite revisit times diminish as the number of and capabilities of satellite systems increase. It is also likely to change as news organizations become more familiar with available systems and formalize image analysis and delivery routines.
9 Of course, third-party arrangements will require licensing arrangements with satellite vendors before imagery may be provided directly to news organizations. For an analysis of the role of third-party advocacy in public affairs, see Manheim (1994).
10 See also the FAS Web site for a chronology of the use of the No-Dong image.
11 A point of clarification is in order. The issue is not that everyone will have access to information. A "digital divide" will remain and may expand (see, among many sources, Hafner, 2000). Rather, because so many actors around the world—including news organizations, NGOs, private citizens, and others—will have access to more information from and about more places in the world than ever before, the number and extent of denied-access areas are shrinking. As a consequence, nearly all who are aware of *tectum* transparency, even in the vaguest way, may behave differently as a result of this knowledge.
12 This new level of transparency also has been referred to as "phase two transparency." See Livingston (forthcoming a).
13 This appears to have been the first of over a dozen airings of the video (see Amanpour, 1999).
14 This image is found on the Web site of the Federation of American Scientists (http://fas.org/irp/imint/kosovo-27.htm).
15 See also M2 Presswire (1995). Newsforce has operated flyaway units transmitting in digital C-band and analogue and digital format in the Ku-band and are powered by portable generators. See M2 Presswire (1998).

References

Amanpour, Christiane, "60 Minutes," CBS News, May 16, 1999.
Brender, Mark, interview at Space Imaging, Washington, D.C., February 7, 2000.
Brockhouse, Gordon, "Video Technology: the Digitization of the Video Industry Is

Very Good News for Retailers," *Computer & Entertainment Retailing*, April 1998, pp. 12–14.

Business Wire, "Television ENG Crews Deploy Sony's Betacam SX Field Editor for Coverage Around the World," February 5, 1998. Available through Nexis.

Carter, Bill, "CBS Is Divided Over the Use of False Images In Broadcasts," *The New York Times*, January 13, 2000, p. C1.

Cellular Telecommunications Industry Association, Annualized Wireless Industry Survey Results: December 1985 to December 1998, 1998, http://www.wow-com.com/images/1298datasurvey2.gif.

The Daily Yomiuri, "No Letup in Strong Sales of Video Cameras," Tokyo, December 1, 1998, p. 13.

Daniel, Donald C., and Katherine L. Herbig, eds., *Strategic Military Deception*, New York: Pergamon Press, 1981.

Dewar, Michael, *The Art of Deception in Warfare*, London: A David & Charles Military Book, 1989.

Dobbs, Michael, "Letter from Yugoslavia," *The Washington Post*, no date, p. C1.

Dubno, Daniel, "Satellites Change How We See the Earth," CBS News Web site, October 13, 1999a.

——, Telephone interview, October 13, 1999b.

Electronic Media, "Technology Lightening Load for Fly-Aways," July 12, 1993, p. 42.

Hafner, Katie, "We're Not All Net-Connected, Yet," *International Herald Tribune*, January 31, 2000, p. 11.

Hall, Lee, "Digital Special Report First Steps to the Future: Digital Video Opens News Vistas," Electronic Media, March 9, 1998, p. D6.

Hays, Peter L., and Roy F. Houchin, II, "Commercial Spysats and Shutter Control: The Military Implications of U.S. Policy on Selling and Restricting Commercial Remote Sensing Data," prepared for the Institute for National Security Studies, USAF Academy, Colo., 1999.

Herring, Hubert B., "Seeing Is Not Believing," *The New York Times*, January 16, 2000

Holey, Brett, Interview, New York, May 3, 1999.

Isby, David C., "US Funds Offensive and Defensive Space Systems," *Jane's Missiles & Rockets*, Vol. 3, No. 11, November 1, 1999.

Keeter, Hunter, "Civilian Satellites Need Protection," *Defense Daily*, Vol. 202, No. 23, May 3, 1999.

Kirby, Kathleen A., Wiley, Rein & Fielding, counsel for RTNDA, Joint Comments on the Radio-Television News Directors Association and National Association of Broadcasters in the Matter of Licensing of Private Remote Sensing Space Systems, Docket No. 951031256-5259-01, April 1, 1998.

Kuczynski, Alex, "On CBS News, Some of What You See Isn't There, *The New York Times*, January 12, 2000, p. A1.

Kurtz, Howard, "On the Web, Newspapers Never Sleep," *The Washington Post*, September 7, 1999, p. E1.

Livingston, Steven, "The New Information Environment and Diplomacy" in Evan Potter, ed., *Cyberdiplomacy: Managing Foreign Policy in the 21st Century*, Montreal, Quebec, and Kingston, Ontario: McGill-Queen's University Press, forthcoming a.

——, "Transparency and the News Media," in Bernard I. Finel and Kristin M. Lord, eds., *Power and Conflict in an Age of Transparency*, New York: St. Martin's Press, 2000 b.

——, presentation at a conference sponsored by the Carnegie Endowment for International Peace. "No More Secrets: Policy Implications of Commercial Remote Sensing Satellites," May 26, 1999, Washington, D.C. (http://www.ceip.org/programs/transparency/RemoteSensingConf/Agenda.htm)

M2 Presswire, "Faster and Cheaper Outside Broadcast Transmissions with New Mobile Satellite Equipment," December 4, 1995.

——, "Newsforce Teams Up with France Telecom's GlobeCast," January 6, 1998.

Manheim, Jarol B., *Strategic Public Diplomacy & American Foreign Policy: The Evolution of Influence*, New York: Oxford University Press, 1994.

National Academy of Science, *The Evolution of Untethered Communication*, Washington, D.C.: National Academy Press, 1997.

Neal, Terry M., "Satirical Web Site Poses Political Test; Facing Legal Action from Bush, Creator Cites U.S. Tradition of Parody," *The Washington Post*, November 29, 1999, p. A2.

Ricchiardi, Sherry, "Searching for Truth in the Balkans," *American Journalism Review*, June 1999, p. 22.

Windrem, Bob, telephone interview, October 15, 1999.

Woodruff, Judy, Brent Sadler, Amanda Kibel, "Strike Against Yugoslavia: Exclusive videotape of Alleged Mass Killings of Ethnic Albanians; Yugoslavs Say NATO Airstrikes Kill 100 Civilians," CNN Worldview, May 14, 1999.

Woodruff, Judy, Walter Rodgers, and David Lewis, "Strike Against Yugoslavia: State Department Spokesman James Rubin Presents 'Hard Evidence' Alleging Mass Graves in Kosovo," Nexis transcript of CNN Live Event, May 19, 1999; Wednesday 4:39 P.M. Eastern Time.

24
Space Remote Sensing Regulatory Landscape

Bob Preston

This chapter concentrates on the regulation of commercial remote sensing from space. It focuses particularly on operational controls or licensed activity, popularly referred to as *shutter control*. It largely excludes related policy issues of export control and technology transfer,[1] because the remote sensing data product, the U.S. policy's preferred commercial commodity, is not subject to export controls on technical data in any case. This chapter reviews the background and authority for regulation; examines specific issues in recent attempts at rule making; and, finally, explores a framework for acceptable, useful, and one hopes durable regulations. In this context, *acceptable* refers to the interested parties—national security, civil, and commercial. *Useful* refers to goals of policy. *Durable* refers to possible legal challenges.

Background

By way of background, we should review the legal basis for regulation in the Constitution, legislation, and court as well as some relevant legislation aimed at similar purposes.[2]

CONSTITUTION AUTHORITY

The constitutional foundations for the authority to regulate commercial remote sensing from space are the commerce clause (U.S. Constitution art. I, sec. 8), which allows Congress to regulate commerce between the states and with other nations, and the president's executive responsibilities as commander in chief, to make treaties, receive ambassadors, and execute laws (U.S. Constitution art. II, sec. 2). The president's executive authority includes responsibilities that create some tension with constitutional limitations on authority to regulate remote sensing. The sources of constitutional limitations are the first amendment freedoms of speech and press, the fourth amendment protection from unreasonable search and seizure, and the fifth amendment protection of due process and requirement of compensation in public taking of private property. The greatest sources of constraint on regulatory mechanisms for remote sensing are the first amendment freedoms of speech and press.

FIRST AMENDMENT

The fundamental purpose of the first amendment is to preserve the foundation of democratic government in the free exchange of ideas, based on the founding fathers' belief that

> "freedom to think as you will and to speak as you think are means indispensable to the discovery and spread of political truth; that without free speech and assembly, discussion would be futile; that with them, discussion affords ordinarily adequate protection against the dissemination of noxious doctrine. . . ." (*Whitney v. California*, 1927, Justice Brandeis concurring.)

Among the Supreme Court decisions most relevant to the issues of first amendment rights and national security is the landmark case of the Pentagon Papers.[3] However, the court's decision in that case leaves considerable ambiguity as to what suitable controls over remote sensing might be. Understanding the meaning of the terms used requires reference to a large body of prior cases. In search of that understanding, we will review some of that material.

The constitutional protection of speech is not absolute. The time, place, and manner of speech may be regulated to "safeguard the peace, good order and comfort of the community." (*Cantwell v. Connecticut*, 1940.) Illegal speech, such as slander, libel, treason, or obscenity, may be punished. Otherwise legal speech, in the wrong circumstances, may be constitutionally punished. Examples include "fighting words" that have the effect of force (*Chaplinsky v. New Hampshire*, 1942), false warnings that incite panic, or words that harm the operation of the nation's armed forces. As the Supreme Court explained in discussing circumstances of clear and present danger that may justify infringement of free speech: "When a nation is at war many things that might be said in time of peace are such a hindrance to its effort that their utterance will not be endured so long as men fight and that no court could regard them as protected by any constitutional right."[4]

In early interpretations, the test for infringing otherwise protected speech was for circumstances posing a "clear and present danger that they will bring about the substantive evils that Congress has a right to protect." (*Schenck v. United States*, 1919.) Over time, the clear and present danger test was strengthened to "imminent danger" of "serious injury to the State." (*Whitney v. California*, 1927.) In more recent years, the court has, at times, moved the line for punishable speech past clear and present danger. "The line between what is permissible and not subject to control and what may be made impermissible and subject to regulation is the line between ideas and overt acts." (*Brandenburg v. Ohio*, 1969, Justice Douglas concurring.)

Beyond the free exchange of ideas, a fundamental purpose of the first amendment was to ensure the accountability of government. "The Government's power to censor the press was abolished so that the press would remain forever free to censure the government." (*New York Times v. U.S.*, 1971, Justice Black concurring.) Concern for this purpose created a strong judicial preference for deterrent punishment over preventive restraint. "A free society prefers to punish the few who abuse rights of speech *after* they break the law than to throttle them and all others beforehand." (*Southeastern Promotions v. Conrad*, 1975.)[5]

The Supreme Court has permitted legislation seeking prior restraint of speech only under narrow circumstances and only with particular safeguards. Aside from the reasonable restrictions on time, place, and manner of speech mentioned above, it has allowed few cases of prior restraint based on the content of speech. It has recognized that government may "bar from its facilities certain speech that would disrupt the legitimate governmental purpose for which the property has been dedicated." (*Consolidated Edison Company of New York v. Public Service Commission of New York*, 1980, citing *Greer v. Spock*, 1976.)[6]

The court has identified a number of factors in deciding whether prior restraint is constitutional. First is the nature of the interest requiring regulation of speech. The court requires the regulator to "act in patent good faith to maintain the public peace . . . or for equally indispensable ends of modern community life." (*Niemotko v. Maryland*, 1950.) Second is the method used to regulate speech. Particularly when the "prohibition is directed at speech itself, and the speech is intimately related to the process of governing" (*First National Bank of Boston v. Bellotti*, 1978), the court puts the burden on the government to show the existence of a compelling interest and to use means "closely drawn to avoid unnecessary abridgement" (*First National Bank of Boston v. Bellotti*, 1978) with narrow criteria related to the government's legitimate interest (*Kunz v. New York*, 1951). Third is the mode of speech regulated. Fourth is the place of speaking. The last two factors overlap with the time, place, and manner constraints mentioned earlier, but their reiteration here indicates some means the court has considered to restrict adequately the reach of content-based prior restraint.

In addition to the narrow circumstances described above, the court has insisted on procedural safeguards on prior restraint to reduce the danger of suppressing protected speech. It requires "an almost immediate" (*Bantam Books v. Sullivan*, 1963) judicial determination to impose a final restraint, permitting administrative restraint only to preserve the status quo for the briefest time compatible with judicial resolution (*Freedman v. Maryland*, 1965). It places the burden of proof that the speech is not protected on the censor (*Speiser v. Randall*, 1958) and places the burden of instituting judicial proceedings likewise on the censor (*Southeastern Promotions v. Conrad*, 1975).

FIFTH AMENDMENT

Beyond the court's decisions requiring judicial supervision in prior restraint on speech, the fifth amendment requires due process in depriving anyone of property and just compensation when private property is taken for public use. The issue of due process arises because regulation of commercial remote sensing activity arguably may cause economic damage to the parties involved, either directly, through lost sales, or indirectly, through erosion of market confidence in the firm as a reliable supplier. The federal government has broad power to regulate. To discriminate *regulation* from *taking*, the court has judged regulation as taking only when it deprives the owner of economically viable use of the property, where a determination of economically viable use requires ad hoc, factual inquiries that have identified several factors—such as the economic impact of the regulation, its interference with reasonable investment backed expectations, and the character of the government action. . . . (*Hodel v. Virginia Surface Mining & Reclamation Association*, 1981).

Regulations may do economic harm but do not amount to taking unless the owner is left with no economically viable use of his property. So, the loss of a single sale would not amount to taking, but the loss of enough sales to endanger the business case for the enterprise could.

The fifth amendment due process requirements affect regulation via the standards of vagueness. The basic requirement on statutory vagueness is that "No one may be required at peril of life, liberty or property to speculate as to the meaning of penal statues. All are entitled to be informed as to what the State commands or forbids." (*Lanzetta v. New Jersey*, 1939.) In general, the tests on statutes require "words or phrases having a technical or other special meaning, well enough known to enable those within their reach to correctly apply them or a well-settled common law meaning." (*Connally v. General Construction*, 1926.) However, the court applies stricter standards of vagueness in first amendment issues, saying in *Hynes v. Mayor of Oradell* (1976) that the "general test of vagueness applies with particular force in review of laws dealing with speech" and in *Smith v. California* (1959) that

> stricter standards of permissible statutory vagueness may be applied to a statute having a potentially inhibiting effect on speech; a man may the less be required to act at his peril here, because the free dissemination of ideas may be the loser.

Legislation

The current source of legislative authority to regulate commercial remote sensing is the Land Remote Sensing Policy Act of 1992, as amended by the Commercial Space Act of 1998. In the 1992 act, the Congress found that

> land remote sensing data from space are of major benefit in studying and understanding humans' impact on the global environment, in managing the Earth's natural resources, in carrying out national security functions, and in planning and conducting many other activities of scientific, economic, and social importance. (P.L. 105-303, October 28, 1998.)

Congress expressed the view that "commercialization of land remote sensing should remain a *long-term* goal of United States policy" (P.L. 105-303, October 28, 1998) but judged that commercialization was not fully possible for the Landsat earth remote sensing program. The act requires a license issued by the Secretary of Commerce for any private-sector party to operate a space remote sensing system. Among other requirements for operation of such a system, the act requires the licensee to

- "operate the system in such manner as to preserve the national security of the United States and to observe the international obligations of the United States" (15 USC §5622 (b)(1))

- make unenhanced data available to sensed countries on reasonable terms and conditions as soon as they are available (15 USC §5622 (b)(2))[7]
- "notify the Secretary [of Commerce] of any significant or substantial[8] agreement the licensee intends to enter with a foreign nation, entity, or consortium involving foreign nations or entities" (15 USC §5622 (b)(6)).

In issuing licenses for commercial remote sensing, the Secretary of Commerce is authorized to consult with other appropriate U.S. government agencies (15 USC §5621 (1)(a)(1)). The secretary is required to consult with the Secretary of Defense on all matters affecting national security (15 USC §5657 (a)) and with the Secretary of State on all matters affecting international obligations (15 USC §5657 (b)). The Secretary of Defense is given responsibility for determining the conditions necessary to meet national security concerns of the United States. The Secretary of State is given responsibility for determining conditions necessary to meet international obligations and policies of the United States.

The act gives the Secretary of Commerce administrative authority, subject to judicial review of any adverse action (15 USC §5623 (b)), to
- "grant, condition, or transfer licenses" (15 USC §5623 (a)(1))
- seek judicial determination to terminate, modify, or suspend licenses ... and to terminate licensed operations on an immediate basis, if the Secretary determines that the licensee has substantially failed to comply with any provisions of this act, with any terms, conditions, or restrictions of such license, or with any international obligations or national security concerns of the United States (15 USC §5623 (a)(2)) and to seek a warrant to seize any object, record or report under probable cause to believe that such object, record, or report was used, is being used, or is likely to be used in violation of this act or the requirements of a license or regulation issued under it (15 USC §5623 (a)(6))
- investigate any matter relating to enforcement of the act (15 USC §5623 (a)(7)) and issue subpoenas for materials, records, documents or witnesses required for a hearing (15 USC §5623 (a)(5))
- impose, compromise, modify, or remit penalties for violations of license or regulations (15 USC §5623 (a)(3) and (4)).

In addition to the 1992 Land Remote Sensing Policy Act, Congress passed language in the 1997 Defense Authorization Act (P.L. 104-201, 1996) that restricted licensed[9] space remote sensing of Israel to data "no more detailed or precise than satellite imagery of Israel that is available from commercial sources."

Another source of legislative limits on commercial space remote sensing is the espionage statutes, which include prohibitions on gathering and transmitting defense information (18 USC §793), gathering or delivering defense information to aid a foreign government (18 USC §794), photographing defense installations or equipment (18 USC §795), using an aircraft to photograph (18 USC §796), and publication or sale of photographs of defense installations and equipment (18 USC §797). Section 793's prohibitions on gathering and transmitting depend on willful collection or communication with "intent or reason to believe that the information is to be used to the injury of the United States, or to the advantage of any foreign nation." The prohibition on collection includes flying over the installation or equipment. The penalties prescribed include a fine and imprisonment for not more than ten years.[10]

Section 794 steepens the penalties for gathering and delivering when the delivery is to a foreign government. Depending on the circumstances, the penalty may be death or imprisonment for any term of years or life and forfeiture of any property used or intended to be used to commit or facilitate the violation. (This could be a substantial financial penalty for satellite imagery.) This section also includes a prohibition on publishing in time of war with intent that the information is to be communicated to the enemy.

Sections 795, 796, and 797 explicitly authorize prior restraint or censorship of publication of photographs or other graphical representations of military and naval installations and equipment designated by the president. The penalties for violation include fines and imprisonment of not more than one year. The current Executive Order (Executive Order 10104, February 1, 1950) designating those installations and equipment identifies installations, equipment, and documents that are classified, designated, or marked as top secret, secret, confidential, or restricted.

Policy

Current Executive Branch policy toward space remote sensing is contained in Presidential Decision Directive 23 (PDD-23), March 10, 1994. PDD-23 proposes to "expand American jobs and business opportunities by enabling U.S. firms to compete aggressively in the growing international market for remote sensing." It states that the fundamental goal of our policy is to support and to enhance US industrial competitiveness in the field of remote sensing space capabilities while at the same time protecting U.S. national security and foreign policy interests.

To protect national security and foreign-policy interests, it stipulates that, when national security or international obligations or foreign policies may be compromised by commercial remote sensing (defined by the Secretary of Defense or State respectively), the Secretary of Commerce may require the licensee to limit data collection and distribution by the system to the extent required by the situation.

It should be noted that the concurrent purposes of supporting industry and of protecting national security and foreign-policy interests are not phrased in terms of balance or compromise. It is not necessary to view these purposes as conflicting interests if security can benefit from commercial remote sensing. Effective implementation of policy can provide such an opportunity. To the extent that a U.S. commercial remote sensing industry can be regulated in operation, the policy can reduce the threat to U.S. national security or diplomacy posed by others' use of U.S. space remote sensing. It can also enhance U.S. national security by providing useful sensing at lower cost to U.S. forces. To the degree that support of the U.S. remote sensing industry increases U.S. dominance of commercial remote sensing, the policy can reduce the scope of international remote sensing threats that must be addressed by other means, diplomatic or military.

While the policy identified opportunity, successful regulation is necessary to make the opportunity a reality. To date, efforts to produce regulations have been protested, halting, and not entirely satisfactory in terms of acceptability to interested parties, usefulness for the goals of law and policy, and durability to possible legal challenges.

Rule-Making Issues

Through the period of protracted rule-making, the rules in effect for licensing commercial space remote sensing pre-dated the authorizing legislation by five years. The Department of Commerce published a notice of inquiry in December 1995 requesting public comment on the extent of change needed. A public hearing was held in June 1996. The department published draft rules for the 1992 act in November 1997, three and a half years after PDD-23, five years after the law (Department of Commerce, 1997). The department then held a public hearing on the draft rules in April 1998. Congress amended the enabling legislation in October 1998, and added a requirement for the Secretary of Commerce to publish in the *Federal Register* within six months a "complete and specific list of all information required to comprise a complete application for a license." At the end of July, 2000, the Department of Commerce produced an interim final rule, eight years after the authorizing legislation.

Regulatory issues have stymied rule making. One might ask whether this matters. Despite the lack of current regulations, the secretary has issued 13 licenses, two in 1993, four in 1994, three in 1995, and one each year from 1996 through 1999. So far, few licensees have commenced commercial operations,[12] and the operations of none have so far been constrained out of concern for foreign policy or national security. However, that should provide little basis for complacency.

Neither commercial nor government interests are well served by ambiguity in this case. The licenses granted contain conditions of the draft regulations that were the basis for claims of adverse effect and improper authority in comments on the proposed rules. The claimed adverse effects may well be inhibiting the industry. The large proportion of dormant licenses suggests so. Uncertainty about the operation of regulations must at least increase risk to business cases. On the government's side, if the license conditions are challenged successfully on a legal basis, the government may be left with less effective safeguards. Needlessly creating and perpetuating regulatory uncertainty increases regulatory risk to commercial enterprises. More importantly, it deprives those with constitutional authority for executing the country's military and diplomatic missions and with the life and death responsibility for the people involved of effective and durable tools to do the job.

If we examine the current state of regulations on remote sensing and objections raised, we might find a basis for workable, effective rules. We can group the issues raised in public comment on the draft rules into issues of authority and issues of implementation.

Authority Issues

There are two issues we might describe as issues of authority, questioning the legitimacy of the authority claimed. The more fundamental of the two with respect to the governmental interests protected is a constitutional challenge to prior restraint on first amendment grounds. The more limited of the two is an objection to restrictions based on foreign policy.

PRIOR RESTRAINT

In its first amendment challenge, the media perspective equated news gathering and news dissemination, likening remote sensing to photojournalism. The comparison with photojournalism seems apt. While one might question the definition of a remote sensing company as the press, the courts have extended concern for prior restraint beyond the traditional press to cover corporate expression of ideas. (*First National Bank of Boston v. Bellotti*, 1978.)[13] One might question a definition of remote sensing data as speech. Is an earth image or a soil moisture or vegetation index the expression of an idea in the marketplace of ideas? It may be evidence of objective fact in that marketplace, say in debate over wetlands policy or legislation. It may also be evidence of objective fact that a government might wish to suppress to avoid embarrassment and accountability, say of the disturbed earth of a mass grave. In content and purpose, remote sensing data are protected speech. As for form, remote sensing data will surely qualify as speech in the diversity of forms of expression, such as film and pictures, that the courts have ruled protected speech.[14]

The aptness of comparing space remote sensing with photojournalism deserves some additional examination because of the venue of sensing and the distinction between news gathering and news reporting. If we think of the remote sensing satellite as a photojournalist, we still have to note that the state retains international responsibility for its activities (whether privately or government owned) (Outer Space Treaty, art. VI; Principle XIV) and absolute liability for damage caused by it (Outer Space Treaty, art. VII). The state's international responsibility for activities in space is to ensure their conformance with the treaty and international law.

The international obligations most directly relevant to remote sensing are the 15 principles on remote sensing from space that the UN General Assembly adopted in 1986 in resolution 41/65. The most relevant to commercial remote sensing are the requirements to conduct remote sensing in accordance with international law (principle III) and to provide sensed countries nondiscriminatory access to sensed and analyzed data (principle XII). The latter, in particular, would seem unusual to a terrestrial photojournalist.[15] Disclosure of sources might seem an encroachment on first amendment liberties.

The courts and the Congress have recognized a distinction between news gathering (action, which may be regulated to protect society or other rights) and publishing (information and ideas, whose restriction runs counter to the first amendment) in allowing prior restraint on gathering as opposed to publishing information. One example is the espionage statutes. When the government interests and criteria for restraints are similar, we might expect the courts to recognize a similar distinction between gathering and publishing remote sensing data. We will return to this similarity later in attempting to synthesize workable regulations.

The Supreme Court has also recognized the distinction between news gathering and news reporting in areas other than national security. Most of the areas have come to the fore when the press's interests in news gathering have come into conflict with the judicial system and the rights of others to due process or privacy. However, there is also precedent for restraint on news gathering in the context of the Executive Branch's responsibility for conducting the nation's foreign affairs.

For example, the court rejected first amendment claims in permitting the Department of State to restrict the travel of U.S. citizens to areas that "might involve the Nation in dangerous international incidents." (*Zemel v. Rusk*, 1965, pp. 22–23.) A similar case might be made to restrict the action of news gathering via remote sensing for foreign-policy reasons, if the law provided the authority. However, even in that case, the burden would seem to be on the government to show that such a restriction was necessary for the protection of the country or for a similarly compelling foreign-policy reason and was not aimed solely at the restriction of access to information, particularly not information that bears on public interests in the actions of government.

The details of the press's first amendment objections to the proposed rules are found in a media position paper prepared by the Radio-Television News Directors Association and the National Association of Broadcasters (Kirby, 1998). The essence of their complaints was that the proposed regulations lack explicit enough criteria, are not drawn narrowly enough, and lack judicial oversight in application. In the complaint of excessive generality and breadth, the media argument was first with the draft regulation's choice of words describing potential limits on sensing and dissemination (echoing the president's policy) "during periods when national security or international obligations and/or foreign policies *may be compromised*. . . ." [emphasis added]. The complaint challenged the phrase "may be compromised" as being too low a standard on the basis that it did not represent clear and present danger of imminent harm to the state. Further, it challenged the criteria for prior restraint as being overly broad, lacking explicit standards for application.

In offering an alternative proposal for regulatory language, the media position paper cited three criteria that the Supreme Court has applied to content-based prior restraint of speech: that a compelling government interest requiring protection exists, that the regulations will achieve the government's objective, and that the regulations are the least-restrictive means available. In this context, "least restrictive" means are those "no more than is reasonably necessary to protect the substantial government interest." (*Brown v. Glines*, 1980.) That is to say, the regulation must be confined to the government's compelling interest (*NAACP v. Button*, 1963). The degree of confinement necessarily requires judgment in application. The court's requirement is for "precision of regulation" (*Lamont v. Postmaster General*, 1965) —not necessarily proof that the regulator has chosen the tightest possible threshold.[16]

Similarly, the requirement that regulations will achieve the government's objective is a restriction on narrowness of scope in application to the compelling interest. It is not a requirement to prove the sufficiency of the regulations standing alone to accomplish the government's objective. The media position paper suggested that this requirement could be taken to hold regulation futile if there were threats to the government's national security interest beyond the reach of the regulations' jurisdiction, such as the existence of foreign remote sensing entities. However, means besides regulation are available to address such threats, such as military or diplomatic action.

In dealing with foreign entities, the Constitution grants the Executive Branch authority over military and diplomatic means without judicial oversight. First amendment protections do not apply to foreign entities threatening U.S. security. In the 1991 Persian Gulf War to liberate Kuwait, France limited access to the remote sensing products of its SPOT satellite. In a future conflict, France and others might be persuaded to limit sensing or dissemination. But with other, less cooperative, suppliers, the U.S. Executive Branch could use more direct means against foreign remote sensing suppliers threatening national security. It could, for example, dazzle or jam offending satellites. As a further incentive for effective domestic controls, we should note that, in the case of diplomatic action, effective domestic regulations may be a prerequisite to establish credibility, precedent, or custom.

The media position paper proposed language it deemed adequate for defining content-based prior restraint as circumstances in which

> there is clear evidence that such action is necessary to prevent serious and immediate injury to distinct and compelling national

security interests of the United States, or to prevent violations of valid international treaty obligations of the United States. (Kirby, 1998, p. 18.)

We will address the international obligations in the next section. But we ought first to comment on the adequacy of the suggestions for national security interests. Despite the addition of such modifiers as "clear evidence," "serious and immediate injury," and "distinct and compelling," there is some risk of constitutional challenge even to this wording. The phrase national security interests may not be enough.

Military activity has always been one of the reasons the court has cited in describing the exceptional cases in which prior restraint may be permitted.[17] In presenting the dissenting opinion in *Greer v. Spock* (1976), Justice Brennan observed that

the first amendment does not evaporate with the mere intonation of interests such as national defense, military necessity, or domestic security. In all cases where such interests have been advanced, the inquiry has been whether the exercise of first amendments rights necessarily must be circumscribed in order to secure those interests.

"The guarding of military and diplomatic secrets at the expense of informed representative government provides no real security for our Republic." (*New York Times v. U.S.*, 1971, Justice Black concurring.)

When the espionage statutes apply (when the offender has reason to believe that the disclosure will harm the United States or aid another country), the court has held that the term "national defense" is not too broad for due process, that it "is a generic concept of broad connotations, referring to the military and naval establishments and the related activities of national preparedness." (*Gorin v. U.S.*, 1940.) However, the ordinary test for due process is not necessarily enough for prior restraint of first amendment rights. Given the court's decision on the Pentagon Papers, the regulation's language should go out of its way to make specific that the restrictions may not be imposed to prevent political discussion or governmental embarrassment. It could, perhaps, in the interest of specificity, be restricted to instances in which military forces, installations, or operations are protected from imminent harm by the restraint.

Finally, on the topic of prior restraint, the courts have permitted prior restraint on publication without reference to the espionage statutes in one instance. The reasoning in that case is instructive for conditions under which shutter control for national security reasons would be likely to survive constitutional challenge. In the case of *U.S. v. The PROGRESSIVE, INC.* (1979), the U.S. district court for the Western district of Wisconsin enjoined publication of portions of an article titled: "The H-Bomb Secret: How We Got It, Why We're Telling It." Although apparently created without unauthorized access to classified information, the article contained material on the design of hydrogen bombs that the government determined to be classified "Restricted Data.[18] Because the publisher's purpose in printing the article was informing the public policy debate, restricting publication went to the heart of the purpose of the first amendment. The court decided that the objectionable portions of the article were analogous to the publication of troop movements in wartime in their consequences for risk to human life and that their publication would result in direct, immediate, and irreparable damage to the United States. Even with that decision, and with potential consequences for loss of life on the scale possible with nuclear weapons, the court was still reluctant to set the precedent of prior restraint. It offered to appoint a panel of mediators in the hope that the government and the publisher could reach a settlement out of court, but the parties could not agree. A clear test for conditions sufficient to impose shutter control would seem to be the need to prevent imminent risk to human life. Even then, the court's reluctance to sanction censorship (preferring to encourage self-censorship) indicates the difficulty that shutter control should anticipate in judicial review.

FOREIGN POLICY

Another objection offered to the draft regulations is the inclusion of the term *foreign policy* as one of the reasons for imposing prior restraint on remote sensing. The basis for the objection was the absence of the term in the enabling legislation and the vagueness of the term as a content-based prior restraint on the first amendment. The enabling legislation required the commercial remote sensing operator to operate so as "to preserve the national security of the United States and to observe the international obligations of the United States." (15 USC §5622 (b)(1).) It required the Secretary of Commerce to certify that the operator would comply with "any applicable international obligations" (15 USC §5621 (b)(1)) of the United States before issuing a license. It required the Secretary of Commerce to consult with the Secretary of State "on all matters under this chapter affecting international obligations." (15 USC §5657 (b)(1)) The legislation mentions foreign policy in describing the Secretary of State's responsibility "for determining those conditions, consistent with this chapter, necessary to meet international obligations and *policies* of the United States." (15 USC §5657 (b)(1)); [emphasis added]. However, in judging consistency with the remainder of the chapter, foreign policy is mentioned as a condition only in sections pertaining to the Landsat program, its potential successors, and the government-funded remote sensing technology program. It is not mentioned in the sections authorizing licensing or operation of commercial remote sensing activities. The intent of the legislation seems clearly to restrict license conditions and operational constraints to national security and international obligations, limiting the applicability of foreign-policy issues to government-funded remote sensing.

Despite the apparent legislative intent, PDD-23 announced clear intent to include foreign-policy conditions on commercial remote sensing operators. To understand the motivation for foreign-policy conditions, consider a foreign-policy crisis, such as the Tupac Amaru guerrilla takeover of the Japanese ambassador's residence in Peru from December 1996 through April 1997. Commercial remote sensing satellites might have disclosed Peruvian government preparations for the eventual commando raid that freed the hostages, such as the digging of the tunnel under the residence that the commandos used to plant explosives and penetrate the compound. The Peruvian government might have been able to control domestic news-gathering activity that would have disclosed preparations. The conventional media might well have self-censored reporting of such preparations. However, commercial satellite imagery would be available not only to media that might not be willing to forgo reporting on the preparations but also to such customers as the Tupac Amaru leadership in exile in Germany. In a case like this, the U.S. government would have difficulty justifying any constraints on the remote sensing operations on the basis of U.S. national security or international obligation. A foreign-policy–based authority to regulate the operations of commercial remote sensing could enable the U.S. government to respond to a foreign request or to limit the U.S. government's liability for harm done by its country's space activity.

Returning to international obligations as the basis for constraints, the media position paper's recommended language recognized international obligations only in treaties. We should note that the obligations of international law also include custom. For example, one public comment on the proposed rules provided an extensive argument for the regulation to include explicitly implementation of the customary law principle of nondiscriminatory access to space remote sensing data (Gabrynowicz, 1998). Among the customs of international relations that the court has recognized in first amendment deliberations is confidentiality in diplomacy:

> It is elementary that the successful conduct of international diplomacy and the maintenance of an effective national defense require both confidentiality and secrecy. Other nations can hardly deal with this Nation in an atmosphere of mutual trust unless they can be assured that their confidences will be kept . . . it is the constitutional duty of the Executive—as a matter of sovereign prerogative and not as a matter of law as the courts know law—through the promulgation and enforcement of executive regulations—to protect the confidentiality necessary to carry out its responsibilities in the fields of international relations and national defense. (*New York Times v. U.S.,* 1971, Justice Stewart concurring.)

While the term *foreign policy* may be too broad, the term *treaty obligation* is too narrow. Under the aegis of international obligations, the regulations should cover international law and diplomacy. However, here again, the regulations need a similar degree of precision and clear avoidance of suppression of accountability or discourse. They should avoid prior restraint when punishment may deter and the damage of the offending speech may be undone. In the case of prior restraint for international obligation, reasonable specificity might restrict the regulations to instances in which diplomatic personnel, installations, or operations are protected from imminent harm. Returning to the Peruvian hostage example, U.S. governmental assistance to the Peruvian government during the crisis could reasonably be construed as such diplomatic operations and could justify limited restrictions on remote sensing. Such restrictions should, in an instance like this, be limited to embargo or delay in dissemination of sensitive data rather than denial of data collection. Consider, for example, the effect of U.S. government denial of coverage in light of the controversy over the Branch Davidian siege at Waco that continued a half dozen years after the incident. Such a denial of coverage would not only provide fodder for conspiracy theories, it would present the real possibility for the kind of abuse of power that the first amendment's protection is designed to hold accountable.

Implementation Issues

Among the issues raised on how to implement regulations were the nature of judiciary involvement, details of reporting required by licensees, and visibility of Executive Branch processes for review of license applications and foreign agreements and for determining the need for prior restraint in operation.

JUDICIARY INVOLVEMENT

The enabling legislation requires judicial involvement for the Secretary of Commerce "to terminate, modify, or suspend licenses . . . and to terminate licensed operations on an immediate basis" for failure to comply with license, act, security concern, or international obligations (15 USC §5623 (a)(2)). The legislation also requires judicial involvement for warrants to "seize any object, record . . . based on a showing of probable cause" of use in violation of act or license conditions (15 USC §5623 (a)(6)). The legislation permits the secretary to investigate, subpoena materials for investigation, impose civil penalties, and modify or remit those penalties without judicial involvement—all subject to judicial review under 5 USC §701. As we noted earlier, the Supreme Court has consistently required judicial supervision of any final prior restraint on speech and allowed temporary administrative restraint only to preserve the status quo and only as long as necessary for judicial review.

The media position paper's recommended regulatory language proposes to require the Executive Branch to seek an injunction *first* whenever it desires to control collection or dissemination of data, affording the licensee and affected parties adequate notice and opportunity for hearing. This omits the possibility of brief administrative restraint to preserve the status quo pending judicial resolution. We will have more to say about administrative process below in formulating a structure for regulation.

REPORTING

Comments on the proposed rules included a number having to do with reporting, both in license application and operation. Among them were some that are relatively easy to accommodate, such as presumption of confidentiality of material reported by applicants or licensees. A presumption of confidentiality could create some conflict between the commercial interests of licensees and those of other interested parties (for example, the press in hearings before final judicial decisions on prior restraint, but we should expect the courts to balance such interests.) Among the more difficult reporting issues were those having to do with foreign ownership and control, particularly the definition of "significant and substantial agreements."

The original language of the law required notification of all foreign agreements. Had that remained the case, and had the draft regulations not presumed the government's right to prevent entering into an agreement, there might not have been cause for so much adverse comment. Two difficulties arise with the current draft rules. First, and probably most objectionable, is a presumption that the government must approve agreements in advance to prevent those it objects to. The law allows the Secretary of Commerce to amend, suspend, or terminate licenses after judicial determination. It allows the secretary to investigate and impose penalties subject to later judicial review. The legislative intent seems clear that the secretary is not to impose administrative and judicial delay on normal business activity. The presumption of advance notice appeared first in the president's policy. This presumption makes the definition of "significant and substantial" even more critical.

The second difficulty with the draft rules on foreign agreement reporting is drawing a line around "significant and substantial." The notice of draft rule making explained its intent in terms of control of the remote sensing system and of distribution of data. The controls on distribution of data included reporting sales agreements that would create a "major customer or distributor." Inasmuch as sales agreements could incrementally accumulate to create a major customer, this suggests a need to see all foreign sales agreements, particularly since the notion of "major customer" does not seem definable with a single number. All foreign sales agreements might be a very large number of agreements, possibly amenable to automated scrutiny and analysis and perhaps reasonable as a reporting requirement. But it hardly seems reasonable to require a company to wait for government approval before accepting every foreign order for an image. While *substantial* and *major* may have precise meanings in changes of ownership, for which government approval has precedent, it is difficult to see how the words might be defined for customer influence and even more difficult to establish the connection between customer size and national security or diplomatic interest. The interim final rule revised the limitation on agreements creating a "major or customer or distributor" to "distributor arrangements involving the routine receipt of high volumes of the system's unenhanced data." "Involving high volumes of data" seems as vague as "major customer or distributor."

The "significance" of data sales with respect to national security may be completely independent of magnitude, depending instead on an operational context that should be accessible only to national security, intelligence, or law enforcement activities. For example, a pattern of purchases associated with such things as military installations and activities, embassies, or diplomatic or presidential travel might be significant, depending on the identity of the customer, timing of the requests, context of related activities, and information from other sources. Understanding of that kind of significance is likely to require substantial and sustained study as purchasing and data use evolve—not simple prescription a priori in regulation. Reporting and study of this kind are probably merited and reasonable within the police powers of the state, similar to the reporting of financial transactions and the investigation of money laundering. The key to reasonableness, though, is to avoid the constitutional issues of free speech and due process by not inserting government approval in the process of transaction.

MEMORANDUM OF UNDERSTANDING (MOU)

The draft regulations proposed documenting the process of interaction between government agencies in reviewing license applications and deciding the need for operational restrictions on data collection and dissemination in an MOU among the agencies. The draft stipulated that the MOU would identify the president as the final arbiter of disagreements in these interactions and expected that the MOU to be made public before issuance of a final rule. The primary objection to this MOU was that it appeared to be an evasion of rule making through public notice and comment in accordance with 5 USC §553, as required by the enabling legislation (15 USC §5624). Without disclosure of the content of the proposed MOU, there was no basis for judging if its description of process would be substantive with respect to rule making. The interim final rule included in an appendix a fact sheet describing the MOU to reassure industry and its customers. However, the inclusion was "not intended to solicit public comments" on the MOU. Clearly, the government should be able to disclose regulatory procedure without risk of compromising legitimate secrecy.[19] There seems no good reason for incomplete regulation requiring reference to a document describing procedures developed outside normal administrative processes.

Synthesis of a Regulatory Framework

With background and current issues in mind, we can attempt to synthesize a regulatory framework for space remote sensing operations. Before we begin, we should recall the goals of this regulation. Our purpose is to encourage the development of a responsible U.S. commercial space remote sensing industry. The notion of responsibility here is to protect legitimate public interests and promote the same kind of responsibility in international remote sensing activities. The value of a civil remote sensing capability begins with the use of its product for public good, which could include direct use by national security organizations. This use represents probably the largest, most developed, initial customer. But larger benefits are possible for environmental protection, earth resources conservation and exploitation, public safety, and unanticipated uses in the growth of an information-based economy.

The value of a commercial remote sensing industry over government ownership or subsidy of civil systems is in the expectation of greater efficiency with profit incentives and competition, where markets will support competition. It remains to be seen when or whether the market for space remote sensing data will be big enough for commercial viability, let alone competition. Initial financial commitments are encouraging. Current regulatory behavior is not.

A commercial remote sensing industry may have the secondary value of supporting the industrial base for national security space and remote sensing. U.S. domination of international remote sensing may also diminish the likelihood of dangerous or irresponsible international remote sensing. However, that possibility, while a worthwhile interest for regulation to consider, should not be the touchstone. Without domination, a U.S. commercial remote sensing industry may still provide a national security advantage by reducing the cost of access to U.S. sources and by encouraging international responsibility in remote sensing. For the irresponsible, coercive means are also available.

Finally, with respect to regulatory purpose, we have not described commercial interests and national security as competing concerns whose threats to each other need to be balanced. Rather, we have described national security interests in encouraging the growth of a viable, if not dominant, commercial industry. This commonality of interest depends on the existence of workable, durable regulations to allow operation "in such as a manner as to preserve the national security." Without such regulations, commercial and security interests are all too likely to be inimical.

Elements

The discussion above allows us to list several necessary elements of the regulatory framework. First is consistency with its purpose and its legislative and constitutional background. This begins with concern for the welfare of a fledgling commercial space remote sensing industry. It requires transparency and specific delimitation of government interests in regulation. Absolutely key is the absence of any taint of restriction of expression in the marketplace of ideas or any attempt to screen government from accountability or embarrassment.

A second element of the regulatory framework should be the absence of reference to foreign policy as the basis for government controls on commercial space remote sensing. We could retain reference to international obligations and substitute protection of diplomatic personnel, installations, and operations for references to foreign policy. However, any controls on these that approach prior restraint on publication should be subject to the same concerns for precision and narrowness that we will discuss shortly in operational controls.

A third element of our framework concerns reporting. We could reasonably require routine reporting of all (or all above some threshold) foreign sales transactions, but we must not make their completion contingent on government approval. We could, on the other hand, require prior government approval for offers to purchase equity interest or control at thresholds set by a transition of control that would remove U.S. jurisdiction in regulation.

A fourth element that barely deserves mention is completeness. Regulations should not refer to procedures documented outside the visibility of the Administrative Procedures Act. This is not a threat to legitimate secrecy in government or an invitation to encroachment on executive authority. It does not require prior documentation and review of all criteria for executive decision-making. The Executive Branch retains command authority and responsibility for military and diplomatic operations. The courts presume it competent and decline to interfere.[20]

The final element, most critical to successful, durable regulation, is operational controls. The next subsections will review alternatives for operational controls and recommend one.

Alternatives for Operational Controls
ESPIONAGE STATUTES

One comment on the draft rules proposed that the government should not attempt prior restraints on remote sensing at all but rather rely on the possibility of prosecution under the espionage statutes to deter dangerous remote sensing. A similar line of reasoning is evident in some of the opinions in the Pentagon Papers case. (*New York Times v. U.S.*, 1971, Justices White and Marshall concurring.)

The key to culpability under the publication or disclosure limitations of the espionage statutes is purpose and intent. The offender must have reason to believe that the information is to be used to the detriment of the United States or to the benefit of another country. In the case of a commercial remote sensing entity collecting information for a customer who has this purpose and intent, could the commercial entity have reason to believe harmful use? Could it be presumed to know the value of the information sensed with respect to the national defense? The entity might be expected to know the location of publicly disclosed fixed installations that coincide with a customer's tasking request. However, even publicly available knowledge of the whereabouts of every "vessel, aircraft, work of defense, navy yard, naval station, submarine base, fueling station, fort, battery, torpedo station, dockyard, canal, railroad, arsenal, camp, factory, mine, telegraph, telephone, wireless, or signal station, building, office, research laboratory or station or other place connected with the national defense . . . ." (18 USC §793(a)) seems an unreasonable burden to place on a commercial entity, particularly for craft and equipment moving in the course of the operations whose security is to be protected.

Even if it could be presumed to know, a commercial entity should not be presumed competent to make the life and death judgments for which commanders in the Executive Branch are held accountable. Worse, its attempts, however well motivated, could operate at cross purposes to the commanders' intent, denying an opponent visibility of a feint deliberately made visible by the U.S. commander. A commercial entity would necessarily err on the side of caution if the espionage deterrent were credible. That alone would work against the government's interest in encouraging the growth and development of the industry.

Further, should the commercial supplier of remote sensing data be expected to know how a customer intends to use or could extract harmful information from a purchased product or service? For example, without awareness of the details of classified signatures of military targets, how is a supplier to know of their detectability in its products? Presumptions of this kind of knowledge, reasonably needed to protect military or diplomatic forces and activity, do not pass the test of vagueness for due process.

Under the espionage statutes, the government can impose prior restraints on gathering and publishing photographs or similar graphical portrayals of defense installations and equipment.[21] It could invoke this legislative authority in publishing regulations for the space remote sensing industry. If it did, the implementation would require some procedure to inform the industry of designated installations and equipment, beyond the signs at the fence customary now but not readily visible in remote sensing. Specific decisions on timing, area, and circumstances of restriction would have to remain with those in the Executive Branch with the command authority, information, expertise, and responsibility for life and mission needed to make the judgment calls on what information, available when and to whom, will help or harm. If too narrowly and timely designated, the designation itself might represent an undue risk to operational security. If broadly applied, the extensiveness of the prohibition would have the same chilling effect on the industry we noted above, not to mention the possibility of first amendment challenges. Clearly, the espionage statutes alone are not sufficient.

TIME, PLACE, AND MANNER RESTRICTIONS

A general alternative to content-based prior restrictions on publishing remote sensing data is restriction based on the time, place, or manner of otherwise protected speech. This kind of restriction would require the government to establish license conditions defining security concerns a priori. It is hard to imagine how to do this usefully without excessive restriction. Place-of-speech restrictions do not seem applicable here. Time restrictions on the speech might have small utility if a minor delay in data delivery could provide protection without harming the market value or civil utility of the information. But no obviously useful nor reasonably harmless brief delay seems likely that would not also harm legitimate use.

Past de facto restrictions defined by technical parameters, such as spatial resolution were restrictions on manner of speech. That particular restriction has already been overturned in the interest of avoiding restrictions on U.S. competition with international providers. Besides competition, there are instances in emerging remote sensing technologies, such as hyperspectral imaging and synthetic aperture radar, in which no technical performance parameter separates military utility from public good. (See Chapter 5 in this volume.)

ADMINISTRATIVE AND JUDICIAL PROCEDURE FOR PRIOR RESTRAINT

Because deterrence by the espionage statutes and time, place, or manner constraints are insufficient, we are back to the delicate problem of defining acceptable procedures for prior restraint. The background discussion listed the general conditions required in past decisions on constitutionality: imminent danger of substantial harm, narrow specificity of conditions to the harm at risk, final judicial supervision, and briefest delay in administrative restraint consistent with judicial review. In prescribing rules that should pass tests of constitutionality, we might find useful insight from prior cases.

A court decision that observed "The phrase 'prior restraint' is not a self-wielding sword. Nor can it serve as a talismanic test." (*Kingsley Books v. Brown*, 1957, Justice Frankfurter delivering the court's opinion) provides some insight into acceptable conditions of prior restraint. In *Kingsley Books v. Brown* (1957), the court allowed a New York law to authorize municipal executives to seek an injunction preventing the sale or distribution of obscene matter. After prompt trial of the issues, the law permitted seizure and destruction of the material. The law provided further that sale or distribution after the complaint had been served was chargeable with knowledge of the nature of the material. The law in this case has the features of prompt, final judicial determination, after minimal delay from administrative procedures, finding cause to restrict distribution based on narrow criteria. In emulating this example, there is another concern to beware of in prescribing acceptable administrative procedures prior to judicial review. The court found in a later case that administrative measures ostensibly intended to advise of potential legal liability could be unacceptable informal censorship if not subject to appropriate safeguards (*Bantam Books v. Sullivan*, 1963).

The administrative procedures for shutter control should provide the Executive Branch the opportunity to exercise the judgment in diplomacy and security that only it can—making its decisions "directly responsible to the people whose welfare they advance or imperil." (*Chicago & Southern Air Lines, Inc. v. Waterman Steamship Corp*, 1948.) Having determined that a particular instance of space remote sensing activity endangers specific military or diplomatic forces (people and materiel), installations, or operations, the government should use administrative procedures to communicate the jeopardy to the commercial remote sensing entities involved. In reference to the espionage statutes, this communication establishes reason to believe the information collected is to be used to the injury of the United States. It also parallels procedure upheld in *Kingsley Books v. Brown*, but could be seen as informal censorship à la *Bantam Books v. Sullivan* if not supervised by judicial safeguards.

After communication of the government's determination of imminent jeopardy, the administrative procedures should reasonably allow for negotiation with the commercial entities of the least intrusive, most flexible adjustments needed to remove the jeopardy. If the imminence of danger does not permit time for discussion with the commercial entities, the government's selection of remedy should be made from a graduated spectrum of choices. Among possible adjustments might be delay of delivery, excision of minimal data, degradation of data quality or content (for example spatial clarity or spectral content), or obscuration of offending information with naturally consistent artifacts (for example water vapor or aerosols).[22] This flexibility may be helpful to the government's need to protect, aside from its benefit to the commercial entity and consistency with first amendment protections. Outright denial of a customer's request may be counterproductive in some circumstances, supplying an indication by itself and inviting resort to alternative means. If only outright denial is permitted, the government may require greater area or duration of denial to obscure the indication. Outright denial of data gathering is also the most intrusive to the first amendment. Considering the range of flexibility possible, the novelty of the regulations, the rarity of anticipated need, and the critical dependence on effective communication among the parties involved, implementation of regulations like these should involve some routine exercise of the administrative apparatus of shutter controls to educate the participants and improve the process.

Communication of jeopardy and a process of negotiation (if time permits) should support due process for the companies and provide the Secretary of Commerce a basis for investigation, civil penalties, and evidence of probable cause for a warrant for seizure of materials or for judicial determination to terminate, modify, or suspend a license. For any adjustments to remote sensing operations imposed to protect national security or diplomacy, except possibly for a delay shorter than judicial process, the government's procedure should require it to initiate judicial review of the adjustment made.[23] In the process of review, the government should not expect to expose legitimate secrets or create the jeopardy it sought to avoid. It should expect to have to demonstrate the narrow specificity of its purpose and justify the scope of means applied.

Summary

To recap the outlines of our sketch of a regulatory landscape design for space remote sensing, rules should
- omit reference to foreign policy, referring instead to U.S. diplomatic forces (people and materiel), installations, and operations
- be complete, having no unreviewed procedure or MOU

- require reporting but not approval of foreign sales
- provide administrative processes to advise licensees of imminent harm to diplomatic or national security forces, installations, or operations posed by their operations. The administrative process should, time permitting, allow negotiation of the least restrictive palliative means and should probably include immediate governmental institution of judicial review before final imposition of restrictions.

In turning this sketch into complete draft rules and in completing rule making after public notice and comment, the government, industry, press, and public should keep rule durability in mind. All should prefer not to have rules that will someday invite constitutional challenge of the rules or underlying law. No amount of precedent can predict the court's response. There is ample opinion expressed in the precedents cited throughout this chapter to support outcomes that any single interest would abhor. A constitutional challenge to remote sensing rules is likely to occur in circumstances similar to the "great case" of the Pentagon Papers, where passions of the moment, political tides, and patriotic fervor may all combine unpredictably and badly. We would all do well to recall Justice Harlan, dissenting with Chief Justice Burger and Justice Blackmun, in the Pentagon Papers decision. In decrying unseemly haste in that case, he quoted Justice Holmes's dissenting opinion in *Northern Securities Co. v. U.S.* (1904):

> Great cases like hard cases make bad law. For great cases are called great, not by reason of their real importance in shaping the law of the future, but because of some accident of immediate overwhelming interest which appeals to the feelings and distorts the judgment. These immediate interests exercise a kind of hydraulic pressure which makes what previously was clear seem doubtful, and before which even well settled principles of law will bend.

Finally, we should reiterate the call to complete a well-designed regulatory landscape for space remote sensing. In its absence, the current ambiguity provides no security, chills industry, and delays the public good that a vibrant remote sensing industry could bring.

Acknowledgments

The details of that sketch benefited substantially from critical comment by Barbara Cochran, President of the Radio and Television News Directors Association; Kathleen Kirby of Wiley, Rein & Fielding; Carol Schwab, Department of State; Karen Dacres, NOAA; and John Woodward, RAND. None of them will likely be entirely pleased with the final chapter, but all were gracious with their time and thoughtful with their comments. Kevin O'Connell suggested the title of the chapter.

Notes

1. Chapter 5 in this volume (by the same author) provides some observations on export controls and remote sensing policy. Within the U.S. policy's hierarchy of preference (selling remote sensing products; then case-by-case consideration of sales of turnkey systems; and, last, export of technology only on a restricted basis), export controls are the mechanisms available for implementing the hierarchy. However, in an increasingly globalized space remote sensing industry, unilateral export controls are a relatively blunt instrument more likely to isolate U.S. industry than to enhance its competitiveness—in the process diminishing U.S. opportunities to protect U.S. national security and foreign-policy interests.
2. For a survey of recent commercial remote sensing policy issues, see Chapters 2–4 and 6 in this volume. For a purely legal perspective of earlier remote sensing regulations, see Merges and Reynolds (1997). For the effects of international law on commercial remote sensing, see Bourely (1988). For more thorough background on military concerns, commercial markets, and previous regulatory issues, see Chapter 2 and Appendix A of Preston (1994).
3. The government sought an injunction against the publication by the *New York Times* and the *Washington Post* of the contents of a classified study entitled "History of U.S. Decision-Making Process on Viet Nam Policy." The Supreme Court decided that the government had not met the burden of justification needed for prior restraint on publication. The popular name of the case, the Pentagon Papers, was in reference to the classified report authored and released to the press by Daniel Ellsberg of RAND. (*New York Times v. U.S.*, 1971).
4. The speech in question was a circular intended to incite resistance to conscription in wartime. *Schenck v. United States*, 249 U.S. 47, 1919.
5. See also *Near v. Minnesota* (1931): Public officers, whose character and conduct remain open to debate and free discussion in the press, find their remedies for false accusations in actions under libel laws providing for redress and punishment, and not in proceedings to restrain the publication of newspapers and periodicals.
6. *Greer v. Spock* (1976), permitted a ban on all distribution of political leaflets and political speeches on a military installation in the interest of good order and discipline. Another case, *Lehman v. City of Shaker Heights* (1974), permitted a ban on political advertising on a city-operated transit system.
7. This implements the U.N. Principles Relating to Remote Sensing of the Earth from Outer Space.
8. The words "significant or substantial" were added with the Commercial Space Act of 1998.

9   It similarly limited declassification and release of government remote sensing.
10 It is worth noting that the language of this section does not include publication. Justice Black's opinion in the Pentagon Papers case highlights this omission and cites the legislative record to note congressional unwillingness to permit general prior restraint on the press to prevent espionage.
11 Published in draft in March 1986 after the enabling legislation passed in 1984.
12 At the time of this writing, two U.S. suppliers are producing satellite images; only one of them has a high-resolution sensor.
13 The reference here is not to commercial speech, which may enjoy a lower degree of constitutional protection, but to the court's refusal to permit legislation to restrict corporate expression of ideas to issues materially affecting the property or business of the corporation.
14 See, for example, *Joseph Burstyn, Inc. v. Wilson* (1952) and *Superior Films, Inc. v. Department of Education* (1954).
15 The requirement to notify a sensed state might seem like a requirement to disclose confidential sources, a requirement that news gatherers are protected from by some states' shield laws and, to some extent, by the first amendment. However, the news gatherer's privilege against disclosure is not absolute. It does not permit the news gatherer to break the law, impede justice, or infringe other rights. (For a discussion of relevant cases, see Eclavea (2000).
16 For example, the court upheld relatively low dollar thresholds for disclosing campaign contributions over objections that the low thresholds lacked "a substantial nexus with the claimed governmental interests" observing that "we cannot require Congress to establish that it has chosen the highest reasonable threshold. The line is necessarily a judgmental decision. . ." *Buckley v. Valeo* (1976).
17 For example, preventing obstruction to recruiting or publication of sailing dates for troop transports, *Near v. Minnesota* 1931.
18 In the sense of 42 USC §2011 et seq.
19 According to *U.S. v. Marchetti* (1972): If in the conduct of its operations the need for secrecy requires a system of classification of documents and information, the process of classification is part of the executive function beyond the scope of judicial review.
20 The Supreme Court has explained: The President, both as Commander-in-Chief and as the Nation's organ for foreign affairs, has available intelligence services whose reports are not and ought not to be published to the world. It would be intolerable that courts, without the relevant information, should review and perhaps nullify actions of the Executive taken on information properly held secret. Nor can courts sit in camera in order to be taken into executive confidences. But even if courts could require full disclosure, the very nature of executive decisions as to foreign policy is political, not judicial. Such decisions are wholly confided by our Constitution to the political departments of the government, Executive and Legislative. They are delicate, complex, and involve large elements of

prophecy. They are and should be undertaken only by those directly responsible to the people whose welfare they advance or imperil. They are decisions of a kind for which the Judiciary has neither aptitude, facilities nor responsibility and which has long been held to belong in the domain of political power not subject to judicial intrusion or inquiry. (*Chicago & Southern Air Lines, Inc. v. Waterman Steamship Corp.,* 1948.)

21 According to 18 USC §795, Whenever, in the interests of national defense, the President defines certain vital military and naval installations or equipment as requiring protection against the general dissemination of information relative thereto, it shall be unlawful to make any photograph, sketch, picture, drawing, map, or graphical representation of such vital military and naval installations or equipment without first obtaining permission of the commanding officer of the military or naval post, camp, or station, or naval vessels, military and naval aircraft, and any separate military or naval command concerned, or higher authority, and promptly submitting the product obtained to such commanding officer or higher authority for censorship or such other action as he may deem necessary.

22 The last, while perhaps more useful for the government, may also be more satisfactory to the commercial entity, doing less harm to good will with customers.

23 Alternatively, the government might choose to take a position that the operational controls are not prior restraint of publication but rather of news gathering and might wait for a legal challenge to provoke judicial review. Should it do so, it should undoubtedly have documented a case for the imposition of prior restraints sufficient to satisfy the criteria identified for prior restraint of publication. If the government takes this incremental approach, it would seem to run some risk of having the legislative authority to regulate commercial remote sensing challenged and should weigh the consequences of having the authority to regulate overturned.

References
Legislation, Laws, and Treaties
15 Code of Federal Regulations Chapter IX, National Oceanic and Atmospheric Administration, Department of Commerce, Part 960, Licensing of Private Remote-Sensing Space Systems, July 10, 1987.

5 United States Code, Part 1, The Agencies Generally.

—, Ch. 7, Judicial Review.

15 United States Code 82, Land Remote Sensing Policy, Subchapter II, Licensing of Private Remote Sensing Space Systems, §5621, General Licensing Authority.

—, §5622, Conditions for Operation.

—, §5623, Administrative Authority of Secretary.

—, §5624, Regulatory Authority of Secretary.

15 United States Code 82, Land Remote Sensing Policy, Subchapter III, General Provisions, §5657, Consultation.

18 United States Code Part I, Crimes, Chapter 37, Espionage and Censorship, §793, Gathering, transmitting or losing defense information

—, §794, Gathering or Delivering Defense Information to Aid Foreign Government.
—, §795, Photographing and Sketching Defense Installations.
—, §796, Use of Aircraft for Photographing Defense Installations.
—, §797, Publication and Sale of Photographs of Defense Installations.
42 United States Code Chapter 23, Development and Control of Atomic Energy, Division A, Atomic Energy, Subchapter I, General Provisions, §2011, Congressional Declaration of Policy.
Administrative Procedures Act—*see* 5 USC §701 et seq.
Commercial Space Act of 1998.
Land Remote Sensing Policy Act of 1992
Executive Order 10104, February 1, 1950.
Outer Space Treaty—*see* United Nations (1967).
Presidential Decision Directive 23 (PDD-23), March 10, 1994.
Principles—*see* United Nations (1986).
Public Law 104-201, National Defense Authorization Act for Fiscal Year 1997, September 23, 1996.
Public Law 105-303, An Act to Encourage the Development of a Commercial Space Industry in the United States, and for Other Purposes, October 28, 1998.
United Nations, *Treaty on Principles Governing the Activities of States in the Exploration and Use of Outer Space, including the Moon and Other Celestial Bodies*, January 27, 1967. Available at http://www.un.or.at/OOSA/treat/ost/osttxt.html (last accessed May 31, 2000).
United Nations, General Assembly, *Principles Relating to Remote Sensing of the Earth from Outer Space*, resolution 41/65, 1986. Available at http://www.un.or.at/OOSA/treat/rs/rstxt.html (last accessed May 31, 2000).

Court Cases
Bantam Books v. Sullivan, 372 U.S. 58, 1963.
Brandenburg v. Ohio, 395 U.S. 444, 1969.
Brown v. Glines, 444 U.S. 348, 1980.
Buckley v. Valeo, 424 U.S. 1, 1976.
Cantwell v. Connecticut, 1940
Chaplinsky v. New Hampshire, 315 U.S. 568, 1942.
Chicago & Southern Air Lines, Inc. v. Waterman Steamship Corp., 333 U.S. 103, 1948.
Connally v. General Construction, 269 U.S. 385, 1926.
Consolidated Edison Company of New York v. Public Service Commission of New York, 447 U.S. 530, 1980.
First National Bank of Boston v. Bellotti, 435 U.S. 765, 1978.
Freedman v. Maryland, 380 U.S. 51, 1965.
Gorin v. U.S., 312 U.S. 19, 1940.
Greer v. Spock, 424 U.S. 828, 1976.
Hodel v. Virginia Surface Mining & Reclamation Association, 452 U.S. 264, 1981.
Hynes v. Mayor of Oradell, 425 U.S. 610, 1976.
Joseph Burstyn, Inc. v. Wilson, 343 U.S. 495, 1952.

Kingsley Books v. Brown, 354 U.S. 436, 1957.
Kunz v. New York, 340 U.S. 190, 1951.
Lamont v. Postmaster General, 381 U.S. 301, 1965.
Lanzetta v. New Jersey, 306 U.S. 451, 1939.
Lehman v. City of Shaker Heights, 418 U.S. 298, 1974.
NAACP v. Button, 1963.
Near v. Minnesota, 283 U.S. 697, 1931.
New York Times v. U.S., 403 U.S. 713, 1971.
Niemotko v. Maryland, 340 U.S. 268, 1950.
Northern Securities Co. v. U.S., 193 U.S. 197, 1904.
Schenck v. United States, 249 U.S. 47, 1919.
Smith v. California, 361 U.S. 147, 1959.
Southeastern Promotions v. Conrad, 420 U.S. 546, 1975.
Speiser v. Randall, 357 U.S. 513, 1958.
Superior Films, Inc. v. Department of Education, 346 U.S. 587, 1954.
U.S. v. Marchetti, 466 F.2d 1309, 1972.
U.S. v. The PROGRESSIVE, INC., 467 F. Supp. 990, 1979.
Whitney v. California, 1927, 274 U.S. 357, 1927.
Zemel v. Rusk, 381 U.S. 1, 1965.

Other Works

Bourely, Michel G., "Legal Problems Posed by the Commercialization of Data Collected by the European Remote Sensing Satellite ERS-1," *Journal of Space Law*, 16, 1988.

Department of Commerce, National Oceanic and Atmospheric Administration, "Licensing of Private Land Remote-Sensing Space Systems," draft rules, *Federal Register*, Vol. 62, No. 212, 15 CFR Part 960, November 3, 1997, pp. 59,317–59,331.

---, "Licensing of Private Land Remote-Sensing Space systems (Rules and Regulations)," *Federal Register*, Vol. 65, No. 147, 15 CFR Part 960, July 31, 2000, pp. 46, 822–46, 837.

Eclavea, R. P., "Annotation: Privilege of Newsgatherer Against Disclosure of Confidential Sources or Information," 99 A.L.R.3d 37, 2000.

Gabrynowicz, Joanne Irene, Professor of Remote Sensing Law and Policy, University of North Dakota, letter to Charles Wooldridge, Department of Commerce, re: 15 CFR Part 960, March 31, 1998.

Kirby, Kathleen A., Wiley, Rein & Fielding, counsel for RTNDA, Joint Comments on the Radio-Television News Directors Association and National Association of Broadcasters in the Matter of Licensing of Private Remote Sensing Space Systems, Docket No. 951031256-5259-01, April 1, 1998.

Merges, Robert P., and Glenn H. Reynolds, "News Media Satellites and the First Amendment: A Case Study in the Treatment of New Technologies," *High Tech Law Journal*, March 1, 1987.

Preston, Bob, *Plowshares and Power: The Military Use of Civil Space*, National Defense University Press, 1994.

25
New Users and Established Experts: Bridging the Knowledge Gap in Interpreting Commercial Satellite Imagery

John C. Baker

An important new challenge is how to sustain the credibility of commercial satellite imagery in the face of new imagery users who have very limited experience in dealing with satellite imagery. As higher-resolution satellite images become widely available, both inadvertent and deliberate errors in imagery interpretation and analysis are likely to occur. In some instances, such errors are likely to receive substantial public attention. This situation is something of a departure from earlier times, when lower resolution Landsat and Satellite Pour l'Observation de la Terre (SPOT) satellite images, which were largely used by scientists and civilian remote sensing experts, rarely received prominent attention in policy debates.

The rapid growth of new imagery users poses a novel problem for the remote sensing field. The advent of commercial observation satellites, which offer broad public and international access to relatively high-resolution satellite images, is attracting a wider range of satellite imagery users (see Table 25.1). New imagery users, including the news media and nongovernmental organizations (NGOs) are being encouraged by growing access to higher-resolution and more-timely imagery data, improved computers and software for handling imagery data, and lower imagery prices (see Chapters 22 and 23 in this volume). Thus, a new generation of "imagery activists" is emerging that intends to use satellite images for focusing public attention on a broad range of domestic

Table 25.1—Comparison of Traditional and New Users of Overhead Imagery Data

Traditional Users	New Users
Governments • Civil planners (mapping, land management, disaster response, etc.) • Armed forces • Intelligence services • Scientific centers • State and local governments Multinational Organizations • United Nations (UN) agencies (e.g., UN Special Commission, UN High Commissioner for Refugees) • Global change research programs • Regional centers (e.g., Western European Union satellite centre) Business • Resource extraction (e.g., oil, gas) • Resource management (e.g., forestry) Academia and research organizations • Geography and geology departments • Remote sensing programs • Environmental studies Remote sensing industry • Aerial photography firms • Satellite imagery data providers • Value-added, image processing firms • GIS firms Remote sensing professional organizations	News media and information providers • Electronic media organizations • Print media • Trade journals NGOs • Environmental policy • Arms control, nonproliferation, and disarmament • Regional conflict resolution • Humanitarian relief • Human rights Academia and Research organizations • Media studies departments • Security policy studies departments • Archaeology departments • Transportation studies • Agricultural and vegetation studies Business • Utilities (e.g., telecommunications, pipelines, etc.) • Insurance firms (e.g., hazard assessments) • Precision agriculture • Environmental impact assessments Consumers • Real estate • Individuals interested in images of their homes or cultural attractions

and international issues, such as environmental problems, the need to support humanitarian operations, and public awareness of the activities of countries suspected of acquiring weapons of mass destruction. A good example is an analysis of North Korea's missile launch site that the nongovernmental analysts at the Federation of American Scientists (FAS) released, which made use of a high-resolution IKONOS commercial satellite image and received considerable public attention.[1]

Despite their best efforts, the new imagery users face major challenges in establishing their imagery credibility. They generally lack the training, experience, and resources required to make consistently accurate interpretations of overhead images, particularly when compared with more-experienced imagery analysts found at government agencies, commercial remote sensing firms, and some university departments. This situation creates an *imagery credibility paradox* because, even though new users possess relatively limited experience and resources for developing their imagery expertise, their imagery work is likely to attract high public visibility.

In comparison, the substantial imagery work of the traditional users in government and industry with the most experience and expertise will continue to receive limited attention because their efforts are often not intended for public consumption. While recognizing their somewhat divergent interests, the best way to bridge this major knowledge gap in imagery interpretation between the new imagery activists and the more-experienced remote sensing experts is through greater engagement that leads to limited cooperation.

This chapter addresses the challenges of sustaining imagery credibility as new imagery users with much less experience take advantage of higher-resolution commercial satellite imagery for the first time. It begins by briefly reviewing the basic techniques, equipment, and activities associated with imagery interpretation and analysis. Next, it examines how deliberate and inadvertent errors could present problems for interpreting satellite data. Finally, the chapter postulates three models (i.e., indifference, rivalry, and cooperation) for describing the possible relationship that could emerge between the new imagery activists and the expert imagery analysts in government and industry. The case is made that adopting a model of limited cooperation to bolster the imagery interpretation and analysis skills of the new imagery users will help sustain the credibility of commercial and civilian satellite imagery for all producers and users of overhead imagery.

Challenges of Accurately Interpreting Satellite Imagery Data

Overhead images acquired by satellite and airborne platforms have a powerful attraction for many individuals partly because of the truism that "seeing is believing." Ironically, however, accurately interpreting satellite images is not so much an intuitive process as it is one requiring a combination of expertise and experience in manipulating imagery data to discover objects and patterns that are usually ambiguous or not obvious at first glance. Several factors complicate untrained efforts to interpret satellite imagery data: The overhead perspective that satellite images provide is unfamiliar to most individuals; unfamiliar scales and resolutions complicate the ability of individuals to identify objects located on the surface; and sometimes the most revealing data are found in the spectral range beyond the visible bands (Campell, 1996, pp. 14–15, 122–130; Bruginoni, 1996). Hence, imagery data seldom yield correct interpretations that are intuitive to users; rather, substantial imagery interpretation skills and resources are often needed to avoid mistakes and to find the most important objects and patterns embedded in the satellite imagery data. As one remote sensing expert cautions:

> Access to raw imagery data does not, by itself, impart useable [sic] knowledge. Military and commercial users still need training and expertise. They will, also and increasingly, need to acquire timely, high-volume, high-technology systems to exploit the full value of high-resolution remote sensing imagery to solve their diverse information needs. (Haakon, 1999, p. 26.)

Depending on the spatial resolution and spectral bands available, analysts can use satellite imagery data to address various imagery interpretation tasks requiring ever more precision. These tasks usually begin with *detection* (i.e., determining the presence or absence of a particular feature) and *identification* (i.e., specifying a feature or an object within a particular class). With more-detailed imagery data, the analyst can move on to the more demanding tasks of *measurement* (i.e., estimating the physical dimensions or image brightness of a particular object or feature) and *analysis* (i.e., providing a much more-sophisticated appreciation of the qualitative character of an object and its relationship to other objects and features). In undertaking these interpretation tasks, imagery analysts look for the presence or absence of a known set of "observables" in the imagery data that are closely associated with important objects or features. Detecting observables, such as the distinctive shape and size of an aircraft or a ship, gives imagery analysts higher confidence in interpreting what they see in an image scene. In their search for generic observables, or "signatures" that are even more definitive and that are uniquely associated with certain objects, imagery analysts focus on the following image elements that are usually grouped in patterns on a pixel-by-pixel basis within the image:[2]

- Tone or hue—the relative brightness or color of an object or region within the image
- Shape—the general configuration or outline of an object
- Size—the dimensions, surface, height, and volume of an object
- Texture—the apparent roughness or smoothness of an image region
- Pattern—the spatial arrangement of objects into distinctive and recurring forms

- Shadow—the effect created if an object casts a shadow that reveals its size and shape
- Site—the location of an object in relationship to its environment
- Scale—the ratio of image size to an object's actual size
- Association—the interrelationships among observed objects
- Situation—how a feature or object is situated on a particular site.

A broad range of image interpretation techniques, technologies, and information-handling processes are typically available to imagery analysts under the best of circumstances. Table 25.2 identifies some basic imagery-interpretation techniques and tools that experienced imagery analysts will use to look for signatures of important objects or patterns. One of the more-powerful techniques is change detection analysis, which can be used if repeat coverage of an imagery target is available. This technique involves comparing the specific pixel values from two overhead images of the same location acquired on different dates. The pixels of these images are then carefully matched to reveal differences that occurred over the time when the images were collected.

Table 25.2—Imagery Data–Interpretation Techniques, Technologies, and Processes

Imagery-interpretation techniques	• Image reference keys (i.e., catalogs of observables) • Image magnification • Change detection based on repetitive imagery coverage • Stereoscopic viewing of overlapping images • Comparing different imagery types (e.g., optical and radar) • Image enhancement techniques (e.g., contrasting, etc.) • Three-dimensional visualization techniques
Technologies	• Preprocessing to correct geometric and radiometric errors • Light table for viewing film transparencies • Stereoscopes and measuration equipment • Specialized image-processing software • Digital image-processing systems (e.g., personal computers, workstations) • High-quality plotters and color printers
Information handling processes	• Ground reference information (as available) • Collateral information (maps, photographs, reports, etc.) • Multidisciplinary teams of analysts • GIS databases • All-source information or intelligence • Structured review process for imagery interpretation

Imagery analysts must also rely on certain technologies in performing their work with modern digital images.[3] At a minimum, these include high-end personal computers or workstations capable of handling the data-intensive files associated with digital images, as well as good image processing and display software that can be used with a high-quality image display. In addition, many imagery interpreters use specialized equipment, such as stereoscopes, for producing a three-dimensional view of overlapping images to gain a more-accurate measure of the heights of objects in images. However, new imagery users are unlikely to have the same level of specialized equipment, such as light tables and zoom stereoscopes for viewing image transparencies, or the more-advanced imagery-related hardware and software that their counterparts in government agencies and corporations often use.

Finally, information-handling processes are an integral element of ensuring accurate image interpretations. Traditional imagery analysts, particularly those in U.S. government agencies with a strong interest in satellite imagery (such as the National Intelligence and Mapping Agency), have well-developed and well-funded capabilities for bringing together disparate source materials and multidisciplinary expertise. The main benefit is to gain a broader perspective for interpreting imagery data and some invaluable historical context that permits knowledgeable analysts to better recognize the particular significance of what they find or do not find in the overhead images of a particular location. Similarly, experienced civil and commercial users will also attempt to make use of diverse information sources, including any available "ground truth" observations of the imaged location. Geographic Information System (GIS) databases can provide various overlays of thematic information or raw geospatial data to help interpret imagery data. Digital elevation models, which can be merged with overhead imagery to create three-dimensional representations of the earth's surface, offer another important data source for both civil and military applications.

The Problem of Imagery-Interpretation Errors

Despite established techniques for imagery interpretation, errors can occur at various stages in the process of translating imagery data into information products. These stages can be defined as imagery data production, imagery data interpretation, and the resulting analysis based on the imagery-interpretation products. For purposes of analysis, it is useful to distinguish between deliberate errors, which occur when someone deliberately seeks to distort

Table 25.3—Imagery Data and Deliberate Errors

Imagery data production	• Data provider removes observables indicating objects considered too sensitive to make public • Data provider or imagery user tampers with imagery meta-data to alter the reported date and/or location of the image • Imagery user tampers with imagery data either to conceal or to "discover" sensitive objects
Imagery data interpretation	• Imagery user claims to detect and identify observables for objects that the user knows are not present in the imagery data • Imagery user claims *not* to detect or identify any observables for objects that the user knows are present in the imagery data
Analysis based on imagery products	• User disputes the larger implications of the observables detected and identified by imagery interpretation
Potential countermeasures	• Accountability in imagery data processing and good security practices for data providers • "Watermarking" technologies and procedures for validating imagery data • Multiple sources of imagery data to permit double-checking • Third-party review of disputed imagery data, interpretation, or analysis

the imagery data, and inadvertent errors, which arise from mistakes in imagery interpretation and analysis resulting from lack of expertise, inexperience, or limited resources. Minimizing both types of errors is important to avoid undermining the credibility of commercial satellite imagery as an important and reliable information source for policymaking.

DELIBERATE IMAGERY DATA ERRORS

Deliberate errors occur if someone consciously changes or distorts the satellite imagery data to alter how others will perceive and respond to the imagery (see Table 25.3). But why would an individual or an organization seek to inject deliberate errors into the imagery-interpretation process? And what options exist for detecting deliberate errors in commercial satellite imagery?

The problem of ensuring data integrity is complicated by the fact that data manipulation is an integral part of imagery data processing and interpretation. Processing of raw imagery data is usually necessary to address geometric and radiometric distortions that typically occur when overhead images are collected.[4] Once this data manipulation is accomplished, imagery analysts often use false colors to interpret and present multispectral imagery in a way that highlights phenomena of particular interest (e.g., healthy versus unhealthy vegetation). In addition, the scale of natural or manmade features in three-dimensional representations of imagery data is often distorted for effect. However, these routine methods for manipulating digital imagery data are well established, and most authors presenting such imagery products acknowledge when they are presenting false color images or images with distorted scales.

The same digital qualities that facilitate the user's ability to manipulate the digital imagery data for legitimate analytical purposes also open the door for deceptive practices. One way of deliberately distorting imagery data is to alter the contents of the image. "Pixel plasticity" is intrinsic to the digital nature of most satellite imagery.[5] In this sense, a user group might tamper with the imagery data by deleting or obscuring a portion of the imagery data observation that the user would prefer not to have publicly available. Alternatively, the user could insert nonexistent imagery data into the image or unduly enhance some of the data to mislead others. Of course, even film-based overhead images can be convincingly doctored with enough expertise, but digital imagery offers pixel plasticity with greater ease and without any obvious indications of tampering.

Many motives exist for deliberately distorting imagery data. National governments could distort overhead imagery data to protect state secrets or to gain a diplomatic edge over other countries. Political groups might alter satellite images to embarrass or discredit their domestic political rivals. Finally, corporations and even individuals could have strong economic or legal reasons for deliberately distorting imagery data. And although digital imagery data readily lend themselves to being manipulated and changed, such actions involve a certain risk of detection because the altered imagery product can be compared with an original image at some point to reveal any discrepancies. Nonetheless, actors in desperate circumstances, or those unworried about damaging their reputations, are unlikely to be dissuaded by the chance that others could detect their actions.

Fortunately, several countermeasures exist to discourage actors from generating deliberate errors in commercial satellite imagery. One method for checking the integrity of a controversial image is for skeptics to obtain their own copies of the satellite image directly from the data provider and check for any discrepancies through their own imagery interpretation. This double-checking process could occur unilaterally or through some type of third-party review. In addition, satellite imagery data providers might also incorporate digital "watermarking" or digital signatures, which can be encrypted with a special key, into their raw imagery data.[6] This practice would make it much easier to determine whether satellite imagery data had been tampered with, provided that the digital imagery data were available for review.

Given that tampering with imagery data involves a substantial risk of being detected, most actors will probably be more likely to make deliberate distortions in interpreting overhead imagery data and drawing broader substantive conclusions (see Table 25.3). For example, a user might argue that the observables recorded on a particular commercial satellite image are anything but the sensitive items that others might detect and identify. Alternatively, a user could seize upon ambiguous imagery data to declare with unwarranted certainty that observables associated with a particular object or activity have been found in the imagery data.

Ironically, the imagery data provider—the enterprise that collects and distributes the satellite imagery data to other users—is in the best position to alter the incoming satellite imagery data with the least risk of being detected. It would seem unlikely that a remote sensing organization would intentionally distort its own imagery products, given the risk of diminishing the products' perceived credibility and reliability. However, some satellite imagery providers could receive direct orders or indirect pressure from their national authorities to conceal objects or activities that officials consider too sensitive to be made public. India offers an example of how this "self-censoring" process works. Press reports suggest that India has made available its medium resolution (e.g., about 6-m resolution) IRS-1C and -1D satellite images only after processing out—apparently blacking out—defense installations that Indian government authorities consider to be sensitive from a national defense perspective (Jayaraman, 2000). In this sense, national security considerations are considered paramount over the risks of diminishing the credibility of Indian imagery in the eyes of others. Presumably, any entity engaging in this form of self-censorship will try to keep it from becoming publicly known.

Commercial remote sensing firms could come under similar pressures from outside actors or even self-imposed pressures to manipulate their imagery data. Cases will undoubtedly arise when corporate management is pressured to "sanitize" certain imagery products to accommodate the particular sensitivities of important government or private customers. To some degree, the imaging satellite firms might anticipate and avoid these predicaments by refusing to collect satellite images of sensitive locations. Alternatively, firms could withhold from public access any imagery of sensitive locations that their satellites collect. While these tactics circumvent the imagery credibility problems raised by tampering with their own imagery products, such actions also diminish the principle of "open skies," that is, the internationally accepted basis for operating imaging satellites without regard for national borders. In addition, professional and industry norms, which are antagonistic to data tampering, are likely to play an important role in discouraging the employees of satellite data providers from playing a role in producing misleading or incomplete satellite imagery data.

INADVERTENT IMAGERY DATA ERRORS

Most errors in imagery interpretation will probably be inadvertent rather than deliberate. Inadvertent errors, which are unwanted and not aimed at deceiving others, can occur in handling and interpreting imagery data, as well as in analyzing their broader implications. Table 25.4 provides examples of inadvertent errors and identifies some ways of minimizing the chance of such errors occurring.

Inadvertent errors can arise during the production stage, particularly if an inexperienced data reseller or imagery user is performing the steps involved in preprocessing the imagery data to correct for the systematic distortions that occur in collecting imagery data from space-based platforms. However, most new imagery users are likely to rely on more-experienced data providers or value-added firms to perform the essential production and preprocessing tasks on the satellite imagery data.

Inadvertent errors are quite likely to occur in the image interpretation stage. This phase relies heavily on the more-variable human qualities, including judgment, expertise, and experience of individual imagery analysts.[7] Errors can arise when an imagery analyst fails to detect signatures of important objects that are present in the imagery data or misinterprets the imagery data by concluding that observables for a particular object are present in the image when, in fact, they are not. For example, an imagery analyst might mistakenly identify an isolated industrial or agricultural facility as a secret military installation, or vice versa. The inadvertent error in identifying an unknown facility could then be compounded as the user draws unfounded conclusions in analyzing the imagery data.

Interpreting satellite imagery data is a challenging task that typically requires substantial expertise and resources to ensure the most accurate and complete assessments. These qualities are in short supply for most new imagery users (e.g., news media organizations, and NGOs) relative to the more-established imagery users in government and certain commercial sectors. This does not mean that the latter imagery users do not also make imagery-interpretation errors. The main difference is that they have more opportunities to identify and correct erroneous interpretations because of greater resources, highly trained staffs, and longer experience in working with overhead imagery regarding a particular problem. And because the work of governments and private firms with higher-resolution satellite imagery is seldom publicly released, there is less likelihood that any inadvertent errors will receive that same degree of public attention.

The new users of satellite imagery face certain challenges in making accurate imagery interpretation. Perhaps the most daunting challenge is that they possess limited resources and expertise. Few NGOs and news media organizations have dedicated imagery analysts on their staffs. Instead, they are more likely to rely on individuals who are learning imagery-interpretation skills as they work on specific questions. In addition, these new users might rely on outside consultants or consulting firms with imagery expertise, as well as seek informal feedback from government and industry experts to double-check their work.

Time constraints pose another major challenge for new imagery users. In particular, news media organizations are working under intense deadlines that can increase the risk of inadvertent errors occurring because there is not enough time available for additional review of satellite imagery data. Chapter 23 in this volume discusses how *Newsweek* erred by rushing into print a misleading satellite image that purported to show the specific test site facility for India's nuclear tests in May 1998. However, *Newsweek* later admitted that the satellite image it printed was actually acquired years before the Indian test occurred and that the specified nuclear test location also appeared to have been mistaken.[8] Of course, not all news media organizations are unprepared to make good use of higher-resolution commercial satellite imagery as an important element of their new reporting. For example, the *New York Times* (2000) effectively used a set of "before and after" high-resolution IKONOS satellite images of a specific location in Grozny to highlight the urban devastation that Russian military attacks have inflicted during the campaign against the Chechnya rebels. Similarly, CBS News has carefully created a very structured review process for using satellite imagery that permits opportunities for drawing on highly experienced imagery analysts to reduce the chances of airing a satellite image featuring a major error (Dubno, 2000). No news organization is immune from the time pressures of the 24-hour news cycle, but some are better equipped and organized in advance to avoid inadvertent errors in their public presentations of commercial satellite images.

Dedicating more resources to imagery analysis is one way of improving the capabilities of new users for accurately interpreting satellite images (see Table 25.4). Imagery analysts can use additional satellite images of the same location to detect changes, which provides a stronger historical context for understanding what the analysts are viewing in the imagery data. Similarly, imagery analysts can greatly benefit from having available other source materials (e.g., foreign press reports and eyewitness observations) that offer added insights into what is occurring at the location in the satellite image. Yet, despite their obvious desirability, such measures require the imagery users to expend additional resources or to reallocate the distribution of resources within their organizations.

Table 25.4—Imagery Data and Inadvertent Errors

Imagery data production	• Data provider or reseller makes an error in providing imagery meta-data, georegistration, radiometric corrections, etc. • Imagery reseller or user makes an error during data manipulation processes in preparing for imagery interpretation
Imagery data interpretation	• Imagery user makes imagery interpretation errors: - mistakenly detects objects that do not exist (false alarms) - misidentifies objects as something they are not - overlooks or discounts observables of significant objects present in the imagery data
Analysis based on imagery products	• Imagery user reaches unfounded conclusions based on inaccurate imagery interpretation • Imagery user reaches unfounded conclusions despite accurate imagery interpretation
Potential Improvements	• Use a broad range of sources to develop a better context for interpreting imagery data • Rely on a structured review process for imagery interpretations and analyses • Seek more imagery-interpretation training and/or added staff with greater imagery-handling experience • Acquire multiple images to develop a basis for using change detection techniques

Implications of the Imagery Paradox

One of the ironies of the growing public access to commercial satellite imagery is that imagery interpretations made by new users, who possess relatively limited expertise and experience in dealing with overhead images, will probably receive the most public attention. These new imagery users, mainly news organizations and NGOs, are at the beginning of a steep learning curve for interpreting higher-resolution satellite imagery data. This makes them somewhat more likely to generate inadvertent errors. Nonetheless, these "imagery activists" are also strongly focused on making their image interpretations and analyses widely known to the public through television broadcasts, Web sites, and articles in journals and newspapers. In comparison, the work of imagery analysts in government, industry, and university remote sensing departments is likely to receive much less public attention.

A key question, therefore, is what type of relationship will develop between the traditional imagery experts and the growing number of new users of satellite imagery? For purposes of analysis, three model relationships are postulated: mutual indifference, active rivalry, and limited cooperation.

MODEL I: MUTUAL INDIFFERENCE

This model is probably closest to current situation between the new "imagery activists" and the much-more-experienced imagery analysts working in government agencies or commercial firms. In this postulated relationship, each side is generally unfamiliar with the other's work or attempts to ignore it.

Some basic differences between the two types of imagery users, including their distinctive organizational practices and cultures, largely account for this reluctance to engage. For example, a fundamental difference exists in how each side handles imagery. Government and industry analysts treat higher-resolution satellite imagery data as either classified data or proprietary information that is seldom made public. In comparison, most NGOs and news media organizations are dedicated to sharing and publicizing imagery data. Certainly, the classified and proprietary nature of government and industry uses of satellite imagery is a major, but not insurmountable, hurdle to a more-active dialogue. However, cultural and resource disparities add to the gap between the two groups. New imagery users are unlikely to be found at many of the professional development activities associated with the highly trained remote sensing experts because they either lack the funds to cover their participation or because they see little direct relevance to their specific uses of imagery data.

In this model, outreach efforts by the traditional remote sensing community are greeted with indifference or suspicion by the new imagery activists.

Nonetheless, attempts to ignore each other are probably futile, at least in the long run. Imagery interpretation of higher-resolution images by NGOs and the news media is likely to attract considerable attention in some cases and is likely to compel government officials to assess the validity of the activists' imagery work even if they do not publicly respond. Furthermore, as the Kosovo conflict suggests, news media organizations will solicit the independent opinions of the new imagery analysts, who are relatively accessible and willing to go on the record, whenever governments release overhead imagery.

MODEL II: ACTIVE RIVALRY

Things could take a turn for the worse if both sides seek to discredit each other. Conditions exist to encourage the imagery activists to become active rivals with their counterparts in government and industry. This type of negative relationship could arise if both sides take to criticizing the competency of each other's imagery interpretations, at least on highly controversial policy issues.

In Model II, many NGOs and news media organizations will use their own interpretations of higher-resolution satellite imagery to challenge official assessments. These disagreements can take the form of challenging the basis for a specific imagery interpretation or questioning the broader conclusions drawn from the imagery data. Thus, a NGO might acquire and interpret imagery of a particular location that leads it to conclude that the government's imagery analysts misinterpreted the objects found in the imagery data or that government officials drew unwarranted conclusions from the imagery interpretation. For example, the nongovernmental analysts at the FAS have used their own interpretation and analysis of satellite imagery as part of their "Public Eye" program to question the validity of U.S. government intelligence assessments of the North Korean missile program and the severity of the threat that Chinese air power poses to Taiwan (Broad, 2000; King, 2000).

Alternatively, government or industry experts could begin attacking the accuracy of the satellite imagery interpretations of the new imagery activists. Such criticism could take the form of public critiques. More likely, however, would be sharply critical comments leaked to trade journals and newspapers that are attributed to unnamed government officials or stated more openly by industry and former government imagery experts (Eddington, 2000). In this model of active rivalry, any inadvertent imagery-interpretation errors become a target of opportunity for critics to challenge the credibility and competence of the new imagery users.

Unfortunately, a competitive approach to imagery interpretation will probably hurt all parties by eroding public confidence in the reliability of satellite imagery data as a source of information. Drawing the line between legitimate constructive criticism and attacks intended to diminish the reputation of the other party will be difficult at best. Imagery credibility problems that arise will not necessarily be associated in the public's mind with one group or the other. Rather, the public and potential imagery data users could come to doubt the perceived competency of both new and traditional users of satellite imagery in making accurate imagery interpretation for various applications. In addition, foreign governments or groups that are the subject of negative assessments of their activities partly based on overhead imagery could also take advantage of this situation to challenge the credibility of satellite imagery interpretations in their cases.

MODEL III: LIMITED COOPERATION

Both the new imagery users and the more-experienced imagery analysts in government and industry could gain much from developing a cooperative relationship. Their very different purposes and backgrounds work against these disparate groups having a close working relationship. However, they do possess overlapping interests that provide a basis for limited forms of cooperation and even occasional collaboration in producing accurate imagery interpretations. In Model III, each of the primary parties (i.e., new imagery users, government agencies, commercial imagery firms, and professional associations) is willing to engage in limited cooperation to achieve its distinctive aims.

New Imagery Users. For NGOs and news media organizations, the main benefit to limited cooperation is progressing more rapidly up the learning curve in making accurate and complete imagery interpretations. These groups also have a practical interest in learning how to acquire and handle imagery data with greater economy and sophistication. Insights from imagery analysts in industry and government could strengthen the skills of the imagery activists by exposing them to proven image-interpretation techniques and best practices in working with imagery data. Another important aim would be to gain a better appreciation of imagery observables for relevant objects, to help reduce the chances of making inadvertent errors. Experienced imagery analysts could also convey their sense of the limits on drawing firm conclusions based on imagery interpretations, particularly when the overhead imagery data are limited or somewhat ambiguous.

In this limited cooperation model, new users of satellite imagery could receive another important benefit concerning their need for credibility in making use of imagery. In a broader sense, Keohane and Nye (1998, p. 94) have highlighted the importance of developing a credible reputation for providing accurate information against the challenge of an overabundance of information, observing that

> If governments or NGOs are to take advantage of the information revolution, they will have to establish reputations for credibility amid the white noise of the information revolution.

Thus, the imagery credibility of NGOs and news media organizations largely depends on their reputation for producing accurate information. In some cases, governments will be able to exert a major influence by validating or challenging the satellite image interpretations offered by the new imagery activists.

Table 25.5 outlines a range of options for improving the imagery credibility of the new imagery users. It identifies various ways that other key actors (i.e., governments, industry, and professional associations) can encourage

Table 25.5—Methods for Enhancing Commercial Satellite Imagery Credibility

New imagery users	• Allocate more resources to imagery interpretation training • Develop a structured review process that involves outside experts • Share imagery expertise and experience with other organizations • Become better organized for interacting with the commercial imagery providers on imagery needs, costs, and licensing • Publicly disclose imagery-interpretation errors and implement improvement measures
Commercial firms	• Develop training courses and material tailored to new users • Offer discounts on imagery data and image processing software • Provide the option of imagery-interpretation review for own images • Formulate imagery-use licensing arrangements that remove obstacles to broader sharing of imagery data among NGOs
Government agencies	• Declassify documents or create public-use material on basic imagery-interpretation techniques and analysis skill • Develop ground rules that would permit experienced government imagery analysts to interact with NGOs and news media users • Create mechanisms for evaluating public uses of satellite imagery data and for providing constructive feedback to new users
Professional associations	• Develop workshops and conference panels focused more on the interests and training needs of new imagery users • Play a facilitating role in matching the imagery interpretation needs of new users with more experienced users • Promote agreed standards for using commercial satellite imagery and mechanisms for identifying and correcting imagery errors

these new imagery users to improve their capabilities for interpreting overhead images and analyzing the imagery data.

Efforts to strengthen the imagery credibility of new users must begin with "self-help" measures. New users can bolster their expertise in imagery interpretation through more training for existing staff members or by hiring imagery analysts with greater experience with higher-resolution satellite imagery. However, NGOs and news organizations will probably be reluctant to expend their limited resources for satellite imagery purposes, which they largely view as an ancillary tool in their work. Thus, these new imagery users must make concerted efforts to convince their senior management and funding sources, including foundations, that effective use of satellite imagery for their purposes requires sufficient in-house capabilities to avoid inadvertent errors that could harm their organizations' overall credibility.

New users should adopt procedures for strengthening their imagery data interpretation process. One potential improvement is to implement a structured process for reviewing their imagery interpretations and analyses. Although this review process would mainly operate internally, it might also involve some outside reviewers in some cases. Indeed, one of the best ways of enhancing the imagery credibility of the NGOs and news organizations is through sharing imagery expertise and experience with other organizations. This is most likely to occur among like-minded NGOs, as well as between NGOs and individual news organizations. Finally, publicly admitting errors in satellite imagery interpretations would also contribute to the long-term credibility of new imagery users while encouraging them to adopt quality assurance measures.

Industry. Private remote sensing firms, such as satellite image providers or value-added resellers of imagery data, can view the emerging imagery activists as important new customers for commercial imagery products and services. However, given tight budgets and competing needs, the demand for satellite images from NGOs and news media organizations will be smaller than that from other commercial users of imagery data.

Nonetheless, industry still has an important stake in developing closer working relationships with the new imagery users. One compelling reason is the worldwide publicity that these imagery activists will attract through their work with satellite imagery. The visibility of news media organizations and NGOs using imagery will also benefit the commercial sector by dispelling the notion that higher-resolution satellite imagery is mainly of interest to countries for spying on their rivals or to private firms seeking to make a profit. Instead, by demonstrating their industry's readiness to work with various types of imagery activists, the commercial firms can build a reputation as an impartial source that provides overhead imagery to a broad range of potential users.

Commercial firms therefore need to undertake "outreach" efforts focused on these new imagery activists, who need more knowledge on how to acquire the most relevant imagery data and how best to interpret and analyze the resulting imagery data. The data providers have a strong interest in having their imagery products presented to the public without errors. Hence, they have a stake in helping the new imagery users avoid basic mistakes. Providers could even offer to review, on a confidential basis, the imagery interpretations that news media organizations and NGOs make of the providers' satellite images, to help catch any inadvertent errors or overlooked items before the images are publicly released.

Both the satellite imagery data providers and value-added firms need to offer training courses that are more tailored to the distinctive interests, experience levels, and limited resources of the new imagery users. Several commercial software firms, such as ERDAS and the Environmental Systems Research Institute, have sold their image processing and GIS products to NGOs at discounted prices. The satellite imagery data providers should consider adopting similar practices. However, NGOs and other nonprofit imagery users also need to speak with a more-coherent voice in dealing with the imagery data providers and software producers. The commercial firms would benefit from receiving a single, clear message from the new users on salient issues, such as developing licensing arrangements, rather than through piecemeal interactions. This would help both sides reach a mutually satisfactory accommodation concerning the interests of these new users in broadly sharing imagery data and interpretations. Developing licensing arrangements for sharing images that are more compatible with the public orientation of these new users could serve the larger public good while developing an appreciation on the part of the new imagery activists of the importance of the intellectual property rights of the satellite imagery data providers.

Governments. Governments are uniquely positioned to assist the new imagery users become more proficient in using satellite imagery data. Several NGOs and news media organizations are interested in using overhead imagery to highlight foreign weapon developments or humanitarian interventions, such as Kosovo. The United States and a few other governments have a cadre of intelligence and military analysts with extensive experience in interpreting and analyzing overhead imagery on such subjects. Hence, the new imagery activists have much to gain from greater interaction with their more-experienced government counterparts. But why would national governments want to help NGOs and news media organizations, which are potential challengers of official policies, to become more adept at using imagery interpretations and analysis?

Model III recognizes that governments possess an understandable reluctance to engage in substantive discussions with imagery analysts working with NGOs and news media organizations. However, it also postulates that top government officials judge that, on balance, they can realize some important benefits from limited cooperation with imagery activists to help them move up the learning curve for imagery interpretation and analysis. One motive is simply to reduce the nuisance factor for government agencies from having to address public claims based on image interpretation errors that NGOs and news media organizations have made. If inadvertent imagery errors become commonplace, the regular work of government imagery analysts can be significantly diverted to disproving public allegations based on questionable imagery interpretations.

A second, and more-compelling reason, is that no public or private organization possesses a monopoly on data sources, information products, or thoughtful analysis of imagery data in this new information age. Nongovernmental actors often have some unique sources of information, including the ability to acquire potentially important information by directly approaching individuals, research institutes, and news media sources in other countries. Thus, NGOs and news media organizations will occasionally acquire timely imagery data or produce analyses that could be invaluable for government experts, particularly in the context of largely unanticipated events, such as the Indian nuclear tests in mid-1998. One example is a Norwegian institute's analysis of Russia's Novaya Zemlya nuclear test site using a combination of sources, including satellite images and interviews with Russian experts, which has provided a valuable source of information for European imagery experts (see Skorve and Skogan, 1992). Another is the detailed analysis by David Albright and Corey Gay Hinderstein that precisely located the Pakistan nuclear test site (see Chapter 17 in this volume), which drew on Pakistani source material as well as on commercial satellite imagery.

A third reason is that occasions will arise when national governments look to NGOs and other independent sources to help validate their public arguments.

Governments sometimes publicly release satellite or aerial images to provide evidence that a particular country is engaging in unacceptable behavior. For example, in September 1999, the U.S. Department of State released a public report criticizing Saddam Hussein for misusing Iraqi resources to build a large vacation resort for government officials rather than to provide basic human services, as well as undertaking attacks on Kurdish and Shi'a population centers. This report presented overhead images as evidence to support the U.S. government charges (U.S. Department of State, 1999). Similarly, during the Kosovo campaign NATO released overhead images that purported to show civilian massacres (see Chapter 23 in this volume).

Finally, unexpected events, such as the Chernobyl nuclear reactor accident, could also put governments in a position where the judgment of outside imagery experts could be helpful. The credibility of this approach could be strengthened if news media organizations and knowledgeable NGOs could independently confirm these charges through their own interpretation and analysis of overhead images.

Thus, governments should consider several steps that would help new imagery users become more proficient with imagery interpretation and analysis. Table 25.5 outlines several forms of more-productive interaction between government and nongovernment imagery users. Governments can begin by making more documents on basic imagery interpretation and analysis publicly available. This would involve either declassifying certain material or creating unclassified versions for public release. More specifically, the United States could release a broader range of older photographic intelligence reports, such as those that were made available in limited numbers as part of the declassification of the CORONA imaging satellite program in 1995 (see Ruffner, 1995, Part III). With careful attention, basic information on imagery interpretation and analysis can be released without compromising the more-sophisticated and highly specialized sources and methods that government imagery experts need to protect.

Governments should also permit more regular and open engagements between their imagery experts and nongovernment imagery users. Model III postulates that governments encourage a continuing dialogue on basic imagery-interpretation techniques (see Table 25.2). In the case of the United States, the government might develop a set of ground rules that could facilitate the interaction between government imagery analysts and their less-experienced counterparts at NGOs and news media organizations. These ground rules would signal the U.S. government's interest in initiating dialogue on the basics of imagery interpretation and analysis while being careful not to expose sensitive sources and methods used by government intelligence analysts. Finally, governments should also consider establishing mechanisms that would enable them to offer constructive critiques on how well the new imagery users are interpreting the new higher-resolution satellite imagery.

PROFESSIONAL ASSOCIATIONS AND TRAINING CENTERS

Professional associations, such as the American Society for Photogrammetry and Remote Sensing (ASPRS) and its international counterpart, the International Society for Photogrammetry and Remote Sensing (ISPRS), have a particularly important role to play in facilitating the potential interactions between the new imagery activists and the more-experienced imagery analysts in government and industry. These organizations that support the remote sensing profession should initiate outreach efforts aimed at encouraging the new imagery users to become more involved in workshops and conferences that could contribute to their professional development in imagery interpretation and analysis. Similarly, multinational training centers, such as the International Institute for Aerospace Survey and Earth Sciences, located in The Netherlands, can also play a vital role by offering short courses or fellowships aimed at enhancing the imagery-handling expertise of the imagery activists.[10]

Another potentially important role for these remote sensing associations is to promote professional standards in using commercial satellite imagery. Associations can sponsor various activities that expose new imagery users to the best practices concerning imagery interpretation and analysis, as well as encourage new users to adopt high professional standards in their use and presentation of imagery data.

These associations should also take a proactive role in addressing controversial or questionable cases of overhead imagery interpretation that are brought to their attention. One method for reducing inadvertent errors in imagery interpretation would be to give public recognition each year to the best examples of imagery interpretation and analysis using commercial or civilian satellite imagery. At the same time, the associations should be willing to identify and publicly note instances of government or nongovernment imagery users relying on erroneous or highly questionable methods in interpreting and analyzing satellite imagery that was publicly released. This emphasis on maintaining a high degree of commitment to professional standards in imagery interpretation could provide new imagery users with an added incentive to acquire the training and capabilities needed for reducing inadvertent imagery errors.

Conclusions

The advent of higher-resolution commercial observation satellites is creating an unprecedented situation in the satellite remote sensing field. Consistent with the basic nature of the information age, the use of satellite imagery data is rapidly growing beyond the smaller community of highly trained remote sensing experts. Higher resolution commercial satellite imagery has a strong appeal to a broad range of new imagery users. Among the most prominent of these new imagery users are news media organizations and NGOs. Neither group has yet acquired much experience in interpreting and analyzing higher-resolution satellite images. Nonetheless, these imagery activists are dedicated to using satellite imagery data in highly visible ways to address public policy issues. Thus, an undesirable imagery-credibility paradox is created by the major disparity that exists in the levels of imagery expertise of the new imagery activists compared with the more-experienced imagery analysts found in government and industry.

This new environment poses certain challenges to the credibility of commercial satellite imagery arising from the possibility of deliberate and inadvertent imagery data errors. Although deliberate imagery errors are an important matter of concern, a variety of technical and procedures countermeasures exist that provide a reasonable chance of exposing such deceptions in the most important instances. The most problematic case, however, is if the satellite data provider is the source of the data tampering. Despite the risk to their own reputation, the commercial and civilian organizations that collect and distribute satellite imagery could face strong pressures from governments, clients, or their own management to alter higher-resolution imagery data to avoid revealing sensitive locations or activities. The best safeguard against such pressures is to encourage a strong professional commitment against any form of imagery data tampering.

The more likely source of imagery errors is the limited expertise, resources, and experience of many of the new imagery users. Inadvertent errors will probably occur in interpreting higher-resolution imagery data and analyzing the broader implications of what is revealed by the imagery interpretation. Although the traditional imagery users and the new imagery activists could attempt to ignore or even discredit each other's work, the result would be to diminish the overall utility and credibility of commercial satellite imagery as mistakes in imagery interpretation become commonplace. A better alternative is available: limited cooperation between the new imagery activists and the more-experienced imagery analysts in government and industry. Several options exist for providing the new imagery users with ways of moving rapidly up the learning curve in making accurate imagery interpretations. A special role exists for professional associations in the remote sensing field, such as ASPRS and ISPRS, for reducing the hurdles to productive engagements between the traditional and new imagery users.

The expectations for productive interactions between the imagery activists and the more-experienced remote sensing experts must be rooted in realistic expectations given that they approach the use of imagery with significantly different backgrounds and even competing aims at times. Nonetheless, all sides have a common interest in narrowing the knowledge gap between the new and the more-experienced imagery users in producing credible imagery interpretations and analyses. Any debates are better focused on the broader policy implications of imagery interpretations than on whether or not an accurate imagery interpretation has been made. There is no guarantee that adopting the constructive measures discussed in this chapter for enhancing the imagery-interpretation capabilities of new users will preclude imagery errors from occurring. However, unless each party adopts a proactive response to the new challenges created by commercial observation satellites, both inadvertent and deliberate imagery errors are likely to increase over time. Left unchecked, this trend could diminish the potential of satellite imagery to continue as a highly credible and reliable information source for a broad range of potential users.

Acknowledgments

The author is grateful for the thoughtful suggestions and advice on this chapter that he received from the following individuals: Kevin O'Connell; Richard Leghorn; Mark Gabriele; Major Charles Gonty, USAF; Professor John Jensen; Professor Mark Marshall; and Frank Pabian.

Notes

1. The FAS experts' imagery interpretation and analysis led them to conclude that the relatively crude nature of the Nodong missile launch infrastructure raised questions on whether the North Koreans were pursuing the necessary program "to develop a fully reliable missile system." (See Broad, 2000.)
2. *Picture elements* (pixels) are the smallest individual elements of an image that can be used to differentiate distinct objects or patterns. Many of the fundamental works on remote sensing outline the key image elements that are relevant to making accurate photographic interpretations (see Colwell, 1983, pp. 993–995;

Philipson, 1997, pp. 50–58; Lillesand and Kiefer, 2000, pp. 192–195; Jensen, 2000, p. 544; and Campell, 1996, pp. 122–130).

3. For a survey of basic technologies associated with photointerpretation and digital imagery interpretation, see Campell (1996), pp. 135–141, 148, and Philipson (1997), pp. 88–101.

4. Geometric corrections are needed to compensate for distortions that typically occur in satellite imagery because of unintended variations in the altitude, attitude, and velocity of the imaging sensor platform. In comparison, radiometric corrections are required because of variations in the radiance measured for a given object that result from changes in scene illumination, sensor viewing geometry, and atmospheric conditions. (Lillesand and Kiefer, 2000, pp. 471–482.)

5. The author is indebted to Steven Livingston for highlighting this possibility (Livingston, 1999). See also Chapter 23 in this volume.

6. On the utility of watermarking for data validation, see Garfinkel (2000), pp. 199–204.

7. For an analysis of the complex processes associated with human perception and imagery interpretation, see Marshall (1999) and Hanson (1999).

8. The satellite image appeared in *Newsweek* (1998a). In a rather terse correction about three months later, *Newsweek* (1998b, p. 13) noted that the image "was not recorded the week before the blasts, but seven years ago."

9. For example, one former U.S. intelligence community imagery analyst has criticized the FAS analysts for misleading the media and the public by engaging in amateur intelligence analysis and for their lack of formal training as imagery analysts (see Eddington, 2000).

10. This international center, popularly known as ITC, is dedicated to training imagery analysts as part of its education and research mission for promoting geoinformation knowledge and geoinformation management tools (see http://www.itc.nl/).

References

Broad, William J., "Spy Photos of Korea Missile Site Bring Dispute," *New York Times*, January 11, 2000, p. A8.

Bruginoni, Dino, "The Art and Science of Photoreconnaissance," *Scientific American*, Vol. 274, No. 3, March 1996, pp. 78–85.

Campell, James B., *Introduction to Remote Sensing*, New York: The Guilford Press, 1996.

Colwell, Robert N., ed., *Manual of Remote Sensing*, Vol. I, Falls Church, Va.: American Society of Photogrammetry, 1983.

Dubno, Dan, producer on science and technology issues at CBS News, personal communication, January 3, 2000.

Eddington, Pat, "Orbital Snooping: Welcome to Amateur Hour," *Space News*, May 22, 2000, p. 15.

Garfinkel, Simson, *Database Nation: The Death of Privacy in the 21st Century*, Sebastopol, Calif.: O'Reilly & Associates, Inc., 2000.

Haakon, Christopher P., "Commercial Space Imagery for National Defense," *Defense Intelligence Journal*, 8, Summer 1999, pp. 24–32.

Hanson, Steven M., "Results of an Experiment Comparing the Spatial Ability of Imagery Analysts and Non-Imagery Analysts," *Defense Intelligence Journal*, Vol. 8, Summer 1999), pp. 120–134.

Jayaraman, K. S., "Imagery Deal Spurs Indian Policy Review," *Space News*, January 17, 2000, p. 3.

Jensen, John R., *Remote Sensing of the Environment: An Earth Resource Perspective*, Saddle River, N.J.: Prentice-Hall, Inc., 2000.

Keohane, Robert O., and Joseph S. Nye, Jr., "Power and Interdependence in the Information Age," *Foreign Affairs*, Vol. 77, No. 5, September–October 1998.

King, Neil, Jr., "Activists Use Satellite to Challenge View That China's Air Force Menaces Taiwan," *Wall Street Journal*, May 12, 2000.

Lillesand, Thomas M., and Ralph W. Kiefer, *Remote Sensing and Image Interpretation*, New York: John Wiley & Sons, Inc., 2000.

Livingston, Steven, presentation at a conference sponsored by the Carnegie Endowment for International Peace, "No More Secrets: Policy Implications of Commercial Remote Sensing Satellites," Washington, D.C., May 26, 1999.

Marshall, Mark G., "Intelligence," *Defense Intelligence Journal*, Vol. 8, Summer 1999, pp. 93–119.

Newsweek, "Ground Zero", May 25, 1998a, pp. 28–29.

---, "Letters," August 31b, 1998, p. 13.

New York Times, "Week in Review," March 26, 2000.

Philipson, Warren R., ed., *Manual of Photographic Interpretation*, Bethesda, Md.: American Society for Photogrammetry and Remote Sensing, 1997.

Ruffner, Kevin C., *CORONA: America's First Satellite Program*, Washington, D.C.: Center for the Study of Intelligence and the U.S. Central Intelligence Agency, 1995.

Skorve, Johnny, and John Kristen Skogan, *The NUPI Satellite Study of the Northern Underground Test Area on Novaya Zemlya*, No. 164, Oslo: Norwegian Institute of International Affairs, December 1992.

U.S. Department of State, "Saddam Hussein's Iraq," report released on September 13, 1999. Available at http://usinfo.state.gov/regional/nea/iraq/iraq99.htm (last accessed June 9, 2000).

26
Conclusions

John C. Baker, Kevin M. O'Connell, and Ray A. Williamson

High-resolution earth-observation satellites were once accessible only to highly trained specialists working at a few government agencies, corporations, and university remote sensing centers in a handful of countries. Now, however, commercial satellite imagery is becoming increasingly available to a wide range of international actors. The result is a dramatic boost to global transparency: Almost any state, corporation, nongovernmental organization (NGO), or even individual has potential access to overhead imagery or to geospatial data and information derived from it.

As the previous chapters illustrate, the new generation of commercial and civilian observation satellite systems is fundamentally transforming the remote sensing field from a primarily science-based enterprise to one that is much more focused on specific civil and commercial applications. This transition has been made possible not only by the development of high-resolution earth-observation satellites but also by the concomitant, independent introduction of high-capacity, low-cost information tools into the global marketplace for geospatial information.

What should we make of this new information technology, which offers an unprecedented ability to observe human activities and environmental changes occurring most anywhere on earth? The book's contributors offer various interpretations of what the rapid growth in global access to commercial satellite imagery portends for international security, national space policies and programs, commercial markets in geospatial information products, and the political relationship between governments and nonstate actors. The discussion has several broad themes, as discussed below.

An Emerging Information Technology

Remarkable recent advances in satellite-based sensors and other technologies are making possible the development of cheaper, lighter, and more capable satellites than ever before. Yet such advances would have little effect in the absence of the even-more-rapid advances in and reduced costs of telecommunications, computer hardware, software, and other information technologies. These improvements have established the technological foundation for putting the data gathered by satellite to use in myriad new commercial, civil, and security-related applications.

The new generation of commercial observation satellites is notably different from the earlier Landsat and Satellite Pour l'Observation de la Terre (SPOT) satellites that pioneered civilian remote sensing. Improvements brought about by cutting-edge technologies and new manufacturing processes have greatly increased the opportunities for various countries, commercial firms, and NGOs, such as universities, to acquire imaging satellite systems. During the first decade of the new millennium, more than three dozen countries and private firms are expected to exploit the technological advances to build or buy their own earth-observation satellite systems.[1] Many of these will be relatively small imaging systems, fully or partly built by foreign satellite manufacturers, as countries develop their indigenous capabilities for producing and operating imaging satellite systems.

The growing number of commercial and civilian observation satellites is matched by a steady advance in imaging capabilities. This book highlights the importance of satellite sensors capable of collecting data at about 1-m resolution—a major improvement over the 10- to 30-m resolutions that earlier satellites could offer. Using 1-m resolution imagery, even inexperienced analysts can detect significant patterns and objects, including cars, trucks, airplanes, and ships or, alternatively, military vehicles and troop deployments. In addition, the color or multispectral imagery these satellites offer (generally at lower spatial resolutions) can provide analysts with useful information in the visible and infrared regions of the electromagnetic spectrum, useful for monitoring vegetation changes and other developments. Thus, advances in satellite imagery open the door to a range of new imagery users at government agencies, firms, and NGOs.

Users can also expect to take advantage of different imaging technologies as they become more readily available for civil and commercial applications. This book explores two emerging technologies, radar and hyperspectral imagery, which have stimulated interest among users of overhead data. Compared with the existing panchromatic and multispectral images, the newer imaging sensors will offer novel capabilities for determining certain physical characteristics remotely. Radar satellites are a mature technology that, unlike the electro-optical imaging satellites, can collect images despite cloud cover or night. Hyperspectral sensors, which are now being developed for satellites, offer a potentially powerful means of finely discriminating among different classes of objects based on their unique spectral signatures.

Along with better resolution, many new users of civilian and commercial observation satellites attach great importance to timeliness in collecting and distributing the data. More satellites and faster communication systems substantially improve the opportunities for acquiring timely data. High-bandwidth telecommunications networks have dramatically decreased the time required to deliver data to the customer and have made possible the delivery of very large digital image files. Finally, the availability of precise positioning signals from the Global Positioning System satellites substantially improves the ability to locate the satellite imagery accurately and in a timely fashion, thus enhancing its utility.

As important as these developments are to the marketplace for earth-observation data, improved satellite technologies could offer few gains for imagery users without corresponding advances in ground-based enabling technologies and the specific applications that address the needs of diverse users. Such technologies include affordable computing power, improved data-display systems, and data-storage equipment capable of handling large digital imagery files. User-friendly software, some of it originally developed for the entertainment industry, allows image processing and data displays to run on desktop computers or more capable workstations. Finally, painstaking research and field testing have resulted in the development of applications for translating data into accurate information products that are relevant to a broad range of commercial and civil users. That process has been greatly bolstered by the rapid growth in geographical information systems that merge multiple layers of geospatial and socioeconomic data to serve practical information needs. Furthermore, other information technologies, such as the Internet and CD-ROM storage devices, are making satellite imagery data available for the first time to the broader public through exposure to remote sensing and geospatial information in classrooms and workplaces, as well as through multimedia connections to homes. Nevertheless, commercial applications of earth-observation imagery still face certain barriers, such as the needs for higher levels of technical expertise to ensure credibility and for the processing capacity to handle large data sets rapidly.

At the Leading Edge of Global Transparency

One of the primary effects of the growing proliferation of earth-observation satellites of all types is to increase the ability of individuals and organizations in any country to follow certain natural and anthropogenic trends throughout most of the world,[2] whether or not they have physical access to the region. Thus, for the first time, knowledge of many human activities has become widely accessible, or "transparent" to the outside world, even in the absence of access to highly classified national data. Taken by themselves, the data have a relatively small effect on transparency, but when combined with other powerful information technologies, such as the Internet, video, the Global Positioning System, cell phones, and satellite communications, earth-observation satellites ensure worldwide transparency by extending public access to timely data concerning remote or "denied" areas.

These developments already contribute to a more accurate and timely sense of the consequences of natural disasters, such as earthquakes, typhoons, floods, and volcanic eruptions in distant lands. Similarly, high-resolution imagery now provides a means of circumventing denied areas created by closed societies, thus making it possible to pinpoint and observe sites of worrisome military developments, such as nuclear bomb tests or missile deployments. As imagery users become more familiar with these data and how to apply them, global transparency will necessarily increase. Some, including several contributors to this book, applaud these developments, believing that they will lead to a much safer world because the information will help reduce uncertainties and miscalculations. Others, however, raise concerns that, in some cases, increasing global transparency will reveal military vulnerabilities or highlight national environmental degradation that could ultimately promote interstate or intrastate conflict. Certainly, government officials used to exerting control over sensitive internal information will worry about the challenges that could arise from growing public and international access to high-resolution commercial satellite imagery.

In part, the media attention paid to the high-resolution commercial systems reflects these worries and the relative novelty of broad access to these systems on the international scene. Hence, these new systems have become the subject of media attention in audio, video, and print articles focused on their qualities as "spy satellites." Yet this narrow focus overlooks another important element of transparency: the enormous day-to-day utility of data from these satellites for urban planning, for disaster management, and for identifying and following detrimental environmental trends on a variety of geographical scales. In our view, as governments, organizations, and individuals gain experience with the use of these data, most national governments will become more accustomed to the increased transparency that information technologies encourage. Thus, they will necessarily develop policies and practices better suited to operating in a more open environment, such as has been done with the Internet and other information technologies.

Transparency also implies that a wider selection of actors than ever before can use the data these satellites provide. Indeed, the media and NGOs with agendas ranging from environmental protection to human rights and greater political openness have begun to use the data to analyze regional developments and to pressure governments for change. On the worrisome side, rogue states or terrorists can also take advantage of earth-observation data to target their adversaries. Greater transparency may contribute to greater risk in certain regions, particularly where a major asymmetry might exist in access to information sources. On the whole, however, the same transparency that may be exploited by malefactors may also be used against them by revealing their intentions and documenting their actions.

The Role of Nonstate Actors

One of the interesting political characteristics of the late 20th century was the development of a broad variety of nonstate actors, supporting myriad different political, economic, and environmental agendas. The public and, to a great extent, governments have come to rely on a variety of NGOs and the media for independent sources of information on important public-policy issues. Access to new types of commercial and civilian observation satellites will potentially strengthen the influence that such nonstate actors as NGOs, international agencies, and news media organizations have on the policymaking process.

Some of these groups, especially environmental and certain political NGOs, welcome the additional leverage that the information derived from earth-observation data can provide. Environmental NGOs, such as The Nature Conservancy and the World Wildlife Fund, are already making extensive use of such data to support their analysis of environmental change. Arms control and nonproliferation NGOs, such as the Institute for Science and International Security, have used satellite imagery and other data sources to monitor nuclear activities of international concern. Other NGOs, such as humanitarian groups that focus on responding to natural and human-caused disasters, have been slower to adopt the technology, though high-resolution satellite imagery could be used to create or update maps and carry out timely analysis of the effects of regional natural disasters. In all cases, however, high-resolution commercial satellite imagery will enable these nonstate actors to play a more important role by providing them with an independent source of information that they can use to support their activities and, on occasion, to challenge government assessments with their own interpretation of events.

The advent of commercial, high-resolution imagery is likely to spur development of what we might call "imagery activists," individuals and organizations, such as the Federation of American Scientists, which use earth-observation data to pierce governmental secrecy. Until the launch of Space Imaging's IKONOS satellite, these imagery activists had only moderate-resolution (6 to 10 m) imagery to work with. Nevertheless, much can be discovered with such data. As data from high-resolution satellites become more widely available, we can expect a greater number and variety of such imagery activists. However, as many of our authors have discovered, accurate interpretation of satellite imagery requires considerable expertise, experience, and resources. This raises the possible problem that new imagery activists, attracted by the apparent ease of analyzing high-resolution imagery, will rush to the media with inaccurate or incomplete imagery interpretations that will confuse, rather than illuminate, public debate.

Despite the potential utility of earth-observation data for a variety of applications, nonstate actors will nevertheless face several impediments to making effective use of such data:

- Costs of data, hardware, and software. Despite the broader availability of data and rapidly falling costs for computer hardware and analytic software, the costs of purchasing data and analyzing data will place satellite imagery out of the range of the smaller and less—well-financed NGOs.

- Costs of training. Acquiring the skills necessary to carry out accurate imagery interpretation and data analysis requires considerable training, increasing the startup costs to organizations.
- Lack of knowledge about earth-observation data and information. The lack of knowledge about the data and methods among the managers who must make funding and staffing decisions is a subtler barrier. Until such individuals gain sufficient understanding of the benefits and drawbacks of using such data to achieve the goals of their organizations, they are unlikely to allocate the funds necessary to acquire the needed expertise, imagery data, and equipment to ensure that their organization is making effective and accurate use of imagery data.

The ability of new imagery users to use commercial satellite imagery to address important domestic and international problems is likely to erode over time unless users can sustain their own credibility in making use of imagery data. This situation suggests that government, industry, and professional associations need to help new users acquire the necessary skills to make the best and most effective use of such data. Each party has a stake in ensuring that satellite imagery continues to be perceived as a highly credible and reliable information source for a broad range of uses.

Dual-Purpose Implications of High-resolution Imagery

Policymakers must consider that commercial earth-observation satellites are a dual-purpose technology that can be used both for commercial and for military applications.[3] Although data from these satellites offer another means for states to enhance their security through greater transparency, such use could also create or exacerbate security problems. Thus, policymakers must be concerned with identifying conditions under which commercial satellite imagery could pose a significant threat and with identifying options to limit the risks that overhead imagery poses for states or even groups within states.

The new commercial observation satellites blur the traditional distinction between civilian remote sensing satellites and military reconnaissance satellites. Although data from earlier Landsat and SPOT civilian observation satellites could be used to detect very large objects, such as population centers and airfields, data from the most-recent commercial satellites can reveal objects as small as cars, trucks, and private aircraft—or tanks, artillery, and combat aircraft. Similarly, commercial radar satellites will offer the all-weather imaging capability that military planners find especially useful, and the new hyperspectral sensors promise to make it possible to distinguish camouflage from natural vegetation. Further, designers of commercial earth-observation satellite systems are attempting to improve the timely delivery of data.

The new earth-observation satellites have both positive and negative implications for international security. On the positive side, higher-resolution commercial satellites offer new instruments for encouraging security cooperation among wary states. This book includes several examples of how satellite imagery has been or could be used to reduce the risk of conflict, benefits that are often unappreciated by policymakers. Diplomats have made good use of civilian and commercial satellite imagery to help resolve difficult territorial issues during peace negotiations over Bosnia and the Ecuador-Peru border conflict. High-resolution commercial imagery could play a similarly important role in supporting a regional transparency regime to mitigate the persistent territorial disputes in the South China Sea. Finally, access to more-detailed overhead information provides NGOs with an unprecedented ability to monitor activities occurring within countries suspected of nuclear proliferation and to undertake more effective humanitarian relief operations.

Nevertheless, high-resolution commercial satellite imagery can potentially pose a security concern for regional rivals or for the United States and its allies in coalition actions. Rogue states might attempt to exploit this dual-purpose technology to gain an information edge over regional rivals. High-resolution imagery could provide additional information about an adversary's force deployments and could even help identify specific tactical or strategic targets. Countries with the expertise to combine high-resolution commercial satellite imagery with precise geospatial information could gain an advantage in targeting precision-guided weapon systems, such as long-range cruise missiles.

However, this risk has often been overemphasized. Gaining a military or intelligence edge over another country requires much more than access to high-resolution satellite images. Imagery is only one important input in a complex process that translates raw data into information or "actionable" intelligence for the user. Drawing valid conclusions from earth-observation data usually requires skill and expertise in processing, interpreting, and analyzing the data. The information derived from the satellite images must be generated in a timely manner to be useful for many military applications. Even then, the user must possess the capabilities needed to take full advantage of the information for supporting various military operations, such as delivering precision-guided weapons. Nonetheless, the potential for countries to make effective use of imagery data will improve as they gain more experience in handling high-resolution satellite images and as new weapon systems that take advantage of the growing availability of commercial imagery become available on the international marketplace.

The security implications of earth-observation satellites are not confined to countries. Certain nonstate actors, such as terrorists or narcocriminals, will have unprecedented access to overhead imagery. However, the added benefit of satellite imagery data for these nonstate actors is likely to be relatively small, given their access to a broad array of information sources better suited to supporting their illegal activities. On the other hand, high-resolution imagery could be very useful to a government seeking to harm ethnic subgroups within its own country. Some governments could use commercial data to monitor the activities of organized opposition groups within their borders. Repressive governments might even take advantage of timely commercial images that are made publicly available by the news media or humanitarian NGOs as another information source to support "ethnic cleansing" operations against domestic opponents or to launch attacks on refugees.

Some governments, such as that of the United States, have adopted measures to limit the military utility of high-resolution commercial satellite imagery that could fall into the wrong hands. One way is to impose technological restrictions, such as limiting the resolution allowed for a country's commercial or civilian imaging satellite systems, to make the data less useful for military users. Alternatively, shutter controls offer an operational constraint that a government could impose during a crisis or military conflict on any earth-observation satellites that it controls. In such cases, governments could temporarily limit the collection or distribution of high-resolution satellite imagery of a particular territory if a conflict is imminent or ongoing. However, imposing timely shutter controls could be difficult. For example, media groups in the United States contend that the ability of Executive Branch officials to impose shutter controls without judicial review is a form of prior restraint on the freedom of the press and is therefore a violation of the first amendment rights provided in the Constitution. The efficacy of imposing shutter controls and their actual utility for restricting access to satellite imagery will remain uncertain until governments gain experience with the new transparency.

In any case, the effectiveness of either technological or operational constraints is likely to be limited unless other countries are also willing to adopt such measures. Although the United States currently enjoys a substantial lead in earth-observation satellites capable of producing high-resolution data, several other countries are pursuing comparable capabilities. Absent of some form of multinational collaboration, it seems unlikely that unilateral technical or operational constraints will be sufficient to prevent threatening actors from obtaining these data. Countries that pursue open information policies, such as the United States, will be affected relatively little by increasing global transparency. Although high-resolution commercial satellite imagery could make potential adversaries more knowledgeable about U.S. monitoring capabilities and more adept at concealing their activities, it is hard to identify any cases in which potential adversaries can extract a significant military advantage from even the best available commercial satellite imagery.

A Complex Global Market for Earth Observation Data and Information

Assessing the market for imagery and its derived geospatial information products requires examining a variety of factors. The size of the market will affect the international availability of information for observing the natural and human developments, including the activities of governments. A strong market will enhance transparency, while a smaller market will constrain it. Yet the market for earth-observation data and associated geospatial information products is complex to assess, partly because of competing information-age sources (e.g., airborne and ground-based information sources). In addition, governments continue to exert a strong but uncertain influence on the prospects for a commercial remote sensing market both as a major customer for imagery data and as the regulator of access to such data flows within their own countries or to foreign customers.

On the one hand, there is substantial cause for optimism concerning the projected growth in commercial and civilian observation satellites as both governments and commercial firms pursue plans to launch a wide range of new earth-observation satellites. Yet the promise of the new technologies and how they enable users of commercial satellite imagery to understand developments prompt an important question: Is there really a sufficient commercial market for satellite remote sensing to support the substantial private investment already in systems and organizations? Will the market support the central premise of this book, that commercial remote sensing offers an unprecedented boost to global transparency?

26 Conclusions

This book suggests at best some skepticism about initial market expectations: Some believe that the market will be largely made up of defense and intelligence services, while others contend that commercial satellite imagery data will stimulate a nascent commercial market demand. Commercial applications, for example, could draw new users ranging from telecommunication and insurance companies to agricultural enterprises anxious to optimize crop yields and minimize environmental damage. Regardless of which prediction is correct, governments will continue to play the predominant role in this emerging enterprise, both as customers and regulators.

Often lost in the attention given to satellite remote sensing is the fact that the current data market is still dominated by commercial and government aerial data providers that have pursued a variety of sensor and business innovations to stay competitive. For example, some aerial firms have adopted sophisticated digital imaging systems and have begun to capitalize on the satellite imagery model by developing digital imagery archives. These firms maintain very strong customer ties, which will assist their continued competitiveness. In addition, many aerial imaging firms are entering into business partnerships with satellite data provider enterprises, an arrangement that offers a potentially profitable way for both sides to take advantage of their complementary remote sensing capabilities.

Today, governments, especially the U.S. federal and state governments, are still the primary customers for commercial earth-observation data, products, and services. Many national governments possess a strong interest in such data for resource management, environmental assessments, natural disaster planning and response, and even as a tool for supporting foreign policy aims. Furthermore, in most countries, the chief concentration of remote sensing expertise will for some time continue to be within national government agencies or the programs they fund at universities and research centers. In sum, today's remote sensing market is best characterized as a quasi-market, given the strong nonmarket forces at work, including the dominating influence of government policies and funding.

U.S. firms have rapidly become leaders in the worldwide market for sales of high-resolution earth-observation data. Yet, stiff competition from parastatal firms, which enjoy substantial direct and indirect subsidies from their governments, could prove difficult for U.S. commercial firms to overcome. Another constraint on their access to the global market is the internal restrictions on high-resolution geospatial data, including satellite imagery, that still exist in many countries. Most of these restrictions are an unnecessary legacy of earlier times, but in some cases, governments might be tempted to use restrictive data policies as an indirect way to inhibit competition with their own nationally supported earth-observation enterprises.

The unrivaled influence that U.S. government policies and spending programs exert on the prospects for both U.S. and non-U.S. commercial remote sensing has become a very prominent issue. This poses important policy questions concerning the government's proper role in encouraging the commercial success of the U.S. commercial satellite data providers given the major influence the U.S. government exerts in its multiple roles as customer, regulator, patron, and occasional competitor, depending on the policy issue.

Space-based remote sensing will surely occupy some niche within this geospatial environment, although the competition for market access will be high. Whether or not commercial satellite imagery becomes competitive with alternative aerial or ground-based sources of geospatial information will greatly depend on how well satellite firms can tailor their products and services to meet the needs of the rapidly expanding global information economy.

Final Observations

We believe that commercial observation satellites are on the leading edge of global transparency, which is being driven by rapidly changing technology, institutions, and public expectations. The book's diverse chapters generally agree that—for better or for worse—the new generation of high-resolution commercial and civilian observation satellites dramatically improves the capabilities of national governments and international organizations to monitor major human and natural developments throughout the world. Equally important, commercial satellite imagery promises to give nonstate actors, particularly the media and NGOs, an unprecedented ability to acquire timely and precise satellite imagery data and to make independent assessments. Thus, rapidly expanding global access to high-resolution satellite images poses a series of challenges for governments, commercial remote sensing firms, and even the new users of earth-observation data. An increasingly transparent world is uncomfortable for many political, military, and intelligence policymakers, but there is little they can do to halt this inexorable trend toward greater transparency.

IMPLICATIONS FOR GOVERNMENTS

When combined with other powerful information and communication technologies, the new imaging satellites present a major challenge to how states operate by making the environment for both domestic and international affairs much more transparent. Closed societies will confront new challenges to their attempts to conceal preparations for military attacks on neighbors, ethnic-cleansing campaigns, weapon proliferation activities, environmental disasters, and the like. However, even democratic governments will need to make some inconvenient adjustments to their diplomatic practices, military doctrines, and geospatial data policies to accommodate growing global transparency. Furthermore, as a dual-purpose technology, high-resolution commercial satellite images can be used to generate information that can support either offensive or defensive military activities.

Some policymakers, both in the United States and abroad, have been taken by surprise by the development of high-resolution commercial satellite systems, despite detailed coverage by the news media and trade press. While the trend toward greater transparency appears inexorable, national governments possess policies and foster practices that can significantly affect the pace of global transparency. This is a familiar choice: During the Cold War, U.S. and Soviet leaders had to choose between transparency and secrecy in their national security approaches. Given the deadly threats they perceived from each other, both countries opted for making secrecy an integral element of their national security strategies. Hence, relying on global transparency was a "road not taken" for Washington and Moscow, leading to major consequences in the level of perceived military risks and the amount of national resources dedicated to military preparedness and intelligence-gathering activities.

Now, at the start of the new millennium, each national government must decide whether to embrace or resist the spread of global transparency. The editors of this book are convinced that the United States and other open societies will generally benefit from adopting policies that promote global transparency. Such policies will permit nations to realize the diverse civil and commercial benefits that can result from unrestricted access to earth-observation data and resulting information products. One element of these national policies will be decisions that support a diverse constellation of such satellites and a robust ground infrastructure. Equally important are policies that eliminate questionable barriers to public access to geospatial data and information, including high-resolution satellite data.

IMPLICATIONS FOR COMMERCIAL FIRMS

The commercial remote sensing industry, which serves a diffuse market for data and information, must develop a global market for its products and services or fail as a viable component of the geospatial information sector. This sector includes not only the few firms that build, launch, and operate the new commercial observation satellites but also the much larger number of enterprises around the world that provide various value-added services. The new commercial firms confront a major challenge in competing with the many state-owed or largely subsidized remote sensing enterprises.

To succeed in the highly competitive international marketplace for information products, the commercial remote sensing industry must undertake the difficult, long-term steps necessary for developing market demand for their satellite data, information products, and geospatial services. One of their key challenges is to avoid becoming too reliant on government spending. Such dependency could undermine corporate willingness to undertake the painstaking and expensive work required to develop specific imagery applications and information management infrastructures needed to satisfy the information needs of potential commercial customers. At the same time, governments must adopt clear and consistent regulatory policies to give the commercial remote sensing firms, and their potential investors and customers, confidence to proceed with these expensive, long-term projects.

Although commercial remote sensing firms are understandably working hard to earn sufficient revenues to recoup their investments, they also face the challenge of striking a reasonable balance between their market interests and various public remote sensing needs. These difficult issues include how best to license the distribution of satellite imagery data, particularly for use in emergency operations. More effort is required to reach practical arrangements among governments, NGOs, and commercial firms that will ensure that commercial satellite firms will have a continuing stake in collecting and distributing data from natural and humanitarian disasters. Similarly, the commercial firms also need to address the right of sensed states to acquire copies of any civilian or commercial satellite imagery taken of their territories on a nondiscriminatory basis and at a reasonable cost.

IMPLICATIONS FOR NONSTATE ACTORS

The market prospects for commercial remote sensing firms are particularly important for many high-visibility users of satellite images, including the news media and NGOs, in taking advantage of commercial satellite imagery. These new users are expected to purchase far fewer imagery products and services than government and commercial users and will therefore have relatively little weight in encouraging firms to undertake the large investments required to develop and operate commercial observation satellites. Nevertheless, they wield great influence in publicizing the capabilities of commercial satellites.

The primary challenge for these new users of satellite imagery is to establish their credibility in interpreting and analyzing high-resolution commercial satellite data. Their long-term credibility depends on taking some difficult steps, such as adopting rigorous in-house review procedures for imagery interpretation and analysis, seeking outside expertise when needed, and spending their limited resources to acquire additional training and to hire staff members with analytic expertise. These imagery activists should boost their own knowledge in selecting and interpreting satellite imagery by taking advantage of the expertise and experience that already exists among imagery analysts found in government and commercial firms. Professional associations for remote sensing and geographical information systems can play an important role in encouraging greater interaction between the new and established users of satellite data.

In sum, the effects of the increased availability of earth-observation data on world affairs are highly uncertain and depend deeply on other economic, technological, and political trends. We believe that, on balance, greater transparency will provide significant benefits to the global community. Nevertheless, it will be necessary to monitor developments and to devise policies that enhance the positive aspects of transparency and mitigate possible negative ones, a process that will require the policy community to be more agile in dealing with new technologies than it has been. However one views the political and economic effects of wider global transparency, the phenomenon is still developing, both from the standpoint of data from high-resolution earth-observation satellites and from the other information technologies that support these data and make them particularly useful in monitoring and better understanding global developments.

Notes

1 For details on the projected trends in land remote sensing satellites, see Appendix B in this volume.
2 Cloud cover may obscure certain regions much of the year, limiting the utility of optical sensors for monitoring developments on Earth's surface in those regions. However, synthetic aperture radar systems would be of considerable utility in obscured regions.
3 The term *dual-purpose* conveys a meaning slightly different from that of the more common term, *dual-use*, implying that the technology is intended from the start to be used for both civilian and military-intelligence activities.

Appendices A–E

Appendix A	List of Abbreviations	page 576
Appendix B	The Past, Present and Future of the Medium- and High-Resolution Satellite World	page 579
Appendix C	Color Plates	page 589
Appendix D	About the Authors	page 617
Appendix E	Selected Bibliography	page 623

Appendix A
List of Abbreviations

ASPRS	American Society for Photogrammetry and Remote Sensing
AVHRR	Advanced High-Resolution Radiometer
CCD	Charge-coupled device
CCRS	Canada Centre for Remote Sensing
CD-ROM	Compact disk-read only memory
CIA	Central Intelligence Agency
CNES	Centre National d'Etudes Spatiales
CNN	Cable News Network
CSA	Canadian Space Agency
CTBT	Comprehensive Test Ban Treaty
DEM	Digital elevation model
DMA	Defense Mapping Agency
DoD	Department of Defense
DOI	Department of Interior
DTED	Digital Terrain Elevation Data
EDC	EROS Data Center
EOSAT	Earth Observing Satellite Corporation
EOSDIS	Earth Observation System Data and Information System
EROS	Earth Resources Observation System, or Earth Remote Observation System
ERS	European Remote Sensing

ERTS	Earth Resources Technology Satellite
ESA	European Space Agency
ETM	Enhanced Thematic Mapper
FAS	Federation of American Scientists
FF	French francs
FOIA	Freedom of Information Act
GIS	Geographic Information System
GOES	Geostationary Operational Environmental Satellite
GPS	Global Positioning System
GSFC	Goddard Space Flight Center
HRMSI	High Resolution Multispectral Stereo Imager
HSI	Hyper-spectral imagery
IAEA	International Atomic Energy Agency
IDP	Internally displaced persons
IGS	Information gathering satellite
IGY	International Geophysical Year
IR	Infrared
IRS	Indian Remote Sensing
ISRO	Indian Space Research Organization
JPL	Jet Propulsion Laboratory (NASA)
METEOSAT	Meteorological Satellite
MITI	Ministry of International Trade and Industry (Japan)
MOU	Memorandum of understanding
MSI	Multispectral (color) imagery
MSS	Multispectral scanner
NAPP	National Aerial Photography Program
NASA	National Aeronautics and Space Administration
NASDA	National Space Development Agency (Japan)
NATO	North Atlantic Treaty Organization
NGO	Nongovernmental organization
NIMA	National Imagery and Mapping Agency
NOAA	National Oceanic and Atmospheric Administration
NPT	Nuclear Non-Proliferation Treaty
NRO	National Reconnaissance Office
NTM	National technical means
OSI	On-site inspection
OTA	Office of Technology Assessment

PAN	Panchromatic (black and white) imagery
PDD	Presidential Decision Directive
PIDC	Prototype International Data Center
RBV	Return beam vidicon
RSA	Russian Space Agency
RSI	RADARSAT International
RTNDA	Radio-Television News Directors Association
SAR	Synthetic aperture radar
SPOT	Satellite Pour l'Observation de la Terre
STA	Science and Technology Agency
START	Strategic Arms Reduction Treaty
TIR	Thermal infrared
TIROS	Television Infrared Observing Satellite
TLM	Topographic line map
TM	Thematic mapper
TRMM	Tropical Rainfall Measuring Mission
UAV	Unmanned aerial vehicle
UN	United Nations
UNHCR	United Nations High Commissioner for Refugees
USC	United States Code
USGS	United States Geological Survey
USSR	Union of Soviet Socialist Republics
VHSR	Very high spatial resolution
VNIR/SWIR	Visible near infrared/shortwave infrared
WIS	West Indies Space
WMD	Weapons of mass destruction

Appendix B
The Past, Present and Future of the Medium- and High-Resolution Satellite World

William E. Stoney

Land Imaging Satellites

There are two basic classes of sensors used to image the Earth from space: Optical sensors, which record the radiation reflected or emitted from the surface, and Radar/Lidar sensors which measure the surface reflections from emissions from the satellite sensor. The specific measurement parameters of the sensors and their satellite systems can be varied to provide data suited to the needs of a wide variety of users. The three major data features are image resolution, image spectral characteristics, and image field of view.

The laws of optics, orbital mechanics, and last but never least, economics, are such that better features in any one of the above characteristics can generally be had only at the expense of one or both of the others. The ideal solution for the user community would be a satellite that combined several sensors providing coverage of a broad range of variables. However the resulting satellite system becomes large, technically complex, and expensive. As a result, land observing satellite systems tend to carry one major instrument with perhaps one additional auxiliary low resolution survey sensor.

The current and planned land imaging satellites can be placed into four observational classes plus a fifth class consisting of advanced systems designed to explore new sensors. The four non-experimental classes can be labeled environmental daily coverage systems, Landsat-like broad area coverage systems, high resolution pointable imaging systems, and Radar systems. The one currently-planned exception is Japan's ALOS 1 which will carry a radar sensor and two optical sensors, one with a wide field of view and one with high resolution. This Appendix will not cover the environmental daily coverage systems, specifically NASA's EOS series, and the NOAA and European weather satellites. While they are very important in the Earth land observation activities which seek to understand the interactions between land changes and climate and weather, their kilometer resolutions do not make them part of the transparency concerns which are a major focus of this volume.

A Short History

Figure A-1 presents all the land imaging satellites having resolutions of 30 m or better, which have been, are in, or are planned to be in orbit during the years between 1996 and 2005. Note the codes: The solid bars are government funded satellites, the vertical dashed bars represent satellites totally funded by the private sector, and the slanted dashed bars indicate systems funded by a combination of government and private funds. Note also that while Landsat, SPOT, and IRS satellites are government funded, the latter two distribute their data through commercial channels and, in that sense, their operators consider themselves part of the commercial imaging industry.

Figure A-1 is intended as a history lesson as well as a portent of things to come. As is indicated by the number and multi-year length of the open bars, planning satellite launches is an inexact science. A major reason for the number and length of the slips in originally-announced launch dates is due to the increasing number of privately funded systems. While the government systems have had their share of delays, commercial system delays tend to be considerably longer, due undoubtedly to the uncertainties of raising private versus public capital.

But there is a clear trend toward the private sector catching up with government systems—in the beginning of 2000, there were 11 government satellites and one private system. Plans indicate that in three years there will be 12 government satellites and 8 commercially funded systems, and that by 2005 there could be 16 private and 10 government satellites. As is better illustrated in Figure A-2, the private sector's entrance into the satellite data world is due to one factor: the technological and political ability to provide high resolution images commercially.

Figure A-1 also shows that launching satellites still caries risks. In the period shown, nine satellites were launched successfully and are still operating, but five others were lost at launch or very soon after and a sixth failed after only months in orbit. These numbers include the unfortunate commercial system record of four losses and two successes. (Note that these figures do not include the two successful SPIN 2 film return missions.)

Satellite Types

The characteristics of the three satellite types that are the subject of this volume and a separate class of test systems are presented in Table A-1 and Figures A-2 and A-3. Table A-1 provides details of the sensor and satellite measurement characteristics, Figure A-2 presents the satellite schedules, and Figure A-3 shows each sensor's radiometric coverage.

LANDSAT-LIKE

This group is characterized by having frequent, large-area coverage capability, (images are 70 to 320 km wide), and multiple color bands with several beyond the visible and near IR spectral regions (all but one have short wave IR bands and several have thermal IR bands as well). Their multispectral resolutions range from 6 to 30 m and two have panchromatic resolutions of 2.5 m. The number and extent of their color bands enables the measurement of the area, type, condition and change over time of the Earth's vegetative cover. Their wide swaths provide the global temporal coverage required to monitor and evaluate global and regional scale land cover changes. Note the final column in Table A-1 which provides an approximation of the number of times a year the systems could image the entire globe, and remember that on average half the globe is cloud-covered at all times. It clarifies the great difference between the large area coverage systems and the high-resolution pointable systems. These characteristics are also useful for the land management responsibilities of government agencies down to the county level. As shown in Figure A-2, the satellites are all government funded with the exception of the proposed Resource21 system. Resource21 is planning to help farmers identify and react to changes in their crop conditions to maximize their yields. This requires their systems to have four-day repeat coverage and the very high radiometric sensitivity and calibration specifications characteristic of Landsat 7.

HIGH RESOLUTION

These mostly commercial satellites provide high resolution ranging from 2 m to the planned half-meter systems of IKONOS and EarthWatch. This has been accomplished by restricting the width of their images (swaths range from 8 to 30 km), and for those that have color bands, limiting the spectral coverage to the visible and near IR bands. While the narrow swath does restrict their utility for the broad area monitoring tasks of the Landsat type systems, they do serve a wide range of commercial needs because their high resolutions can be used to identify, measure, and track changes in man-made features such as roads, buildings, airports, etc. Equally important is their ability to point off axis, which provides 1- to 2-day repeat capabilities and the measurement of the vertical dimensions by means of stereo images. Except for the two Indian systems and the single Russian film system, all are privately funded. The established use of high resolution aircraft imagery has convinced several in the private sector that satellite imagery of sufficient resolution can be profitably marketed at prices that will both capture some of the aircraft market and create entirely new markets based on the satellite's ability to gather scenes anywhere, anytime, for anyone, clouds willing.

RADAR

Radar satellites have been operating since the early 1990s. They have one very useful characteristic: They record their images through clouds and at night, which ensures acquiring images when they are needed. Their sensitivity to a target's physical characteristics and to its moisture content provides data with an entirely different information content than optical systems measurements of reflected and emitted radiation. Their potential for a wide variety of applications is promising, but as yet undeveloped compared to the optical systems. All civil radar systems flying and currently planned are funded by foreign governments; however, their data are available commercially, and the Canadian RADARSAT program has an aggressive plan to be nearly privately funded and operated by the time a third satellite is needed.

TEST, MULTI- AND HYPERSPECTRAL

These systems, all funded totally or in part by the government, test advanced sensors' ability to increase the value of space data by increasing the range and sensitivity of the radiation measurements. The number and locations of the bands for these systems are compared with the current systems in Figure A-3. Aircraft-mounted sensors have indicated that measurements of the radiances of essentially all the spectrum, for example 200 to 400 bands over the visible and near IR range instead of the 3 to 7 measured by current sensors, provides the additional information content needed to improve identifying natural targets.

The Data

The satellite data in the Figures in this Appendix originally came from a survey MITRE made under a contract with NASA's Goddard Space Flight Center in support of the 1995 ASPRS Land Satellite Information in the Next Decade Conference. Mitretek conducted a second survey for the second ASPRS Land Satellite Information Conference, in 1997. Mitretek has kept the data up to date since then under a contract with NASA's Stennis Space Center in support of its Land Imaging Satellite DataBase.

Only satellites whose data are or will be available to the general public, commercially or otherwise, are included. The Japanese 1-m Information Gathering Satellites (see Chapter 10 in this volume), scheduled for launch in late 2002 (IGS-1 optical and IGS-2 radar), are not listed because the distribution of their data is still under discussion. Should the data become generally available, they would offer strong commercial competition, especially in the radar arena.

At the time of this writing, many both governments and private corporations were discussing and/or announcing plans for remote sensing satellites. RapidEye of Germany announced a commercial four-satellite 6-m system to monitor crops. Korea has launched a 10-m panchromatic, 20-m multispectral system (not noted in the Figures because of the uncertainty of general data availability) and plans to launch Kompsat 2, a 1-m panchromatic, 4-m multispectral system, by the end of 2003. Taiwan has announced plans for Rocsat 2, a 2-m satellite to be launched in 2002. The Europeans have been discussing developing a seven-satellite system, three optical and four radar, having 2- to 3-m resolution. Turkey is selecting a contractor for its high-resolution satellite. India has announced development of the Test Evaluation Satellite (TES), a 1-m "spy satellite" to fly early in 2001.

Updated information will be available through the ASPRS Web page, at www.asprs.org./asprs/news/satellites/.

Table A.1—Data Characteristics of Current and Planned Land Imaging Satellites

SATELLITE	LAUNCH	MS BANDS RES M			PAN BAND RES M			# OF SPECTRAL BANDS				GLOBAL COV	
		>10	4.1-10	<=4	<=1	1.8-3	>2.5	VNIR	SWIR	MWIR	TIR	>7X/YR	<6X/Y
HIGH GLOBAL REPEAT (LANDSAT-LIKE)													
Landsat 5	1984.250	30						4	2		1	23	
SPOT 1	1986.016	20					10	3				14	
SPOT 2	1990.100	20					10	3				14	
IRS-1C	1995.930	23					5	3	1			15	
IRS-1D	1997.790	23					5	3	1			15	
SPOT 4	1998.230	20					10	3	1			14	
Landsat 7	1999.288	30					15	4	2		1	23	
CBERS-1	1999.790	20					20	4	2		1	14	
IRS-P6	2001.100	23	6					4,3	1			15	3
CBERS-2	2001.790	20					20	4	2		1	14	
SPOT 5	2002.120		10			2.5		3	1			14	
ALOS-1	2002.500		10			2.5		4					5
Resource 21 - A	2004.800		10					4	1			45	
Resource 21 - B	2004.950		10					4	1			45	
HIGH RES, PAN & MS													
SIE IKONOS-2	1999.731			4	1			4					1.5
OrbView-4	2001.496			4	1			4					1
EW QuickBird-2	2001.624			4	1			4					2.5
OrbView-3	2001.748			4	1			4					1
ISI Eros-B2	2002.915			4	1			4					2
ISI Eros-B3	2003.414			4	1			4					2
ISI Eros-B4	2003.916			4	1			4					2
SIE IKONOS-Blk-2	2004.000			2	0.5			4					1.5
ISI Eros-B5	2004.414			4	1			4					2
EW Blk. 2	2004.500			2	0.5			4					1.5
ISI EROS-B6	2004.915			4	1			4					2
HIGH RES, PAN ONLY													
SPIN-2	1998.242					2							
ISI Eros-A1	2000.951					1.8							1.6
ISI Eros-A2	2001.674					1.8							1.6
Cartosat-1	2002.000					2.5							3.4
ISI Eros-B1	2002.414					1.8							2
Cartosat-2	2003.000				1								3.4
MULTI-HYPERSPECTRAL													
Terra (ASTER)	1999.956	15						3	6		5	7.5	
MTI	2000.170	20	5					7	3	2	3		1.6
EO-1	2000.931	30						115	115				1.9
OV-4 Warfighter	2001.247		8					100	100				0.6
NEMO	2002.500						5	105	105				3.4
ARIES	2003.748						10	32	32				1.9
RADAR													
ERS-2	1995.300						10					12.3	
Radarsat 1	1995.750						8.5						6
Envisat	2001.500						10					12.3	
ALOS-1	2002.500						10					8.7	
Radarsat 2	2003.329					3							6

Appendices A–E

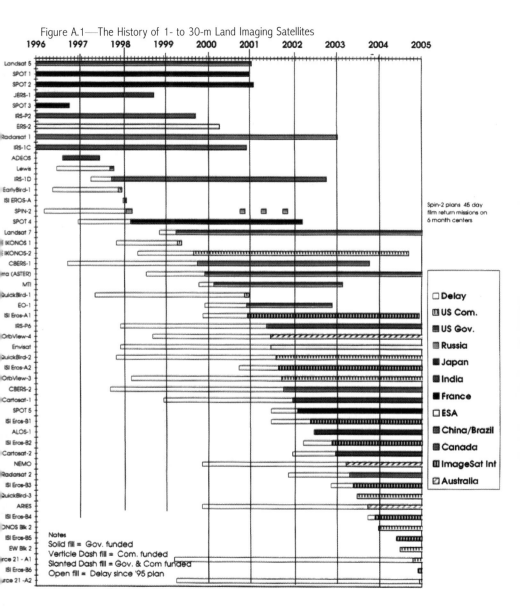

Figure A.1—The History of 1- to 30-m Land Imaging Satellites

Figure A.2—The Next Five Years

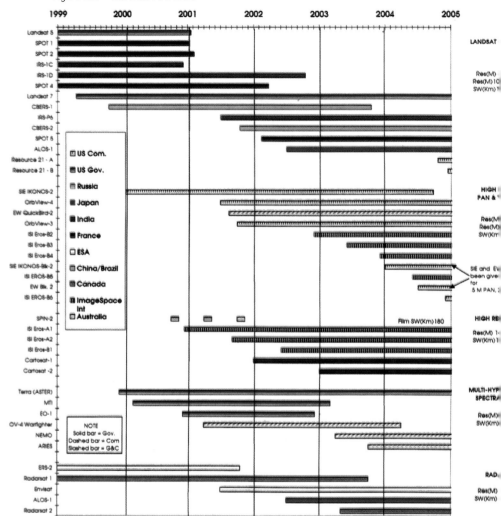

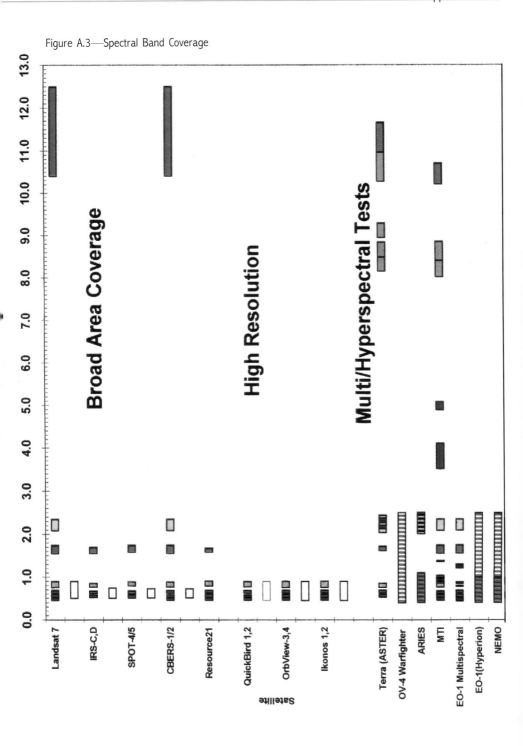

Figure A.3—Spectral Band Coverage

Appendix C
Color Plates

Figure 4.1—Digital Orthoimage with Road Network Overlay, Austin, Texas. This image contains overlays urban transportation features on a Texas Strategic Mapping Program (StratMap) digital orthoimage of part of the city of Austin. (Image courtesy Drew Decker, Texas Natural Resources Information System, [www.stratmap.org])

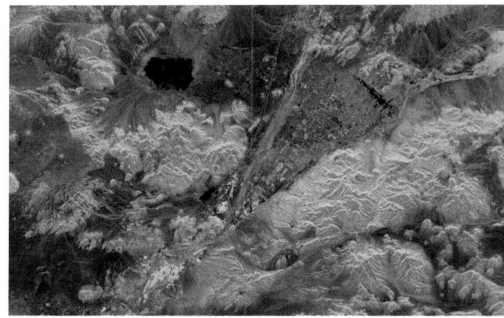

Figure 5.2—False Color SAR Image: Barstow, California, and Vicinity, April 12, 1994. This image was acquired by Spaceborne Imaging Radar-C/X-Band Synthetic Aperture Radar onboard the space shuttle Endeavour; north is toward the upper left. This image of part of the Mojave Desert reveals evidence of human activities. Barstow is the pinkish area in the lower left center. The V-shaped blue area in the center of the image is the Manix Basin, which includes the bed of the Mojave River. The orange circular and rectangular patches within the basin are irrigated fields. Radar data have been used to evaluate the effects of irrigation practices on soil stability and land degradation in this desert region. Sparsely vegetated areas of sand and small gravel appear blue in this image, while rocky hills and rougher gravel deposits appear mostly in shades of orange and brown. The dark patch in the upper left is the very smooth surface of Coyote Dry Lake. Sources: Image and caption, JPL; image also at http://www.jpl.nasa.gov/radar/sircxsar/barstow.html.

Appendices A–E

Figure 5.3—SAR Interferogram: A Portion of the Rutford Ice Stream in Antarctica. Sources: Image and caption, JPL; Goldstein et al. (1993); image also at http://southport.jpl.nasa.gov/scienceapps/dixon/index.html.

Figure 5.4 (below)—Notional Hyperspectral Data Cube. A hyperspectral data set is often portrayed as a cube as a strictly notional representation. Here, the top layer looks like a color picture of the scene. Spectral information content is portrayed by the rainbow of colors underneath, intended to indicate different spectra at each picture element. At any point in the scene, the actual data look more like Figure 5.5, a sampling of the spectrum of the materials sensed in a single picture element.

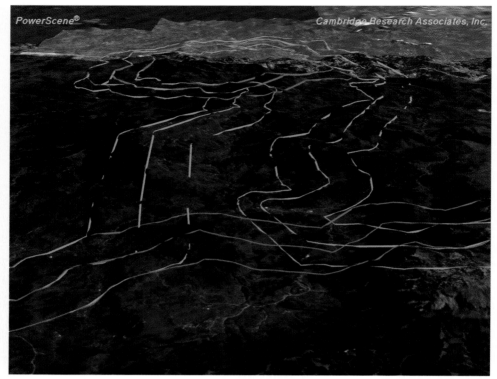

Figure 14.5—Negotiated Lines and Buffer Zones, Gorazde Corridor, Bosnia-Herzegovina. Aligning a proposed all-weather road within a militarily defensible corridor linking the Muslim-Croat sector (foreground) to the Muslim enclave of Gorazde (toward horizon), and then determining construction feasibility of the road, led to a diplomatic breakthrough in the Dayton Proximity Peace Talks. Initial alignment of the road, and determination of military defensibility, occurred on the PowerScene™ visualization system, using SPOT panchromatic imagery draped over Digital Terrain Elevation Data Level 1 data from the Defense Mapping Agency.

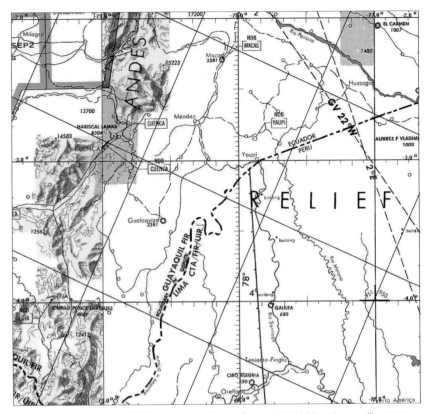

Figure 15.2—Small-Scale NIMA Aeronautical Chart. This 1:1,000,000-scale map illustrates the lack of detail near the border.

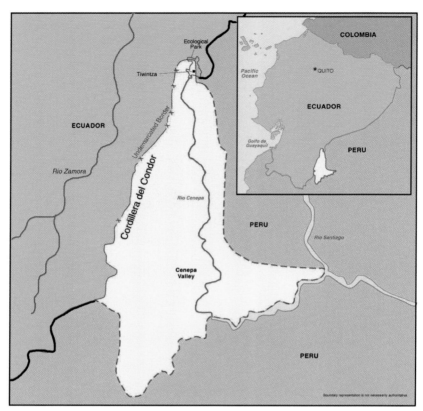

Figure 15.3—Cenepa River Valley. This 150-km section of undemarcated border was the principal zone of contention between Peru and Ecuador.

Figure 15.5—PowerScene Screen Capture of the Northern End of the Cordillera del Condor. Existing monuments and rivers are flagged along with lines in contrasting colors representing the differing interpretations of the border. Eventually, a compromise line was chosen in this area.

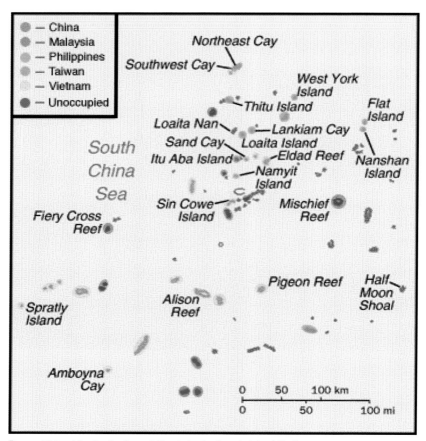

Figure 16.1—Islands, Reefs, and Shoals in the Spratly Island Region

Appendices A–E

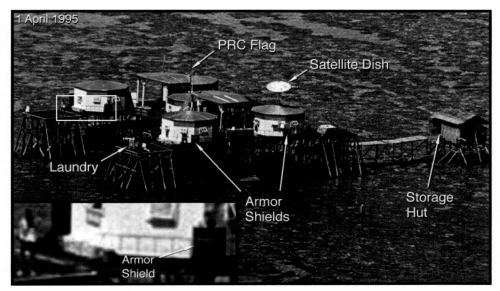

Figure 16.4—Aerial Photograph, Taken a Few Months After China Took Control of Mischief Reef, Showing One of Four Building Clusters China Constructed and Occupied. The inset shows an individual standing on a platform and a second, possibly armed, individual positioned behind one of the armor shields. Source: Aerial image, Agence France-Presse.

Figure 16.5—Aerial Photograph Showing Another Chinese Building Cluster on Mischief Reef. As in Figure 16.4, the huts were deliberately set apart, which suggests the storage of items that need to be kept away from the living quarters for safety or hygienic reasons. The inset shows an individual with binoculars watching the observation aircraft. Source: Aerial image, Agence France-Presse.

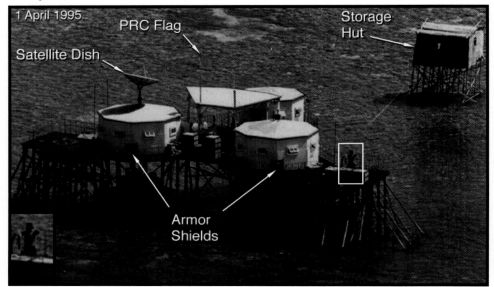

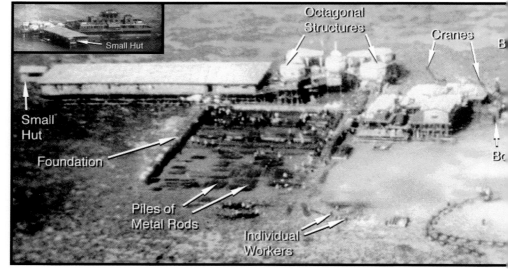

Figure 16.6—Aerial Photograph of Construction at a Chinese-Occupied Area on Mischief Reef. The photograph, released by the Philippine Department of National Defense on November 10, 1998, shows significant construction activity at one of the four Chinese-occupied areas on Mischief Reef. Taken from an aircraft at 300 feet, January 15, 1999, the insert shows the five-story building nearing completion. Source: Aerial image, Agence France-Presse.

Figure 16.7—Aerial Photograph Showing New Construction Activity at a Second Chinese Occupied Area on Mischief Reef. It was also released by the Philippine Department of National Defense on November 10, 1998. In the background, the construction workers can be seen erecting a large truss structure. The large black object on the center platform is probably a weapon system. Source: Aerial image, Agence France-Presse.

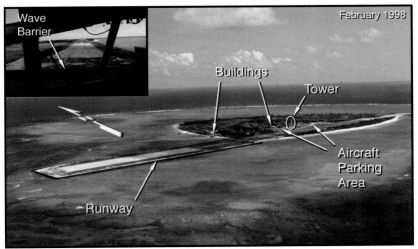

Figure 16.17—February 1998 Aerial Photograph of Thitu Island. This image was acquired from a C-130 transport plane. (The C-130 propellers can be seen in another photograph that was taken by the photographer; see Dahlby, 1998, p. 9.) In addition to the runway, the image shows buildings among the lush vegetation and the aircraft parking area. An observation tower next to the taxiway can also be seen. The inset is a close-up view of the runway, which appears to consist of grass, coral, and concrete. The photograph was taken from the C-130 cockpit while on landing approach from the west. Notice the wave barrier at the end of the runway. The same barrier can be clearly seen in the RADARSAT-1 image in Figure 16.16.

Figure 17.4—Before, during, and after the tests on May 28, 1998 at Pakistan's first nuclear test site; looking from west to east. Source: Video, Pakistan Television.

Appendices A–E

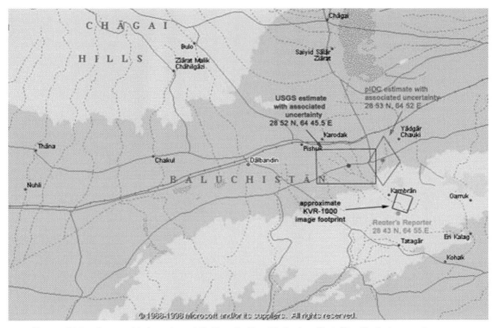

Figure 17.5—Geographic Location of Pakistan's May 28 Nuclear Test Site. Preliminary seismic estimates and associated error ellipses are as of mid-June 1998.

Figure 17.8—Pakistan's First Nuclear Test Site. Sources: video, Pakistan Television; annotations, ISIS and Frank Pabian.

Figure 17.9—Instrumentation Bunker, Forward Recording Site. Inset photo shows recording camera on an oscilloscope. Sources: video, Pakistan Television; annotations, ISIS and Frank Pabian.
Figure 17.10—Concealed Tunnel Portal at Pakistan's First Nuclear Test Site. Source: annotations, ISIS and Frank Pabian.

Figure 17.11—Top: IRS-1C 5-m resolution image. Bottom: Geologic Map. Sources: Image, Space Imaging, Inc.; map, Government of Canada (1958).

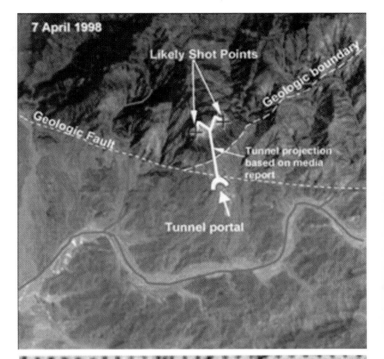
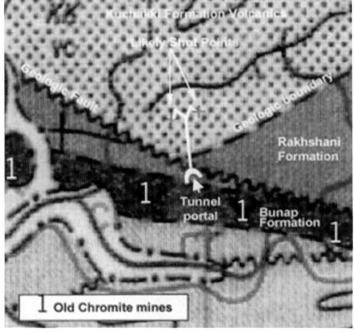

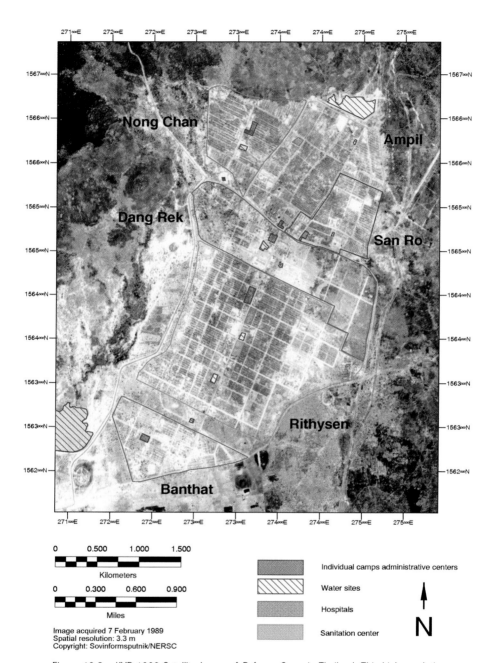

Figure 19.2—KVR-1000 Satellite Image of Refugee Camp in Thailand. This high-resolution (3.3-m) multispectral image combines panchromatic and near-infrared channels to show Thailand's now-demolished Site 2 refugee camp, which was once home to over 140,000 people. Source: Image, ©1998 Sovinformsputnik.

Appendices A–E

Figure 19.3—Detail View of Refugee Camp shown in Figure 19.2. Closer examination of the camp in Figure 19.2 reveals its key internal features, including its water site and dense housing structures. Source: Image, ©1998 Sovinformsputnik.

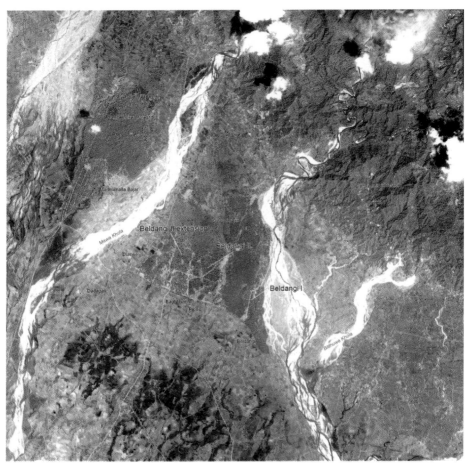

Figure 19.4—High-Resolution IKONOS Image of Beldangi Refugee Camps, Nepal, May 15, 2000. This image of three refugee camps in Nepal shows the surrounding environment of the camps, including local villages, the road network, rivers, and forest cover. Sources: Image, ©2000, Space Imaging, Inc.; processing, Satellus for ENVIREF.

Appendices A–E

Figure 19.5— Detail View of Refugee Camps shown in Figure 19.4. Closer examination of the refugee camps shown in Figure 19.4 illustrates the planning utility of combining 1-m panchromatic and 4-m multispectral image channels to reveal details in nearly natural colors. Individual housing structures, camp layout, and even individual trees can clearly be seen in the image. Sources: Image, ©2000, Space Imaging, Inc.; processing, Satellus for ENVIREF.

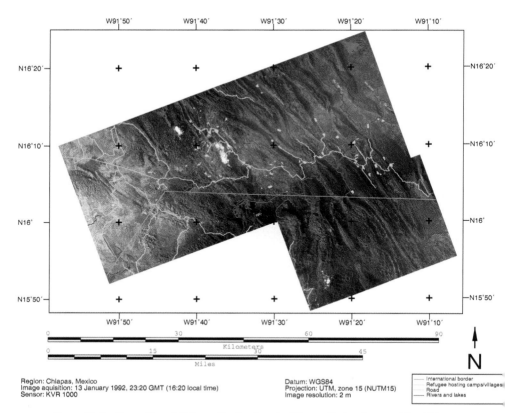

Figure 19.6—Space Map of Refugee Hosting Area in Chiapas, Mexico, January 13, 1992. This composite image consists of three KVR-1000 high-resolution (2-m) images. The high-resolution imagery data were used to derive the overlay data indicated in the color annotations representing road networks, rivers, and the international border with Guatemala. Source: Image, ©1998 Sovinformsputnik.

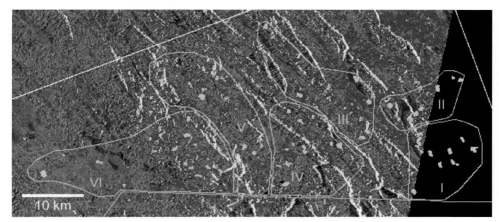

Figure 19.8—Deforestation over time (1993–1997) in Chiapas, Mexico, as recorded in satellite imagery. The 1993 synthetic aperture radar (ERS-1) image, in shades of gray, forms the background. The overlay information is derived from higher-resolution satellite images. Green areas were forested in 1993. Red areas indicate where the forest disappeared between 1993 and 1997. The Light blue polygons indicate the extents of the refugee camps and villages, as derived from the higher-resolution satellite images. Light blue lines (I–VI) indicate refugee-administrative regions, as described in UNHCR field reports. Source: Image, ©1998 European Space Agency.

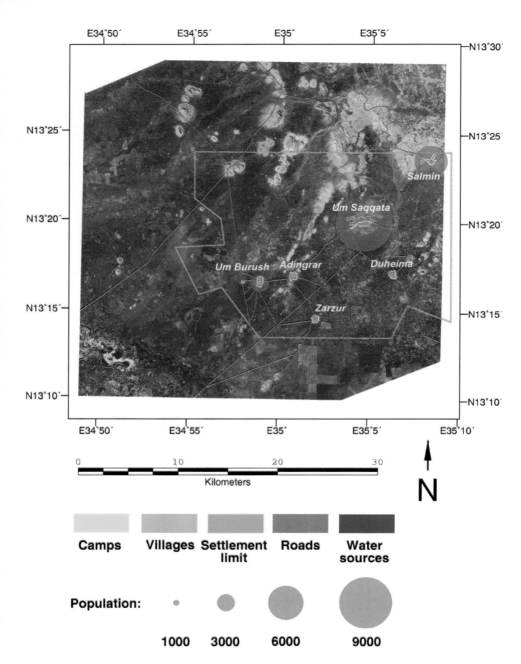

Figure 19.9—Space Map of Qala en Nahal Refugee Settlement Plan, Sudan, January 20, 1992. Satellite image of Qala en Nahal refugee settlement plan (within the green line) with descriptive color overlay. The background is provided by a KVR-1000 high-resolution (2-m) satellite image. The image indicates that the refugee settlement is located in a relatively flat area, with some nearby population centers, and without substantial water resources. Source: Image, ©1998 Sovinformsputnik.

Figure 19.11—IKONOS image of the Ifo Refugee Camp, Dadaab Settlements, Kenya, May 25, 2000. The image combines the 1-m resolution panchromatic and 4-m resolution multispectral data. Note the relatively rigid camp structure with rectangular "greenbelt" reforestation activities on the outskirts of the camp, as well as the block structure of refugee housing. Sources: Image, ©2000 Space Imaging; processing, Satellus for ENVIREF.

Figure 19.12—Detail View of Refugee Camp Shown in Figure 19.11. The larger buildings in the lower left quadrant of the image are camp administration units. Note the trees planted as part of the reforestation scheme inside the blocks of refugee housing. Sources: Image, ©2000 Space Imaging; processing, Satellus for ENVIREF.

Figure 19.13—Satellite image of Kosovar Refugee Hosting Area, Albania and Macedonia. This image map was made on short notice using a Landsat satellite image as background. An already existing Landsat image was combined with GPS data on refugee camp locations and on information on road networks and hydrology to produce a simple map, which was disseminated to relief agencies. Sources: Image, ©1999 Eurimage; GIS information, UNHCR and USAID; camp locations, UNHCR; image analysis; Nansen Center/Satellus.

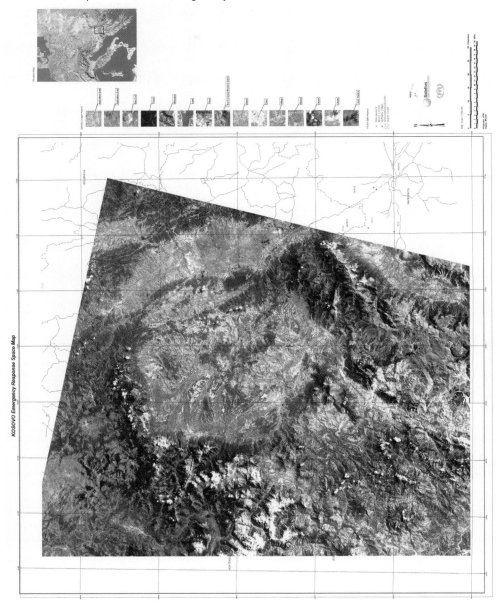

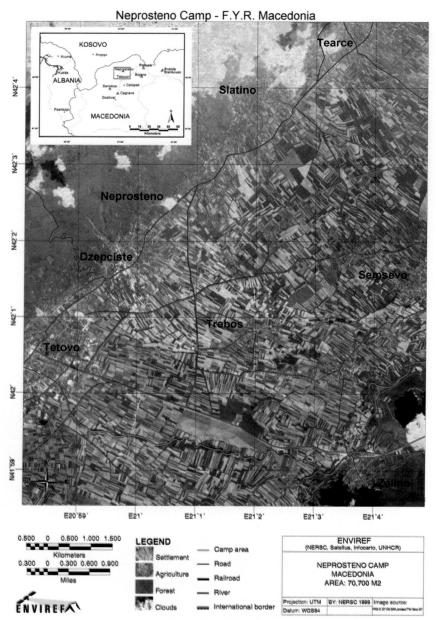

Figure 19.14—Satellite image of Neprosteno refugee camp. This image combines a Landsat (30-m resolution) image with a higher resolution IRS-1D (5.8 m resolution) satellite image, acquired on May 27, 1999, along with an overlay of relevant geographical information system (GIS) data. The area of the Neprosteno refugee camp (outlined in green) was easily derived using standard image-processing tools. The image shows agricultural areas and villages located near the camp. Sources: Image, ©1999 Eurimage; GIS information, UNHCR and USAID; camp locations, UNHCR; image analysis; Nansen Center/Satellus.

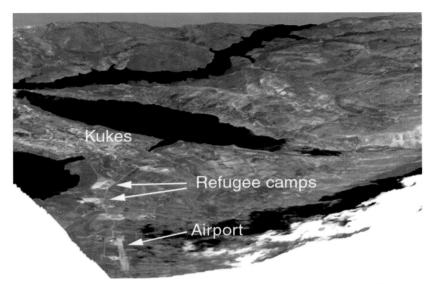

Figure 19.15—Three-dimensional perspective of Kukës refugee camps in Albania. This view was created by overlaying an IRS-1D panchromatic image (5.8-m resolution) over a DIGITAL Terrain Mode derived from two tandem-phase ERS radar satellite images. Note the location of the airport and where refugee camps were set up, given the local terrain. Sources: Image, ©1999 Eurimage and European Space Agency.

Appendix D
About the Authors

David Albright is president of the Institute for Science and International Security (ISIS). His research organization has been a leader in assessing how commercial and civilian satellites could be used by nongovernmental organizations to monitor potential nuclear weapon facilities and test sites of foreign countries.

John C. Baker is a technology policy analyst at RAND. His work is concerned with commercial observation satellites and other dual-use space technologies. Previously, as a staff member at the Space Policy Institute at George Washington University, he directed the Institute's South China Sea Remote Sensing project. He has also been a research associate at the International Institute for Strategic Studies, Pacific-Sierra Research Corporation, and the Brookings Institution.

Dr. Adam Bernstein is a physicist working in the Radiation Detection Department at the Livermore branch of Sandia National Laboratories. He is a specialist in the development and use of technologies for treaty verification and cooperative monitoring, with an emphasis on nuclear arms control.

Dr. Einar Bjorgo works for the United Nations High Commissioner for Refugees (UNHCR) Mapping Unit as an associate information officer for geographic data. He received his Ph.D. in remote sensing from the Nansen Environment and Remote Sensing Center at the University of Bergen, Norway. While at the Nansen Center, he initiated the U.S. Institute of Peace–sponsored ReliefSat and European Commission—supported ENVIREF projects on the use of high-resolution satellite imagery in refugee relief operations. He is the author of several international publications on applications of commercial high-resolution remote sensing technology in humanitarian relief operations.

Dr. Michel Bourbonniere is associate professor of International Law at the Royal Military College of Canada. He is a graduate of the McGill University Institute of Air and Space Law. He is a member of the International Institute of Space Law; a fellow of the Center for Hemispheric Defense Studies, National Defense University; and a member of the American Society of International Law. He has acted as consultant for the Canadian Space Agency on the RADARSAT-2 project. Professor Bourbonniere has published numerous articles in Canada, the United States, and in Europe dealing with air, space, and armed conflict laws.

Yahya Dehqanzada was, at the time this book was written, a RAND researcher specializing in commercial observation satellite issues and substantially contributed to producing this book. He now works for the U.S. government. He was previously with the Project on Transparency at the Carnegie Endowment for International Peace and was a coauthor (with Ann Florini) of *Secrets for Sale: How Commercial Satellite Imagery Will Change the World* (2000). He has also served as an intern at the State Department and the National Security Council.

Dr. Ann Florini is a resident associate at the Carnegie Endowment for International Peace. She wrote one of the earliest scholarly articles on the security implications of the worldwide proliferation of imaging satellites and has since published widely on the policy implications of growing access to satellite imagery. She currently directs two efforts at the Carnegie Endowment that address broad issues of global governance: the Project on Transparency and the Project on Transnational Civil Society.

Deborah Foster is a research analyst at the Strategic Assessment Center of the Science Applications International Corporation (SAIC), where she specializes in assessing trends in foreign use of space-based capabilities. She previously worked at the Center for Strategic and Budgetary Assessments.

Dr. Mark D. Gabriele is an expert on the Treaty on Open Skies and has written a book on its policy implications for the United States. His other work involves international transparency and studies of the collection and interpretation of early warning information as a stabilizing factor in international relations. He received his Ph.D. in Public Policy Analysis from the RAND Graduate School and is currently a policy analyst at RAND.

John Gates is an employee of the National Imagery and Mapping Agency (NIMA) in Bethesda, Maryland, and has held previous positions with NIMA and its predecessor agencies in Panama, Peru, and Chile. As a participant in the Ecuador-Peru border negotiations following the 1995 conflict, Mr. Gates provided technical support and analysis to representatives of the guarantor nations and the deliberating parties. Mr. Gates holds Bachelor and Master of Science degrees from the State University of New York, College of Environmental Science and Forestry in Syracuse, New York.

Dr. Vipin Gupta is a senior member of the technical staff at the Livermore branch of Sandia National Laboratories, where he is currently working on assignment in the Advanced Concepts Group. He is a pioneer in the application of civilian and commercial satellite imagery to international security problems. Along with his colleague, Dr. Adam Bernstein, he provided the satellite imagery analysis for the South China Sea Remote Sensing project.

Dr. Louis Haeck is an adjunct professor at the *Ecole Polytechnique*, University of Montreal, and at the Royal Military College of Canada. Professor Haeck received his Ph.D. from McGill University in space law, where his thesis examined the military uses of outer space by the Canadian forces. He was awarded a NATO fellowship and obtained postdoctoral research scholarships from the Department of National Defence. He serves as a consultant for the military attaché at the Canadian Space Agency and has published many legal articles in various professional reviews in Europe and in the Americas.

Dr. Gregg Herken is a historian and the military curator for the Smithsonian Institution's National Air and Space Museum. He organized the museum's "Secret No More?" workshop in May 1998 that focused attention on the broader political and security implications of higher resolution commercial observation satellites.

Greg Hilgenberg is currently with Litton/TASC. Previously, he was a researcher at RAND, working on space policy and military issues. He holds a graduate degree in science, technology, and public policy from the Elliott School of International Affairs at the George Washington University. At RAND, he focused primarily on commercial remote sensing, microsatellites, and Air Force space policy issues.

Corey Gay Hinderstein is a policy analyst at the Institute for Science and International Security (ISIS). She has been a main contributor to the institute's continuing nonproliferation research on using civilian and commercial satellite imagery for monitoring nuclear-related sites.

Dr. Dana Johnson is national security analyst at RAND, where she focuses on military space policy issues. She has coauthored studies on air and space integration issues for the U.S. Air Force, as well as analyses on space support to the warfighter and the national security community.

Dr. Richard Johnson is an associate with Booz-Allen & Hamilton, Inc., and chief engineer of the Engineering Edge Alliance, a consortium of firms providing systems engineering and integration support to the National Imagery and Mapping Agency (NIMA). At the time of the Dayton peace talks, then–Colonel Johnson was Commander of the U.S. Army Topographic Engineering Center at Fort Belvoir, Virginia, and was deployed to Dayton as the night-shift supervisor for mapping support to the negotiators.

Beth E. Lachman is a science and technology policy analyst at RAND. Her current research focuses on analyzing market trends of geospatial technology industries and the application of such technologies to address environmental issues and urban systems. Her recent work includes analyzing trends in the commercial remote sensing and GIS marketplaces, developing an indicator system within a GIS to support decisionmaking within the Greater Yellowstone Ecosystem, and analyzing public-private partnerships for geospatial data sharing.

Richard Leghorn is the former president of the Itek Corporation and a pioneer in aerial and satellite reconnaissance. He has contributed first-person insights to the book regarding the evolution of early U.S. postwar efforts to acquire a remote sensing capability and early thinking on transparency as national policy.

Dr. Karen Litfin, associate professor of Political Science at the University of Washington, is an expert on global environmental politics and nongovernmental organizations. Her recent research includes work on how various NGOs are making use of satellite imagery.

Dr. Steven Livingston is associate professor of Political Communication and International Affairs at the George Washington University. He is an expert on assessing the influence of the news media on U.S. foreign and defense policy.

His current research focuses on how the news media are taking advantage of emerging information technologies, including commercial observation satellites.

Kevin M. O'Connell is a senior international policy analyst and manager of intelligence community programs at RAND. Prior to joining RAND, he served for more than a decade in various government positions, including on the intelligence community management staff and on the National Security Council staff in the White House Situation Room. His research projects at RAND focus on space commercialization, including assessments of the prospects for commercial remote sensing markets and international opportunities for imagery cooperation.

Dr. Xavier Pasco is an expert on space policy and American politics. He directs the research program at the *Fondation pour la Recherche Strategique* in Paris. He has written extensively about French military and civilian satellite programs and is a coauthor of *Espace et Puissance* (1999).

Bob Preston is a former U.S. Air Force officer who is now with RAND. Much of his Air Force career was spent working on space policy and technology issues. In addition, he is the author of *Plowshares and Power: The Military Uses of Civil Space* (1994), which includes an assessment of commercial observation satellites and their implications for U.S. security policy and policymaking.

Isabelle Sourbès-Verger is a geographer and senior researcher at the *Centre National de la Recherche Scientifique* and the *Fondation pour la Recherche Strategique* in Paris. Her research focuses on comparative space policies in the world with a special interest in national strategy for space occupation. Most recently, she coauthored *Atlas de Géographie de l'Espace* (1997), to be translated and updated in English as *Cambridge Encyclopedia of Space* (2000), and she is also coauthor of *Espace et Puissance* (1999). She has written different French and English articles about civilian and military uses of space.

Dr. Gerald Steinberg is professor of politics and directs the Program on Conflict Resolution at Bar-Ilan University in Israel. He has written extensively on arms control, verification, and space imaging issues, including a study assessing the security implications of commercial observation satellites: *Dual Use Aspects of Commercial High-Resolution Imaging Satellites* (1998).

William E. Stoney is currently principal engineer for Mitretek Systems, supporting NASA's Stennis Space Center's Commercial Data Buy and Land Imaging Satellite Data Base programs. He was director of NASA's Earth Observation

Program from 1972 through 1978, during which Landsats 1, 2, and 3 were launched, the Thematic Mapper sensor was defined and developed, and NOAA's TIROS and GEOS satellites and sensors were developed and launched. After leaving NASA, he worked for RCA and General Electric, supporting NASA in the development of the EOS program and for MITRE and now Mitretek on the current and future Landsat systems.

Kazuto Suzuki is writing his doctoral thesis on European space policy and is also studying Japanese space programs in association with Isabelle Sourbes-Verger and Xavier Pasco. He is currently a researcher at the Sussex European Institute at the University of Sussex and has been a part-time lecturer at the School of Policy Science at Ritsumeikan University since April 2000. He received his master's degree in international relations from Ritsumeikan University in Kyoto.

George J. Tahu is a space policy analyst at NASA headquarters in Greenbelt, Maryland. He is a graduate of the George Washington University's Space Policy Institute and the International Space University, and previously worked with the ANSER Corporation as a specialist on Russian space activities, including remote sensing programs.

John Weikel is an employee of the National Imagery and Mapping Agency (NIMA) in Reston, Virgina, and has been the requirements officer and production manager for the SOUTHCOM Customer Operations Team for the past 10 years. In this capacity, Mr. Weikel initiated the use of RADARSAT to produce many of the early products used by the Military Observer Mission in Ecuador and Peru following the 1995 conflict, and he oversaw the production of the final treaty maps. Mr. Weikel holds a Bachelor of Arts degree in geography from Miami University in Oxford, Ohio.

Dr. Ray A. Williamson is a research professor of space policy and international affairs at the George Washington University's Space Policy Institute, which is part of the Elliott School of International Affairs. He served for 16 years with the Office of Technology Assessment (OTA) of the U.S. Congress, where he directed their space-related studies and several studies on imaging satellite policies. Dr. Williamson directs the Space Policy Institute's Dual-Purpose Space Technologies project. He has written extensively on earth observation and remote sensing policy issues.

Appendix E
Selected Bibliography

Abiodun, Adigun Ade, "Remote Sensing in the Information Age," *Space Policy*, Vol. 14, No. 4, November 1998, pp. 229–238.

Albright, David, Corey Gay (Hinderstein), and Frank Pabian, "New Details Emerge on Pakistan's First Nuclear Test Site," *EOM: The Magazine for Geographic, Mapping, Earth Information*, Vol. 7, No. 12, December 1998–January 1999, pp. 18–21.

Albright, David, and Corey Gay (Hinderstein), and Khidhir Hamza, "Development of the al-Tuwaitha Site: What If the Public or the IAEA Had Overhead Imagery?" *ISIS Issue Brief*, Washington, D.C.: Institute for Science and International Security, April 26, 1999.

Albright, David, and Corey Hinderstein, "The Age of Transparency," *Imaging Notes*, Vol. 15, No. 2, Thorton, Colo.: Space Imaging, Inc., March–April 2000.

—, "Evidence of Camouflaging of Suspect Nuclear Waste Sites," David Albright and Kevin O'Neill, eds., *Solving the North Korean Nuclear Puzzle*, Washington, D.C.: Institute for Science and International Security, ISIS Press, 2000.

Amato, Ivan, "God's Eyes for Sale," *Technology Review*, Vol. 102, No. 2, March–April 1999, pp. 37–41.

American Society for Photogrammetry and Remote Sensing (ASPRS), *Land Satellite Information in the Next Decade*, Conference Proceedings, Bethesda, Md.: American Society for Photogrammetry and Remote Sensing, 1995.

Anselmo, Joseph C., "Commercial Imagery Aids Bosnian Peace Mission," *Aviation Week and Space Technology*, February 5, 1996, p. 60.

Apt, Jay, and Michael Helfert, and Justin Wilkinson, *Orbit: NASA Astronauts Photograph the Earth*, Washington, D.C.: The National Geographic Society, 1996.

Babington-Smith, Constance, *Air Spy: The Story of Photo Intelligence in World War II*, New York: Ballantine Books, 1957.

Baker, James D., *Planet Earth: The View From Space*, Cambridge, Mass.: Harvard University Press, 1990.

Baker, John C., *Trading Away Security? The Clinton Administration's 1994 Decision on Imaging Satellite Exports*, Pew Case Studies in International Affairs, Washington, D.C.: Institute for the Study of Diplomacy, Georgetown University, 1997.

Baker, John C., and David G. Wiencek, "Sat-images Could Be Spratlys Salvation," *Jane's Intelligence Review*, Vol. 11, February 1999, pp. 50–54.

Baker, John C., and Ray A. Williamson, "The Implications of Emerging Satellite Information Technologies for Global Transparency and International Security," in Finel and Lord (2000), pp. 221–225.

Bankes, Steve, and Carl Builder, *Seizing the Moment: Harnessing the Information Technologies*, Santa Monica, Calif.: RAND, N-3336-RC, 1992.

Behling, Thomas G., and Kenneth McGruther, "Satellite Reconnaissance of the Future," *JFQ: Joint Forces Quarterly*, No. 18, Spring 1998, pp. 23–30.

Bell, Trudy E., "Remote Sensing," *IEEE Spectrum*, March 1995, pp. 24–31.

Berner, Steve, "Proliferation of Satellite Imaging Capabilities: Developments and Implications," in Sokolski (1996), pp. 95–129.

Bernstein, A., *Monitoring Large Enrichment Plants Using Thermal Imagery from Commercial Satellites: A Case Study*, Albuquerque, N.M.: Sandia National Laboratories, SAND2000-8671, May 2000.

Bjorgo, Einar, *Use of Very High Spatial Resolution Satellite Sensor Imagery in Refugee Relief Operations*, Berger, Norway: University of Bergen and Nansen Environmental and Remote Sensing Center, Ph.D. Thesis, ISBN 82-995329-06, July 1999.

—, "Aiding Refugee Relief," *Imaging Notes*, Vol. 14, No. 6, Thorton, Colo.: Space Imaging, Inc., November–December 1999, pp. 20–21.

—, "Very High Resolution Satellites: A New Source of Information in Humanitarian Relief Operations," *Bulletin of the American Society for Information Science*, Vol. 26, No. 1, October–November 1999.

—, "Using Very High Spatial Resolution Multispectral Satellite Sensor Imagery to Monitor Refugee Camps," *International Journal of Remote Sensing*, Vol. 21, No. 3, February 2000.

Black, Todd J., "Commercial Satellites: Future Threats or Allies?" *Naval War College Review*, Vol. 52, No. 1, Winter 1999, pp. 99–114.

Brauer, Doug, "Restrictions on Remote Sensing Satellite Imaging: Is Israel the Exception or the Precedent?" *Swords and Ploughshares: A Journal of International Affairs*, Vol. 8, No. 1, Fall 1998, pp. 63–76.

Brin, David, *The Transparent Society*, Reading, Mass.: Perseus Books, 1998.

Bruginoni, Dino, "Aerial Photography: Reading the Past, Revealing the Future," *Smithsonian*, Vol. 14, No. 12, March 1984, pp. 150–161.

—, *Eyeball to Eyeball: The Inside Story of the Cuban Missile Crisis*, New York: Random House, 1990.

—, "The Art and Science of Photoreconnaissance," *Scientific American*, Vol. 274, No. 3, March 1996, pp. 78–85.

Burrows, William E., *Deep Black: Space Espionage and National Security*, New York: Random House, 1986.

———, *The New Ocean: The Story of the First Space Age*, New York: Random House, 1998.

Campbell, James B., *Introduction to Remote Sensing*, New York: The Guilford Press, 1996.

Colwell, Robert N., ed., *Manual of Remote Sensing*, Vol. I, Falls Church, Va.: American Society of Photogrammetry, 1983.

Commission on Earth Studies, Space Studies Board, National Research Council, *Earth Observations From Space: History, Promise, Reality*, Washington, D.C., National Academy of Sciences, 1995.

Conway, Eric D., and The Maryland Space Grant Consortium, *An Introduction to Satellite Image Interpretation*, Baltimore, Md.: The Johns Hopkins University Press, 1997.

Crocker, Chester A., Fen Osler Hampson, and Pamela Aall, eds., *Herding Cats: Multiparty Mediation in a Complex World*, Washington, D.C.: United States Institute of Peace Press, 1999.

Dailey, Brian, and Edward McGaffigan, "US Commercial Satellite Export Control Policy: A Debate," in Sokolski (1996), pp. 169–180.

Davis, Richard, "The Foreign Policymaking Role of Congress in the 1990s: Remote Sensing Technology and the Future of Congressional Power," *Congress & the Presidency*, Vol. 19, No. 2, Autumn 1992, pp. 175–195.

Day, Dwayne A., John M. Logsdon, and Brian Latell, eds., *Eye in the Sky: The Story of the Corona Spy Satellites*, Washington, D.C.: Smithsonian Institution Press, 1998.

Dehqanzada, Yahya A., and Ann M. Florini, *Secrets for Sale: How Commercial Satellite Imagery Will Change the World*, Washington, D.C.: Carnegie Endowment for International Peace, 2000.

Doyle, Stephen E., *Civil Space Systems: Implications for International Security*, United Nations Institute for Disarmament Research, 1994.

Dunn, Lewis A., and Richard Davis, et al., *Implications of Commercial Satellite Imagery On Arms Control: A Conference Report*, McLean, Va.: Science Applications International Corporation, Center for Global Security and Cooperation, April 5, 1999.

Eisenbeis, Kathleen M., *Privatizing Government Information: The Effects of Policy on Access to Landsat Satellite Data*, Metuchen, N.J.: The Scarecrow Press, Inc., 1995.

Estes, John E., and John R. Jensen, "Development of Remote Sensing Digital Image Processing and Raster GIS," in Foresman (1998), pp. 163–180.

European Space Agency, Committee on Earth Observation Satellites (CEOS), *Towards an Integrated Global Observing Strategy: 1997 CEOS Yearbook*, 1997.

Ferster, Warren, "Private Spacecraft Imagery Evolves as Treaty Tool," *Space News*, October 20–26, 1997, p. 3.

Finel, Bernard I., and Kristin M. Lord, "The Surprising Logic of Transparency," *International Studies Quarterly*, Vol. 43, June 1999, pp. 335–336.

Finel, Bernard I., and Kristin M. Lord, eds., *Power and Conflict in the Age of Transparency*, New York: St. Martin's Press, 2000.

Florini, Ann M. "The Opening Skies: Third Party Imaging Satellites and US Security," *International Security*, Vol. 13, Fall 1988, pp. 91–123.

---, "The End of Secrecy," *Foreign Policy*, Vol. 111, Summer 1998, pp. 50–63.

Florini, Ann, and Yahya Dehqanzada, "Commercial Satellite Imagery Comes of Age," *Issues in Science and Technology*, Vol. 16, No. 1, Fall 1999, pp. 45–52.

Foresman, Timothy W., ed., *The History of Geographic Information Systems: Perspectives from the Pioneers*, Upper Saddle River, N.J.: Prentice Hall PTR, 1998.

Frelinger, David, and Mark Gabriele, *Remote Sensing Operational Capabilities: Final Report*, Santa Monica, Calif.: RAND, MR-1172.0-OSTP, October 1999.

Gabbard, C. Bryan, Kevin M. O'Connell, George S. Park, and Peter J.E. Stan, *Emerging Markets of the Information Age: A Case Study in Remote Sensing Data and Technology*, Santa Monica, Calif.: RAND, DB-176-CIRA, 1996.

Gabriele, Mark David, *The Treaty on Open Skies and Its Practical Applications and Implications for the United States*, Ph.D. Thesis, Santa Monica, Calif.: RAND Graduate School, RGSD-143, 1998.

Gabrynowicz, Joanne I., "The Promise and Problems of the Land Remote Sensing Act of 1992," *Space Policy*, No. 9, November 1993, pp. 319–328.

Gaddis, John Lewis, "Learning to Live with Transparency: The Emergence of a Reconnaissance Satellite Regime," in John Lewis Gaddis, *The Long Peace: Inquiries Into the History of the Cold War*, New York: Oxford University Press, 1987, pp. 195–214.

George, Hubert, "Remote Sensing of Earth Resources: Emerging Opportunities for Developing Countries," *Space Policy*, Vol. 14, 1998, pp. 27–37.

Gertz, Bill, "Crowding In on the High Ground," *Air Force Magazine*, Vol. 80, No. 4, April 1997, pp. 38–42.

Glackin, David L., and Gerard R. Peltzer, *Civil, Commercial, and International Remote Sensing Systems and Geoprocessing*, El Segundo, Calif.: The Aerospace Press, 1999.

Graham, Mary, "High Resolution, Unresolved," *The Atlantic Monthly*, July 1996, pp. 24–28.

Grundhauser, Larry K., "Sentinels Rising: Commercial High-Resolution Satellite Imagery and Its Implications for US National Security," *Airpower Journal*, Vol. 12, No. 4, Winter 1998, pp. 61–80.

Gupta, Vipin, "New Satellite Images for Sale," *International Security*, Vol. 20, Summer 1995, pp. 94–125.

Gupta, Vipin, and Frank Pabian, "Investigating the Allegations of Indian Nuclear Test Preparations in the Rajasthan Desert: A CTB Verification Exercise Using Commercial Satellite Imagery," *Science and Global Security*, Vol. 6, 1996, pp. 101–189.

—, "Viewpoint: Commercial Satellite Imagery and the CTBT Verification Process," *The Nonproliferation Review*, Vol. 5, No. 3, Spring–Summer 1998, pp. 89–97.

Haakon, Christopher P., "Commercial Space Imagery for National Defense," *Defense Intelligence Journal*, Vol. 8, No. 1, Summer 1999, pp. 24–32.

Hall, Cargill R., "From Concept to National Policy: Strategic Reconnaissance in the Cold War," *Prologue: Quarterly of the National Archives*, Summer 1996, pp. 107–125.

—, "Post War Strategic Reconnaissance and the Genesis of Project Corona," in McDonald, ed., *Corona Between the Sun & the Earth: The First NRO Reconnaissance Eye in Space* (1997), pp. 25–58.

Hall, Keith R., "The National Reconnaissance Office: Revolutionizing Global Reconnaissance," *Defense Intelligence Journal*, Vol. 8, No. 1, Summer 1999, pp. 7–13.

Harris, Ray, *Satellite Remote Sensing: An Introduction*, London: Routledge & Kegan Paul, 1987.

Harris, Raymond, *Earth Observation Data Policy*, Chichester, UK: John Wiley & Sons, 1997.

Hays, Peter L., and Roy F. Houchin II, "Commercial Spysats and Shutter Control: The Military Implications of U.S. Policy on Selling and Restricting Commercial Remote Sensing Data," unpublished paper, October 1999.

Henry, Ryan, and C. Edward Peartree, eds., *The Information Revolution and International Security*, Washington, D.C.: The CSIS Press, 1998.

Heppenheimer, T. A., "Eyes in the Skies," *Invention and Technology*, Vol. 15, No. 4, Spring 2000, pp. 10–16.

Holbrooke, Richard, *To End a War*, New York: Random House, 1998.

Iannotta, Ben, "Setting the Rules for Remote Sensing," *Aerospace America*, April 1999, pp. 34–38.

Jasani, Bhupendra, and Toshibomi Sakata, *Satellites for Arms Control and Crisis Monitoring*, Oxford: Oxford University Press, 1987.

Jayaraman, K.S., "Imagery Deal Spurs Indian Policy Review," *Space News*, January 17, 2000, pp. 3, 20.

Jensen, John R., *Introductory Digital Image Processing: A Remote Sensing Perspective*, Upper Saddle River, N.J.: Prentice Hall, 1996.

—, *Remote Sensing of the Environment: An Earth Resource Perspective*, Upper Saddle River, N.J.: Prentice Hall, 2000.

Johnson, Dana J., Max Nelson, and Robert J. Lempert, *U.S. Space-based Remote Sensing: Challenges and Prospects*, Santa Monica, Calif.: RAND, N-3589-AF/A/OSD, 1993.

Johnson, Dana J., Scott Pace, and C. Bryan Gabbard, *Space: Emerging Options for National Power*, Santa Monica, Calif.: RAND, MR-517, 1998.

Johnson, Rich, "Drawing the Lines in Bosnia," *The Military Engineer*, February–March 1996, pp. 52–55.

Keohane, Robert O., and Joseph S. Nye, Jr., "Power and Interdependence in the Information Age," *Foreign Affairs*, Vol. 77, No. 5, September–October 1998, pp. 81–94.

Klotz, Frank G., *Space, Commerce, and National Security*, New York: Council on Foreign Relations Press, 1998.

KPMG, *The Satellite Remote Sensing Industry: A Global Review*, Arlington, VA: KPMG Peat Marwick, Space and High Technology Practice, 1998.

Krepon, Michael, "Spying from Space," *Foreign Policy*, Vol. 75, 1989, pp. 92–108.

Krepon, Michael, Peter Zimmerman, Leonard Spector, and Mary Umberger, eds., *Commercial Observation Satellites and International Security*, New York: St. Martin's Press, 1990.

Krepon, Michael, and Amy Smithson, eds., *Open Skies, Arms Control, and Cooperative Security*, New York: St. Martin's Press, 1992.

Lauer, Donald T., Stanley A. Morain, and Vincent V. Salomonson, "The Landsat Program: Its Origins, Evolution, and Impacts," *PE&RS*, Vol. 63, No. 7, July 1997, pp. 831–838.

Lenczowski, Roberta E., "The Military as Users and Producers of Global Spatial Data," in David Rhind, ed., *Framework for the World*, Cambridge: GeoInformation International, 1997, pp. 85–110.

Lillesand, Thomas M., and R. W. Kiefer, *Remote Sensing and Image Interpretation*, New York: Wiley & Sons, 2000.

Litfin, Karen T., "The Status of the Statistical State: Satellites and the Diffusion of Epistemic Sovereignty," *Global Society*, Vol. 13, No. 1, 1999, pp. 95–116.

Liverman, Diana, Emilio F. Moran, Ronald R. Rindfuss, and Paul C. Stern, eds., *People and Pixels: Linking Remote Sensing and Social Science*, Washington, D.C.: National Academy Press, 1998.

Livingston, Steve, and Todd Eachus, "Humanitarian Crises and U.S. Foreign Policy: Somalia and the CNN Effect Reconsidered," *Political Communications*, Vol. 12, No. 4, 1995, pp. 413–429.

Logsdon, John M., ed., *Exploring the Unknown: Selected Documents in the History of the U.S. Civil Space Program*, Vol. III: *Using Space*, Washington, D.C.: U.S. Government Printing Office, 1998.

Logsdon, John M., and Russell J. Acker, eds., *Merchants and Guardians: Balancing U.S. Interests in Global Space Commerce*, Washington, D.C.: The George Washington University, Space Policy Institute, May 1999.

Mack, Pamela E., *Viewing the Earth: The Social Construction of the Landsat Satellite System*, Cambridge, Mass.: The Massachusetts Institute of Technology Press, 1990.

Mack, Pamela E., and Ray A. Williamson, "Observing the Earth From Space," in Logsdon (1998), pp. 155–177.

Macauley, Molly, *Economic Aspects of Earth Observation Policy Development*, Washington, D.C.: Resources for the Future, May 1992.

Macauley, Molly K., and Timothy J. Brennan, *Enforcing Environmental Regulation: Implications of Remote Sensing Technology*, Washington, D.C.: Resources for the Future, 1996.

Mahnken, Thomas G., "Why Third World Space Systems Matter," *Orbis*, Vol. 35, Fall 1991, pp. 563–579.

Mathews, Jessica T., "Power Shift," *Foreign Affairs*, Vol. 76, No. 1, January/February 1997, pp. 50–66.

McDonald, Robert A., ed., *Proceedings: Space Imagery and News Gathering for the 1990s: So What?* Bethesda, Md.: American Society for Photogrammetry and Remote Sensing, 1991.

——, *Corona Between the Sun & the Earth: The First NRO Reconnaissance Eye in Space*, Bethesda, Md.: American Society for Photogrammetry and Remote Sensing, 1997.

——, "NRO's Satellite Imaging Reconnaissance: Moving from Cold War Threat to Post–Cold War Challenges," *Defense Intelligence Journal*, Vol. 8, No. 1, Summer 1999, pp. 55–91.

Merges, Robert P., and Glenn H. Reynolds, "News Media Satellites and the First

Amendment," *High Technology Law Journal*, Vol. 3, 1987, pp. 1–32.

Morain, Stanley A., "A Brief History of Remote Sensing Applications, with Emphasis on Landsat," in Liverman et al. (1998), pp. 28–50.

National Academy of Public Administration, *Geographic Information for the 21st Century: Building a Strategy for the Nation*, Washington, D.C.: National Academy of Public Administration, 1998.

Nye, Jr., Joseph S., and William A. Owens, "America's Information Edge," *Foreign Affairs*, Vol. 75, 2, March–April 1996, pp. 20–36.

Oberg, James, "Spying for Dummies," *IEEE Spectrum*, Vol. 36, No. 11, November 1999, pp. 62–69.

Pace, Scott, Kevin M. O'Connell, and Beth E. Lachman, *Using Intelligence Data for Environmental Needs: Balancing National Interests*, Santa Monica, Calif.: RAND, MR-799-CMS, 1997.

Pace, Scott, Brant Sponberg, and Molly Macauley, *Data Policy Issues and Barriers to Using Commercial Resources for Mission to Planet Earth*, Santa Monica, Calif.: RAND, DB-247-NASA/OSTP, 1999.

Philipson, Warren R., ed., *Manual of Photographic Interpretation*, Bethesda, Md.: American Society for Photogrammetry and Remote Sensing, 1997.

Preston, Bob, *Plowshares and Power: The Military Uses of Civil Space*, Washington, D.C.: National Defense University Press, 1994.

Richelson, Jeffrey T., *America's Secret Eyes in Space*, New York: Harper & Row, Publishers, 1990.

Roberts, Margaret A., *U.S. Remote Sensing Data from Earth Observation—Law, Policy and Practice*, Washington, D.C., American Institute of Aeronautics and Astronautics, 1997.

Rostow, W. W., *Open Skies: Eisenhower's Proposal of July 21, 1955*, Austin: University of Texas Press, 1982.

Ruffner, Kevin C., ed., *CORONA: America's First Satellite Program*, Washington, D.C.: Center for the Study of Intelligence, U.S. Central Intelligence Agency, 1995.

Simpson, Christopher, ed., *Remote Sensing: A Legal Primer and Resource Book*, Washington D.C.: American University School of Communication, 1996.

Smith, D. Brent, "Overcoming Impediments to International Cooperation in Space-based Earth Observation," *Space Times*, Vol. 37, No. 2, March–April 1998, pp. 4–9.

Sokolski, Henry, ed., *Fighting Proliferation: New Concerns for the Nineties*, Washington, D.C.: U. S. Government Printing Office, 1996.

Spires, David N., *Beyond Horizons: A Half Century of Air Force Space Leadership*, Washington, D.C.: U.S. Government Printing Office, 1998.

Stares, Paul B., *The Militarization of Space: U.S. Policy, 1945–84*, Ithaca, N.Y.: Cornell University Press, 1985.

Steinberg, Gerald M., *Satellite Reconnaissance: The Role of Informal Bargaining*, New York: Praeger, 1983.

——, *Dual Use Aspects of Commercial High-Resolution Imaging Satellites*, Ramat Gan, Israel: Bar-Ilan University, The Begin-Sadat Center for Strategic Studies, February 1998.

Strebeck, John, "Organizing National Level Imagery and Mapping," *JFQ: Joint Forces Quarterly*, No. 18, Spring 1998, pp. 31–37.

Tahu, George J., John C. Baker, and Kevin M. O'Connell, "Expanding Global Access to Civilian and Commercial Remote Sensing Systems and Data: Implications and Policy Issues," *Space Policy*, Vol. 14, No. 3, August 1998, pp. 179–188.

Uhlir, Paul F., "From Spacecraft to Statecraft: The Role of Earth Observation Satellites in the Development and Verification of International Environmental Protection Agreements," *GIS Law*, Vol. 2, No. 3, Winter 1995, pp. 1–15.

U.S. Congress, House of Representatives, Joint Hearing before the Committee on Science, Space, and Technology and the Permanent Select Committee on Intelligence, *Commercial Remote Sensing in the Post–Cold War Era*, 103rd Congress, 2nd session, Washington, D.C., U.S. Government Printing Office, 1994.

——, House of Representatives, Committee on Science, Subcommittee on Basic Research, Hearing, *Remote Sensing Applications as a Research and Management Tool*, 105th Congress, 2nd session, Washington, D.C., U.S. Government Printing Office, 1998.

——, Office of Technology Assessment, *Civilian Satellite Remote Sensing: A Strategic Approach*, Washington, D.C.: U.S. Government Printing Office, September 1994.

——, Office of Technology Assessment, *The Future of Remote Sensing from Space: Civilian Satellite Systems and Applications*, Washington, D.C.: U.S. Government Printing Office, July 1993.

——, Office of Technology Assessment, *Remotely Sensed Data: Technology, Management, and Markets*, Washington, D.C.: U.S. Government Printing Office, September 1994.

——, Office of Technology Assessment, *Verification Technologies: Cooperative Aerial Surveillance in International Agreements*, Washington, D.C.: U.S. Government Printing Office, July 1991.

U.S. Department of Commerce, National Oceanic and Atmospheric Administration, "Licensing of Private Land Remote-Sensing Space Systems (Proposed Rules)," *Federal Register*, Vol. 62, No. 212, November 3, 1997, pp. 59, 317–59, 331.

——, National Oceanic and Atmospheric Administration, "Licensing of Private Land Remote-Sensing Space Systems (Rules and Regulations)," *Federal Register*, Vol. 65, No. 147, July 31, 2000, pp. 46,822–46,837.

U.S. Senate, Select Committee on Intelligence, Hearing, *Commercial Imagery*, 103rd Congress, 1st session, Washington, D.C.: U.S. Government Printing Office, 1995.

Wagner, Caroline S., *International Agreements on Cooperation in Remote Sensing and Earth Observation*, Santa Monica, Calif.: RAND, MR-972-OSTP, 1998.

Wilkie, David S., and John T. Finn, *Remote Sensing Imagery for Natural Resources Monitoring: A Guide for First-Time Users*, New York: Columbia University Press, 1996.

Williamson, Ray A., "The Landsat Legacy: Remote Sensing Policy and the Development of Commercial Remote Sensing," *PE&RS*, Vol. 63, No. 7, July 1997, pp. 877–885.

Winnefeld, James A., Preston Niblack, and Dana J. Johnson, *A League of Airmen: U.S. Air Power in the Gulf War*, Santa Monica, Calif.: RAND, MR-343-AF, 1994.

Wright, Robert, "Private Eyes," *The New York Times Magazine*, September 5, 1999.

Wriston, Walter B., "Bits, Bytes, and Diplomacy," *Foreign Affairs*, Vol. 76, No. 5 September–October 1997, pp. 172–182.

Index

A

Advanced Earth Observing Satellite (ADEOS), 208, 214-215
Advanced Land Observing Satellite (ALOS), 214-215, 580
Advanced Very High Resolution Radiometer (AVHRR), 141, 176, 404
  use for environmental monitoring during Gulf War, 103-104
Aerial Images, 157, 175, see also TerraServer
Aerial photography and imagery, 38, 74, 460, 466-468, 477
  aerial reconnaissance, 18-22, 507, 597-599
  alternative to satellite imagery, 15, 60, 73-74, 84, 97, 326, 467, 568
  dominant role in remote sensing market, 10, 569, 582
  early postwar development of air reconnaissance, 19-20
  technological advances in, 60, 90, 583
Africa,
  Great Lakes region, 473-474
de Aguiar, Captain Braz Dias, 313
Air Force, see U.S. Air Force
Albania, 404, 418-422, 477, 613, 615
Albright, Madeleine, 285, 290n, 433
Alenia (Aerospazio), 269, 438
Algeria, 468
American Society for Photogrammetry and Remote Sensing (ASPRS), 553, 555, 583
Amnesty International, 463, 479
Antarctica, 267, 277

Anti-satellite weapons, 94-95, 127, 495
Applications, see Remote sensing
Aranha, Oswaldo, 312
Argentina, 439
  guarantor nation for Peru-Ecuador peace treaty, 311-312
Arms control and nonproliferation, 119
  monitoring with imaging satellites, 109, 111, 272, 361-362, 381-383, 468-469
ASEAN Regional Forum, 290n
Asymmetries in access to satellite imagery, 566
  national security implications, 451-455, 459-461, 566
  privacy implications, 456-461
Australia, 135, 196, 266
  acquiring imaging satellites, 10
  remote sensing program, 255
Autometric, 487-488, 497n
Axworthy, Lloyd, 285, 290n

B

Bahrain, 435
Ball Aerospace and Technologies Corporation, 152-153
Balloons,
  means for overhead observation, 18, 20, 32n
Bangladesh, 196
Barre, Raymond, 192
Belgium, 191-192
Bissell, Richard, 23
Biodiversity, 465, 468
Bjorgo, Einar, 471, 474

Boeing Corporation, 157
Bosnia-Herzegovina, 9, 111, 227, 249, 259, 300-303, 307, 437, 475, 477, 517, 566
Brazil,
    acquiring imaging satellites,10, 439
    Brasilia Declaration, 323
    guarantor nation for Peru-Ecuador peace treaty, 311-312
Brender, Mark, 486
Browne, VADM Herbert, 495
Brunei, 327, 329-330
Bush, George, 43
    administration, 5
    Open Skies Treaty proposal, 30, 455

C

Camber Corporation, 299
Cambridge Research Associates, 199, 318
Camouflage, 92, 94, 107, 333, 367-368, 373-375, 377, 400-401, 479, 566, *see also* Concealment
Canada, 3, 49, 58, 76n, 85, 98n, 135, 137, 439, 442, 466
    agreements and disagreements with United States, 264-265, 272, 276-286, 289n, 290n
    applications radar imaging data, 267-270, 272-276
    commercial imagery sales, 4, 39, 46
    government regulations on satellite imaging, 76n, 130n, 264-266, 270-272, 267-280, 284-285
    data policy for satellite imagery data, 265-266, 270-272, 276, 278
    history of remote sensing program, 263-265, 268, 274-275, 276-277
    international partnerships, 264-265, 266, 276-282
    ground receiving stations, 266, 272
    maritime monitoring missions, 267-270, 247
    RADARSAT-1, 76n, 130n, 151-152, 199, 238, 263-278, 286n, 287n, 288n, 317-318, 319, 320, 321, 405, 582
    RADARSAT-2, 76n, 96, 99n, 156, 162n, 263-265, 277-282, 284, 286-287, 288n, 442
    remote sensing commercialization, 277-278, 286
    remote sensing organizations, 264-265, 270-272
    Sentry mobile receiving station, 273-274, 288
Cartography, 143, 197, 198, *see also* maps
    automated cartography, 298
Chairman of the Joint Chiefs of Staff, 315
Chernobyl nuclear reactor, 552
    disaster captured on satellite imagery, 470
Chiapas, Mexico, 404, 411-415, 608-609
Chile,
    guarantor nation for Peru-Ecuador peace treaty, 311-312

China, *see* People's Republic of China (PRC)
China-Brazil Earth Resources Satellite (CBERS), 237
Churchill, Winston,
    "Iron Curtain" speech, 18
Civilian observation satellites,
    diverse national motives for acquiring, 62-63
    eroding distinction with commercial observation satellites, 4-5, 565
    projected growth in numbers of satellites, 579-587
    smaller imaging satellites, 5, 74, 560
Clark, LTG Wesley Clark, 300
Clinton, William J., 280, 283
    administration, 5
Cloud cover, *see* Weather
CNES, *see* France
Cohen, Eliot, 227
Cold War, 4, 14, 136, 17-31, 56, 73, 226, 452, 464, 478, 490, 571
    effect of end on commercial imaging satellites, 5, 31, 68
    superpower monopoly on high-resolution imaging satellites, 135, 437, 455
Commercialization of satellite remote sensing,
    ambivalence of governments, 71, 159-162, 519
    differing national conceptions, 62, 146-150, 159-162, 192-195, 197-199, 212-213, 277-278
    distinguishing commercialization from privatization, 50n, 56
    incentive to assist new imagery users, 549-550
    interest of private firms in satellite remote sensing, 1, 56-57, 58, 68-71, 150-158, 561, 572
    global market for satellite imagery, *see* Market
    need to adopt new business models, 2, 71-73
    U.S. government support for, 56-57, 85, 146-150, 159-162, 519-520
Commercial observation satellites, 106-107, 580-582, *see also* Chapters 14, 15, and 16
    asymmetrical access to, 433, 562, 566
    competition from parastatal enterprises,10, 62-63, 73-75, 569
    costs of, 443-445, 446n
    diminishing barriers to acquiring, 441-446
    diverse national motives, 62, 74-75, 443
    projected growth in numbers of satellites, 3, 107, 560, 568, 581
    potential security benefits, 101, 107-111, 226, 566
    potential security risks, 57-58, 59, 101, 115-125, 199-202, 226-227, 440, 515, 566-567
    sales of turnkey systems, 62, 443, 480
    technical setbacks, 11n, 70, 580-581
    used to help settle territorial disputes, 2-3, 101, 109, 111, 296,
Comprehensive Test Ban Treaty (CTBT), 361, 362, 378
    CTBT Organization, 370, 377, 469

monitoring use of commercial satellite imagery, 362, 377, 378n
Concealment, 7, 92, 94, 127-128, 377
  India's concealment of nuclear testing, 362, 367-368
  military activities in the South China Sea, 337
  Pakistan's concealment of nuclear testing, 362, 369-370, 375, 602
Congress, *see* U.S. Congress
Coolidge, Charles A., 25
Coral Cay Conservation, 466
Core Software Technology (CST), 176, 232-233, 239n
CORONA satellite program, 38, 139
  concern over "freedom of space," 24-25
  declassification of satellite imagery, 5, 31, 160, 331, 469, 552
  national reconnaissance capability, 4
  origins, 23-25

D

Data policy for satellite imagery, 40-41, 47, 49, 571
  difference between U.S. policy and other nations, 49
  nondiscriminatory access to satellite imagery, 143, 515, 572
  restrictions on imagery data access, 4, 59, 438-440, 559, 569-570
Davies, Merton E., 23, 32
Dayton peace talks concerning Bosnia, 296, 453, 566
  Implementation Force (IFOR), 302
  use of imagery to support negotiations, 9, 111, 297-310, 592-593
Deception, 92, 368, 540
  pixel plasticity, 496, 540
  used to avoid detection, 94, 368, 377, 479
  uses of satellite imagery, 105
  use in warfare, 94, 127-128, 495-496
Defense Mapping Agency, see U.S. Defense Mapping Agency
Desert Storm, *see* Gulf War
Dewar, Michael, 495-496
Digital,
  "digital divide," 497n
  digital elevation model (DEM), 107, 110, 267, 318, 322, 473
  digital images, 3, 47, 69, 171, 172, 569
  digital image processing, 47-48
  digital maps used in Dayton talks, 298
  digital ortho-image, 64, 65, 67
  digital terrain models, 422
Digital Terrain Elevation Data (DTED), 309n, 592
Diplomacy, 566
  use of satellite imagery to support negotiations, 9, 566

Dobbs, Michael, 494
Drug trafficking, 118, 270, 276
Dual-purpose technology, 101, 125, 565-567
  defined as, 439, 573
Dual-use technology, 1, 2, 9, 13, 15, 202, 272-276, *see also* Dual-purpose technology
  potential military and security benefits, 173, 226, 566
  potential security risks, 8, 226-227, 240-241, 276, 288n, 565
Dubno, Daniel, 485, 489-490

E

EAGLE VISION program, 106, 112
Earth Observing Satellite (EOSAT) Corporation, 42-43, 151, 175, *see also* Landsat
Earth Observing System (EOS), 136, 140-141, 275-276, 580
Earth Observation System Data and Information System (EODIS), 49, 146
Earth Resources Technology Satellite (ERTS), *see* Landsat
Earth Science Enterprise, *see* National Aeronautics and Space Administration (NASA)
Earth Search Sciences, Inc. (ESSI), 156
  NEMO satellite program, 156-157
EarthWatch, 50, 150, 152-154, 159, 232, 582
  EarlyBird satellite, 147
  international partnerships, 153
  origins as WorldView, 152-153
  QuickBird satellites, 147, 153-154
Ecuador, 255
  Ecuador-Peru peace treaty, 296, 311
  map signing ceremony in St. Louis, 312, 326
  use of satellite imagery to support peace negotiations, 9, 111, 307-308, Chapter 15, 566, 594-595
Egypt, 230
  remote sensing program, 229, 239
Einaudi, Luigi, 311, 319
Eisenhower, Dwight, 14, 32n
  administration, 21-26
  and U-2 aerial overflights, 22, 26-27, 33, 34n
  decision to pursue satellite reconnaissance, 22-23
  Open Skies proposal of 1955, 21-26, 32, 455
Electro-optical imaging satellites, 5, 7, 11n, 85, 122, 573n, 579
Environment, 569
  deforestation, 7, 413-415, 466, 480, 490, 609, 612
  environmental management, 90, 325, 415, 417-418
  environmental monitoring, 7, 72, 90, 267-270, 277, 288n, 425, 468, 559, 564, 581, 590
  environmental problems and disasters, 7, 268-269, 404, 406, 467, 562

Montreal Protocol, 466
U.S. Endangered Species Act, 466
Environmental Systems Research Institute (ESRI), 155, 467, 474, 478, 550
EOSAT Corporation, 151-152, 188, 207, 228, 238-239, 255
ERS-1 and -2 satellites, 45, 84, 151-152, 189-190, 199, 207, 254, 288, 404-405, 422, 426, 609, 615
  ground receiving stations, 190
ERDAS, 309, 310n, 399, 467, 474, 478, 550
Ethnic cleansing,
  use of satellite imagery to carry out, 567
  use of satellite imagery to provide evidence, 437, 477
Eurimage, 176, 178
Europe, 5, 61, 85, 116, 122, 503
European Space Agency (ESA), 45, 49, 151, 187, 189-190, 192, 254, 287
European Union, 201, 309n,
  Common Foreign and Security Policy, 201
Export controls, 94-96, 501, 526n-527n

F

Federal Emergency Management Agency (FEMA), 159
Federation of American Scientists (FAS), 426n, 469, 497n-498n, 534, 546, 564
  assessment of North Korean missile program, 489, 555n
  Public Eye project on overhead imagery, 288n, 469, 546
First amendment, see U.S. Constitution
France, 3, 21, 56, 58, 439-440, 443, see also SPOT
  Centre National d'Etudes Spatiales (CNES), 44, 187, 188, 190, 191, 195, 198
  commercial imagery sales, 4, 194-198, 204n
  commitment to global transparency, 188, 200, 202
  data policy for satellite imagery data, 192, 197, 201
  different approach to commercialization, 136, 187-188, 192-193, 194, 195-199
  Earth observation organizations, 188-190, 193-196, 203n
  Helios military reconnaissance satellite, 45, 188, 200, 201-202
  history of remote sensing program, 188-197, 203n, 204n
  proposal for international satellite monitoring agency, 188
  restrictions on Iraqi access to imagery, 104, 512
  security concerns about high-resolution imagery, 199-202
  use of French satellite imagery in Gulf War, 102-104

G

Gabrynowicz, Joanne I., 287n
Gejdenson, Rep. Sam, 281
Geographic information system (GIS), 10, 54, 59, 63-65, 70, 71, 76n, 77n, 152, 155, 168, 203, 274, 407, 415, 422, 466-467, 475, 480, 589
  as enabling technology for imagery users, 47, 63-65, 69, 158, 561
  definition, 76n
  how used with remote sensing data, 64, 300, 613, 614
  smart GIS data sets, 64, 67
  use to support Dayton negotiations, 298, 300, 305, 309
Geolocation accuracy, see Imagery characteristics
Germany, 31, 515
  remote sensing program, 191, 255, 445, 583
Global Disaster Information Network (GDIN), 426n
Global Positioning System (GPS), 10, 59, 76, 94, 125, 130n, 227, 228, 304, 323, 419, 424-425, 472, 561, 562, 613
  differential GPS, 325-326
Global transparency, 1, 13-14, 17, 54, 206, 272, 295, 429-430, 480-481, 559
  ability to image denied areas, 7, 30
  bilateral transparency, 451, 461
  causes of, 6-7
  characteristics of, 1
  constraints on, 75
  contribution of imaging satellites, 1, 6-10, 25, 74-75, 135, 161-162, 463, 480, 559, 562, 570
  contribution of information technologies, 1, 6-7, 32, 49, 486, 562
  implications of, 75, 121-122, 404
  information asymmetry, 430, 451-452, 458-460
  market implications, 74-75, 161-162
  negative aspects of, 8-9, 225-227, 240-241, 450-452, 562
  positive aspects of, 21, 446, 562, 573
  two schools of thought concerning, 450
  tectum transparency, 491-495, 497n
  unilateral transparency, 430, 454-455, 461
Governments,
  changing relationship with nonstate actors, 463-464, 479-480
  dominant role in remote sensing, 480, 580-581
  multiple roles in remote sensing marketplace, 10
  ownership or subsidies to satellite imaging programs, 10
Guatemala, 411-412
Gulf War, 15, 130n, 228, 259, 512
  Coalition Forces, 103-106
  news media use of imagery, 470
  risk of detecting "Hail Mary" flanking maneuver, 98n, 105, 129n, 436

Index

stimulates international interest in imaging satellites, 56, 434
use of civilian and commercial satellite imagery in, 44, 101-106, 129n, 173, 273, 434, 436
Gupta, Vipin, 228, 488

H

Hall, Cargill, 24
Holbrooke, Richard Amb., 297-298, 300, 309n
Holey, Brett, 487
Hostage rescue operations, 116, 287n, 344
  in Peru, 515-516
  risk of exposing operations, 119
Human rights, 438, 478, 562
  Large-scale atrocities, 430, 437
  satellite images to detect mass graves, 437, 477
  using imagery against internal groups, 117, 474
Human Rights Watch, 476, 479
Humanitarian emergencies, 3, 8, 110, 116, 296, 566, 572
  ENVIREF, 408, 426n
  internally displaced persons (IDP), 275, 403
  OpenEyes, 426n
  relief operations, 270, 405-407
  ReliefSat, 408, 426n
  use of satellite imagery, 274-275, 404-425, 566, 604-615
Hyperspectral imaging, 6, 15, 58, 86, 425, 459, 468, 487, 565-566
  challenges of licensing, 96-97
  development of hyperspectral imaging satellites, 90, 107, 154, 156, 561, 583, 587
  imagery data characteristics, 86-87, 98n, 591
  potential military applications, 107, 154, 156, 566

I

IKONOS, 4, 33n, 34n, 70, 107, 139, 151-152, 199, 239, 256, 377, 404, 410-411, 417-419, 426n, 469, 476, 486-487, 534, 582
  first commercial imaging satellite, 4, 107n, 149, 152, 241, 480, 485, 564
  satellite imagery, 4, 74, 606, 611-612
  Space Imaging, 50, 70, 74, 139, 147, 150, 159, 232, 255, 377, 486, 489, 491, 497n, 564
Image interpretation, 88, 560
  change detection, 537
  equipment for, 537-538
  ground truth, 538
  key steps, 536-537
  necessary skills, 120-121, 535-538, 561, 565, 566
  observables, 536, 541, 547
  signatures, 97, 482n, 536
Image processing,
  geometric corrections, 556n
  radiometric corrections, 556n

ImageSat International, 231-234, 235-236, 237, 440, *see also* Israel
  EROS commercial observation satellites, 229, 231-234, 235-237
  origins as West Indian Space (WIS), 233
  satellite operating partners (SOP), 233-234
Imagery activists, 18, 431, 533, 547-550, 553-554, 564, 573
  challenges facing new imagery users, 8, 431, 479, 542-544, 564-565
  new users of satellite imagery, 407, 411, 431, 533-534, 560-561, 565
Imagery characteristics,
  Field-of-view, 579
  geolocation, 93, 106, 561
  image elements, 536-537,
  price, 41, 47, 60, 169, 423-424, 470, 473, 476
  resolution, 84, 86-90, 98
  timeliness, 98, 423-424
Imagery credibility, 8, 565
  challenge of imagery interpretation and analysis, 407, 411, 535-536, 537, 564-565, 573
  deliberate errors in imagery interpretation, 431, 538-542, 554
  inadvertent errors in imagery interpretation, 431, 539, 542-544, 554
  means for enhancing, 540-542, 548-554
  paradox of, 534-535, 554
Imagery data users,
  alternative relationships between new and established users, 545-555
  new users, 433, 534-535
  traditional users, 534-535
Imaging spectrometry, 79, 85-91, 96-98n, 459
India, 2, 58, 137, Chapter 12, 146, 439, 442, 451-452, 468
  border conflicts with Pakistan, 258-259, 439, 451
  "censorship" of satellite imagery, 541
  civilian applications of satellite imagery, 249-251, 259
  commercialization of remote sensing satellites, 255
  data policy for satellite imagery data, 49, 255-256
  history of remote sensing program, 247-256
  Indian Remote Sensing Satellite (IRS), 151, 200-201, 253-255, 257-258, 376, 404, 419-423, 541, 580, 603, 614, 615
  military interest in satellite imaging, 256-259
  missile and rocket launch program, 247-248, 256, 258n-259n
  nuclear testing, 296, 361-368, 481n, 488-489
  restrictions on access to satellite imagery, 255-256, 441
  space-related organizations, 200-201, 247, 259n-260n
  use of commercial imagery to monitor its nuclear test sites, 362-368, 377-378
  uses for satellite imagery data, 250-252

Indonesia, 328
Information age, 1, 479-480
  civilian 'system of system' technologies, 486, 490
  imaging satellites as part of, 1, 10,
  information technologies, 6, 14, 48, 59, 430, 562, 573
  proliferation of video camera technologies, 491-492
  tectum transparency, 491-495, 513n
  telecommunication advances, 491-494
Institute for Science and International Security (ISIS), 381, 469, 481n
International Atomic Energy Agency (IAEA), 382
International Institute for Aerospace Survey and Earth Science (ITC), 553, 556n
International Seismological Center (ISC), 363
International Society for Photogrammetry and Remote Sensing (ISPRS), 553, 555
International Traffic in Arms Regulations (ITAR), 283-285
International War Crimes Tribunal, 437
Internet, 478, 562, *see also* World Wide Web (WWW)
  contribution to global transparency, 6, 49, 490-491, 494
  means for transferring imagery data, 1, 6, 48, 157, 172, 177, 424, 480
Intelligence-gathering by imaging satellites, 84, 92, 110-111, 92, 95
  national technical means (NTM), 119
Iran, 181, 225-228, 230, 242n, 435
  remote sensing program, 240
  missile programs, 240
Iraq, 102-105, 128, 225-228, 230, 235, 424, 435-436
  access to satellite imagery in Gulf War, 102-105, 512
  remote sensing program, 180-181, 239-240
  U.N. Special Commission, 195
  use of satellite images to support military operations, 226, 436
Israel, 3, 10, 58, 116, 136, 439-440
  commercialization of remote sensing satellites, 229, 231-234, 235-237
  concern over threat from high-resolution imagery, 137, 227, 234-235, 236-237, 240-241
  data policy for satellite imagery data, 237-238
  desire to offset military satellite costs, 238
  Earth observation organizations, 231-233, 237
  EROS commercial imaging satellites, 150, 229, 231-234, 235-237
  history of remote sensing program, 229-237
  images of Dimona nuclear complex, 226
  Israel Aircraft Industry (IAI), 231-233, 235, 237, 242n
  Ministry of Defense (MOD) role in shaping policy, 231-232, 235-237
  Offeq military imaging satellite program, 229-231, 234, 237
  U.S.-Israeli relations, 227, 229-230, 234-237
  U.S. restrictions on imaging Israel, 150, 236-237, 506
Iyengar, P. K., 365-366
Izbica, Kosovo, 492-493

J

Japan, 3, 85, 136, 255, 266, 268, 290n, 439
  civilian remote sensing satellites, 151, 208, 211, 213-215, 580
  data policy for satellite imagery data, 210-213
  Earth observation organizations, 206-211, 220
  effect of Taepodong missile launch, 136, 205
  focus on technology development, 205-210
  history of remote sensing program, 213-218
  Information-Gathering Satellites, 215-220, 583
  National Space Development Agency (NASDA), 206-215, 217-221, 221n
  "peaceful purposes" commitment, 211
  remote sensing commercialization, 206-210
Johnson, Lyndon B.,
  public comment on satellite reconnaissance, 30
Johnson-Freese, Joan, 286

K

Karadzic, Radovan, 437
Kennedy, John F.,
  administration, 28-29, 34n
Kenya, 415-419, 425, 611-612
Khrushchev, Nikita, 22, 26-27, 33n
Khrushchev, Sergei, 33n
Kosovo conflict, 110, 116, 255, 477, 493, 546, 613
  humanitarian relief operations, 404, 418-422
  use of imagery to detect atrocities, 477, 492-493
Kuwait, 102, 103, 104, 237, 424, 434
  oil well fires in Gulf War, 103, 273

L

Land Remote Sensing Commercialization Act of 1984, 42-43, 56
Land Remote Sensing Policy Act of 1992, 5, 45, 56, 146, 235, 505-506, 508
  amended by Commercial Space Act of 1998, 505
Landsat program, 5, 7, 37-50, 69, 95, 106, 107, 136, 139, 176, 191-192, 194, 207, 233, 255, 317, 419, 421-425, 465, 467-468, 472, 477, 480, 514, 533, 560, 565, 580-582, 613
  contribution to global transparency, 7, 30, 142
  data policy, 40-41, 47, 49, 143, 145-146
  emphasis on continuity of data, 41-42, 45, 142-143, 145-146
  Enhanced Thematic Mapper (ETM), 46, 144-145
  Enhanced Thematic Mapper Plus, 46, 144-145
  EOSAT, *see* Earth Observing Satellite Corporation
  EROS Data Center (EDC), 40, 47, 48, 143
  Earth Resources Technology Satellite (ERTS), 11n, 39, 276

Index

failed privatization effort, 14, 37, 41-48, 49-50, 521
first civilian remote sensing satellite, 3, 139, 144
ground stations, 40, 143, 454
history of, 37-50, 142-146, 434, 505
imagery characteristics, 49, 86, 144
Landsat-1, -2, and -3, 11n, 39, 142
Landsat-4, and -5, 40-41, 45, 47, 142, 151
Landsat-6 planned mission, 42-43, 46, 142
Landsat-7, 45-47, 140, 142-143, 145, 162n, 582
Multispectral Scanner (MSS), 39, 144-145
model for other national programs, 187-188
remote sensing applications, 142-143
technology development, 144-145
tension between public and private benefits, 43
Thematic Mapper (TM), 39, 144, 145
use in Gulf War, 44, 98n, 101-106, 109, 128
Leghorn, Richard S.,
role in developing open skies concepts, 18-21, 25-27, 33n
Libya, 225, 229
LIDAR (Light Detection and Ranging), 579
Litton/Itek, 235
Lockheed Martin, 151, 232, 264

M

McBride, George, 313, 325n
Macedonia, 404, 418-420, 477, 613, 614
MacDonald Dettwiler, 162n, 264-265, 277-278, 442
Malaysia, 327, 329-330, 332, 351, 353, 445n
Manley, John, 280, 442
Maps, 300-301, 305, 315-317, 318, 323-325, *see also* Cartography
digital mapping, 298, 302, 304, 307-309
meta-data, 307
support equipment, 299-300, 318-319
Topographic Line Maps (TLMs), 102, 317-318
World Geodetic System 1984 (WGS 84) datum, 316-317, 325n
Market, 13, 47-49, 53-78, 95-98, 568-569
global market for satellite imagery, 2, 9-10, 15, 38
government roles in market for imagery, 2, 71, 85, 569
market connection with transparency, 74-75, 568
market projections, 9-10, 68-71, 569
potential market segments, 72
remote sensing as a 'quasi-market,' 569
spatial technology marketplace, 54, 59, 63-67, 72-73, 76n, 559, 568, 572
value-added industry for satellite imagery, 40, 47-48, 53, 70, 72, 157-158
Marine Corps, *see* U.S. Marine Corps
Matra, 106, 193-194, 199
Media, *see* News media
Meyers, Gen. Richard, 280
Microsatellites, 74, 443-446

Disaster Monitoring Constellation, 445n
UoSat-12, 443
RapidEye, 445
Tactical Small Satellite Concept Demonstrator, 445
Middle East, 116, 136, 225-226, 228-230, 238, 240, 443, *see also specific countries*
Military applications of satellite imagery, 101-129
bomb damage assessment, 103
limitations of commercial satellite imagery, 104
military maps and nautical charts, 102
mission planning, 103
reconnaissance, 102, 110, 112
targeting support, 103
Military Observer Mission Ecuador-Peru (MOMEP), 311-312, 315, 317
Military reconnaissance satellites, 1, 18, 102, 226
as National Technical Means (NTM), 159-160, 226, 490
clear distinction with civilian imaging satellites, 4
early postwar development of, 18-20
Mitretek, 583
Missiles,
imagery use for targeting cruise missiles, 108, 125, 130n, 566
Scud missiles, 230
Mission to Planet Earth, *see* NASA
Mongolia, 240
Mowry, Clayton, 280, 447n
Mozambique, 270
Multispectral remote sensing systems, 3, 6, 107, 144-145, 167, 257, 444-445, 487, 579, 581
multispectral imagery, 4, 67, 80, 86, 98n, 109, 116, 173, 181, 191, 408, 410, 423, 425n-426n, 560, 561, 583, 607
pan-sharpened imagery, 107

N

Narco-criminals, 270, 276
access to satellite imagery, 118, 567
National Aeronautics and Space Administration (NASA), 71, 203n, 256, 259n, 583
Commercial Remote Sensing Project (CRSP), 158
Earth Science Enterprise programs, 141, 158, 580
Mission to Planet Earth, 37, 71, 141
role in developing Landsat satellites, 39-41, 45-47
role in RADARSAT program, 264-265, 276-277, 282
science data purchase, 151-152, 158
use of earth observation data, 37, 49
National Association of Broadcasters (NAB), 511
National Oceanic and Atmospheric Administration (NOAA), 43, 198, 255
environmental monitoring satellite program, 140, 580
role in licensing of commercial remote sensing satellites, 130n, 148-150

role managing the Landsat satellites, 40-41, 43-44
role in RADARSAT program, 264-265, 277, 282
National Imagery and Mapping Agency (NIMA), 71, 113-114, 129n, 142, 158-160, 203, 298-299, 309n, 538, *see also* U.S. Defense Mapping Agency (DMA)
   Commercial Imagery Program, 113, 158-159
   role in supporting diplomatic negotiations, 298-299, 311, 318, 319, 315, 317n
National Reconnaissance Office (NRO), 28, 71, 113-114, 160
   Future Imagery Architecture program, 71
National Security Council, 29
   role in Landsat program decisions, 39
National Space Council, 43
Natural disasters, 403-404, 569, 572
   potential use of satellite imagery, 85, 270, 288n, 490, 562, 564
Natural Heritage Network, 465
Nature Conservancy, The, 465
Naval EarthMap Observer (NEMO) satellite program, 156-157
Navy, *see* U.S. Navy
Nepal, 408, 410-411, 425, 466, 606-607
News media, 3, 13, 31, 72, 429, 430, 509, 533, 545-552, 562, 567, 570, 572
   ABC, 470
   CBS, 470, 485, 487-488, 496, 543
   CNN, 487, 492
   concern over restrictions on satellite imagery access, 443, 509, 511-512, 517
   fly-a-way units for transmitting reports, 493, 498n
   interest in satellite imagery, 75
   time constraints on use, 90, 486-489
   NBC, 487, 490
   New York Times, 469-470, 527n
   Newsweek, 481-482, 488-489, 497n, 543, 556n
   Time, 488
   24-hour news cycle, 489, 493
   US News and World Report, 488
   Washington Post, 493, 527n
Nongovernmental organizations (NGOs), 3, 8, 13, 64, 296, 429, 430, 463-379, 533, 544-552, 559, 562, 570, 572
   as users of satellite imagery, 1, 208, 75, 111, 399-401, 463-464, 561, 564-565, 572
   as "imagery activists," 8, 74, 471, 563
   changing relationship with governments, 8, 433
   imagery credibility as new users, 8, 400
   imagery used for public education, 387
   limitations on imagery use, 321, 399-400, 470-471
   limited sharing of resources, 470

NGOs concerned with arms control and non-proliferation, Chapter 17, Chapter 18, 464, 468-471, 480, 489, 564
NGOs concerned with environmental issues, 463-468, 480, 489, 564
NGOs involved in humanitarian assistance, Chapter 19, 471-475, 480, 482n, 564
NGOs concerned with human rights problems, 438, 463-464, 475-480
Nonstate actors, 295, 430, 464, 562-565, 567, 570
   as satellite imagery users, 2, 480, 562, 564-565
   changing relationship with governments, 2, 560, 564
North Atlantic Treaty Organization (NATO), 30, 274, 287, 297, 437, 477, 493
North Korea, 435, 479
   attack on South Korea, 18, 20
   imagery used to monitor missile facilities, 288n
   imagery used to monitor nuclear facilities, 387-392, 469
   probable nuclear weapon facilities, 383-384, 387-392
   Taepodong missile launch, 205
Norway, 135, 177, 266
Novaya Zemlya, 551
Nuber, Major Gen. Philip, 299
Nuclear Non-Proliferation Treaty (NPT), 382
Nuclear weapons production facilities, *see also* Chapter 18
   Iraq, 392-394, 396-398
   North Korea, 383-384, 387-392, 395
   Pakistan, 384-385
   United States, 386-387
   satellite monitoring of, 3
Nuclear weapon testing, *see also* Chapter 17
   India nuclear testing, 361-368
   Pakistan nuclear testing, 361-362, 369-377, 379

O

Observables, *see* imagery interpretation
Ocean monitoring satellites, 140, 154, 165, 176, 177, 253, 267-270, 274
Offeq imaging satellite, *see* Israel
Open skies, 542, *see also* Chapter 21
   legitimacy of space-based observation, 450, 530
   Open Skies proposal, 1955, 14, 32, 33n, 455
Open Skies Treaty, 1992, 30, 450, 453, 455
Open source information,
   use for corroborating imagery interpretation, 361, 493, 538, 551
Openness, 4, 24-28, 49, 568, *see also* Global transparency
   declassified satellite imagery, 5, 31
   political openness, 562
   "road not taken" during Cold War, 14, 571

Orbital Imaging Corporation (Orbimage), 50, 159 232
 Eyeglass program, 235, 238
 international partnerships, 154-155, 199, 277, 279
 hyperspectral sensor (Warfighter) program, 154, 155
 Orbital Sciences Corporation, 140, 277, 279, 442
 Orbview-2 satellite, 140, 154
 Orbview-3 satellite, 154-155
 Orbview-4 satellite, 154-155
 RADARSAT-2 role, 154-155, 442
 SeaWiFS (Sea-viewing Wide Field of View Sensor), 154
Orthophotos/orthoimages, 64-66, 589
Outer Space Treaty, 11n, 510

P

Pace, Scott, 287
Pakistan, 230, 296, 439
 nuclear testing, 139-140, 369-377, 469, 551
 use of commercial imagery to monitor nuclear activities, 369-373, 377-378, 384-385, 469, 600-603
Panchromatic remote sensing, 84, 88, 96, 107, 167, 191, 232-233, 253, 445, 581, 583, 607
 panchromatic imagery, 4, 80, 98n, 109, 116, 173, 181, 325, 408, 410, 423, 425n-426n, 561
Pan-sharpened multispectral imagery, 46, 86,174
Parastatal providers of imagery, 62-63, 75, 569
PDD-23, *see* Presidential Decision Directive-23
Peacekeeping, 275
People's Republic of China (PRC), 196, 240, 276, 284-285, 439, 442
 presence in the South China Sea, 327, 330-342, 345-346, 348-351, 353, 355, 359n
 remote sensing satellites, 445
Persian Gulf, 44, 102-103, 116, 225, 229, 487
 imagery of oil spills during Gulf War, 113-114
Peru,
 Ecuador-Peru peace treaty, 296, 311
 map signing ceremony in St. Louis, 312, 325
 use of satellite imagery to support peace negotiations, 9, 116, 307-308, Chapter 15, 566, 594-595
Perry, William, 227
Philippines, 327, 329-335, 337, 339, 346-349, 351, 353, 358, 359n
Pike, John, 469
Pixels, 86, 96, 555n-556n
 "pixel plasticity," 496, 540
 subpixel analysis or detection, 96
PowerScene™ three-dimensional visualization software, use for mission planning, 303
 use to support diplomatic negotiations, 299, 300-302, 303-304, 309, 318-319, 322, 592, 595

Precision-guided weapon systems, 123, 135, 566
Presidential Decision Directive-23 (PDD-23), 31, 122-123, 126, 147-149, 234, 242n, 280-281, 507-508, 515
 encourages U.S. commercial satellite firms, 57, 58, 96, 130n, 147, 519-520
 regulatory guidelines for commercial imaging satellites, 5, 57, 58, 148-149
Privacy, 92, 430, 457-458, 461-462
 effect of satellite imagery, 459-461
Prototype International Data Center (PIDC), 370
Public diplomacy, 464

Q

Quarles, Donald, 25
QuickBird satellite program, *see* EarthWatch

R

Radar, 6, 7, 178, 206, 209, 266, 561, 579-580
 advantages compared with electro-optical sensors, 11, 19, 82-84, 258, 266, 317, 425, 565, 582
 interest in commercial applications, 58
 interferometry, 83-84, 98n, 309n, 591
 maritime monitoring capabilities, 82-85, 268-270
 military utility of, 84-85, 109, 122
 phase history data, 83, 96
 radar image features, 80-85
 synthetic-aperture radar (SAR), 15, 80, 98n-99n, 142, 155-156, 254, 258, 266, 278, 459, 573n, 590
Radar imaging satellites,
 Almaz, 166, 167, 178-179
 ALOS, 213-215, 217, 219, 580
 ERS-1, -2, 189, 404-405, 426n
 ENVISAT, 189
 JERS-1, 151-152, 213-215, 222n
 RADARSAT-1, 76n, 130n, 152, 238, 263-278, 286n, 287n, 288n, 317-318, 319 320, 321, 405, 582
 RADARSAT-2, 96, 99n, 263-265, 277-282, 284, 286-287, 288n, 442
 Seasat, 405
RADARSAT, *see* Canada
RADARSAT International (RSI), 238, 264-266, 272, *see also* Canada
Radio-Television News Directors Association (RTNDA), 440, 511
RAND, 118
 role in early space system development, 18, 20, 32n
Rao, P. V. Narishmha, 362
Raytheon, 151, 232, 239
RDL, *see* Research Development Laboratories Space Corporation
Reagan, Ronald,
 administration, 41

Regulations on imaging satellites, 70-80, 256, 263, 276, 278-288, 460-461, 501, 504, 567-568, 572
    challenges facing technology restrictions, 90, 93, 97-98, 567
    debated constitutionality of restricted access, 440, *see also* Chapter 24
    governments as regulators, 2, 57-58, 59, 94-98, 520-521, 568
    unfeasibility of unilateral "shutter controls" 439-440, 489-490, 494-495
    operational controls, 93, 97-98, 501
    "shutter controls," 9, 31, 58, 97, 75, 104, 122-123, 126, 130n, 148-149, 278, 280-281, 286, 424, 430-431, 439-440, 470, 480, 485-486, 489-490, 495-496, 497n, 501, 514, 524-525, 567
    U.S. regulations on commercial imaging satellites, 9, 91-98,122-123, 130n, 146-150, 234, 235-237, 275-278, 430, 439-440, 507-508, 518, 526-527
Remote sensing,
    applications for using imagery data, 43, 72, 84-85, 90, 109-111, 250-251, 267-270, 272-276, 405-407, 466-467, 468, 471-472, 561, 569
    definition, 76n, 79-80
Research Development Laboratories (RDL) Space Corporation, 147, 155-156
    RADAR-1 program, 156
Resolution, 4, 5, 96, 98n-99n, 266, 535, 579
    improving resolution of commercial observation satellites, 4, 5, 113,116-117, 124, 582
    measure of imagery quality, 80-81
    required levels for detection and identification, 119, 435
    spatial resolution, 87, 93, 96,
    spectral resolution, 86-89, 93, 96-97
Resource21, 157, 581
    imaging satellite program, 157, 581
Russia, 1, 19, 30, 56, 58, 85, 136, 230, 239, 439, 470, 582, *see also* Chapter 8
    Almaz radar imaging satellite, 166-167, 178-179, 180
    Chechnya, 476
    Chernobyl nuclear accident observed on imagery, 470
    civilian remote sensing satellites, 167-169, 176-179
    commercial satellite imagery sales, 4, 5, 31, 56, 167-168, 170-172, 173-176, 182-183
    data policy for satellite imagery data, 166, 168, 174, 178, 182-183
    Earth observation organizations, 168-171, 173-177
    history of remote sensing program, 165-179
    impediments to commercial imagery sales, 166-167, 168, 171-172, 179-180, 183

    KVR-1000 satellite images, 174-176, 183n, 363-365, 404, 408-409, 469, 472, 479, 604-605, 608, 610
    Mir space station as imaging platform, 165, 166-167, 175-176
    Novaya Zemlya, 551
    partnerships with Western firms, 166, 172, 175-176, 177-179, 182
    planned imaging satellites, 175-182
    Priroda institute, 171, 173, 175-176, 183n
    Resurs-F imaging satellite, 166-167, 173, 183
    Resurs-O imaging satellite, 166-167, 172, 176-177, 181, 183n
    sale of declassified imagery, 136, 165, 173-175
    Sovinformsputnik, 171, 174, 176, 180, 183n
    Soyuzkarta, 166, 173, 470
    SPIN-2 project, 155, 157, 166, 175, 184n, 373, 404, 412-417, 581
Rwanda, 472, 474-475

S

Saddam Hussein, 103, 105, 240, 552
Sarabhai, Vikram, 257
Satellites,
    camera-pointing technologies, 107-108, 191, 581
    legitimacy of space observation, 24-30
    unconstrained by political borders, 1, 7, 21, 476, 490
    vulnerabilities to attack, 94, 130n-131n, 494-495
Satellite imagery,
    advantages compared to aerial platforms, 20-21
    civilian user requirements, 4
    civilian utility of satellite imagery, 82-85, 90, 92-93
    commercial applications of satellite imagery, 93
    ground-based data collection as alternative, 60
    military user requirements, 4, 565-566
    military utility of satellite imagery, 85, 90, 91-92, 434, 566-567
    potential users, 1, 463-479, 533-534
    price of, 60, 423-424, 470, 473, 476
    shortcomings compared to aerial platforms, 20-21, 60
Saudi Arabia, 102, 104, 181, 196, 436
    controversy over EIRAD, 235, 238
    remote sensing program, 229, 238, 255, 266
Schriever, Bernard A., 20, 33n
SeaWiFS (Sea-viewing Wide Field of View Sensor), *see* Orbital Imaging Corporation (Orbimage)
Secrecy, 17-18, 468, 474, 562, 571
    censorship of imagery data, 255-256
    classified satellite imagery, 4, 562
    closed society, 22, 490, 562, 571
    denied areas or territory, 7, 18, 85, 476, 489-490, 562
    national security purposes, 378

Index

tension between openness and secrecy, 31-32, 571
Serbia, 227, 306, 437, 477, 479
Shutter controls, *see* Regulations on imaging satellites
Shuttle Radar Topography Mission (SRTM), 141-142
Signatures, *see* Imagery interpretation
Singapore, 135, 196, 266
Small satellite systems, 124, 233, *see also* Microsatellites
Somalia, 472, 475
Sources and methods of intelligence collection, 92, 115, 119-120, 464, 478, 553
South China Sea, 111, 296, 596-599, *see also* Chapter 16
  comparing aerial and satellite monitoring, 331-348, 355-358
  Mischief Reef, 329, 331-346, 355, 597-598
  risk of military conflict, 330-331
  Spratly Islands, 329-331, 596
  Thitu Island, 329, 332-333, 346-348, 599
  use of satellite imagery to monitor, 331-359
South Korea, 255, 266
  acquiring imaging satellites, 10, 240, 439, 443-445
  Korea Advanced Institute of Science and Technology (KAIST), 443-444
Soviet Union, 4, 17-30, 68, 571
Space control, 94, 125-127, 279, 512, *see also* Anti-satellite weapons
Space Imaging, 50, 70, 74, 139, 147, 150, 151-152, 159, 232, 564, *see also* IKONOS
  EOSAT acquisition, 151
  international partnerships, 151-152, 255
  Pacific Meridian acquisition, 152
Spain, 201
Spatial technologies, 63, 497n
Spectral features of satellite imagery, 107
SPIN-2 satellite images, 175, 184n, 373, 404, 412-417
SPOT Image Corporation, 44, 45, 69, 192, 195-196, 198-199, 204n, 207
SPOT satellite program, 7, 8, 14, 30, 106, 107, 136, 274, 176, 200, 207, 255, 288n, 299, 301, 404-405, 423-425, 465, 467-470, 472, 480, 510, 512, 533, 560, 565, 580, 592, *see also* France
  contribution to global transparency, 7, 30
  ground receiving stations, 196, 201
  history of, 68, 191-197, 203n, 434, 436
  international partners, 191-193, 198-199
  leader in commercial imagery sales, 44, 56, 68, 194-195, 196-198, 203, 204n
  SPOT-1, 191-193, 195, 200, 201
  SPOT-2, 191-195, 201
  SPOT-3, 191-192, 194
  SPOT-4, 191-192, 194, 197
  SPOT-5, 191, 194

use in the Gulf War, 101-106, 128, 195
Spratly Islands, *see* South China Sea
Sputnik satellite, 5, 13, 22-24
Strategic Arms Limitation Treaty (SALT), 30
Stereoscopic viewing of imagery data, 7, 191, 267, 537-538
  equipment, 537-538
  stereo image pair viewing, 80, 318
Sudan, 404, 415-418, 424, 472, 610
Surprise attack, 113
  1958 Geneva conference on, 23-24
  attack on Pearl Harbor, 18, 33n
  military concern with, 17-18, 20, 21, 32-33n, 91-92, 490
Surrey Satellite Technology LTD, 444-445
Sweden, 177, 191
Synthetic aperture radar (SAR), *see* Radar
Syria, 181, 225, 229-230, 241n

T

Taiwan,
  acquiring imaging satellites, 10, 439, 583
  presence in the South China Sea, 327, 330-332, 346, 348
Technology,
  advances in imaging satellite technologies, 65, 69, 84, 124, 80-91, 443-444, 560
  enabling technologies for imagery users, 6, 47-48, 63, 65, 69, 90, 480, 559-562, 570
Teran, Edgar, 322
TerraServer, 74, 159, 473
Terrorists, 115, 225, 227, 229, 482n, 562
  access to satellite imagery, 2, 101-102, 117-118, 567
  imagery support to counter-terrorism, 119
  targets of terrorism, 118
Texas Strategic Mapping Program (StratMap), 59, 66-67
Thailand, 240, 255, 266, 404, 408-410, 604-605
Thematic Mapper (TM), *see* Landsat
Thermal infrared remote sensing, 7, 382
Three-dimensional views of imagery data, 73, 76n, 110, *see also* PowerScene™
  Fly-throughs to support negotiations, 300-301, 306-308, 318-319
  use for mission planning, 303
  use in Gulf War, 103
  use to support negotiations, 9, 296, 303, 318, 592, 595
Timeliness, 267, 274, 561
  improving timeliness of commercial imagery, 274, 561, 566
  problem of weather conditions, 61, 108
  timeliness of nonmilitary satellite imagery, 107-109, 122
  revisit time capabilities, 61, 107, 581, 584

Tiven, Ken, 492
Topologically Integrated Geographic Encoding and Referencing System (TIGER), 66-67
Transparency, *see* Global transparency
Truman, Harry S., 20
Turkey, 181, 228, 583

U

U-2 reconnaissance aircraft, 477
   U.S. overflights of Soviet Union, 22, 26-28, 33n-34n
Ukraine, 30, 177
Ultraspectral imaging, 86, 98n
Unilateral transparency, *see* Global transparency
United Arab Emirates (UAE),
   remote sensing program, 229, 235, 238-239
United Kingdom, 21, 266, 444-445
United Nations, 270, 426n, 436, 472, 474
   Committee on the Peaceful Uses of Outer Space (COPUOS), 248, 257n
   High Commissioner for Refugees (UNHCR), 403, 408, 415
   principles for remote sensing, 270, 453, 460, 510, 527n
   U.N. Arms Registry, 226
   UNHCR Mapping Unit, 405
   U.N. Security Council, 437
   UNISPACE III, 441, 446n
United States, 2-3, 17-34, 71, 311-312, 438, 440, 442, 479, 567, 569
   civilian remote sensing satellites, 37-47, 140-146, *see also* Landsat
   commitment to 'open skies,' 24-30, 450, 453, 455
   data policy for satellite imagery data, 145-150, 160
   Earth observation organizations, 39-46, 113-114, 129n, 140-146, 158-160
   history of remote sensing program, 18-20, 27-29, 37-47, 136, 140-160
   military reconnaissance satellites, 18, 22-24, 27-29, 139, 160
   satellite remote sensing commercialization, 41-50, 56-63, 76n, 139, 146-160
   U.S. government, 56, 58-59, 71, 570-571
U.S. Air Force, 422, 436
   EAGLE VISION program, 106, 112
   Space Aggressor Team assessments, 435-436
   use of satellite imagery in Gulf War, 103
U.S. Army, 288n, 299
   Army Space Command, 495
   Topopgraphic Engineering Center (TEC), 298-299, 304-305
   use of satellite imagery in Gulf War, 103
U.S. Central Intelligence Agency (CIA),
   role managing CORONA satellite program, 23
U.S. Congress,
   legislation encouraging commercialization, 5, 42-43
   restrictions on imaging Israel, 150, 235-236
U.S. Constitution, 457, 567
   U.S. constitutional rights, 75
   first amendment issues, 9, 31, 440, 470, 485, 502, 514, 567
   fourth amendment issues, 457
   fifth amendment issues, 504-505
   issue of 'prior restraint' in limiting access to satellite imagery, 503, 509-510, 513, 517, 523
U.S. Defense Mapping Agency, 102-103, 298-299,
   see also National Imagery and Mapping Agency (NIMA)
U.S. Department of Commerce, 42, 95, 140
   role licensing U.S. commercial remote sensing satellites, 148-150, 506, 508-509, 514, 517, 525
U.S. Department of Defense (DoD), 275-276, 506, 508
   role shaping Landsat program, 45
U.S. Department of the Interior, 39, 46
U.S. Department of State, 95-96, 286, 299, 325, 442, 492-493, 506, 508, 511, 514, 552
U.S. Geological Survey (USGS), 46-47, 143
U.S. Marine Corps, 359
   use of satellite imagery in Gulf War, 103
U.S. Navy,
   Naval Research Laboratory (NRL), 156
   use of satellite imagery in Gulf War, 103
U.S. National Spatial Data Infrastructure, 76n
U.S. Southern Command, 315, 317-318
U.S. Space Command, 114-115, 279, 495
   Global Partnership initiative, 114-115
U.S. Supreme Court, 457, 502-503, 511, 517, 528n
   Pentagon Papers case precedent, 513, 521, 526
USSR, *see* Soviet Union

V

Value-added imagery products and firms, 15, 63-64, 70, 157-158, 161, 272, 467, 487, 497n, 549
Verification, Research, Technology, and Information Centre (VERTIC), 468
Very small aperture terminal (VSAT), 258
Vietnam, 327, 329-332, 346, 348, 351, 353

W

Wassenaar Arrangement, 441-442, 446n
Watermarking, 539, 540-541, 556n
Weather,
   cloud cover problems for satellite imaging, 61, 108, 122, 313, 316, 405, 419, 423, 479, 573n, 581
Weather satellites, 9, 43
   Defense Meteorological Satellite Program (DMSP), 141

Index

Geostationary Operational Environmental Satellite (GOES), 140-141, 404
Meteorological Operational Weather Satellite program (METOP), 141
Meteorological Satellite (METEOSAT), 404
Polar-Orbiting Operational Environmental Satellite (POES), 140-142
Television Infrared Observing Satellite (TIROS), 38, 139
West Indian Space, *see* ImageSat International
Western European Union (WEU), 201
  WEU Satellite Centre, 201-202, 534
Wilderness Society, 467
Wilson, Steve, 239n
Windrem, Bob, 487, 490
Woolsey, James, 228
World Resources Institute, 466
WorldView, 5, *see also* EarthWatch
World Wide Web (WWW), 6, 59, 67, 179, 276, 424, *see also* Internet
World Wildlife Fund, 467, 564

Y

Yeltsin, Boris, 184
Yugoslavia, 297, 472

Z

Zaire, 275, 474
Zakheim, Dov, 230
Zambia, 472

The Four Health Segments

Second youngest segment

The Disillusioned are the second youngest Health segment with a median age of 63 years. Fifty-five percent of the Disillusioned are 65 or younger (see Figure 14–3).

Least married, most divorced

Fewer Disillusioned (50 percent) as compared to the over-50 population (58 percent) are married, making them the least married of the Health segments. In addition, significantly more (10 percent) are divorced or separated. One-third are widowed (see Figure 14–4).

Among least educated

Along with the U.S. population over 50 and the Proactives, the Disillusioned have a median of 12 years of education. Fewer of the Disillusioned (24 percent) have attended four years of college or more as compared to the over-50 population (26 percent) (see Figure 14–5).

Struggling financially

Along with being one of the least educated segments, the Disillusioned are also the poorest of the Health segments with a median annual pre-tax household income of under $10,000. More of the Disillusioned (51 percent) have incomes of less than $10,000 as compared with the over-50 population (30 percent). Only 11 percent have median incomes of over $50,000 a year (see Figure 14–6).

Excluding their homes, this segment also has the lowest level of assets of the four Health segments, a median of $20,779 compared to a median of $43,743 for the national sample. In sharp contrast, the Optimists have a median level of assets of $67,361.

Living in the South

Like the other three Health segments, the largest percentage of the Disillusioned (35 percent) live in the South census region. However, 25 percent of the Disillusioned live in the Northeast and another 24 percent live in the Midwest (see Figure 14–7).

Not working for pay

Although they are in the worst financial situation of the four Health segments, the Disillusioned are less likely than the mature population to work for pay. Fewer of the Disillusioned work either part or full time (56 percent) as compared to the U.S. population over 50 (59 percent) (see Figure 14–8).

More homemakers before retirement

Thirteen percent of the Disillusioned have spent their lives working as homemakers, compared to only 10 percent of the over-50 population and 5 percent of the Faithful

Patients. In addition, significantly more Disillusioned work or did work as owner/entrepreneurs, laborers, farmers, and service workers.

Opting against retirement

We have seen that fewer of the Disillusioned work for pay. This segment is also the least likely of the four Health segments to retire with no intention of returning to work. Compared to the over-50 population (51 percent), fewer of the Disillusioned have retired (46 percent) (see Figure 14–9).

More working after retirement

Of those Disillusioned who have retired, significantly more have returned to work (41 percent) as compared to the U.S. population over 50 (29 percent) (see Figure 14–9).

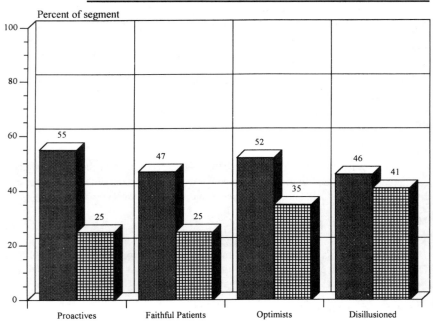

Figure 14–9 Rate of Retirement and Employment Since Retirement, 50+ Health Segments

■ Retired with no intention of returning to work ⊞ Employed since retirement

More Proactives have retired with no intention of returning to work. Far more Disillusioned have been employed since retirement.

The Four Health Segments

Return to homemaking, clerical positions

After retirement, 28 percent of those in this segment became homemakers. More of the Disillusioned (24 percent), in comparison to the over-50 population (13 percent), took clerical jobs after retiring. Nine percent took professional/technical positions.

Avoid volunteering

Only 36 percent of the Disillusioned volunteer their time, compared to 45 percent of the over-50 population. Their rate of volunteerism is in dramatic contrast to that of the Proactives, 52 percent of whom volunteer.

Demographic summary

Predominantly women, the Disillusioned are the least married and the most divorced of the four Health segments. They are also the poorest, in terms of both income and assets, and one of the least educated of these segments. More Disillusioned live in the South than in any other region, although many can also be found in the Northeast and Midwest.

The Disillusioned, who work in a variety of positions such as homemaker, entrepreneur, clerical worker, and farmer, are the least likely of the Health segments to retire and the most likely to return to work after retirement. However, in comparison to the over-50 population, fewer Disillusioned work for pay or do volunteer work.

FIFTEEN

HEALING THE MATURE MARKET

A HUGE MARKET

Question: "Where does the 800-pound gorilla sleep?"
Answer: "Anywhere he wants to."

In U.S. healthcare, the mature market is the 800-pound gorilla. And keeping mature America healthy has become a national crisis. As this population continues to grow, the stress placed on the national economy will become even more severe than it is today. According to estimates by the Health Care Financing Administration, U.S. healthcare expenditures will reach $1.6 trillion by the year 2000, compared to $666.2 billion in 1990. Health expenditures could account for 17 percent of Gross Domestic Product (GDP) in 2000, up from 12.4 percent of GDP in 1990.

Healthcare provided to mature Americans, 26 percent of the population, absorbs 56 percent of these healthcare expenditures. A person aged 55 to 59, for example, has 50 percent higher medical expenses than a person in his or her late forties and more than double the expenses of a person in his or her late thirties. People 65 and older spend four times as much on healthcare as younger people.

In this chapter we will again demonstrate that the mature market is not one monolithic entity. Some of the Health segments take better care of themselves; others use a greater share of healthcare products and services. Products and services can be designed and marketed for all of these segments, taking into account not only their illnesses but also their attitudes regarding health.

This chapter explores topics ranging from the one essential in promoting hospital services to segments that are heavy users of prescription painkillers. We will look at which segments use a Health Maintenance Organization for their Medigap insurance and those which suffer most from diseases such as high blood pressure.

Fundamentally, marketers and public sector planners have to understand how the Health segments differ. Knowing what each segment prefers and needs can lead to greater marketing efficiencies and success.

Market potential

For the pharmaceutical industry, people over 50 are a prime market. They spend $30 billion a year on prescription and over-the-counter medications. Those under 50 account for expenditures of only about $10 billion.

The marketing of medical products and services is changing in ways that were unthinkable a few years ago. For example, in the past marketers of pharmaceuticals concentrated their efforts primarily on doctors. Today medical marketing is changing and direct-to-consumer advertising is booming, even for prescription drugs.

MOST PREVALENT ILLNESSES

Circulatory disease most prevalent

The illness that affects many of those who are over 50 is high blood pressure. This affliction costs more than $15 billion in healthcare expenditures annually and affects 35 million people. Of those over 50, 35 percent report having this disease. In terms of the 50+ study, Optimists (28 percent) report the lowest incidence of high blood pressure. Since this is a disease with few apparent symptoms, and this segment tends to avoid doctors, the Optimists probably have high blood pressure and don't know it.

High blood pressure affects those over 65 more than those 65 and younger, regardless of their segment. In our study, females in all segments reported a higher incidence of high blood pressure than males.

Heart disease and stroke

Heart disease and stroke affect 11 percent of adults over 50. Examining these conditions by age, both the Proactives and Faithful Patients over 65 have much higher rates of heart disease and stroke than those 65 and younger. However, a difference between age groups is only slightly true of the Optimists and Disillusioned.

Looking at heart disease by gender, more males than females in three of the four Health segments have this disease. This difference is reversed in the Disillusioned segment, where more females (17 percent) have heart disease than do males (10 percent).

Arthritis second biggest problem

Arthritis is almost as prevalent as high blood pressure in the mature population, with 31 percent of those over 50 complaining of this disease. As with high blood pressure, Optimists, with a 26 percent incidence, claim to be the least afflicted with arthritis.

While heart disease is found in more males than females in three of the four Health segments, arthritis plagues more females than males in these same segments. Only among male and female Disillusioned do the same rates of this disease occur.

Females have higher cancer rates

About one in ten adults over 50 reports having cancer now or having had it in the past. Rates of cancer are highest among the Proactives (14 percent) and lowest among the Optimists (6 percent). For all segments, females have higher rates than males.

Examining the rates of this disease in each segment by age, only in the Proactive segment do those over 65 (17 percent) have much higher rates of cancer than those 65 and younger (9 percent).

Poor vision and hearing

Both poor vision and hearing are problems for almost one in five persons in the mature population. Older Proactives and Faithful Patients have a higher incidence of poor vision compared to the younger subsegments. However, almost twice as many Disillusioned 65 and younger report poor vision as compared to those who are over 65.

Although one in five persons over 50 claims to have a hearing impairment, only about one in fourteen actually wears a hearing aid. Those over 65 are most likely to have one.

Hearing-aid manufacturers have long concerned themselves with the fact that most people with hearing problems refuse to use one of their devices. While hearing aid use is more prevalent among those over 65, the rates of wearing one vary greatly among the over-65 component of the Health subsegments.

More over-65 Proactives (16 percent) wear hearing aids, for example, as compared to over-65 Optimists (6 percent). We contend that this difference is due far more to the segments' divergent attitudes about taking care of their health than to the lack or presence of an actual physical impairment.

Hearing-aid companies, such as Dahlberg, Inc., which control both the marketing and distribution of their products, should target the Proactives. Doing so will make their advertising and direct marketing even more effective. For example, although Dahlberg advertises on television, Proactives prefer newspapers, magazines, and radio. While Dahlberg makes extensive use of direct mail, we know that the Proactives are at best average in making purchases via this route.

Digestive problems afflict many

Digestive tract problems, such as frequent indigestion and constipation, afflict almost one in ten mature adults. The highest rates of frequent indigestion and constipation are reported by the Disillusioned (16 and 11 percent respectively).

About one in six of those over 50 reports having hemorrhoids, but more Faithful Patients (20 percent) do so than any other segment. Despite proclaiming themselves free of most medical problems, 18 percent of male Optimists report having hemorrhoids.

Overall, Optimists report substantially lower incidence of digestive-tract maladies. As we have pointed out with the other diseases, it is difficult to say whether Optimists are simply more fortunate in their good health than those in other segments.

In our chapter on wellness and prevention, we discuss the fact that the Optimists are, at best, only average in terms of a positive view of life. We suspect that their low incidence of reported diseases derives not from feelings of well-being, but from the fact that they ignore their problems and are, in fact, just as sick as others in this population.

VISITING THE DOCTOR

The Faithful Patients, as their name suggests, are faithful to their physicians. More of them (17 percent) visit their physician one to four times a month, compared to those over 50 (12 percent). Fewer such physician visits are made by the Proactives (13 percent) and Disillusioned (11 percent). The Optimists, true to their attitudes toward health, report the fewest physician visits. Only 5 percent of them visit a doctor one to four times a month (see Figure 15-1).

The Optimists, as we showed earlier in this chapter, reported fewer illnesses than the other segments. Their avoidance regarding their health keeps them from visiting doctors and finding out what their problems are in time to take corrective action. For example, 18 percent of the Optimists 65 and younger report having high blood pressure, half the national rate. However, that rate more than doubles to 38 percent when one considers only Optimists over 65 who report this disorder. The latter percentage is close to the national rate. We suspect that many of the Optimists did have high blood pressure when they were younger but, because they avoided going to a doctor, they didn't know it.

Public health professionals need to understand that the Optimists exist. They must find ways to convince them of the necessity of periodic checkups to identify problems as early as possible.

PAYING FOR CARE

No coverage at all

While only 4 percent of the mature market has no health insurance coverage at all, lack of coverage is concentrated in the Disillusioned segment. The low-income Disillusioned are more than three times as likely to have no coverage as compared to those over 50. In addition, major differences exist between the sexes and age groups within the Disillusioned segment with respect to the lack of healthcare coverage.

More female Disillusioned (16 percent) have no coverage as compared to male Disillusioned (10 percent). While the Disillusioned over 65 are all covered by Medicare, the Disillusioned 65 and younger are left with substantially no coverage. Almost one in four Disillusioned (24 percent) 65 and younger, six times the rate of the over-50 population, has no coverage.

Major sources of healthcare insurance

The most frequently cited sources of healthcare coverage are Medicare (48 percent) and a major-medical policy (46 percent). Slightly more than one in five of the mature

Healing the Mature Market

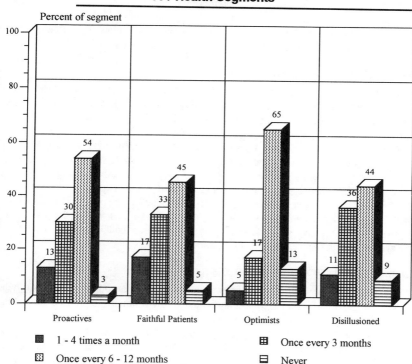

Figure 15-1 Frequency of Physician Visits, 50+ Health Segments

Optimists visit their physicians far less frequently than do the other Health segments. In contrast, more Faithful Patients visit their physicians one to four times a month.

market belongs either to a Health Maintenance Organization (HMO) or Preferred Provider Organization (PPO). More of the Faithful Patients (25 percent) are members of HMOs or PPOs than are Optimists (23 percent) and Proactives (21 percent). Only 14 percent of the Disillusioned belong to a managed-care organization.

We see, then, that more of the segment least interested in wellness and prevention, the Faithful Patients, belong to an HMO or PPO, and the two segments most interested in such matters are least likely to be members. It's no wonder that HMOs and PPOs encounter difficulties in motivating lifestyle changes among their older members.

Since the Faithful Patients are such heavy consumers of medical services, managed-care organizations need to be aware of them. Those in this segment, as we have seen, do not take responsibility for their own health. Instead, they rely on the medical profession to take care of them when their bodies malfunction.

Faithful Patients strain system

The public sector also needs to be aware of the Faithful Patients, because they will produce the greatest strains on the system. As disproportionate healthcare consumers at their present median age of 60, one shudders to contemplate the burden that the Faithful Patients will place on our resources when they reach 80.

Medigap insurance for the over 65

The vast majority of consumers of healthcare over 65 are covered under the federal Medicare program (88 percent). In order to obtain procedures and services not covered by Medicare, the majority of persons over 65 now purchase a so-called Medigap policy.

More of those over 65 have a major medical policy (35 percent) as their Medigap policy than other health insurance options. In contrast, only 13 percent of the Disillusioned have such a policy. One in ten of those over 65 has supplementary coverage from an HMO, while 6 percent use a PPO for Medigap coverage.

Medicaid, which covers the very poor, provides healthcare coverage for 13 percent of those over 65 who are on Medicare and for 16 percent of the Disillusioned (see Figure 15–2). A small percentage (4 percent) of the over-65 population receives coverage through the Veterans Administration.

Long-term care insurance option

According to the U.S. Department of Health and Human Services, a person who was 65 in 1990 had a 40 percent chance of entering a nursing home. Of those who enter a nursing home, one in ten will remain there for five years or more.

In 1990 the U.S. spent $60 billion a year on long-term care. Of this amount, only 2 percent was covered by long-term care insurance. Two government programs, Medicare and Medicaid, covered 53 percent of such care at a cost of $32 billion per year.

The above statistics illustrate the importance of increasing acceptance and use of long-term care insurance. And the situation will only get worse. By 2005, more than one out of every three persons in the United States will be over 50. At that time, government benefits for each retiree will be provided by taxes from only three workers.

When it comes to the purchase of long-term care insurance by those over 50, the Proactives are far more interested than the other three Health segments in using such insurance.

Of the eleven ways in which respondents could select to pay for long-term nursing home care, 21 percent of the Proactive segment would use an insurance policy to cover such care, compared to 16 percent of the over-50 population and only 5 percent of the Disillusioned.

The Proactives are planners who approach life in an orderly fashion. For example, more of the Proactives (35 percent) have made out a living will as compared to the population over 50 (29 percent). In reference to nursing home care, more Proactives, compared to those over 50, have thought about how they would pay for long-term

Figure 15–2 Medigap Coverage, Over 65, 50+ Health Segments

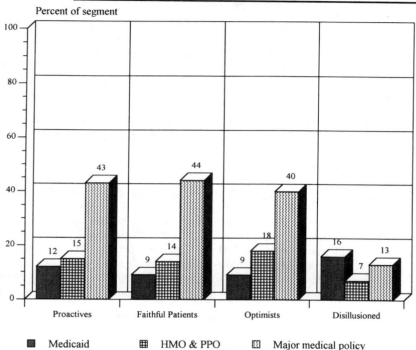

On a multiple-response question listing six sources of healthcare coverage, more over-65 Faithful Patients and Proactives on Medicare report using major medical policies to provide supplemental Medigap coverage.

nursing home care (84 percent versus 80 percent) and fewer of them "don't know" how they would pay for such care (23 percent versus 29 percent).

Our belief that attitudes, rather than demographics, distinguish interest in long-term care insurance parallels the conclusions of a major recent industry study completed by the Health Insurance Association of America (HIAA). Moving beyond that conclusion, our study has identified the Proactives as the segment most interested in such insurance. This finding is significant for both the marketing of long-term care policies by insurance companies and governmental and social planning. Reaching the Proactive segment, the one that is attitudinally most open to the long-term care insurance message, is crucial to the effective and efficient marketing of this product.

CHOOSING A HOSPITAL

Doctors make the difference

By far, the over-50 population selects a hospital based on its ability to attract the "best doctors." This characteristic was chosen first among 10 attributes by 55 percent of the mature market. Such a resounding vote for best doctors being the primary reason for selecting a hospital underscores the need for hospital marketers to stress this facet of their services in their promotions to those over 50. Promoting the latest equipment, compassionate care, or friendly nurses will not do as good a job in attracting the mature market.

For most of the segments, there is little difference between older and younger or male and female subsegments in regard to reasons for choosing a hospital. The one exception is the Disillusioned. More male Disillusioned (56 percent) are consistent with the other segments in wanting the "best doctors" than females in this segment (48 percent).

A greater percentage of female Disillusioned are more concerned with costs than with getting the best doctors. While "reasonable cost" is in second place for those over 50 as the most important reason for selecting a hospital (14 percent), this characteristic was chosen by an overwhelming 26 percent of female Disillusioned. In contrast, only 14 percent of the male Disillusioned selected reasonable cost as most important. More likely to be widowed and poor, the Disillusioned female is in the subsegment most likely to have no medical coverage. This segment's choices regarding what is important in selecting a hospital undoubtedly reflect their desperate situation.

PRESCRIPTION PHARMACEUTICALS

The segments differ greatly in their levels of use of prescribed pharmaceuticals. More of the Faithful Patients (32 percent) take three or more prescribed drugs per day as compared to those in the population over 50 (24 percent). Far fewer of the Proactives (24 percent), Disillusioned (27 percent), and Optimists (10 percent) take this number of prescription drugs daily (see Figure 15-3).

By avoiding doctors, Optimists not only don't find out what is wrong with them, but they also surrender access to pharmaceuticals they might actually need.

High blood pressure drugs most used

The most widely used pharmaceuticals are drugs for high blood pressure, such as Dyazide or Lasix. These drugs are used by 27 percent of the mature market, but by 35 percent of the Disillusioned (see Figure 15-4).

As one would expect, people over 65 are far more likely to take high blood pressure medications than those 65 and younger. However, there are wide variations among the segments. For example, 23 percent of the Disillusioned 65 and younger take high blood pressure medicine as compared to 29 percent of those over 65. This spread is very narrow compared to that encountered in the Optimist segment. A mere 13 percent

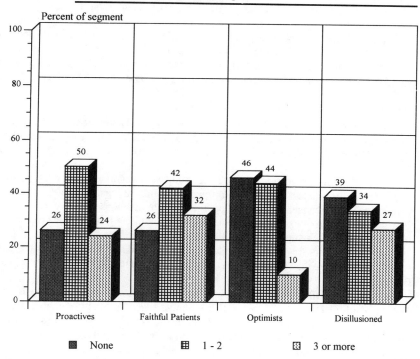

Figure 15-3 Use of Prescription Drugs, 50+ Health Segments

Compared to the other Health segments, more Faithful Patients use three or more prescription drugs daily. Far more Optimists use no prescription drugs.

of the younger Optimists take this kind of medicine compared to 34 percent of the older Optimists.

Mature males and females appear to be taking high blood pressure medications at about the same rate, although there is some variation. Fewer male Optimists (19 percent) as compared to female Optimists (28 percent) take this kind of medication. Since 27 percent of the mature market takes high blood pressure medication, it can be seen that female Optimists are consistent with the national norm, while males in this segment are not.

Drugs for arthritis used extensively

The second most widely used prescription pharmaceuticals are drugs for arthritis, such as Naprosyn or Motrin, used by 16 percent of the mature market. The Disillusioned and the Faithful Patients are higher users of these pharmaceuticals than the other two Health segments (see Figure 15-4).

In the use of medications for arthritis, variations also exist between younger and older subsegments. The widest variation again is between older and younger Optimists. Far fewer Optimists (7 percent) 65 and younger take these medications, as compared to older Optimists (17 percent).

It appears that Optimists wait as long as possible before committing to any regular pharmaceutical usage. This interpretation of their behavior parallels their attitudes. Taking medicine only when it is "absolutely necessary" is the statement with which the Optimists agree most strongly in the attitudinal portion of the 50+ study.

Use of pain killers

Mild-to-moderate prescription drugs for pain, such as Darvocet and Tylenol with codeine, are used regularly by 13 percent of the mature market. Twice as many Faithful Patients (18 percent) use these drugs, as compared to Optimists (8 percent) (see Figure 15–4).

Usage of these medications among older and younger members of the various segments varies widely. For example, fewer Faithful Patients 65 and younger (13 percent) regularly consume such medications in contrast to those who are over 65 (29 percent). The level of usage of such medications among Faithful Patients over 65 is more than twice the national norm (13 percent).

Even among the drug-avoiding Optimists, use of mild-to-moderate prescription drugs for pain more than doubles from those 65 and younger (5 percent) to those over 65 (11 percent).

Females generally are more likely to be consumers of painkillers than males. Female Faithful Patients (23 percent) are almost twice as likely as male Faithful Patients (12 percent) to use these drugs. Although fewer of them consume such drugs as compared to the other segments, female Optimists (9 percent) are also more likely than their male counterparts (7 percent) to be users.

Use of estrogen replacement therapy

While half of females over 50 report not having any menopausal symptoms, differences exist between females in the four Health segments. These differences are particularly dramatic if one considers only females who are between 50 and 65. For example, 53 percent of the female Optimists 65 and younger report having symptoms, in comparison to 71 percent of the Proactives in this subsegment.

Overall, one in five females over 50 is using estrogen replacement therapy. However, 33 percent of Proactive females between 50 and 65 are doing so. This rate compares to 15 percent of the female Optimists and 25 percent of the Faithful Patients between 50 and 65 who are on estrogen replacement therapy. Obviously, Proactive females between 50 and 65 are the very best target for pharmaceutical companies, such as Wyeth-Ayerst and Mead Johnson, that market estrogens.

When one considers the *preventive* aspects of estrogen therapy in regard to osteoporosis and heart disease, it is not surprising Faithful Patients aren't using it that much. As we have seen, Faithful Patients do not have a commitment to prevention. In

Healing the Mature Market 229

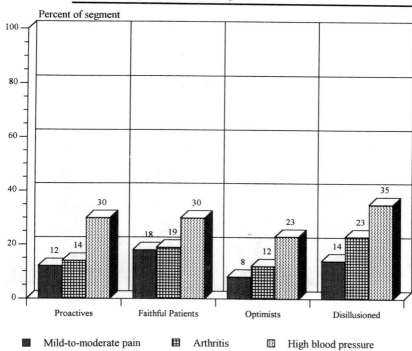

Figure 15-4 Types of Prescription Drugs Used, 50+ Health Segments

On a multiple-response question listing 28 illnesses for which prescription drugs could be taken, more Disillusioned report prescription drugs for high blood pressure as well as for arthritis.

fact, more female Faithful Patients 50 to 65 (22 percent) report having menopausal symptoms and not doing anything about them.

Besides taking estrogen replacement therapy, more female Proactives (19 percent) between 50 and 65 are also exercising more as a way of dealing with menopausal symptoms as compared to females 50 to 65 in the other Health segments. Again, the Proactives are taking charge of their healthcare (see Figure 15-5).

Segmentation explains major differences

One can see that there are substantial differences in the use of prescription pharmaceuticals between age groups and genders, as most studies of the mature market would identify. However, the Health segmentation yields major new insights by identifying the consumption patterns of groups as different as the Faithful Patients and Optimists. By understanding the psychological dynamics of these segments, health

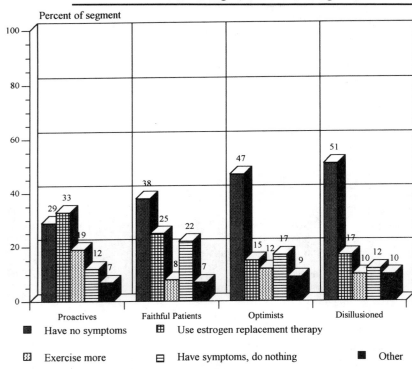

Figure 15–5 Dealing with Menopausal Symptoms, 65 and Younger, 50+ Health Segments

On a multiple-response question listing five ways of dealing with menopausal symptoms, far more Proactives aged 50-65 take estrogen replacement therapy or do more exercise. More Faithful Patients do nothing to treat their symptoms.

professionals and pharmaceutical manufacturers can plan better appeals to motivate them to take the medications they require.

OVER-THE-COUNTER DRUGS

Turning to over-the-counter (OTC) drugs, one can see again the disparity in consumption patterns among the segments. More Faithful Patients (10 percent) take three or more of these kinds of medications daily as compared to the Optimists (3 percent). Six percent each of the other two Health segments consume three or more OTC drugs daily.

Only 46 percent of the mature market reports taking no OTC drug daily. The segment with the lowest usage rate is the Optimists, which should be no surprise in

Healing the Mature Market

view of our previous discussion of this segment. Fifty-four percent of them don't take any OTC on a daily basis (see Figure 15–6).

The Faithful Patients, rather than taking care of themselves, rely on outside "quick fixes," such as OTC drugs and, as we have shown, on their doctor. The Optimists, true to their perceived good health and aversion to medicine, are low consumers of OTC drugs.

A study completed by the public relations firm Ruder Finn, Inc., New York, indicates that older Americans appear more likely to be influenced by their doctors to take OTC drugs than any other source. Seventy-four percent of adults 45 years and older cite a physician's opinion to buy a drug as a determining factor in purchasing it. Since the Faithful Patients are the most likely to visit a doctor and have a perceived need for medical support, it follows that they are overconsumers of OTC drugs.

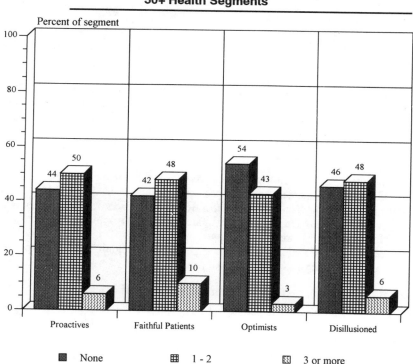

Figure 15–6 Number of Over-the-Counter Drugs Used, 50+ Health Segments

More Faithful Patients use three or more OTC drugs daily. More Optimists use no OTC drugs daily.

Painkillers, antacids heavily relied on

Within the mature market, vitamins, painkillers, and antacids are the most frequently used categories of OTC drugs. While vitamins are used regularly by 38 percent of the over-50 population, they are taken by more Proactives (42 percent) than Optimists (31 percent). Painkillers and antacids are used by 22 percent of this population. Overconsumers of painkillers, more Faithful Patients (29 percent) and Disillusioned (26 percent) report regular use of such drugs (see Figure 15-7).

With the exception of the Disillusioned, the younger versus older Health subsegments have the same level of consumption of painkillers and antacids. In the case of the Disillusioned, far more of them who are over 65 (32 percent) consume painkillers regularly than those who are younger (22 percent).

More females in both the Faithful Patients (35 percent) and Optimists (28 percent) subsegments use painkillers than their male counterparts (21 and 16 percent respectively). However, more male Disillusioned (31 percent) use these drugs than female Disillusioned (23 percent). There are no differences in use of painkillers by male and female Proactives—17 percent of both males and females in this segment regularly use them.

Antacids taken as frequently as painkillers

With 22 percent of the over-50 population reporting regular usage, antacids rank with painkillers in frequency of use. Along with painkillers, more Faithful Patients (28 percent) and Disillusioned (29 percent) also consume antacids than the over-50 population (see Figure 15-7).

The female Disillusioned are by far the heaviest consumers of antacids. Thirty-seven percent use these pharmaceuticals on a regular basis. Fewer male Disillusioned (17 percent) use antacids regularly. In the other segments, consumption of antacids doesn't differ between sexes.

Use of cold medicines differs

Cold medicines and antihistamines are two over-the-counter drugs that have high use in the mature market. Fourteen percent of people over 50 use cold medicine regularly. The Disillusioned (19 percent) are the heaviest users of this medicine. While the Faithful Patients (16 percent) are moderate users, Proactives (12 percent) and Optimists (10 percent) are by far the lowest users of cold medicines (see Figure 15-7).

Compared to the population over 50 (8 percent), the Faithful Patients (12 percent) are by far the heaviest users of antihistamines.

Females in two segments overconsumers of laxatives

Laxatives are used on a regular basis by 11 percent of the mature market. However, far more female Faithful Patients (18 percent) and female Disillusioned (22 percent) overconsume laxatives compared to their male counterparts. There is no difference in laxative usage by males and females in the other segments.

Figure 15–7 Types of Over-the-Counter Drugs Used, 50+ Health Segments

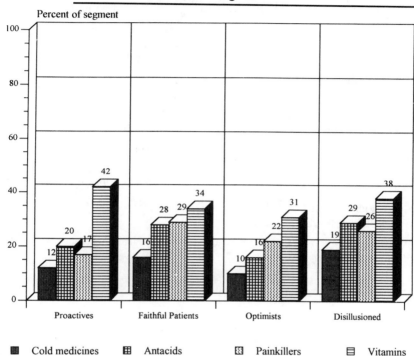

On a multiple-response question listing 11 types of OTC drugs, more Faithful Patients and Disillusioned regularly use painkillers and antacids. While more Proactives regularly use vitamins, more Disillusioned are regular users of cold medicine.

CONCLUSION

In every respect, from rates of specific diseases to which pharmaceuticals they take, the Health segments define mature consumers far more clearly than a mere examination of behaviors or demographic characteristics.

Knowing that because of their attitudes younger Proactive females are far more likely to be open to estrogen replacement therapy, pharmaceutical manufacturers can pursue them more directly. For this segment, behaviors parallel attitudes. Faithful Patients, overconsumers of over-the-counter drugs, many pharmaceuticals, and physician visits, represent a lucrative market for product and service providers and a challenge for those attempting to implement managed care. Can the Faithful Patients be motivated to take better care of themselves?

Awareness of the four Health segments and their patterns of healthcare use will help marketers and planners to position their products and services to one or more of the segments. Rather than simply targeting all those over 50, specific segments should be selected by their attitudes toward a product or service.

REFERENCES

"Doctors Influence Consumer OTC Purchases," *Quirk's Marketing Research Review*, June/July, 1991. Source of Ruder Finn, Inc. study.

Koretz, Gene, "American Medical Tab Is Growing Faster than Ever...," *Business Week*, June 8, 1992, p. 22.

Koretz, Gene, "... And That Is Bad News for an Aging Population," *Business Week*, June 8, 1992, p. 22.

Manly, Lorne, "Prescription for Success," *Adweek*, May 18, 1992, pp. 24-30.

Shafto, Marjorie, and Gerald L. Hunt, *The Body at 40*, New York: Putnam Publishing Group, 1987, p. 190.

"Who Buys Long-Term Care Insurance," study by the Health Insurance Association of America (HIAA), 1992.

SIXTEEN

CULTIVATING WELLNESS

REDUCING HEALTHCARE COSTS

Chronically ill, a 75-year-old man lies in the intensive care unit of a large metropolitan hospital. Tethered to this earth by a tangle of tubes, only pulsing lights witness his waning life. Another man, the same age, slaloms down the slopes at Vail, cutting his way through the snow. What accounts for the difference? Not smoking, exercising, eating a low-fat diet, great genes—or good luck?

Today the answers to these questions are relevant not only for individuals, but also for those who pay for much of the elderly's healthcare costs: business and government. In 1988 Medicare outlays were $81.6 billion. The total amount spent for personal healthcare in 1987 for those over 65 totaled $162 billion.

Every discussion of how to cure the ailments that beset the U.S. healthcare system focus on external problems. Paper shuffling must be reduced; immense bureaucracies must be slashed. Very little, if any, attention is given to individual responsibility for health and a reduction in the use of expensive healthcare services through prevention. Our four Health segments illustrate both the barriers to and the possibilities for individual responsibility for health.

Individuals can take action

From exercising vigorously to reducing the intake of fats, much can be done by the individual to take responsibility, to reduce illness, to create a healthy and active old age. Many studies have shown that our stereotypes of old age are based not on the normal aging process, but on disuse and abuse.

One program instituted by the Bank of America illustrates that a "demand reduction" program can work as a way of improving health while reducing costs. The Bank's program, limited to retirees, provided information, assessment, and support. The result: a reduction in actual claims of $60 per patient during the first year among the group in the demand reduction program.

If we multiplied these savings across all of those in the U.S. population over 50, a savings of almost $4 billion would result. And these savings do not even take into

235

account an improved quality of life or continued savings in the years ahead because of better health.

What motivates healthy living

This chapter explores the Health segments to learn what can be done by providers of healthcare services to motivate various segments to take better care of themselves. Marketers interested in wellness products and services will realize from this chapter which segments are their best targets. Most importantly, this chapter brings together aspects of healthful eating, exercise, and preventive care so marketers and program providers can see how these facets of health interact within the specific Health segments.

HOW HEALTHY ARE THEY?

Asked to rate their current state of health, half of those persons over 50 in the U.S. population said that they were either in excellent or very good health. Slightly more Proactives made this claim (55 percent), but many more Optimists (68 percent) thought that they were either in excellent or very good health. At the other extreme, 17 percent of the U.S. mature population says that its health is either fair or poor. Many more Faithful Patients (27 percent) and Disillusioned (26 percent) agree with this description (see Figure 16–1).

Thinking positively

Besides an assessment of physical health, our 50+ survey also looked at indications of mental well-being. For example, when asked whether they agreed or disagreed with the statement that "the best years of their lives are now and in the future," one in five of the U.S. population over 50 strongly agreed, but one in four of the Proactives did so. In contrast to the Proactives, fewer Optimists (17 percent) strongly agreed with this statement.

How one feels toward life is seen in the Health segments' reaction to the statement, "I feel like a successful person." While 29 percent of the population over 50 strongly agreed with this statement, more of the Proactive (38 percent) did so than the Optimists (29 percent) or the Disillusioned (16 percent).

We see, then, that the Proactives, more of whom think they are in excellent or very good health, have an upbeat, positive view of life. Although the Optimists in great numbers also report that their health is excellent or very good, not as many of them share this positive perspective.

Considering death and suicide

More frequently than those over 50, the Faithful Patients and Disillusioned think about death and dying and consider suicide. For example, at a rate higher than would be expected, more Faithful Patients (15 percent) sometimes feel so overwhelmed by life that they consider suicide as compared to the population over 50 (12 percent).

Cultivating Wellness

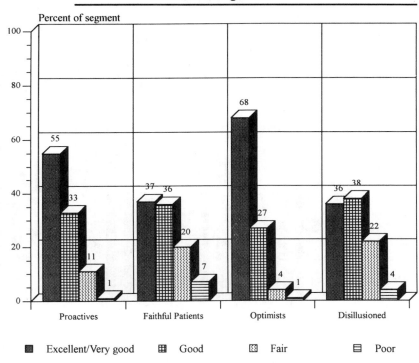

Figure 16–1 Current State of Health, 50+ Health Segments

More Optimists rate their current state of health as "excellent" or "very good." Faithful Patients in greater numbers rate their health as "poor."

When asked how often they thought about death and dying, more of those in the Disillusioned segment (15 percent) have such thoughts "always" or "frequently" (see Figure 16–2).

Disease as a motivator

While many of both the Proactives and Optimists report being in very good health, the Proactives have actually suffered from more disease and have embarked on a course emphasizing wellness and prevention. In contrast, the Optimists have either been very lucky in avoiding disease or, because they avoid doctors, may not know that they have a disease such as high blood pressure. For example, far more of the Proactives (78 percent) believe strongly in getting a yearly physical exam from their doctor. In contrast, only 26 percent of the Optimists hold this view.

At rates either equal to or slightly higher than the population over 50, the Proactives report having had or now having high blood pressure, arthritis, poor hearing, hemor-

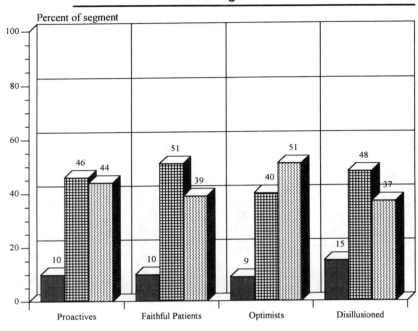

Figure 16–2 Thinking about Death and Dying, 50+ Health Segments

Far more Optimists "rarely" or "never" think about death and dying. In contrast, more Disillusioned "always" or "frequently" think about this subject.

rhoids, heart disease, and, perhaps most importantly, cancer. Because of their encounters with these illnesses, it isn't surprising that far more Proactives (59 percent) agree strongly that "as they grow older, their health becomes a greater concern for them." The number of Proactives who hold this view contrasts dramatically with the Optimists (22 percent).

Fewer Optimists, as compared to the total U.S. population over 50, report having the diseases listed in the paragraph above. In terms of one example, cancer, one in ten of the over-50 population now has or has had this disease. However, only 6 percent of the Optimists are in this category. Half of the Optimists agree strongly with the statement that they "almost never get sick." Fewer Proactives (38 percent) hold this view.

The fundamental positions toward life, wellness, and healthcare that the Health segments hold show up again and again in the ways in which each segment cares for itself.

YOU ARE WHAT YOU EAT

Healthful eating a goal for all

The National Cancer Institute's recently launched campaign encourages us all to eat five or more servings a day of fruits or vegetables. According to the Institute's own survey, however, only about one in five of us actually does so.

Overall, 63 percent of the over-50 population reports eating fresh vegetables daily. More of the Proactives do so (72 percent) as compared to either the Faithful Patients (53 percent) or Optimists (58 percent). Almost two-thirds of the Disillusioned, as poor as they are, do manage to eat fresh vegetables daily. While 60 percent of the Proactive segment eats fresh fruit daily, about 15 percent fewer of each of the other three Health segments do so.

In many cases, the Disillusioned are second only to the Proactives in their consumption of foods that are considered healthful. For example, slightly more than half of each of these two segments eats cereal daily. In contrast, only 38 percent of the Optimists do (see Figure 16-3).

Unlike the Proactives, however, the Disillusioned appear to consume more fat. Cheese is one of the Disillusioned's downfalls. About three-quarters of them report eating cheese at least once a week or more. More Disillusioned (43 percent) also report eating butter once a week or more as compared to the Proactives and Optimists (33 percent each) (see Figure 16-4).

Bad habits die hard

The Faithful Patients lead in the consumption of take-out or delivered pizza with 37 percent eating this dish at least once a month or more. More than the other segments, the Faithful Patients also eat red meat at least once a week (86 percent). Of the four Health segments, more of the Faithful Patients consume lunch meat and brownies or muffins from a mix. Canned soups and canned vegetables are an integral part of the Faithful Patients' diet.

Making a good effort

The Proactives say they are concerned about eating a more healthful diet, and of the four Health segments they are making the best effort to do so. To begin with, more Proactives (55 percent) agree strongly that they are more concerned about nutrition now than when they were younger in contrast to far fewer of the Optimists (24 percent). However, we have to note that all of those in this segment haven't become followers of Dr. Dean Ornish, nor have they submitted themselves to the rigors of the Pritikin regime.

The Proactives appear to have reduced the number of eggs they eat, but 50 percent of them still eat eggs once a week or more. Fewer of them eat red meat than the other segments, but almost three-quarters of them still consume it at least once a week or more (see Figure 16-4). Eating cookies more than once a week is something the Proactives (56 percent) do more than the over-50 population (53 percent). While the Proactives eat more dried fruit than the other segments, as many of them as those in the over-50 population also eat frozen pizza and candy.

Figure 16–3 Healthful Foods Eaten Daily, 50+ Health Segments

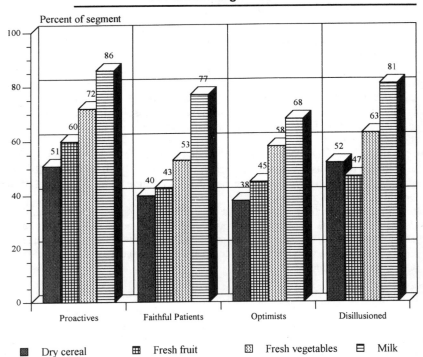

On a multiple-response question on the frequency of eating 36 foods, more Proactives report daily consumption of fresh fruits, vegetables, and milk. More of both the Proactives and Disillusioned consume dry cereal every day.

Segments differ on healthful eating

While 38 percent of the Faithful Patients agree strongly that they are more concerned about nutrition now than before, they have difficulty putting such concerns into action.

For example, fewer Faithful Patients (7 percent) agree strongly that they are careful to eat a balanced diet in contrast to the Proactives (52 percent). A basic tenet of a more healthful diet is fat reduction. But far more Proactives (61 percent) agree strongly that they avoid foods high in fat as compared to Faithful Patients and Optimists (19 and 18 percent respectively).

In rating 15 food attributes, the Proactives rate almost every attribute, from low-in-fat to free-of-preservatives, high-in-vitamins to nutritious, as extremely important.

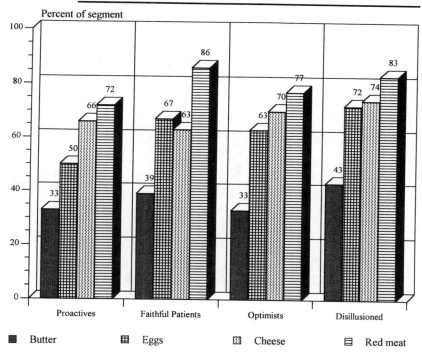

Figure 16-4 High-Fat and High-Cholesterol Foods Eaten, 50+ Health Segments

On a multiple-response question on the frequency of eating 36 foods, more Disillusioned report eating eggs and cheese once a week or more. Greater numbers of Faithful Patients eat red meat this frequently.

While the low-fat aspect of food is important to almost three-quarters of the Proactives, it's only that important to slightly more than one-third of the Optimists (see Figure 16-5). The Disillusioned and Proactives share some concerns about food attributes. For example, equal percentages of Disillusioned and Proactives rate food being high in vitamins (39 percent each) (see Figure 16-6) and being free of pesticides (29 and 30 percent respectively) as extremely important. In fact, more Disillusioned (20 percent) than Proactives (15 percent) rate low-in-lactose as an extremely important food attribute.

Optimists favor salty foods

Because they aren't concerned about eating low-salt products, more of the Optimists are overconsumers of salty snacks such as chips and popcorn. One out of seven of them eats such snacks daily as compared to fewer than one out of ten of the mature

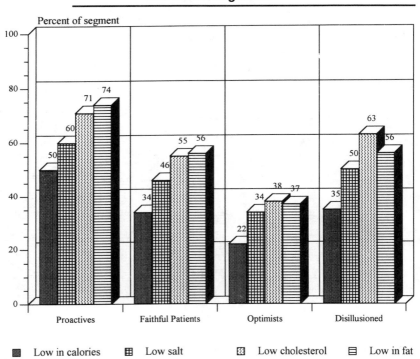

Figure 16–5 Importance of Food Attributes, 50+ Health Segments

On a multiple-response question on the importance of 15 food attributes, far more Proactives rate low in calories, low salt, low cholesterol, and low in fat as "extremely important."

population. Unconcerned about low-fat as a food attribute, the Optimists also relish ice cream and, of all the Health segments, more of them consume ice cream at least once a week or more (see Figure 16–7).

In contrast to the Optimists, the Proactives are well intentioned and, for the most part, are eating a more healthful diet than the other segments. However, the Faithful Patients are eating poorly, and they know it. While 68 percent of the Proactives agree strongly or somewhat that they couldn't be doing more to stay healthy, only 26 percent of the Faithful Patients hold this view. But who or what will motivate the other Health segments to improve their diets, exercise, and take care of themselves?

Exerting a positive influence

We asked respondents which of seven changes to their diets they had made on their own or on the advice of their doctor. These included reducing fat, cholesterol, sugar, and calories. With only one exception, an average of 18 percent more of those in all

Cultivating Wellness

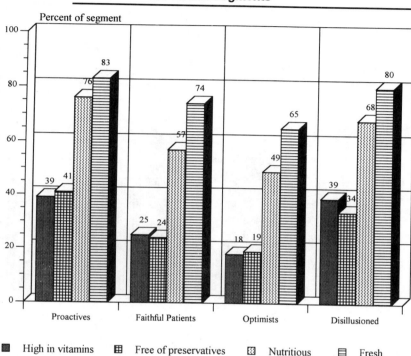

Figure 16–6 Importance of Food Attributes, 50+ Health Segments

On a multiple-response question on the importance of 15 food attributes, more Proactives rate free of preservatives, nutritious, and fresh as "extremely important." More of both the Proactives and Disillusioned rated high in vitamins as "extremely important."

the segments changed specific things they ate on their own, rather than on the advice of a physician. More in only one segment were influenced to make one dietary change through the advice of a physician, rather than on their own.

Through the advice of their doctor, more Faithful Patients had reduced their intake of cholesterol as compared to those who had reduced it on their own. It should be pointed out that an almost equal percentage of Disillusioned reduced their cholesterol intake on the advice of a physician as compared to those who did so on their own (40 and 41 percent respectively).

That only one Health segment believes that they made one dietary improvement far more on the advice of their physician than on their own suggests that physicians to those over 50 are not exerting a great deal of influence on their mature patients' diet. Mature patients are clearly acting on their own perceptions and knowledge in making food choices.

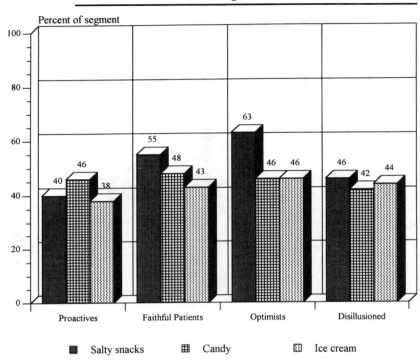

Figure 16–7 Eating Less than Healthfully, 50+ Health Segments

On a multiple-response question on the frequency of eating 36 foods, far more Optimists report eating salty snacks once a week or more. Fewer Proactives eat ice cream this frequently.

As we have seen in the description of the Faithful Patients' attitudes, they rely very much on their doctors and will do what they tell them. This data supports the idea that this segment could be motivated by its doctors to at least reduce the intake of cholesterol. However, self-motivation will influence the other segments to make changes to achieve a healthier lifestyle.

ADOPTING HEALTHY HABITS

Avoiding sun damage

One recent survey shows that about half of both men and women are concerned about too much exposure to the sun. In addition, that study showed that younger people are almost as concerned as older people about skin cancer due to overexposure to the sun.

Cultivating Wellness

Although one out of seven persons over 50 regularly uses a sunscreen, slightly more than one out of five Proactives does so. In contrast, only one in ten Disillusioned regularly uses a sunscreen. This is another example of the Proactives' belief that the actions they take can influence their health. However, we don't believe that the Proactives use sunscreens solely to protect themselves from skin cancer. Their use of rejuvenating creams and cosmetics suggests that they want to look as young and attractive as possible. Besides reducing the risk of skin cancer, sunscreens also curtail the aging effects of the sun's rays on the skin and its supporting structure. In using sunscreens, the Proactives appear motivated by both beauty and health benefits.

Reducing threat of lung cancer

According to the American Cancer Society, every hour of every day 17 people in the United States die of lung cancer, the number one cause of cancer deaths. About three out of four people say that by not smoking they are reducing their risk of cancer.

In the U.S. population, 19 percent of those over 50 currently smoke cigarettes, cigars, or a pipe. While one in seven Proactives smoke, this percent is lower than for any other Health segment. In contrast, one in four of the Disillusioned currently smokes (see Figure 16–8). Just as the Proactives are now eating more healthful foods, 42 percent of them used to smoke, but have given it up.

In the spirits

By not drinking alcoholic beverages we can reduce the risk of cancers of the mouth, throat, and liver. In terms of wine, beer, and hard liquor, the segment that has most dramatically reduced its consumption of alcoholic beverages is the Disillusioned. Fewer Disillusioned (18 percent) drink hard liquor than any other Health segment.

Overall, more Optimists currently drink alcoholic beverages than any other segment. More than one in three Optimists still drinks both hard liquor and beer, and 42 percent drink wine (see Figure 16–8). Nor have significant numbers of Optimists given up alcoholic beverages.

EXERCISE FOR LIFE

Obtaining aerobic benefits

Of the entire over-50 population, 67 percent exercise, but not as part of their job. This exercise could include such activities as bowling, shuffleboard, and leisurely walking. While all of these activities are beneficial, they are not aerobic. Of those over 50 in the population who exercise, 44 percent are exercising aerobically. But what of the others? It's aerobic exercise that offers the greatest benefits to the cardiovascular system.

Of all the Health segments, more Proactives, regardless of age or gender (see Figure 16–9), exercise. Of those Proactives who exercise aerobically, one in three does so three or more times a week. Perhaps inspired by videos of dancing grandmothers, more Proactives do aerobic dance (18 percent) as compared to the Optimists (7 percent).

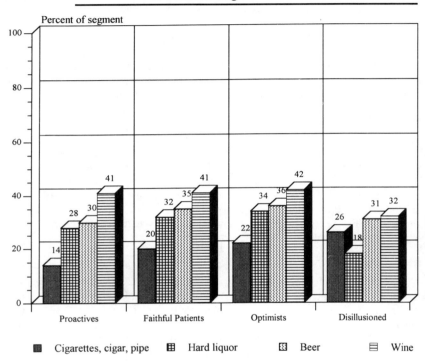

Figure 16–8 Use of Tobacco and Alcohol Products, 50+ Health Segments

On a multiple-response question on the use of six tobacco and alcohol products, more Optimists, compared to the other segments, report currently using alcohol products. More Disillusioned currently use cigarettes, cigars, or pipes.

More Proactives also use exercise machines and do aerobic walking in comparison to the Optimists.

While these statistics are heartening, we must still note that even among the Proactives one-quarter do not exercise aerobically even once a week. And yet this segment believes that it is doing all it can to stay healthy.

While only half of the Faithful Patients exercise at all, of those who do, 72 percent exercise aerobically at least once a week. In terms of achieving the minimum recommended level of at least three aerobic exercise sessions a week, however, only 20 percent of the Faithful Patients do so. For example, bicycling is preferred by the Faithful Patients, but slightly less than a quarter of them bike once a week.

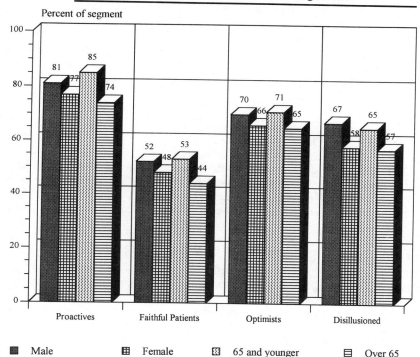

Figure 16-9 Exercise, but Not as Part of Job, Males and Females, 65 and Younger and Over 65, 50+ Health Segments

More males and those 65 and younger in all segments exercise outside their jobs. Across all subsegments, however, more Proactives exercise outside the workplace.

Not breaking a sweat

While more of the Optimists and Disillusioned exercise than do the Faithful Patients, these two segments do aerobic exercise the least. About a third of them do no aerobic exercise. Leisurely walking is an exercise that more of the Disillusioned do as compared to the other Health segments.

A pattern of exercising, but not doing so aerobically, is also seen in the fact that one-quarter of the Optimists play golf once a week, far more than the population over 50 (17 percent).

Gender and age affect exercise

While it's true that exercising decreases with age, the drop in the level of participation by those over 65 differs by sport and also by segment. For certain sports, such as golf, participation doesn't decline with age. In the case of male Optimists, almost three times as many golf who are over 65 as compared to those who are 65 and younger.

When examined by gender and segment, the 50+ study shows that more males than females are exercising. But as many females in certain segments are exercising as are males in certain sports. For example, female Proactives and Optimists play tennis as frequently as their male counterparts. Participation in other sports is more concentrated in one gender over another. Weightlifting and using Nautilus equipment are popular with more male Optimists (14 percent) than females (9 percent) while more female Proactives (13 percent) than male (5 percent) participate in aerobic dance.

PROACTIVE MEDICINE

Seeing a doctor

Two of the Health segments represent the extremes in terms of visits to a physician. Overutilization is seen among the Faithful Patients (17 percent), more of whom see their doctor at least once a month as compared to the over-50 population (12 percent). Underutilization, if not outright avoidance, is seen in the Optimists segment. More of the Optimists (43 percent) see their doctor once a year or never in contrast to 31 percent for those over 50.

Reliance on prescription and OTC drugs

While Chapter 15 in this section on the Health segments discusses pharmaceutical and over-the-counter (OTC) drug consumption in detail, reliance on medications is linked to taking responsibility for one's own health.

An indication of how different segments consume healthcare resources is seen in the fact that more Faithful Patients (29 percent) take OTC painkillers regularly than the over-50 population (22 percent), while more Proactives (42 percent) pop vitamin pills than those over 50 (38 percent).

Rather than taking better care of themselves, it's clear that the Faithful Patients rely on prescription and OTC drugs. Almost a third of the Faithful Patients take more than three prescription drugs a day, something done by slightly less than a quarter of the over-50 population.

The Optimists, as their attitudinal profile supports, dislike taking drugs and almost half of them take no prescription medication or OTC drugs.

As with the number of visits to the doctor, the OTC and prescription drug consumption patterns of the Faithful Patients and Optimists represent polar opposites. The Faithful Patients may be taking too many prescriptions, thus endangering their health through drug interactions. Or they may be relying on prescriptions rather than taking preventive measures. In contrast, the Optimists, by avoiding physicians and medications, may not be treating their diseases.

CONCLUSION

How many millions, if not billions, of dollars could be saved by encouraging the Faithful Patients to eat more healthfully, exercise more often, and take care of themselves? Besides encouraging preventive behaviors among the Faithful Patients, those who market healthcare products and services or supervise corporate wellness programs can do more. Already there are indications that such programs can be effective.

Reinforcing the Proactives' positive behaviors and supporting the struggling Disillusioned in their quest for better health will result in an immense return on whatever investment is made. As we have seen, the Disillusioned are interested in some aspects of healthful eating and could be motivated to do even more.

Whether in the public or private sectors, there is a great need and opportunity for effective programs that offer nutritional information, promote exercise, and devise effective smoking-cessation programs.

REFERENCES

Bowe, Claudia, "The Up Generation: A Louis Harris Survey for Lear's," *Lear's,* March 1989, pp. 66, 68.

Cope, Lewis, "Save Your Life," *Star Tribune,* Oct. 29, 1992, Section E, p. 1.

Minnesota Poll sponsored by Minneapolis *Star Tribune* and WCCO TV.

SEVENTEEN

REACHING THE HEALTH SEGMENTS

A NEW HORIZON

The friendly female pharmacist leans over the counter as she hands the customer his prescription. Has he tried, she asks, the newest version of the medication he takes? An informative discussion ensues. The episode occurs not in real life, but on network television.

From attractive middle-aged couples touting the benefits of prostate testing, to crisply dressed professional women pondering the need for estrogen therapy, advertising of prescription medications has joined the ranks of promotions for cough remedies and antacids.

Marketers of healthcare products and services are counting on instilling or reviving brand loyalties. In the past, generalizations describing brand loyalty, particularly among mature consumers, often disregarded the possibility that new loyalties can develop, even toward mature products. Clearly, for certain never-before-advertised healthcare products and services, the development of new loyalties is an exciting possibility.

In this chapter we'll explore the dimensions that brand loyalty to specific medical or health-related products or services could take. Strategies on how to reach specific Health segments will be developed addressing every type of media. Direct marketers will learn which Health segments are most open to appeals and which segment spends the most on direct mail purchases.

BRAND LOYALTIES WILL DEVELOP

In Chapter 13 on reaching the Self segments we saw that the brand loyalty displayed by mature consumers varies according to both segment and product category. There we defined *brand loyalty* as the purchase of only one or two brands in a specific category. Because certain Health segments are more brand loyal than others, opportunities don't exist equally among all mature consumers. As we will see, this new loyalty can best be achieved through promotions and media preferred by the targeted Health segment.

Pharmaceutical companies address consumers

For example, the development of brand loyalty could be of great importance to pharmaceutical companies that have only recently begun to advertise directly to the consumer. At the present time, consumers may know the medication they take as Zantac or Mevacor, but far fewer of them are familiar with the company that manufactures these drugs. If pharmaceutical companies begin now to build brand loyalty by targeting specific segments, strong preferences among a specific segment could be achieved in the years ahead.

Cutting through clutter

In the case of over-the-counter (OTC) drugs, the opposite situation exists. Consumers have been bombarded with television messages for years touting the benefits of one painkiller or cough medicine over another. Mature consumers may have been so inundated with messages that they have difficulty distinguishing one product from another. Perhaps they have even stopped listening to the advertisements.

An attempt to cut through the clutter and boredom surrounding a mature OTC painkiller is seen in Bristol Myers' use of target marketing to sell Bufferin. Featuring Angela Lansbury as its celebrity spokesperson in television advertising, Bufferin also built a database of users through responses to coupon drops. Those in the database received a quarterly newsletter, *Active Lifestyles*. A mini-catalog included in the newsletter allowed recipients to order merchandise at a discount—if they submitted a Bufferin proof-of-purchase with their order.

This promotion should have appealed to Faithful Patients who are 22 percent of those over 50. Far more of the Faithful Patients have arthritis as compared to the population over 50. In addition, more Faithful Patients (29 percent) take OTC painkillers as compared to the over-50 population (22 percent).

In this chapter, we'll also see that the Faithful Patients spend more on direct mail purchases than any other Health segment. Finally, the Bufferin promotion should have been successful because the Faithful Patients are heavy viewers of network television.

Proactive most brand loyal

On a diverse group of eight products, more of the Proactives, 41 percent of the over-50 population, show themselves to be brand loyal when contrasted to the other three Health segments. The next most brand loyal Health segment, the Disillusioned, is the poorest of the Health segments.

The two remaining segments, the Faithful Patients and the Optimists, are, at best, average in their loyalty to specific brands and, most typically, not at all brand loyal. In fact, the Faithful Patients' lack of brand loyalty parallels their attitudinal preference for generic rather than prescription drugs. While lacking brand loyalty, however, more of the Faithful Patients as compared to the Proactives pay attention to advertising.

While the products that we asked about are not in healthcare categories, the variation among the segments indicates that marketers can promote brand switching more easily with some segments than others (see Figure 17–1).

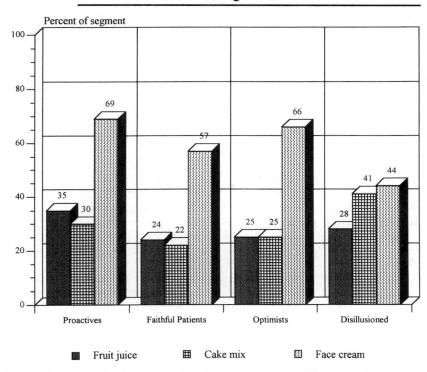

Figure 17–1 Brand Loyalty, 50+ Health Segments

On a multiple-response question on the number of brands bought in eight different product categories, more Proactives who are category users buy only one brand of two of the three products illustrated. More Disillusioned who are cake mix category users are brand loyal.

Loyalty a complex issue

The values and benefits that a brand represents shape the mature population's degree of brand loyalty. Each of our segments values brands to a greater or lesser degree, and there are additional variations in each segment's loyalty to specific products.

For example, the product category to which more of the Proactives are brand loyal as compared to the over-50 population is face cream for dry skin (see Figure 17–1). Besides being loyal to a brand of face cream, the Proactives are also overconsumers of night creams and moisturizers.

Whether using treatment products by Estée Lauder or Elizabeth Arden, this segment is composed of true believers. Their overconsumption of such products, combined with their loyalty to one or two brands of cosmetics, has significant financial implications for cosmetics companies.

Creating brand loyalty for health products

How loyal will the mature market be to specific providers or brands of healthcare services, wellness programs, pharmaceuticals, or hearing aids? And how can such loyalty be cultivated? The answers to these increasingly important questions have wide-ranging implications for many companies and providers.

Only one segment, the Proactives, show themselves to be consistently brand loyal to products they have perhaps purchased for over half a century. This segment brings together a commitment to wellness and prevention with a proclivity toward brand loyalty. Smart marketers of newly advertised health-related products and services can create entirely new loyalties within this segment.

ATTITUDES SHAPE PROMOTIONS

All promotional efforts can be integrated using the attitudinally based Health segments. Because we've linked our 50+ attitudinal segments to behaviors, demographics, and media usage, promotions can be designed with the Health segments in mind. Appropriate images can be used which tie in to that segment's activities and concerns.

The process of structuring all promotional efforts becomes far more efficient and integrated by identifying and targeting the attitudinal segment(s) most interested in a product or service. Healthcare products and services will be more successfully promoted to mature consumers once companies realize that these consumers have different needs and attitudes. For example, a wellness program sponsored by a pharmaceutical company will appeal strongly to the Proactives, less so to the Disillusioned, but not at all to the Faithful Patients or Optimists.

Buying media using attitudinal segments

Our 50+ study shows that the Proactives are consumers of magazines, newspapers, and radio programming at a higher rate than the population over 50. The Faithful Patients and Disillusioned are heavy consumers of network and cable television. No particular medium appeals to the Optimists, who are at best average consumers of the media and more typically underconsumers.

TELEVISION VIEWING

Faithful Patients prefer networks

Of all the Health segments, the Faithful Patients watch the most hours of television per week. More of this segment (53 percent) watches 11 or more hours of network television per week as compared to those over 50 (52 percent) and the Optimists (47 percent) (see Figure 17–2).

Reaching the Health Segments

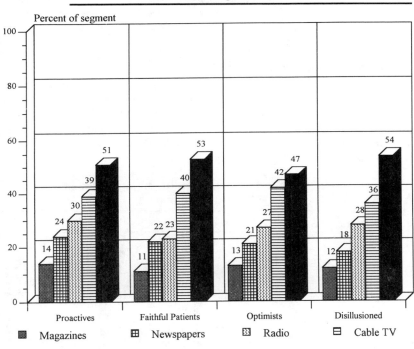

Figure 17–2 Media Usage, 50+ Health Segments

On a multiple-response question on the frequency of use of six different types of media, the Proactives are shown to be heavy media users. More of them spend 11 or more hours a week using magazines, newspapers, and radio. More Disillusioned and Faithful Patients spend 11 or more hours a week watching network television.

Cabling up

The Optimists are heavy consumers of cable television. More of them watch cable 11 or more hours per week (42 percent) as compared to the population over 50 and the Disillusioned (36 percent each) (see Figure 17–2).

What the segments watch

Of the 13 television program options presented in the 50+ questionnaire, the three types of television shows watched by more of those over 50 either several times a week or more are news and current affairs (86 percent), game shows (47 percent), and talk shows (38 percent).

Faithful Patients prefer comedies, talk shows

Advertising targeted to the Faithful Patients should be broadcast during *Golden Palace*. More Faithful Patients watch situation comedies (36 percent) several times a week or more as compared to the Optimists (26 percent) and to the population over 50 (17 percent) (see Figure 17-3).

Proactives selective in viewing

In almost all types of television programming, fewer of the Proactives are frequent viewers as compared to those over 50. Although television is not their preferred medium, more Proactives watch sports programs several times a week or more (41 percent) as compared to those in the over-50 population (36 percent) (see Figure 17-3). In addition, almost a quarter of the Proactives watch educational shows several times a week, slightly more than the other segments.

Figure 17-3 Television Programming Watched, 50+ Health Segments

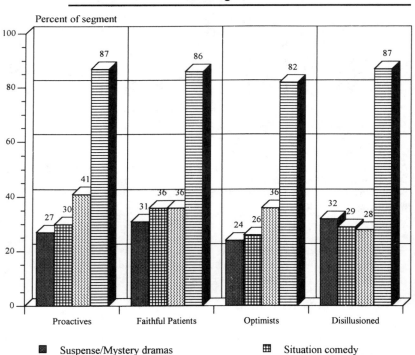

On a multiple-response question on the frequency of watching 13 types of television programs, more Proactives watch sports several times a week or more. Faithful Patients in greater numbers view situation comedies at this level.

NEWSPAPER READERSHIP

Proactives reading more

In terms of time devoted to reading the newspaper, more Proactives (24 percent) report spending 11 or more hours a week reading a newspaper compared to the over-50 population (22 percent). Fewer Disillusioned (18 percent) spend this much time weekly reading a newspaper (see Figure 17-2).

Proactives looking for entertainment

Specific sections of a newspaper attract certain segments. For example, a local hospital needs to know that in placing an advertisement in the entertainment section of a paper it would be more apt to reach Proactives. Besides that section, the Proactives are reading international news, opinions and commentary, editorials, and letters to the editor at rates higher than the mature population as a whole.

Health food manufacturers should note that more Proactives are also regularly reading the food (57 percent) and coupon (62 percent) sections of newspapers as compared to those over 50 (50 and 58 percent respectively) (see Figure 17-4). In fact, far more Proactives (42 percent) strongly agree that their coupon use has increased as compared to those over 50 (36 percent).

Disillusioned not reading papers

For each of the 19 sections of the newspaper listed in the questionnaire, from cartoons to community news, food to entertainment, dramatically fewer Disillusioned are newspaper readers. More of those over 50 (39 percent) regularly read cartoons, as compared to the Disillusioned (28 percent).

Advertising attracts Faithful Patients

The sections read on a daily basis by more of the Faithful Patients than the other Self segments are limited to advertising, advice columns, cars, and community news (see Figure 17-4). Because they display brand loyalty in only a very few categories, it isn't surprising that more Faithful Patients would read newspaper advertising sections.

Optimists are sports fans

More of the Optimists (50 percent) read the sports pages than those over 50 (41 percent). In addition, more of them read obituaries and local and international news.

SELECTING MAGAZINES BY SEGMENT

Measuring magazine readership

Unlike other measures of readership, our 50+ study defined readership more narrowly. The term *reader,* as we have used it here, refers to either the purchaser or

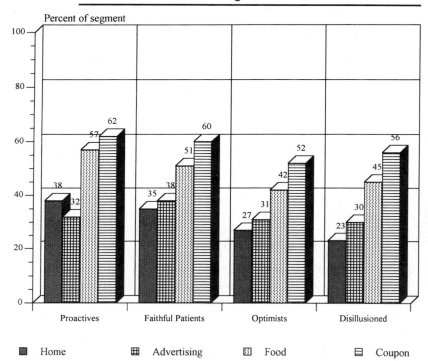

Figure 17–4 Newspaper Sections Read, 50+ Health Segments

On a multiple-response question listing 20 sections of the newspaper, far more Proactives than the other segments regularly read the home, food, and coupon sections. More Faithful Patients regularly read advertising.

subscriber of a magazine, or a respondent who lives in a household where the magazine is regularly available through subscription or purchase.

Reaching the Proactives

Although many magazines offer very similar demographics, our 50+ study shows that some magazines are better than others in reaching specific 50+ segments. A company developing a workout studio specifically for mature consumers would do well to target the Proactives. Such a company should go beyond demographics and select media that clearly offer more direct access to the Proactives, the segment most committed to exercising and staying in shape.

Magazines the Proactives read at a statistically significant level include *Health Magazine* and *Prevention.* However, more Proactives than we would expect also read *Money, Vogue,* and *Gourmet.* From such magazines, media planners will have to make

Reaching the Health Segments

the most cost-effective selections. They must balance cost per thousand with the ability to reach prime prospects—defined as a specific Health segment with a stated interest in their product or service.

Women an important market

Because women are responsible for the purchase of the majority of health-related products and services, it is especially important to reach them. Good bets in planning media campaigns to reach more of the female Proactives are *Good Housekeeping* (29 percent) and *Family Circle* (32 percent). *Prevention* reaches 14 percent of the female Proactives as compared to 9 percent of the over-50 population as a whole. However, only 7 percent of male Proactives read *Prevention*.

Targeting the Faithful Patients

In targeting Faithful Patients, the overconsumers of healthcare services who are not taking care of themselves, specific media can also be used. One-quarter of female Faithful Patients read *TV Guide* in contrast to one-fifth of the mature population. Other publications that the Faithful Patients regularly read at a statistically significant level include *Good Housekeeping, People,* and *Life* (see Figure 17–5). Male Faithful Patients favor *National Geographic*. One out of four reads it regularly as compared to one out of five of those over 50.

Multiple magazine readers

More of the Proactives (31 percent) fall into the Magazine Publishers of America's (MPA) definition of a magazine reader: a person who regularly reads five magazines. About a quarter of the Faithful Patients and Optimists qualify as magazine readers under this definition.

Two magazines read by considerable numbers of Proactives who are multiple magazine readers are *Better Homes & Gardens* (59 percent) and *Good Housekeeping* (53 percent). *TV Guide* has more Faithful Patients (38 percent) among its readers as compared to those over 50 (31 percent). Optimists (63 percent) prefer *Better Homes & Gardens* more than the population over 50 (57 percent). In greater numbers, Optimists who are multiple magazine readers (15 percent) also read *Travel & Leisure*.

More Disillusioned (74 percent) who read multiple magazines read *Reader's Digest* than those over 50 (69 percent). *Family Circle* is also read by far more Disillusioned (68 percent) as compared to the national norm (50 percent).

SENIOR OR MAINSTREAM PUBLICATIONS?

Companies that want to advertise to the mature market have to decide whether to approach it in mainstream publications or in magazines that target only the mature market.

Figure 17-5 Magazines Read, 50+ Health Segments

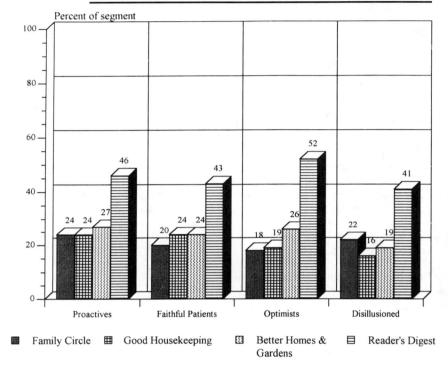

On a multiple-response question listing 39 magazines, more Optimists regularly read *Reader's Digest*. More Proactives regularly read *Family Circle*.

The case for Modern Maturity

In Chapter 13 we discussed the immense circulation numbers of *Modern Maturity*. A very impressive 52 percent of the U.S. population over 50 reports reading *Modern Maturity* regularly.

Some Health segments believe *Modern Maturity* is doing a better job than others. To begin with, far more of the Optimists and Proactives (56 percent each) read the publication regularly as compared to the Disillusioned (44 percent). Two-thirds of those over 50 who receive the publication believe that it delivers useful health information. However, more of the Faithful Patients (73 percent) and Proactives (67 percent) share this view.

Targeting mature consumers and others

Another option is to select a publication targeted to health or wellness that has a base of older subscribers. A new publication, *Rx REMEDY,* located in Westport, Connecticut, targets persons over 55. At the present time its rate base is 1.5 million readers. The magazine produces authoritative and informative copy on healthcare issues and products for its readers. Advertisers focused on, but not limited to those offering healthcare products and services, can capitalize on its ability to target readers cost-effectively.

TUNING IN TO RADIO

While the Optimists are average in the amount of time per week that they devote to listening to radio, the Faithful Patients are actually below average. In contrast to these two segments, more Proactives (30 percent) spend 11 or more hours a week listening to radio (see Figure 17–2). At a significant level this segment finds a number of types of radio programming appealing. These include easy listening (54 percent), news (48 percent), talk shows (26 percent), and classical music (24 percent) (see Figure 17–6).

BOOKS REACH DISILLUSIONED

Although books do not typically carry advertising, they could be offered in a promotional effort focusing on the Disillusioned. More of the Disillusioned are book readers as compared to the other segments. For example, far more of the Disillusioned (30 percent) read books 11 or more hours a week, than do the Proactives (25 percent).

Because they read and are committed to healthier living, the Disillusioned would find a health-information library sponsored by the government or a nonprofit organization appealing. With the lowest income of all the Health segments, the Disillusioned are the least able to afford to buy such books.

IMAGES FOCUS ON ACTIVITIES

Proactives staying active, healthy

Advertising targeted at a specific 50+ Health segment should show them engaged in some of their favorite activities. For the Proactives more than the other Health segments, some of these activities include gardening (45 percent), physical fitness and exercise (35 percent), and health foods (17 percent). As compared to the over-50 population, the Proactive segment enjoys a wide range of activities and a higher level of participation in many of these.

Disillusioned inactive

In contrast, the Disillusioned participate in only two activities at rates greater than the population over 50. Of those in this segment, more (16 percent) designate health

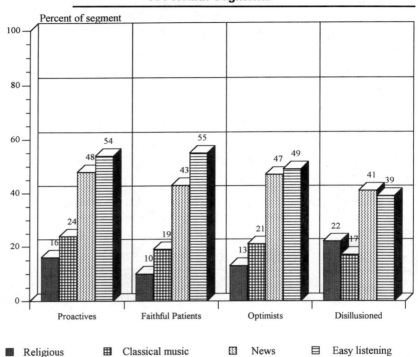

Figure 17–6 Radio Programming Listened to, 50+ Health Segments

On a multiple-response question listing 11 types of radio programming, more Proactive and Faithful Patients prefer easy listening. More Proactives listen to classical music, while more Disillusioned tune in to religious programs.

foods as an activity as compared to those over 50 (13 percent). Sewing is another activity in which more Disillusioned participate than the mature population.

Participating in sports

Optimists favor activities related to sports. For example, more Optimists watch sports on television (51 percent) and play golf (17 percent) in contrast to the over-50 population (43 and 12 percent respectively). Optimists (28 percent) also list do-it-yourself projects as a regular activity.

Faithful Patients enjoy offspring, pets

The Faithful Patients spend time recording videos and playing records and tapes. Another favorite activity of theirs is entering sweepstakes, something that more of them (22 percent) do in contrast to the over-50 population (18 percent).

More of the Faithful Patients are also involved on a regular basis with a child (38 percent) or grandchild (44 percent) as compared to those over 50 (32 and 41 percent respectively). Besides their offspring, far more of the Faithful Patients are involved with pets (31 percent) as compared to the mature population (24 percent).

Tying activities into ads and promos

The Faithful Patients have more grandchildren under 12 than the other Health segments. More of the Faithful Patients also report regular involvement with a pet. With this information in mind, the image in an advertisement of a grandparent playing with an appealing young grandchild and a puppy would find a receptive audience among the Faithful Patients.

THE DIRECT MAIL OPTION

Correlating attitudes in list purchases

In dealing with the present, affluent, mature market that has never before existed, past behaviors and demographics will not predict new needs or motivators. The reliance that too many direct marketers place on history to create catalogues and target potential new customers is reflected in the stacks of inappropriate, me-too catalogues we all find in our mailboxes.

Databases for direct marketing can be structured using attitudes that correlate with behaviors and demographics. We use statistical methods and sophisticated, cutting-edge technologies to find relationships between these various facets. With the information we've provided, attitudes can be incorporated into the process of selecting and merging and purging lists. The first question to be answered, however, is how do the Health segments react to direct marketing? Which segment(s) is the best target for such offers?

Reading contents carefully

Overall, female Faithful Patients demonstrate a higher level of interest in direct mail offers than any other Health subsegment. More of them (11 percent) as compared to those over 50 (8 percent) open direct mail and read it carefully. In addition, more female Faithful Patients (11 percent) put such offers away to be read later as compared to the population over 50 (8 percent). Clearly, if the direct mail offering is kept and put away, it can be used at a later time.

Who buys through direct mail?

Overall, the same percentages of Optimists, Faithful Patients, and Proactives (42 percent) made a purchase through direct mail over the past three months. In contrast, far fewer of the poorer Disillusioned made such a purchase (34 percent). A closer look at the Health subsegments, however, shows that more of the female Faithful Patients (48 percent) had made a direct mail purchase over the past three months.

Spending by segment

Although the same percentage of Faithful Patients, Optimists, and Proactives had made a purchase, the amount spent by the Faithful Patients ($100.67) is far above average. Those over 50 had spent a median of $59.83 in purchases over the past three months through direct mail. The Proactives ($57.60) and Optimists ($55.20) had spent below this average (see Figure 17–7).

Female Faithful Patients spend more

The Faithful Patients, with a median expenditure of $100.67, are clearly spending the most. And within the Faithful Patient segment, females spend more ($105.01) as compared to males ($73.84).

For the other three segments, spending on direct mail declines from 14 to as much as 52 percent for those over 65 as contrasted to those 65 and younger. However, spending by the Faithful Patients on direct mail remains unchanged regardless of age.

More of the Faithful Patients also purchase the greatest number of items. While 5 percent of those over 50 bought from six to over nine items in the past three months through direct mail, one out of every five Faithful Patients had done so.

Whether it's buying through direct mail, the telephone, or a television shopping channel, the pattern is repeated again and again. More of the Faithful Patients respond to such selling situations, and they make more purchases as compared to the other segments. However, their expenditures are far greater than the other Health segments only on direct mail and television shopping channel purchases.

CREATING TRULY TARGETED MAILING LISTS

Companies can target a specific Health segment. Those that offer pharmaceuticals through the mail will find a good target in the Faithful Patients who take the greatest number of drugs and who agree that generic drugs are as good as branded ones. A direct marketer of wellness products such as Comfortably Yours should consider targeting the Proactive segment.

Using media, car ownership lists

As we have already seen, each of these segments prefers specific magazines from which mailing lists can be purchased. Additionally, lists could be selected by income, sex, and census region. If the Faithful Patients are targeted, the region would be the South where 37 percent of them live. The list could then be merged and purged using information such as that given below to significantly increase the likelihood of locating Faithful Patients.

Because lists based on car ownership are used so frequently in direct mail efforts, it's important to know that more of the Faithful Patients (96 percent) own cars as compared to the population over 50 (93 percent). Two cars more popular with the Faithful Patients than the over-50 population are Lincolns and Cadillacs. More of the cars driven by the Faithful Patients (25 percent) are also 1990 to 1992 models as compared to those driven by the Proactives (23 percent).

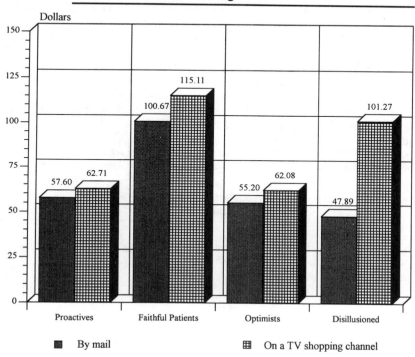

Figure 17-7 Median Expenditures on Direct Marketing, 50+ Health Segments

Compared to the other three Health segments, the Faithful Patients have spent far more on direct marketing purchases in the last three months.

Modeling facilitates targeting

In working with marketers, whether they use direct mail or mass media, we use a variety of statistical techniques and other technologies to locate our attitudinally based segments.

Figure 17-8 shows a "tree diagram" produced by KnowledgeSeeker from FirstMark Technologies in Ottawa. This analysis examines a large battery of demographics and behavioral measures, such as magazine readership in this case, and divides the respondents to focus on meaningful characteristics.

This program examines all of the data and splits the sample to magnify differences between subgroups that are made up of either a high or low percentage of the target segment. Each split is then subdivided, again choosing variables to highlight significant differences expressed in percentages of segment members.

By examining the variables that form the tree, marketers are able to better choose the media with which to communicate to the target segments. For example, while 23 percent of the population over 50 is made up of Optimists, by limiting the selection to males who read *TV Guide* and are college graduates, we increase our probability of reaching the Optimists to 34.1 percent or another 11.1 percent (see Figure 17–8).

Activities can be mapped

Another type of analysis that would help marketers in list and media selection is seen in figure 17-9. The map in this figure was produced using a perceptual mapping program called MAPWISE from MarketACTION Research in Peoria, Illinois. This program takes data from a crosstabulation and presents the relationships in graphic form.

Figure 17-9 shows activities in which the segments regularly participate. The activities shown are those that best correlate with each segment. In interpreting the

Figure 17–8 KnowledgeSeeker "Tree Diagram," Optimists

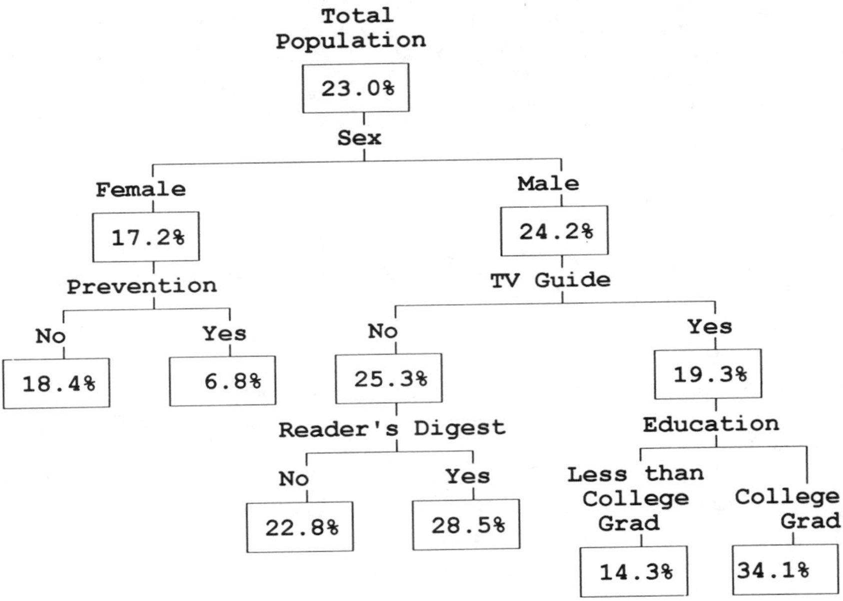

```
                        Total
                      Population
                        23.0%
                         Sex
           ┌─────────────┴─────────────┐
         Female                       Male
         17.2%                       24.2%
       Prevention                   TV Guide
       ┌────┴────┐              ┌──────┴──────┐
      No        Yes            No            Yes
     18.4%     6.8%           25.3%         19.3%
                          Reader's Digest   Education
                          ┌─────┴─────┐   ┌─────┴─────┐
                         No         Yes  Less than   College
                                         College     Grad
                        22.8%      28.5%  Grad
                                         14.3%      34.1%
```

More Optimists can be reached by limiting the selection to males who read *TV Guide* and are college graduates.

Reaching the Health Segments

map, note that when a segment is shown to be close to an activity, it means that this segment participates in that activity at a high level. For example, the Proactives are highly involved in physical fitness and community and civic events.

Conversely, when a segment and an activity are shown at a distance from one another, the segment participates in that activity at a low level. Thus we see that fewer Optimists are involved in gardening (see Figure 17-9).

CONCLUSION

Those over 50, 26 percent of the population, consume 56 percent of all health-related products and services. Opportunities exist for the marketers of healthcare products and services to reach each of the Health segments. The potential is particularly great for products and services that have previously not been advertised directly to the consumer.

Our 50+ study shows that brand loyalty varies by each of the 50+ Health segments and according to specific categories of products. By linking an attitudinal segment's media and brand preferences with its interest in the product, marketers can increase the efficiency of their advertising and direct marketing.

Figure 17-9 Perceptual map by MAPWISE, 50+ Health Segments

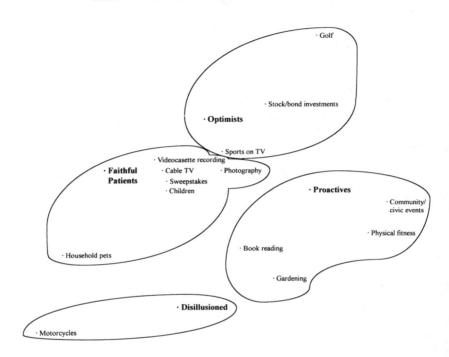

The Proactives are more highly involved in physical fitness and community/civic events compared to the other segments. More Optimists watch sports on television, while more Faithful Patients enter sweepstakes. The only activity in which more Disillusioned participate than other segments is motorcycle riding.

PART FOUR

The Food Segments: Attitudes toward Eating and Nutrition

EIGHTEEN

THE THREE FOOD SEGMENTS

As the senior population grows, marketers at major food companies, supermarkets, and restaurant chains must consider how they will cope and prosper with the mature market's changing food needs. Marketers will have to decide which trends should be developed or abandoned. How important is convenience to seniors? Does food packaging really have to be redesigned with the elderly in mind? From fat content to digestibility, smaller portions to promotions, scores of issues will face food marketers.

In completing our segmentation process, respondents sorted 50 statements into piles according to their levels of agreement or disagreement. These statements included the following:

- I'm willing to pay more for easy-to-prepare foods.
- I am trying to cut down on the amount of salt I consume.
- Eating at restaurants is too expensive.
- If foods that normally go in my freezer could be packaged in a way so that I could store them safely in my cupboard, I would buy those foods in the new package.
- I find that a lot of food containers are too difficult for me to open.
- When I go to the grocery store, I stick with my shopping list.

From the segmentation methodology described in Chapter 1, the following three Food segments developed: the Nutrition Concerned, the Fast & Healthy, and the Traditional Couponers.

THE NUTRITION CONCERNED SEGMENT

At 46 percent of the over-50 population, the Nutrition Concerned are the most health-conscious of the three segments when it comes to their eating habits (see Figure 18–1).

Figure 18–1 Segment Size, 50+ Food Segments

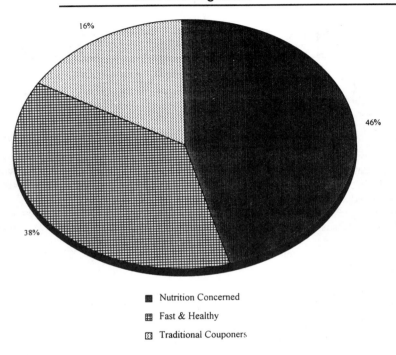

- Nutrition Concerned
- Fast & Healthy
- Traditional Couponers

The three Food segments include the Nutrition Concerned (46 percent), the Fast and Healthy (38 percent), and the Traditional Couponers (16 percent).

Attending to food content, labels

Compared to the other segments, the Nutrition Concerned believe more strongly that how you feel is influenced by what you eat. Their view of this relationship may explain why the Nutrition Concerned put so much energy into eating healthfully. Eating regular meals and not skipping meals are of great importance to only the Nutrition Concerned. They are also the only segment that still cooks most of its meals, not just when their families are together.

In addition, the Nutrition Concerned are more concerned about nutrition now than when they were younger. They're paying more attention to the content of the food they eat. Because of their concern with the content of food, the Nutrition Concerned believe that reading labels is important when trying new products. They're the only segment to agree with this statement.

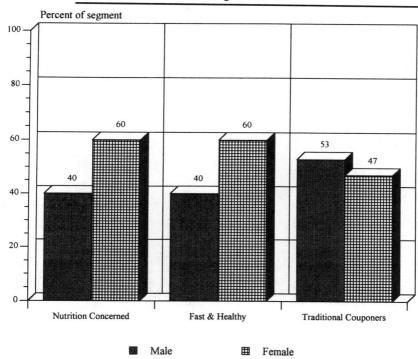

Figure 18–2 Gender, 50+ Food Segments

Of the three Food segments, only the Traditional Couponers segment has more males than females.

Serious about healthy eating

More than the other two segments, the Nutrition Concerned are interested in avoiding fats and eating enough fiber, a concern that has become more important as they've grown older. They are also the only segment concerned about eating hormone-treated red meat.

New food products using low-calorie fat substitutes are more appealing to the Nutrition Concerned than to the other segments. In addition, they keep their eyes open for products that use artificial sweeteners instead of sugar.

Food products specially formulated to provide optimum nutrition for older people would find the Nutrition Concerned receptive. The Nutrition Concerned, along with the Fast & Healthy, think it's important to take a daily vitamin supplement. All three segments are committed to eating fresh fruits and vegetables and are trying to cut back on their salt consumption.

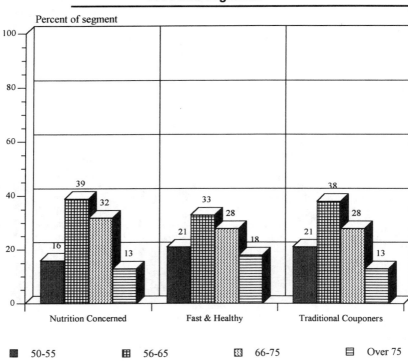

Figure 18–3 Age, 50+ Food Segments

The same percentage of Fast & Healthy and Traditional Couponers are in the 50-55 age group.
The Traditional Couponers and Nutrition Concerned have lower representation in the over-75 category.

Little interest in convenience

Those in this segment are the only ones who are not using more convenience foods now than before and are unwilling to pay more for such foods. The Nutrition Concerned lack interest in purchasing complete pre-cooked meals made from fresh food that they could heat and serve at home. This segment, compared to the others, is negative toward the increasing use of the microwave oven to cook or reheat packaged foods.

Packaging no concern

The Nutrition Concerned have neutral attitudes toward all statements related to food packaging, suggesting that perhaps other issues in the Marketer® sort, namely nutrition, are of much greater importance and interest to them.

The Three Food Segments

Figure 18–4 Marital Status, 50+ Food Segments

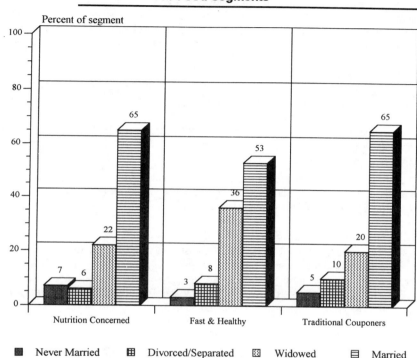

While more of the Nutrition Concerned and Traditional Couponers are married, a far greater percentage of the Fast & Healthy is widowed.

Passing up fast food

With their concern for eating nutritiously, the Nutrition Concerned believe fast food restaurants do not serve food that is healthy for older people.

The cost of eating out may also be of some importance to the Nutrition Concerned. They agree that they would be more likely to eat at restaurants offering age-related discounts.

Stick to shopping lists

The Nutrition Concerned strongly agree that they stick to their shopping lists when they go to the store.

Swayed by advertising

The Nutrition Concerned are more swayed by advertising than the other two segments. They admit that they do pay some attention to advertising when selecting food products. The Nutrition Concerned segment is most likely to look favorably on a food product that is endorsed by a well-known figure.

Brand loyal

In relation to brand loyalty, the Nutrition Concerned, along with the Traditional Couponers, find themselves relying on their favorite brands more than before.

DEMOGRAPHICS—THE NUTRITION CONCERNED

Majority are women

Sixty percent of the Nutrition Concerned are women, which is about the same as that for the over-50 population (58 percent). In contrast, only 47 percent of the Traditional Couponers are women (see Figure 18–2).

Few in youngest age category

The Nutrition Concerned, along with the Fast & Healthy, have a median age of 63 years, compared to a median of 64 years for the U.S. population over 50. However, far fewer of the Nutrition Concerned are in the study's youngest age category. Compared to 21 percent of the Fast & Healthy and Traditional Couponers, only 16 percent of the Nutrition Concerned are aged 50 to 55 (see Figure 18–3).

More married, more never married

Far more Nutrition Concerned (65 percent) are married as compared to the over-50 population (58 percent) and the Fast & Healthy (53 percent). However, more of the Nutrition Concerned (7 percent) have never been married as opposed to all those over 50 (5 percent). In addition, only 22 percent of the Nutrition Concerned are widowed in comparison to 30 percent of the over-50 population (see Figure 18–4).

Highly educated

The Nutrition Concerned have a median of 13 years of education, compared to a median of 12 years for the total U.S. population over 50. Just as 26 percent of the over-50 population have attended four years of college or more, so have the Nutrition Concerned (see Figure 18–5).

More incomes over $50,000

The Nutrition Concerned have a median annual pre-tax household income of $34,308, higher than the median of $25,670 for the over-50 population. The Nutrition Concerned's income, highest of the three Food segments, is higher than that of the Traditional Couponers, but only by a couple of hundred dollars.

The Three Food Segments

Almost a third (32 percent) of the Nutrition Concerned have incomes of over $50,000. The number of Nutrition Concerned with these incomes compares to 26 percent for the Fast & Healthy, 24 percent for the Traditional Couponers, and only 23 percent of the U.S. population over 50 (see Figure 18-6).

Living in the Northeast

More of the Nutrition Concerned (34 percent) live in the Northeast census region as compared to the over-50 population (22 percent). In addition, 26 percent live in the South. The remainder is almost equally divided between the West and Midwest (see Figure 18-7).

Least likely to work

The Nutrition Concerned are the least likely of the Food segments to work for pay. Only 60 percent, compared to 66 percent of both the Traditional Couponers and Fast & Healthy, work either part time or full time (see Figure 18-8).

Professionals, clericals, managers

A quarter of the Nutrition Concerned have spent their lives working in professional or technical positions. In addition, as compared to the over-50 population (12 percent), more of the Nutrition Concerned (15 percent) had careers as managers or executives. Sixteen percent were employed as clerical workers.

Homemakers, professionals after retirement

Of those Nutrition Concerned who returned to work after retirement, 29 percent became homemakers and 19 percent entered professional or technical jobs. In addition, a small number took either managerial (7 percent) or clerical (7 percent) positions.

Volunteer less than others

The Nutrition Concerned are the least likely of the Food segments to do volunteer work. Only 42 percent of the Nutrition Concerned volunteer their time each week in comparison to 45 percent of the U.S. population over 50.

Demographic summary

The Nutrition Concerned have a median age of 63 years, an educational level of 13 years, and an annual pre-tax household income of $34,308, the highest of the Food segments, but only by a few hundred dollars. Compared to the national over-50 population, more Nutrition Concerned are either married or have never been married. The Nutrition Concerned are more likely to be found in the Northeast than in any other region.

The Nutrition Concerned, the majority of whom had careers as professionals, clerical workers, and managers, are the least likely of the Food segments to currently work full time or to volunteer.

THE FAST & HEALTHY SEGMENT

The Fast & Healthy, 38 percent of the population over 50, are extremely interested in convenience foods and new packaging ideas, but are also concerned about health and nutrition (see Figure 18–1).

Using more convenience foods

Convenience foods have the greatest appeal for this segment, and they are using them more than ever before. Of the three segments, only the Fast & Healthy are both interested in convenience foods and willing to pay for them. And they are the only segment that has even a moderate level of interest in pre-cooked, complete meals purchased at a grocery store.

Cooking less, skipping meals

The Fast & Healthy, who report cooking fewer full meals now than they used to, tend to cook only when they have the family together. They are also the only segment that is skipping more meals now than before.

Becoming more concerned about nutrition

Although the Fast & Healthy are highly interested in convenience foods, they are also more concerned about nutrition than when they were younger, and are paying more attention to food content. This heightened interest in nutrition may be due to their belief that what they eat is an important factor in how they feel.

Their commitment to eating healthful foods is, however, at a far less intense level than that of the Nutrition Concerned. For example, compared to the other segments, namely the Nutrition Concerned, the Fast & Healthy have no interest in reading food labels. If the Fast & Healthy don't know what is in a food product, how will they know if it is good for them or not?

Healthful eating important

The Fast & Healthy are trying to cut down on their consumption of fat and salt and are concerned about getting enough fiber. They also want their nutrition fast and so they rely, as do the Nutrition Concerned, on popping a daily vitamin pill.

Committed to eating foods that are good for them, the Fast & Healthy try to eat lots of fruits and vegetables. They are also interested in foods with low-calorie fat substitutes and products that use artificial sweeteners. In addition, they, like the Nutrition Concerned, would buy food products specially formulated for the nutritional needs of older people.

Although they are obviously concerned about healthy eating, the Fast & Healthy are not frightened of eating hormone-treated red meat.

Rely on their microwave ovens

The Fast & Healthy strongly believe that microwave ovens are safe and are the only segment interested in more conveniently packaged foods for this appliance. In addi-

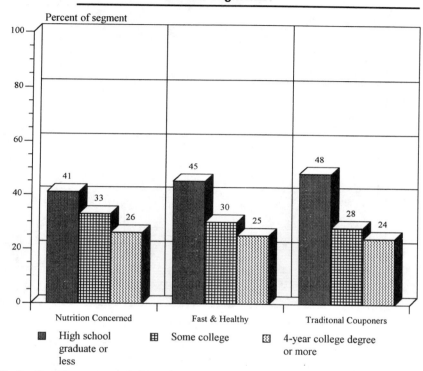

Figure 18–5 Level of Education, 50+ Food Segments

The Traditional Couponers are the least educated of the three Food segments; more Nutrition Concerned have a college degree.

tion, much more than the other segments, the Fast & Healthy think it's great to be able to heat packaged foods in their microwave ovens.

Receptive to packaging innovations

Of all the segments, the Fast & Healthy appear to be the most receptive to innovations in packaging. Paralleling their concern with convenience, they are the only segment interested in foods packaged in individual or smaller servings. They also claim that they would buy shelf-stable foods that are normally frozen.

The Fast & Healthy is the only segment to disagree that a lot of food containers are too difficult to open.

Figure 18–6 Annual Pre-Tax Household Income, 50+ Food Segments

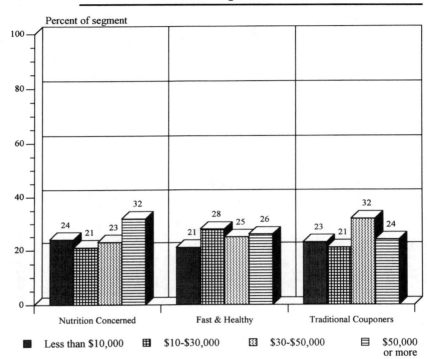

Almost a third of the Nutrition Concerned have annual pre-tax household incomes of $50,000 or more. Half of the Fast & Healthy have incomes under $30,000.

Shopping is a breeze

This segment has fewer problems with grocery shopping than do the Traditional Couponers. In relation to their physical experience within the supermarket, it is not important to the Fast & Healthy that a store have clerks who can carry their packages to the car. They don't find it difficult to move their carts through supermarket aisles, or to get in and out of the parking lot.

Dining out enjoyable, affordable

Dining in restaurants is more appealing to the Fast & Healthy than to the other segments. They are the only segment that would prefer meeting friends at a restaurant than having the bother of cooking for them at home. In addition, they find restaurant chairs and seats to be comfortable.

In relation to costs, the Fast & Healthy don't believe that restaurants are too expensive. Of all the segments, the Fast & Healthy would not select a restaurant just because it offered discounts to seniors. This segment is also least open to being served by people of their own age.

When it comes to nutritional concerns, the Fast & Healthy don't find eating at restaurants to be a problem. They believe they will be able to stay on their diets while eating out. The Fast & Healthy would tend to agree more than any other Food segment that fast food is good for older people.

Don't rely on lists, coupons

When shopping, the Fast & Healthy are impulse buyers. They are the only segment that doesn't stick to a grocery list. In addition, they do not find themselves using coupons more than before.

Not affected by endorsers, brand loyalty

The Fast & Healthy do not tend to believe well-known people endorsing food products. When compared to the other two Food segments, the Fast & Healthy exhibit less brand loyalty than before. Compared to the other segments, however, they do pay some attention to advertising when selecting food products.

DEMOGRAPHICS—THE FAST & HEALTHY

Predominantly women

Sixty percent of the Fast & Healthy are women, compared to 58 percent of the over-50 population (see Figure 18–2).

More Fast & Healthy aged 50 to 55

The Fast & Healthy, along with the Nutrition Concerned, have a median age of 63 years, compared to a median of 64 years for the over-50 population. In comparison to 18 percent of the national sample, 21 percent of the Fast & Healthy are between the ages of 50 and 55 (see Figure 18–3).

Least married, most widowed

The Fast & Healthy are the least married and the most widowed of the three Food segments. Significantly fewer of the Fast & Healthy (53 percent) are married as compared to the U.S. population over 50 (58 percent). In contrast, almost two-thirds of the Nutrition Concerned and the Traditional Couponers are married. On the other hand, the Fast & Healthy are the most widowed. Compared to 30 percent of the over-50 population, 36 percent of the Fast & Healthy are widowed. But fewer Fast & Healthy (3 percent) have never been married, compared to the over-50 population (5 percent) and especially the Nutrition Concerned (7 percent) (see Figure 18–4).

Education level on par with average

The Fast & Healthy have the same median level of education, 12 years, as does the national over-50 population. Virtually the same percentage of Fast & Healthy (25 percent) have attended four years of college or more, compared to the U.S. population over 50 (see Figure 18-5).

Least wealthy Food segment

Although the Fast & Healthy's median annual pre-tax income of $30,759 is higher than that of the over-50 population ($25,670), it is the lowest of the three Food segments. Still, more of the Fast & Healthy (26 percent) have incomes of over $50,000 as compared to people over 50 (23 percent) (see Figure 18-6).

Living in the South, Northeast

The Fast & Healthy are concentrated in the South and Northeast. Significantly more Fast & Healthy (33 percent) live in the South as compared to the other two Food segments. In addition, 29 percent of the Fast & Healthy live in the Northeast, far more than the national over-50 population (22 percent). However, fewer of the Fast & Healthy live in the Northeast as compared to the Nutrition Concerned and Traditional Couponers (see Figure 18-7).

Working more

Along with the Traditional Couponers, more of the Fast & Healthy work than does the population over 50. More Fast & Healthy (66 percent) work at either part-time or full-time jobs as compared to the over-50 population (59 percent) (see Figure 18-8).

From professionals to homemakers

According to the Fast & Healthy, 26 percent made careers in professional or technical positions, 17 percent in clerical jobs, and 14 percent in managerial or executive positions. In addition, more Fast & Healthy (13 percent) spent their lives working as homemakers as compared to the over-50 population (10 percent) and the Traditional Couponers (5 percent).

Returning to homemaking, clerical positions

Of those Fast & Healthy who have been employed since retiring, 25 percent became homemakers and 16 percent became clerical workers. In addition, 13 percent took professional or technical positions when returning to work after retirement.

Most likely to volunteer

The Fast & Healthy are the most likely of the three Food segments to volunteer. Far more Fast & Healthy (52 percent) volunteer their time as compared to the U.S. population over 50 (45 percent) and the other two Food segments.

Figure 18–7 Residence within Census Regions, 50+ Food Segments

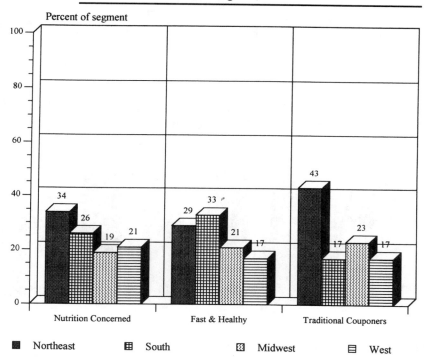

Far more Traditional Couponers reside in the Northeast as compared to the other segments. More Fast & Healthy are found in the South.

Demographic summary

The Fast & Healthy, 60 percent of whom are women, have a median age of 63 years. Of all the Food segments, the Fast & Healthy are the least likely to be married and the most likely to be widowed. They have a median of 12 years of education. Their median annual pre-tax household income of $30,759, although higher than that of all those over 50, is the lowest of all the Food segments. More Fast & Healthy live in the South than in any other region.

Having careers as professionals, clerical workers, managers, and homemakers, the Fast & Healthy are more likely than the U.S. population over 50 to work either full time or part time. In addition, the Fast & Healthy volunteer their time far more than do the other two Food segments.

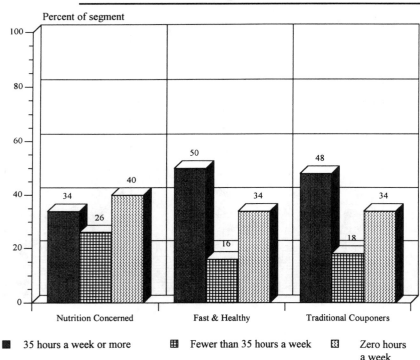

Figure 18–8 Hours of Paid Employment Per Week, 50+ Food Segments

Compared to the Nutrition Concerned, more of the Fast & Healthy and Traditional Couponers work full time.

THE TRADITIONAL COUPONERS SEGMENT

The Traditional Couponers, at 16 percent of the over-50 population, are the smallest Food segment (see Figure 18–1). They are concerned with saving money and strongly committed to coupons. The Traditional Couponers aren't worried about eating healthy foods, nor are they interested in convenience. Compared to the other segments, the Traditional Couponers have the greatest problems with physical comfort in restaurants, negotiating supermarket carts, and opening food containers.

Lack nutritional concern

As they have grown older, only the Traditional Couponers of all the Food segments have not become more concerned with nutrition than they used to be. Of the three segments, they are the least likely to pay attention to the contents of foods they eat, or to believe that what they eat can determine how they feel.

Healthful eating not a priority

The Traditional Couponers are unconcerned about eating enough fiber or cutting down on fat. They are not interested in foods made with a low-calorie fat substitute or products that use artificial sweeteners. In addition, they are not concerned about eating red meat from hormone-treated cattle.

Since nutrition is a minimal concern for them, it is not surprising that only the Traditional Couponers do not feel it is important to take vitamin supplements. Of all the Food segments, they are the least likely to be interested in specially formulated foods meet the nutritional needs of older people.

Although they have very little concern for nutrition, the Traditional Couponers, like the other segments, are trying to cut down on salt and to eat more fruits and vegetables.

Interested in shelf-stable foods

Only this segment prefers buying food packaged so that it can be stored in the cupboard for a long time. However, both the Fast & Healthy and Traditional Couponers are interested in foods that are typically frozen that have been repackaged so that they are shelf stable. In addition, believing that microwave ovens are safe to use, this segment likes being able to purchase packaged foods that they can heat up in that appliance.

The Traditional Couponers are the only segment that finds it difficult to open some food containers.

More convenience, fewer meals

The Traditional Couponers agree they are using more convenience foods and cooking fewer meals. However, they would not like to have gourmet meals delivered hot to their homes.

Restaurants too expensive

The Traditional Couponers favor restaurants that offer special discounts to people over a certain age. They are also the only Food segment to agree that eating in restaurants is too expensive.

Although nutrition is less important to the Traditional Couponers than to the other segments, they agree far more strongly that fast food restaurants do not serve the kind of food that is healthy for older people.

Whether related to nutrition, cost, or appetite, only the Traditional Couponers would appreciate restaurants serving smaller portions.

Coupon use a must

The main interest the Traditional Couponers have in grocery shopping is being able to use coupons. They are the only segment which agrees, and agrees very strongly, that they are using more coupons for discounts and for grocery shopping as they get older. In fact, an increased use of coupons in grocery shopping is the statement with which the Traditional Couponers have the highest level of agreement.

Rely on "favorite brands"

The Traditional Couponers' use of coupons, however, doesn't appear to conflict with their reliance on their "favorite brands," which has been increasing as they have grown older.

Advertising ineffective

Of the three segments, the Traditional Couponers report paying the least attention to advertising for food products. They, along with the Fast & Healthy, do not believe product endorsements made by well-known people.

DEMOGRAPHICS—THE TRADITIONAL COUPONERS

More males than females

The Traditional Couponers segment, 53 percent of which are male, has significantly more males than the other two Food segments (40 percent each) and the over-50 population (43 percent). This is the only Food segment in which males outnumber females (see Figure 18–2).

Youngest Food segment

The Traditional Couponers have a median age of 62 years, slightly younger than the U.S. population over 50 which has a median age of 64.

While more of the Traditional Couponers are in younger age categories, fewer are in older age categories. In comparison to 18 percent of the population over 50, 21 percent of the Traditional Couponers are aged 50 to 55. In addition, only 13 percent of the Traditional Couponers are 75 or older, compared to 18 percent of the Fast & Healthy and 23 percent of the over-50 population (see Figure 18–3).

Least widowed, most divorced

The Traditional Couponers are the least widowed and the most divorced of the three Food segments. Fewer of the Traditional Couponers (20 percent) are widowed as compared to those in the total over-50 population (30 percent). In terms of widowhood, the contrast between the Traditional Couponers and the Fast & Healthy (36 percent) is particularly dramatic.

Compared to 7 percent of the over-50 population, 10 percent of the Traditional Couponers are divorced. However, more Traditional Couponers (65 percent) are also married as compared to the population over 50 (58 percent) (see Figure 18–4).

Average education

Like the Fast & Healthy, those in the Traditional Couponers segment have a median of 12 years of education. Compared to 26 percent of the over-50 population, 24 percent of the Traditional Couponers have attended four or more years of college (see Figure 18–5).

The Three Food Segments

Financially secure

The Traditional Couponers have a median annual pre-tax household income of $34,119, only about $200 less than the Nutrition Concerned, but far more than the median income of $25,670 of those over 50. In this segment, 24 percent have annual incomes of $50,000 or more (see Figure 18–6).

Many Northeasterners

Two-thirds of the Traditional Couponers can be found in the Northeast or Midwest. Far more of the Traditional Couponers (43 percent) live in the Northeast as compared to the over-50 population (22 percent). In contrast to the Traditional Couponers, fewer of the Nutrition Concerned (34 percent) and the Fast & Healthy (29 percent) live in the Northeast.

On the other hand, only 17 percent of the Traditional Couponers live in the South, compared to 26 percent of the Nutrition Concerned, 33 percent of the Fast & Healthy, and 35 percent of the U.S. population over 50. Twenty-three percent of the Traditional Couponers can be found in the Midwest, while only 17 percent reside in the West (see Figure 18–7).

More work full time

Sixty-six percent of the Traditional Couponers work either part or full time, compared to 59 percent of the U.S. population over 50 and 60 percent of the Nutrition Concerned (see Figure 18–8).

Professionals, skilled craftsmen, managers

Twenty-six percent of the Traditional Couponers made careers in professional or technical positions. Significantly more of those in this segment (13 percent) spent their lives working as skilled craftsmen as compared to the over-50 population (6 percent). In addition, 12 percent were managers or executives.

Return to work as homemakers, professionals

Of those Traditional Couponers who have been employed since retirement, 21 percent became homemakers, 18 percent took professional or technical jobs, and 10 percent work in sales positions.

Volunteer at average rate

On par with the 45 percent of the over-50 population that volunteers, 44 percent of the Traditional Couponers give some of their time to volunteering.

Demographic summary

The youngest Food segment, the Traditional Couponers segment has significantly more males than do the other two segments. The Traditional Couponers are the least widowed of the Food segments, and, although also the most divorced, more of them are married as compared to the national over-50 population. The Traditional Couponers have the same level of education as the over-50 population and have a median annual income of $34,119, higher than the median for the U.S. population over 50. More of those in this segment live in the Northeast than in any other region.

More likely to work either part time or full time than the over-50 population, the Traditional Couponers had careers as professionals, skilled craftsmen, and managers. Those in this segment volunteer at about the same rate as does the national population of people over 50.

NINETEEN

FOOD TRENDS AND PURCHASES

THREE FOOD CARTS

Imagine that on your next trip to your local supermarket you assess the contents of three carts, each one pushed by someone over 50. The first cart is filled with a dozen containers of nonfat yogurt, two giant boxes of shredded wheat cereal, a turkey breast, a box of artificial sweetener, and apples, carrots, and lettuce.

Several low-fat, low-salt frozen dinners, a jar of instant coffee, a tube of frozen biscuit dough, two bags of frozen vegetables, a box of frozen oat-bran waffles, and three prepared entrees from the deli take up space in cart number two.

The third cart is loaded with three bags of corn chips, a quart of ice cream, a pound of butter, a sirloin steak, two six-packs of beer, pickled beets in glass jars, a package of bologna, and a box of muffin mix.

The contents of each cart are representative food choices of each of our 50+ Food segments. The first cart, filled with good-for-you foods, belongs to someone in the Nutrition Concerned segment. The Fast & Healthy person with cart number two has focused on convenience-oriented food items that are also positioned as healthful. Snacks and traditional fare represent many of the food choices of the Traditional Couponer pushing cart number three.

This chapter examines some of the possibilities and challenges that the growing mature market holds out to those producing, marketing, and distributing foodstuffs. Their current purchases, reflecting existing trends—from individual packaging to healthful eating, convenience to organically grown—are analyzed.

We also explore some of the Food segments' unmet needs, needs that food marketers and distributors can profitably address. From microwave oven usage to their level of alcohol consumption, the specifics of each Food segment's consumption are defined here.

WHY PAY ATTENTION TO THE MATURE MARKET?

Mature population continues to grow

Both their numbers and expenditures on food compel companies that manufacture or sell food products to pay more attention to mature adults. Those 45 to 64 include the first wave of the Baby Boomers, a group that will begin to turn 50 in 1996. Because of the Boomer phenomenon, the 50 to 64 age group is predicted to grow by 25 percent over the next ten years, while those under 50 will grow by only 3.5 percent.

Those food and grocery store marketers who address the needs and interests of mature adults now will be better able to focus on the concerns of the Boomers as they age.

High expenditures on food at home

According to the Family Economics Review, an individual over 50 spent approximately $2,115 in 1992 on food eaten at home on a "moderate-cost plan." An individual 20 to 50 spent about $2,200—virtually the same amount.

An argument can also be made for targeting those who are now over 65 and spend an average of $1,892 per year on food at home. Before dismissing this age group, consider the fact that they spend $600 more per year than households headed by someone under 25. Although both of these populations represent smaller-sized households dominated by single adults and fewer couples and children, per capita spending in the over-65 age group is still comparatively high.

50+ segments spending more on groceries

Average weekly expenditures on food by persons over 50 for 1989 were placed by the U.S. Bureau of Labor Statistics at $52.43. Although not directly comparable, our 50+ survey shows weekly median expenditures for groceries by the national sample at $67.49. The Fast & Healthy, more of whom are not married as compared to the other two segments, estimate that they spend a median of $64.53 on groceries per week. The remaining two segments, the Nutrition Concerned and Traditional Couponers, spend similar amounts per week: $70.47 and $70.04.

Per capita spending, then, by the Fast & Healthy, more of whom live alone, is actually higher than that of the other two Food segments, more of whom are married. Because of a projected increase in the number of older people living alone, per capita spending for food among this age group will grow. One-person households spend more buying smaller packages of food items and single-serve packages, both of which are more expensive.

REPRESENTATIVE OF CURRENT FOOD TRENDS

The food concerns and selections of our three Food segments represent the major food trends that exist today in the United States: healthy, good-for-you eating; convenience; traditional; ethnic; and diet. But the purchase of specific types of foods is largely concentrated in one or another of the 50+ segments.

Segments don't differ on demographics

Demographic characteristics, such as household income or age, however, don't distinguish our 50+ Food segments. For example, one characteristic of those who eat healthful foods is reading food product labels. Doing so is important to 59 percent of the population over 50. But this averaged score masks a dramatic difference between the Nutrition Concerned and the Traditional Couponers.

It's significant that 74 percent of the Nutrition Concerned strongly agree that it's important to read food product labels, in contrast to only 30 percent of the Traditional Couponers. This example shows how strongly divergent positions are lost when they are averaged, something that occurs in most traditional types of marketing research.

The difference between the Nutrition Concerned and the Traditional Couponers on the importance of reading labels on food products can't be explained by demographics. The education, age, and income of these two segments are virtually identical. The difference between the two segments on this point supports this book's basic premise: attitudes, not demographics, are often better at explaining the consumption patterns that exist in a market.

Relating attitudes to behaviors

The linkage between attitudes and behaviors is often questioned, and the relationship is frequently dismissed as tenuous. However, an example from our 50+ study focusing on use of artificial sweeteners or fat substitutes shows a clear relationship between attitudes and behaviors.

Only the Nutrition Concerned and the Fast & Healthy express a strong interest in such products. When it comes to their behaviors, the use of artificial sweeteners is clearly concentrated in these two segments. About a third of these two segments use artificial sweeteners daily. The majority of Traditional Couponers (62 percent) never use this product.

HEALTHY EATING ATTRACTS NUTRITION CONCERNED

The general trend toward healthy eating

Is the American diet healthier today than it was ten years ago? In general, food consumption patterns suggest that it is. Of the ten foods for which per capita consumption increased between 1978 and 1988, five are fruits and vegetables. For example, the consumption of fresh broccoli increased by 232 percent between 1978 and 1988.

In contrast, half of the ten foods that suffered the largest declines in consumption were red meats and dairy products. The consumption of whole milk declined by 34 percent, that of beef declined by 18 percent.

One pattern suggests that many consumers are seeking freshness and nutrition and, at the same time, reducing fat. However, we also note that the consumption of cheese increased by 46 percent and that of nonfat dry milk declined by 23 percent. Do these contradictory patterns suggest that we're eating out of both sides of our mouths?

If we didn't have the 50+ Food segments, we might be swept into agreeing with a statement by Carole Sugarman reprinted in *Supermarket Business* that such sales data shows "some revealing—and schizophrenic—trends...." Sugarman notes that sales of snack foods more than doubled during the 1980s at the same time that sales of yogurt soared from $675 million to $1.1 billion.

These trends are not "schizophrenic." They don't exist in equal strengths simultaneously within all consumers in general. Our 50+ data and the studies we have completed for clients consistently show dissimilar patterns of consumption concentrated in various motivational segments. When traditional market research averages these drastically different patterns of consumption across all consumers, they are submerged and lost.

Seniors more focused on nutrition

Patterns of consumption do show a greater interest in fresh fruits and vegetables among some consumers. Studies also support the idea that more Americans over 50 are buying healthful foods than are younger adults. The idea that concern with healthy eating is also disproportionately concentrated on those over 50 is supported by several studies.

A study for the Food Marketing Institute in 1988 found that 72 percent of people over 65 say that nutrition is a primary concern. In contrast, the 19- to 24-year-old group had the least amount of interest in nutrition.

A Gallup study conducted in 1989 for *Advertising Age* showed that healthful foods were bought by more than half (53 percent) of those over 55. This figure contrasts with consumers aged 18 to 34, only 39 percent of whom had bought such foods.

Committed to healthful eating

In further defining the interest in healthful eating from consumers in general to those over 50, our 50+ study shows that a very strong commitment to healthful eating is concentrated in only one of our Food segments.

Although 43 percent of our total respondents strongly agree that they are much more concerned about nutrition now than when they were younger, a far larger percentage of the Nutrition Concerned (58 percent) agree with this statement. At 39 percent, the Fast & Healthy are actually below those over 50 in their level of agreement.

Avoiding some foods

This general concern for healthful and nutritious eating is seen when the specific contents of foods are examined. For example, of the Nutrition Concerned, 67 percent strongly agree that they are "concerned about eating too much fat." That percentage compares to 43 percent of the Fast & Healthy and only 6 percent of the Traditional Couponers.

These attitudes are reflected in what each segment consumes. The Traditional Couponers are clearly the heaviest consumers of red meat, butter, and cheese. For

Food Trends and Purchases

example, of this segment, 18 percent eat meat daily. In contrast, only about one in ten of the Nutrition Concerned and Fast & Healthy eats meat this often (see Figure 19-1).

Good-for-you food

Recent dietary studies have shown that almost half of all Americans don't eat fruit daily and almost a quarter don't eat vegetables on any given day. These statistics are about the same as those reflected in the population over 50 as determined by the 50+ study. However, it's important to note that the Nutrition Concerned are eating more fruits and fresh vegetables than the population as a whole.

Following their stated commitment to healthy eating, more of the Nutrition Concerned report eating fresh fruit daily (58 percent) as compared to all those over 50 (51 percent). But among the Fast & Healthy, only 47 percent eat fresh fruit daily.

Figure 19-1 High-Fat and High-Cholesterol Foods Eaten, 50+ Food Segments

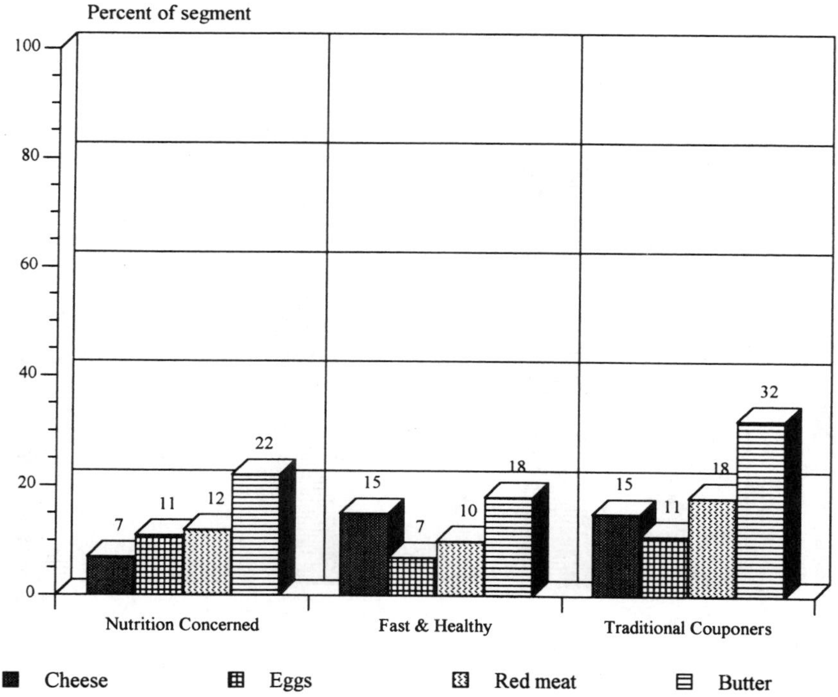

Segment	Cheese	Eggs	Red meat	Butter
Nutrition Concerned	7	11	12	22
Fast & Healthy	15	7	10	18
Traditional Couponers	15	11	18	32

On a multiple-response question on the frequency of eating 36 foods, more of the Traditional Couponers report consuming foods higher in fat and cholesterol daily as compared to the Nutrition Concerned and Fast & Healthy.

Not only do the Nutrition Concerned eat more fresh fruit daily, more of them also consume fresh vegetables. Of this segment, 76 percent eat fresh vegetables daily compared to 53 percent of the Fast & Healthy and 46 percent of the Traditional Couponers (see Figure 19–2).

Research from the Economic Research Service (ERS) reports that in households where the head was 45 years or older, 36 percent more was spent per person on cereals and bakery products than in younger households. While these average statistics should be heartening to Kellogg's and General Mills, our research shows that consumption of cereals is concentrated in one segment: the Nutrition Concerned.

More than half of the Nutrition Concerned (53 percent) eat dry cereal daily, compared to the Fast & Healthy (43 percent) and Traditional Couponers (32 percent). Their heavy consumption of fiber could be having some healthful benefits. Interestly,

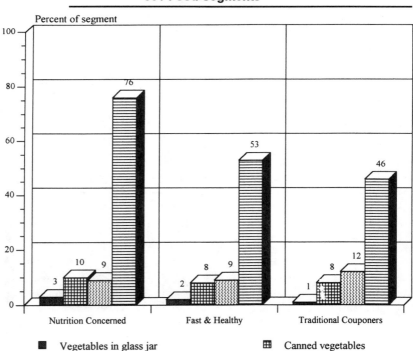

Figure 19–2 Vegetables Consumed, 50+ Food Segments

On a multiple-response question on the frequency of eating 36 foods, the Nutrition Concerned report eating far more fresh vegetables daily than the other two Food segments. However, all three segments consume processed vegetables at similar levels.

fewer of the Nutrition Concerned, as compared to the population over 50, report being bothered by frequent constipation or hemorrhoids.

Having turned away from red meat, it isn't surprising that more of the Nutrition Concerned (57 percent) are eating seafood at least weekly as compared to the other two segments (42 percent each). More of the Nutrition Concerned (87 percent) also eat chicken at least weekly as compared to those over 50 (81 percent).

What motivates the Nutrition Concerned?

The Nutrition Concerned, unlike the other Food segments, have a proactive mindset. They believe they can take actions that will affect their state of health. For example, more than the other two Food segments they believe that "what I eat is an important factor in how I feel." The Nutrition Concerned agree far more strongly than the other segments that they are careful to eat a balanced diet. Healthful eating may contribute to the fact that more of the Nutrition Concerned (44 percent) agree strongly that they "feel much younger than they really are" as compared to people over 50 (37 percent).

Their proactive mindset is also shown in the dramatic rate at which the Nutrition Concerned exercise. Whereas 67 percent of the over-50 population says that it exercises, but not as part of their jobs, 78 percent of the Nutrition Concerned do so. Interestingly enough, while the types of exercise that the Traditional Couponers favor include golf and shuffleboard, the Nutrition Concerned prefer jogging and aerobics. Clearly, the Traditional Couponers, unlike the Nutrition Concerned, don't do fat-burning, aerobic exercises.

When we further defined exercise as getting within the range of one's target heart rate and working hard for at least 20 minutes each time, 28 percent of the Nutrition Concerned exercised four or more times a week at that level as compared to 14 percent of the Traditional Couponers. It isn't surprising, then, that more of the Nutrition Concerned (54 percent) rate their current state of health as "excellent" or "very good" as compared to those over 50 (50 percent).

While the Nutrition Concerned have been touched by illness, they aren't necessarily motivated to eat healthfully now because of overwhelmingly bad health. For example, more of them, as compared to the national sample, have had cancer in the past. However, the Nutrition Concerned were about the same as the over-50 population—or lower—in their rate of the remaining 36 diseases included in the study.

Data from the 50+ study supports the findings of Dr. Philip Garry, a pathologist at the University of New Mexico School of Medicine, who conducted the New Mexico Aging Process study. Garry studied 300 older adults between the ages of 65 and 93 and concluded that the process of aging may be slowed down through behavioral changes.

"You really can slow aging down through eating well, exercising properly, keeping positive, and staying active," says Garry. These are all behaviors that the Nutrition Concerned practice. "We don't see much change," he continues, "in them [participants in his study] due to aging . . . they are going on in life like they did when they were much younger."

What influences better eating?

At the suggestion of their doctors, about one-third of the sample have reduced cholesterol, fat, and sodium in their diets. Not surprisingly, the Nutrition Concerned have done so far more than have the Traditional Couponers.

It is important to note another sign of proactivity among the Nutrition Concerned. More of them have improved their diets on their own rather than under the instruction of their physicians. For example, 37 percent of the Nutrition Concerned reduced fat in their diets because their doctor told them to do so as compared to 63 percent who did it on their own. Far fewer Traditional Couponers have taken such action on their own (see Figure 19-3).

Figure 19-3 Influencers on Improving Diet, 50+ Food Segments

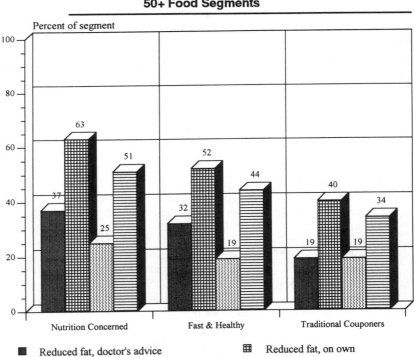

On a multiple-response question listing six food characteristics, more of the Nutrition Concerned have reduced fat and calories in their diets on their own as compared to those who did so on the advice of a doctor.

SNACKING POPULAR WITH TRADITIONAL COUPONERS

High-fat, high-cholesterol diet

As we've seen, more of the Traditional Couponers eat red meat and butter than those in the other segments. In addition, fewer of them eat such items as fresh or frozen vegetables, fruits, cereal, and fruit juice. Compared to the Nutrition Concerned, their diet is high in fat and cholesterol, and loaded with sugar and salt from the snacks they consume.

An important snack food market

A day in the life of a Traditional Couponer is punctuated by the consumption of sugar- and fat-intensive snack foods. For example, more Traditional Couponers (58 percent) eat candy weekly compared to the Nutrition Concerned (39 percent). Salty snacks, such as chips and popcorn, are eaten on a weekly basis by more of the Traditional Couponers (54 percent) as compared to the Nutrition Concerned (43 percent) (see Figure 19-4).

CONVENIENCE IMPORTANT TO FAST & HEALTHY

Cooking less

Although they are more concerned about their health now than when they were younger, the Fast & Healthy are really driven by an interest in convenience. Forty-one percent of those in this segment agree strongly that they are using more convenience foods now compared to when they used to cook from scratch.

Of the Fast & Healthy, 43 percent strongly or somewhat agree that they are skipping more meals now than before, compared to 28 percent of people over 50. In contrast, only 14 percent of the Nutrition Concerned agree with this statement.

Not only are the Fast & Healthy skipping meals, they are also cooking fewer meals now than when they were younger. Although 53 percent of the over-50 population strongly agree with this statement, 83 percent of the Fast & Healthy do so. Only 31 percent of the Nutrition Concerned either strongly or moderately agree. The extent to which the Fast & Healthy are simply not cooking is seen in their responses to how many meals they had prepared for themselves and/or their spouse over the past three days. Nine or 10 meals could be expected to have been prepared over this period. However, almost three-quarters of the Fast & Healthy report preparing fewer than five meals. In contrast, 57 percent of the Nutrition Concerned report cooking six to ten-plus meals over the three-day period, something only 27 percent of the Fast & Healthy did.

Exploding some food myths

A common idea about the eating habits of older people is that they miss meals. Meals are supposedly not eaten because mature consumers are depressed and lonely, cannot get to the store, or are too poor to buy food. From the 50+ study we see that the Fast & Healthy are missing more meals now than they did before and cooking

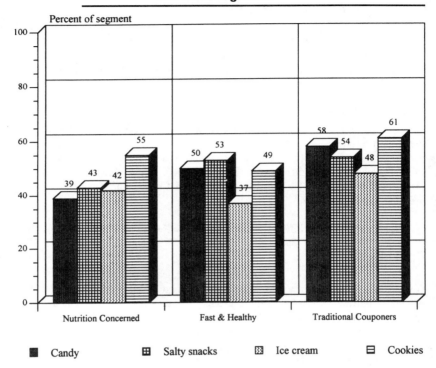

Figure 19-4 Snack Foods Eaten, 50+ Food Segments

On a multiple-response question on the frequency of eating 36 foods, far more of the Traditional Couponers report consuming snack foods once a week or more than the other two Food segments. Even though they are underconsumers of snacks, more than half of the Nutrition Concerned eat cookies once a week or more.

fewer meals as well. While it isn't surprising that the Fast & Healthy do so, their reasons differ significantly from any links to poverty or depression.

It's true that there are far more widowed persons in the Fast & Healthy segment than one would normally expect. However, they are not impoverished. This Food segment's annual pre-tax median household income of $30,759 is higher than the median of $25,670 for all persons over 50. What the Fast & Healthy do far more than the other segments is eat meals in a restaurant. Over a year's time, they eat 68 restaurant meals, as compared to only 44 restaurant meals eaten by those in the Nutrition Concerned segment.

The Fast & Healthy engage in a number of activities. In the realm of travel, the Fast & Healthy are most apt of all three Food segments to have gone on a cruise, taken an escorted tour, and traveled by air to get to a vacation destination.

Food Trends and Purchases 299

Our data shows that one reason the Fast & Healthy are so committed to convenience foods and skip meals is that they are very busy people. Far more of the Fast & Healthy are working full time (50 percent) as compared to the Nutrition Concerned (34 percent).

Not only are more of them working full time, but more of them are engaged in unpaid volunteer work. Compared to both the Nutrition Concerned (42 percent) and the Traditional Couponers (44 percent), far more of the Fast & Healthy volunteer (52 percent).

We see no evidence that the Fast & Healthy are a depressed segment. The Fast & Healthy don't think about death and dying more than the population over 50. In terms of being so overwhelmed by life that they would consider suicide, fewer of the Fast & Healthy consider this alternative as compared to all of those over 50.

Convenience foods appealing

Although the Fast & Healthy crave convenience, some types of convenience foods are clearly more appealing to them than to the other segments.

The Fast & Healthy's reliance on convenience foods is seen in their heavy use of frozen dinners. About one-third of the Fast & Healthy eat a frozen dinner at least once a week, as compared to 6 or 7 percent of the other two segments (see Figure 19-5). In fact, fully half of the Nutrition Concerned say they never buy frozen dinners.

Whether the product is instant coffee or frozen vegetables, the Fast & Healthy are committed to healthy eating—as long as it is convenient. Whereas the vast majority of the Nutrition Concerned eat fresh vegetables daily, more of the Fast & Healthy eat frozen or canned vegetables at least once a week.

Use of microwave

Various studies have reported that 75 to 80 percent of American households have a microwave oven. The fact that 83 percent of the over-50 population owns a microwave oven shows that seniors are quite willing to accept new technology. In addition, more of the Fast & Healthy (90 percent) have one than those over 50.

Given the fact that they prefer to cook from scratch and are least interested in convenience, it isn't surprising that the Nutrition Concerned have the lowest ownership of a microwave oven of our Food segments.

Compared to figures from other national studies, more respondents in our 50+ study are using their microwave ovens for a greater variety of tasks as compared to U.S. adults in general. This comparison underscores the fact that the vast majority of those over 50 will use time-saving technologies such as the microwave oven. The three primary uses of their microwave ovens by our respondents are reheating leftovers (99 percent), heating coffee, tea, or water (88 percent), and defrosting (94 percent).

These percentages are higher for the same tasks than those from a study by The Roper Organization on U.S. owners of microwave ovens as a whole. In some areas the 50+ respondents are using their microwave ovens in ways that demand increasing levels of skill and a more sophisticated knowledge of this appliance.

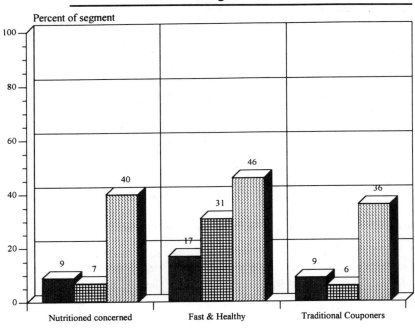

Figure 19–5 Convenience Foods Eaten, 50+ Food Segments

On a multiple-response question on the frequency of eating 36 foods, more Fast & Healthy report using convenience foods once a week or more than the other two Food segments.

For example, in the Roper study only 29 percent of the sample used their microwave ovens to cook meat or fish. In contrast, 57 percent of those over 50 do so. Not surprisingly, the segment that uses this appliance the least extensively is the Traditional Couponers (see Figure 19–6).

ALCOHOL CONSUMPTION DIFFERS BY SEGMENT

The Fast & Healthy drink alcoholic beverages the least as compared to the other two segments. Whereas both the Nutrition Concerned and Traditional Couponers drink hard liquor in similar numbers, each of these segments favors either beer or wine.

Food Trends and Purchases

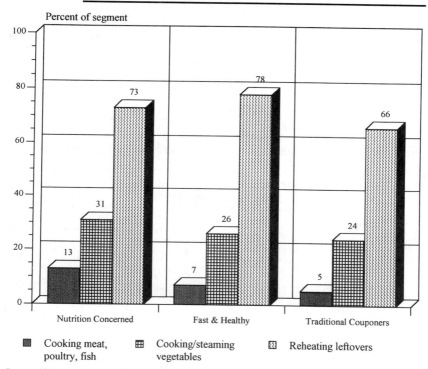

Figure 19–6 Use of Microwave Oven, 50+ Food Segments

On a multiple-response question on the frequency of using the microwave for 10 purposes, more Fast and Healthy report regularly using their microwaves for reheating leftovers. More of the Nutrition Concerned are using their microwaves in less conventional ways.

Nutrition Concerned prefer wine

More of the Nutrition Concerned (47 percent) drink wine, as compared to the over-50 population (39 percent) (see Figure 19–7). Some of them count wine among their favorite activities, along with gourmet cooking and fine foods.

Drinking hard liquor at all ages

Hard liquor is drunk by 28 percent of people over 50, but by 34 percent of the Nutrition Concerned and 36 percent of the Traditional Couponers (see Figure 19–7). Almost three-quarters of the Fast & Healthy don't drink hard liquor now or never did drink it.

The consumption of hard liquor does not decline with age across all segments. What does happen is that the percentage of Fast & Healthy who consume hard liquor declines substantially between those 65 and younger (35 percent) and those over 65 (19 percent). But for the two remaining Food segments, the consumption of hard liquor remains virtually unchanged. For example, although 33 percent of the Nutrition Concerned between 50 and 65 drink it, 35 percent over 65 do so. Just about the same percentages of Traditional Couponers in both age groups consume hard liquor.

Again we see that marketers working with averaged data would assume a general decline in hard liquor consumption among those over 65. Clearly, the same percentages of Nutrition Concerned and Traditional Couponers continue drinking hard liquor and marketers should recognize and address these segments.

CONCLUSION

We have seen that the over-50 population is, in general, more interested and committed to healthful eating than are younger consumers. Food trends such as

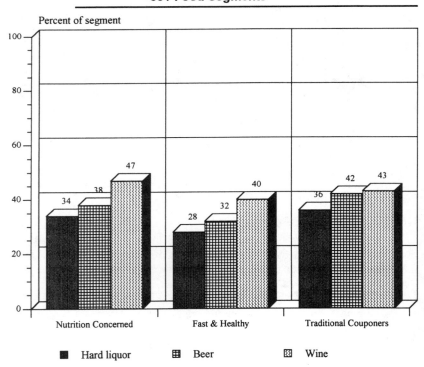

Figure 19–7 **Use of Alcohol Products, 50+ Food Segments**

On a multiple-response question listing six tobacco and liquor products, more Nutrition Concerned report consuming wine, while more Traditional Couponers consume beer. The Fast and Healthy are the least likely of the Food segments to drink any type of alcoholic beverage.

nutritious eating or snacking are evident in mature consumers as well as in the general population. Our study shows that these trends are concentrated in three attitudinal segments: the Nutrition Concerned, the Fast & Healthy, and the Traditional Couponers. These attitudinal or motivational segments are not defined by demographics. We have, however, seen that their behaviors and attitudes correlate.

REFERENCES

Carlson, Ellen, "Get Cooking with the Microwave," *St. Paul Pioneer Press Dispatch,* March 8, 1989, pp. 1C-2C.

Economic Research Service, *Food Consumption, Prices, and Expenditures,* SB-804, U.S. Department of Agriculture, May 1990.

Family Economics Review, U.S. Department of Agriculture, 1992.

Erickson, Julie Liesse, "Wary Consumers Want More Health Ad Info," *Advertising Age,* December 4, 1989, p. 12. Source of Gallup study on purchase of healthful foods.

"Feeding Frenzy," *Newsweek,* May 27, 1991, pp. 46-53.

Freedman, A. M., "The Microwave Cooks Up a New Way of Life," *Wall Street Journal,* Sept. 19, 1989, p. B1.

Greeley, Alexandra, "Nutrition and the Elderly," *FDA Consumer,* October 1990, pp. 25-28. Source of Philip Garry's study on the effects of behavioral changes on the aging.

Menchin, Robert, *The Mature Market,* Chicago, IL: Probus Publishing Company, 1989, p. 137. Source of study by the Food Marketing Institute on concern about nutrition.

Putnam, Judith Jones, "Food Consumption," *National Food Review,* July-September 1989, pp. 1-9.

Sugarman, Carole, "How Are We Doing?" *Supermarket Business,* September 1991, pp. 30-36.

"Who Zaps Their Food?" *American Demographics,* May 1991, p. 16. Source of study by The Roper Organization on U.S. owners of microwave ovens.

TWENTY

NEW FOOD PRODUCTS AND SERVICES

FULFILLING NEEDS OF THE MATURE

Several years ago seniors were given the opportunity to reach for jars of pureed food on supermarket shelves. This product, one of the few ever positioned solely for seniors, flopped.

While seniors have special interests and needs, this example makes it clear that positioning foods just for seniors, especially when the focus is on a disability, is clearly not the way to go. Marketers should, however, consider positioning some food products specifically to seniors and creating new products that reflect their needs.

We believe that a product which a segment within the mature market finds appealing may very well find a market within younger populations. Once limited to a narrow niche, products such as decaffeinated coffee and bran cereal have found a far larger and wider market. New food products, appealing to segments within the immense mature market, remain to be discovered.

In this chapter we will take a close look at exactly what each of the Food segments wants in new products. Food attributes will be analyzed by segment, giving food marketers the information they need to create new products that meet each segments' needs. Awareness and trial by the segments are explored, showing that attitudes often parallel behaviors. The importance of packaging to certain Food segments is discussed as well. Finally, we try to separate myth from reality in terms of the services offered to mature customers by grocery stores.

Appeal of new food products

In creating and marketing new food products for the mature market, food manufacturers and retailers must relate to a variety of interests. Depending on the specific 50+ segment, these include improved labeling; good-for-you foods; modified packaging; reduced additives, pesticides, and calories; and enhanced aromas. Most of these interests or concerns are also shared by segments within the under-50 population.

Asked about the importance of 15 health-related attributes in the foods they eat, the 50+ Food segments told us that four attributes are extremely important: fresh (77 percent), nutritious (65 percent), low cholesterol (60 percent), and low in fat (59 percent).

Three emerging trends

In examining these four attributes and the remaining 11, however, three trends emerge. First of all, it is obvious that the Traditional Couponers have only marginal interest in healthy eating. Even when it comes to a generally accepted attribute such as "nutritious," far fewer of them consider it important as compared to the other two segments (see Figure 20–1).

Secondly, the Nutrition Concerned are far more committed to good-for-you eating than are the Fast & Healthy. While more of the Nutrition Concerned agree that certain more mainstream food attributes such as "low salt," "high in vitamins," or "high in

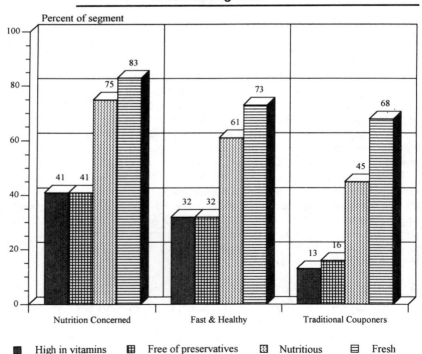

Figure 20–1 Importance of Food Attributes, 50+ Food Segments

On a multiple-response question on the importance of 15 food attributes, more Nutrition Concerned rated high in vitamins, free of preservatives, nutritious, and fresh as "extremely important."

New Food Products and Services

fiber" are extremely important, fewer of the Fast & Healthy do so (see Figures 20-1, 20-2, and 20-4).

The third and final trend shows the Nutrition Concerned defining healthful eating in greater detail and more broadly than do the Fast & Healthy. For example, more of the Nutrition Concerned than the Fast & Healthy rate certain food attributes, such as "grown organically with no pesticides," as extremely important. Fully one-third of the Nutrition Concerned hold this view as compared to one-quarter of the Fast & Healthy (see Figure 20-3).

Besides pesticides, the Nutrition Concerned want foods to be free of preservatives. This attribute is important to more of the Nutrition Concerned (41 percent) as compared to the Fast & Healthy (32 percent) (see Figure 20-1).

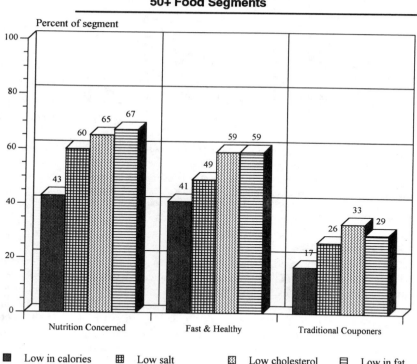

Figure 20-2 Importance of Food Attributes, 50+ Food Segments

■ Low in calories   ▦ Low salt   ▩ Low cholesterol   ☰ Low in fat

On a multiple-response question on the importance of 15 food attributes, more Nutrition Concerned rated low in calories, low salt, low cholesterol, and low in fat as "extremely important." Low in calories was "extremely important" to almost as many Fast & Healthy.

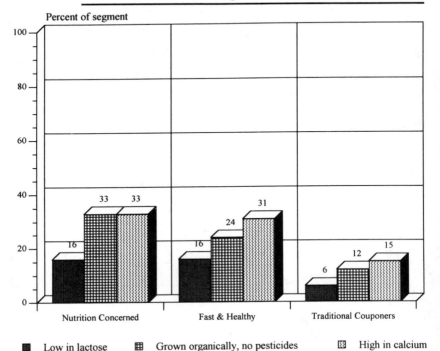

Figure 20–3 Importance of Food Attributes, 50+ Food Segments

On a multiple-response question on the importance of 15 food attributes, the same percentage of Nutrition Concerned and Fast & Healthy rated low in lactose as "extremely important." However, more Nutrition Concerned rated grown organically with no pesticides and high in calcium as extremely important food attributes.

Fear food additives

The Nutrition Concerned's fear of food additives also includes hormone-treated red meat. While a quarter of those over 50 strongly agree that they are frightened about eating red meat that comes from hormone-treated cattle, one-third of the Nutrition Concerned strongly agree with this statement. In contrast, only 9 percent of the Traditional Couponers strongly agree with it. The question for growers and food manufacturers has always been whether those who share these concerns are willing to pay more for foods free of pesticides, additives, and hormones.

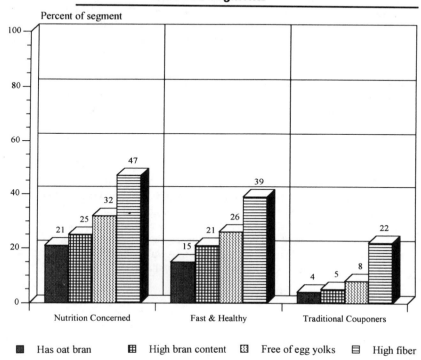

Figure 20-4 Importance of Food Attributes, 50+ Food Segments

On a multiple-response question on the importance of 15 food attributes, far fewer Traditional Couponers rated oat bran, high bran content, free of egg yolks, and high fiber as "extremely important" food attributes.

LOW-FAT, GOOD-FOR-YOU EATING

Good-for-you foods, appealing to both the Nutrition Concerned and the Fast & Healthy, should find a market with 84 percent of all seniors. With this level of interest from the majority of mature food buyers, it's not surprising that sales of fresh cauliflower increased by 174 percent from 1978 to 1988 or that consumption of whole milk declined by 34 percent over the same time span.

Explosion of low-fat foods

The interest that segments such as the Nutrition Concerned and Fast & Healthy have in low-fat products have no doubt contributed to their explosive growth. Rosanna Mentzer Morrison and Judith Jones Putnam, agricultural economists with the Economic Research Service (ERS) at the U.S. Department of Agriculture, believe consumers' desire for low-fat foods "presents challenges and opportunities for the food industry to produce and promote reduced-fat products."

In responding to consumers' concerns about low-fat foods, 1,024 such products were introduced in 1990, double the number of 1989. In 1990 sales of low-fat foods reached an estimated $18 billion. By 1995 they are projected to more than double to $41.4 billion, according to Experience, Inc., a Minneapolis-based international agribusiness consulting firm.

One example of a new low-fat product is ConAgra's Life Choice, a line of meatless, low-fat frozen dinners with 1.5 grams of fat and 235 calories. Another manufacturer, Health Valley, offers over 80 fat-free products ranging from cookies to chilies, fruit bars to corn puffs.

The example of low-fat beef

The development of low-fat ground beef, introduced in 1991, illustrates what can be done to make a high-fat product more acceptable to the Nutrition Concerned and Fast & Healthy. Three-quarters of these segments are eating red meat at least once a week, and ground beef accounts for 44 percent of all beef sales. In 1990, the beef industry sold about 6.75 billion pounds of ground beef, averaging 20 percent fat. Reducing the fat content of ground beef is, then, a significant achievement.

Today scores of products ranging from ConAgra's Healthy Choice Extra Lean Ground Beef to McDonald's McLean Deluxe are available to the American consumer.

Reduce fat, but zap up aromas

While focusing on the good-for-you aspects of foods, food marketers and manufacturers concerned with the mature market can consider increasing their products' aromas. Convenience foods with intense and pleasing aromas should attract more mature consumers than those without them.

The ability to taste is linked with how well a person is able to detect smells. Because our sense of smell declines with age, our ability to taste the foods we eat decreases as well. "The ability to taste remains fairly intact with age," says James Weiffenbach, research psychologist with the National Institutes of Health in Bethesda, Maryland. "It's the decline in the ability to smell that seems to be responsible for 'taste' complaints," he says.

But when fat, for example, is removed from the ground meat sold in a supermarket, the product loses flavor. To make up for flavors lost in processing, new flavors must be added so that all consumers find the product appealing. However, mature consumers may require additional flavors. For them the product must also emit such intense and delicious aromas that it will taste good when eaten.

HEALTHFUL SNACKS

Fast & Healthy most attractive target

As food manufacturers continue to develop low-fat, low-calorie microwavable snacks, they can position them specifically to the Fast & Healthy. The 50+ study shows

New Food Products and Services

the Fast & Healthy cooking fewer meals, skipping meals, and snacking through the day. In terms of developing snack products specifically for the Fast & Healthy, vitamin-enriched snacks, discussed later in this chapter, will have specific appeal for this segment.

Fast snacks appeal to Fast & Healthy

Combining snacking with microwave-oven usage is something we could expect from the Fast & Healthy, more of whom have microwave ovens and favor using them. Specifically, we know that more of the Fast & Healthy (42 percent) use their microwave ovens to make popcorn than do people over 50 (33 percent). It's not surprising then that this segment had tried a number of snack products for the microwave that were new when we completed our 1989 50+ study.

For example, more of the Fast & Healthy (50 percent) had tried Orville Redenbacher Butter-flavored Popcorn, as compared to those over 50 (37 percent). Lack of awareness hadn't kept the other Food segments from purchasing these popcorn products. Whereas 60 percent of the Nutrition Concerned knew about the Orville Redenbacher product, only 34 percent had tried it.

Traditional Couponers prefer salty snacks

As we've seen, almost half of those over 50 eat a salty snack at least once a week or more, with the Traditional Couponers leading in consumption. Fifteen percent of them eat such snacks every day. The Traditional Couponers favor a variety of salty snacks, but are not as favorably disposed to microwaved popcorn products as are the Fast & Healthy.

A product like the highly successful Sunchips, a brand of salty snacks from Frito-Lay, may appeal to both the Traditional Couponers and the Fast & Healthy. In developing Sunchips, "The aging baby boomers were a very significant factor," notes Dwight Riskey, Frito-Lay marketing vice president. "We were looking for new products that would allow them to snack. But we were looking for 'better-for-you' aspects in products."

VITAMIN-ENRICHED FOODS

Need for low-calorie, high-vitamin products

As Senauer, Asp, and Kinsey point out in *Food Trends and the Changing Consumer*, "Since the need for calories diminishes with age but the need for vitamins and minerals apparently does not, the need for nutrient-dense foods and low-calorie foods will increase."

Seeking vitamins, minerals

Foods enriched with vitamins and minerals to provide optimum nutrition to seniors would be very appealing to the Fast & Healthy and Nutrition Concerned who both agree that they would buy such products. These segments think it's important to take a daily vitamin supplement.

A good-for-you entree that is low in fat and full of vitamins would appeal to the Nutrition Concerned and Fast & Healthy, which constitute 84 percent of the market. Because we don't suggest that such products be positioned as only for seniors, other groups of consumers, such as pregnant women or health-conscious younger people, may find these entrees or other products delivering all necessary vitamins appealing. Currently, however, General Mills' Total cereal is one of the few mainstream products on grocers' shelves to tout its delivery of all recommended vitamins.

GETTING IT OFF THE SHELF

In 1988 when Hormel introduced Top Shelf, a line of shelf-stable, meat-based meals, one supermarket manager was surprised that most of his sales of this line were made to residents of a nearby retirement community. Although Top Shelf was positioned for a far younger crowd, these results don't surprise us.

Convenience, long shelf-life attractive

Top Shelf fulfills a number of the Traditional Couponers and Fast & Healthy's interests. It is convenient to serve, microwavable, and packaged in individual portions. In addition, these two 50+ Food segments are also interested in a product that they can store for "a long time" in their cupboards. Shelf-stable products offer both portability and longer shelf-life compared to frozen meals.

In fact, the Traditional Couponers and Fast & Healthy say that if given the choice between frozen foods and shelf-stable substitutes, both would choose the shelf-stable item. The Nutrition Concerned have no interest in shelf-stable products, preferring to cook their meals from scratch.

These attitudes are supported by data from our 1989 50+ study. At that time, we found that while 23 percent of the total sample had tried Top Shelf, 31 percent of the Fast & Healthy had done so.

In 1988 sales of shelf-stable, microwavable foods reached $250 million. They are projected to reach $696 million by 1993. No doubt the Fast & Healthy and Traditional Couponers, 54 percent of the mature market, will contribute significantly to the increase in sales of shelf-stable microwavable foods.

READY-TO-EAT CONVENIENCE FOODS

Fast & Healthy only target

Marketers offering complete cooked meals at a grocery store should look to the Fast & Healthy as a prime target, actually the only viable target in the mature population for this type of product. More of the Fast & Healthy (34 percent) are interested in such meals as compared to those over 50 (19 percent). These numbers are in sharp contrast to the Traditional Couponers, only 5 percent of whom are strongly open to this concept. The Fast & Healthy's desire for ready-to-eat foods fits well with the marketing of new products such as Freshtables, a line of complete salads and stir-fry vegetables from Pillsbury's Green Giant division.

Opportunities for delis

The deli section of supermarkets, long the bastion of high-fat, high-sodium, and high-calorie offerings, can present the Fast & Healthy with healthful meals and operators with the opportunity for increased sales. For example, a chain in Dayton, Ohio, Dorothy Lane Markets, has developed a highly successful line of foods labeled "Healthy Alternatives."

Instead of cannibalizing its existing deli sales, Dorothy Lane Markets has found that Healthy Alternatives has resulted in additional sales, proof that unmet needs do exist in the marketplace, including among the Fast & Healthy. This segment is, after all, the only one willing to "pay more for the convenience of easy-to-prepare foods."

PACKAGING: DEVELOPING NEW OPTIONS

Three populations will benefit from innovations that will make food packaging easier to open and use: seniors, the disabled, and children. Many seniors and disabled have always prepared their own meals, and now children are increasingly doing so. These groups often lack the manual strength and dexterity to open current packaging. Indeed, many within the general population would agree that making packaging more accessible or easier to open would be helpful.

Some seniors may find it difficult to position their finger tips precisely on a thin strip of plastic in order to exert enough pressure to zip open a can of frozen orange juice. Punching open a box of detergent along a line that is dotted, but not perforated, requires strength. Some food packages are difficult to open, but how many seniors find them to be a serious problem?

Traditional Couponers frustrated

In the 50+ study, only the Traditional Couponers agree that "a lot of food containers are too difficult for them to open." In contrast, the Fast & Healthy, who disagree with this statement, aren't having problems opening food containers.

The fact that more Traditional Couponers have arthritis (33 percent) as compared to people over 50 (31 percent), may explain their difficulties with food containers. But not only do more Traditional Couponers have arthritis, fewer of them take a prescription drug for this condition (12 percent) as compared to those over 50 (16 percent).

Appealing as well as functional design

Besides packaging that is easier to open and grasp, seniors would benefit if the typeface, design, and colors of packaging were improved with their needs in mind. According to the 1989 *Statistical Abstract of the United States,* 13 percent of all seniors have some visual impairment. In our 50+ study, 19 percent of the over-50 population report poor vision.

For these seniors, reading food labels and instructions becomes difficult if the typeface is too small or if the colors used in the label offer little contrast. And food labels will certainly be read by those in the largest segment, the Nutrition Concerned.

Fast & Healthy want individual packages

It isn't surprising that only the least-married Fast & Healthy agree that they would like to see more foods packaged in individual servings and are "particularly interested" in finding new foods packaged in smaller portions.

SENIOR SERVICES—ARE THEY NEEDED?

Facing intense competition and reduced profit margins, retail food stores of all types and sizes are searching for products and services that will tie consumers, including those over 50, to them. When retailers suggest services for the mature market, home delivery, extra help in bagging groceries, and faster check-out lines are frequently mentioned.

Not interested in home delivery

As retailers themselves struggle with how to provide home delivery using the right formula, our 50+ study doesn't show a need for this type of service. None of our Food segments agreed that they would use the services of a grocery store that offered home delivery of items ordered by phone. Our conclusions are echoed by those from a study conducted in 1989 for the *Wall Street Journal.* That study found that 68 percent of the respondents, who were either retirees or their spouses, would be unwilling to pay someone to do their shopping for them.

Shopping for themselves

Our Food segments don't report problems in negotiating their way around a supermarket. No segment reported that it had difficulty in moving a cart around supermarket aisles. In fact, the Fast & Healthy denied this was a problem. The Fast & Healthy also responded negatively to the idea that they found parking lots at supermarkets "too difficult to get into." No Food segment agreed with this statement.

In terms of additional support or services that would make their shopping experience easier, no segment affirmed that it needed such help. We didn't find a segment interested in a store that has clerks who carry packages to the car.

In a study conducted in one midwestern state, only 10 percent of the elderly reported difficulty in grocery shopping. Senauer, Asp, and Kinsey, writing in *Food Trends and the Changing Consumer,* state that "some studies found that the elderly responded positively to businesses that provided comfort needs . . . other studies found these amenities to be less important."

OPEN TO NEW PRODUCTS

When Top Shelf was introduced, supermarket managers reported that "shoppers are extremely skeptical that a non-refrigerated meat-based product can have a shelf life of 18 months." But still mature shoppers bought it. A type of product that is described as "likely to meet a far greater degree of consumer resistance and skepticism than any other category of microwavable food" was embraced by many mature consumers.

Besides underscoring the acceptability of a shelf-stable entree to certain segments within the mature market, we think that it is important to note this radically different food product, which many younger consumers look at skeptically, was purchased by those over 50.

Two Food segments adventurous

In our 1989 50+ study, we asked respondents whether they had tried 22 products that were new at that time. With the exception of one product, all were available in supermarkets. This data reveals a consistent pattern with the Nutrition Concerned trying good-for-you products, the Fast & Healthy buying convenience foods and snacks, and the Traditional Couponers resisting new food products even if they were aware of them. In fact, the only new product that the Traditional Couponers had bought in significantly larger numbers was Michelob Dry beer.

Examples of good-for-you products, such as Campbell's Low-Sodium Soups and Quaker Oat Bran, had been purchased by significantly more Nutrition Concerned than those in the other two Food segments. For example, while 34 percent of the population over 50 had tried Campbell's Low-Sodium Soups, 41 percent of the Nutrition Concerned had done so.

The low rate of trial among the Traditional Couponers can't be blamed on lack of awareness. Although 48 percent of the Traditional Couponers were aware of this product, only 18 percent had tried it.

Convenience foods that were new at the time, such as Top Shelf, Pillsbury's microwave cake mixes, and General Mills' Suddenly Salad, were clearly preferred by the Fast & Healthy. Suddenly Salad had been tried by 26 percent of the Fast & Healthy, but by only 17 percent of people over 50. The Nutrition Concerned's lack of interest in convenience foods is illustrated by the fact that only 10 percent of them had tried Suddenly Salad, even though 45 percent of them were aware of it.

These illustrations support the idea that certain seniors are open to new food products—if these products meet their needs and interests. Marketers need to discard the stereotype that portrays seniors as unwilling to try new products.

CONCLUSION

In creating foods for the mature market, manufacturers will be able to reposition existing foods or mass market products for seniors. On the other hand, they will have to pay increasing attention to food issues that revolve around portions, aromas, nutritional content, healthfulness, and eating patterns.

Whether manufacturers or retailers of foods, marketers will also have to realize that, just as with other populations, not all seniors are interested in healthful eating or a new snack product. At the same time, products created for the mature market may very well find more universal appeal.

REFERENCES

Carlson, Eugene, "'Graying' Market May Not Be So Golden," *Wall Street Journal*, December 27, 1989, p. B1. Source of study conducted for the *Wall Street Journal* on services desired by retirees or their spouses.

Chase, R. A., "Minority Elders in Minnesota," Wilder Research Center, St. Paul, MN, March 1990. Source of study on difficulties with grocery shopping.

Dagnoli, Judann, and Julie Liesse Erickson, "The Looming Battle for the Center of the Plate," *Advertising Age*, November 13, 1989, pp. S-10-S-12.

Greeley, Alexandra, "Nutrition and the Elderly," *FDA Consumer*, October 1990, pp. 25-28.

"Hormel: Top Shelf Not a Top Seller," *Grassroots Monitor*, December 2, 1988, Volume V, No. 46.

"Impaired Hearing and Sight and Prevalence of Selected Conditions of Persons 65 Years Old and Over: 1984," *Statistical Abstract of the United States*, 1989, Table 185, p. 115.

Lawrence, Jennifer, "The Sunchip also Rises," *Advertising Age*, April 27, 1992, pp. S-2-S-6.

Liesse, Julie, "ConAgra Gets Healthier," *Advertising Age*, August 17, 1992, p. 4.

Litwak, David, "The Service Deli: Can Its Upswing Survive the Downturn?" *Supermarket Business*, February 1992, pp. 45-53.

Litwak, David, and Nancy Maline, "Competition by Class of Trade," 9th Annual Product Preference Study, *Supermarket Business*, March 1992, pp. 25-26, 35-38, 64.

Pici, Anne, "Nutrition: A Deli Menu Option that Means Business!" *Supermarket Business*, June 1992, p. 49.

Putnam, Judith Jones, "Food Consumption," *National Food Review*, April–June 1989, pp. 1-9.

Senauer, Ben, Elaine Asp, and Jean Kinsey, *Food Trends and the Changing Consumer*, St. Paul, MN: Eagan Press, 1991.

Smith, Rod, "Beef Industry Writing New Story as It Rolls Out Low-Fat Ground Beef," *Feedstuffs*, January 17, 1992, p. 16.

U.S. Department of Health and Human Services, *Surgeon General's Workshop*, 1988, pp. B2-B3.

Zbytniewski, Jo-Ann, "What's in a Name?" *Progressive Grocer*, August 1992, p. 115.

TWENTY-ONE

DINING OUT

AN IMPORTANT RESTAURANT MARKET

That today's seniors represent an important force in restaurant sales cannot be denied. Research from *Modern Maturity* reports that each week 26 million persons over 50 eat at a restaurant.

Although marketers at many restaurant chains have acknowledged that seniors represent a major part of their business, few of them are interested in actively pursuing them. Taking such an important market for granted appears to be an extremely short-sighted position. For restaurants, the mature market is important now and will become even more so as Baby Boomers age.

Senior restaurant patronage increasing

It is true that the average number of restaurant visits drops after the age of 55. However, the number of restaurant or foodservice meals eaten by seniors over the past five years has increased at a higher rate as compared to such meals eaten by the entire U.S. population. From 1986 to 1991, the average number of restaurant meals eaten by the U.S. population increased 3.7 meals to 3.8 meals per week. But for those 55 to 64, the increase occurred at twice that rate. The increase is even more dramatic for those 65 plus. In 1986 they ate an average of 1.8 restaurant meals per week, which, by 1991, had jumped to 2.4.

Restaurants that are serious about the mature market should get to know the 50+ segments better and focus on their needs. Unfortunately, attempts by restaurants to market to seniors often stop with discounts on Tuesdays.

This chapter points out that restaurants can do far more. We will show restaurant owners that not all segments want the same thing when they eat out. From vegetarian entrees to the opportunity to socialize, this chapter shows how each of the Food segments has different interests and needs. Promotions, foods, and services should be selected with specific Food segments in mind in order to return the greatest profit.

FAST & HEALTHY FREQUENT VISITORS

Frequency of eating out

Although adults over 50 eat out with great frequency, eating out is very much concentrated in one of our 50+ segments, the Fast & Healthy. In each of five restaurant categories we measured—fast food, coffee shop, cafeteria, steak house, and casual dining—the Fast & Healthy eat out more frequently than do the other two Food segments. During a week's time, over half of the Fast & Healthy have eaten out as compared to only one-third of the Traditional Couponers and Nutrition Concerned.

Preferred restaurant types

A restaurant offering casual dining is favored by all of the Food segments. We defined "casual dining" as a restaurant that is "mainly for lunch and dinner. It is more expensive than a coffee shop and offers an extensive and varied menu." The Fast & Healthy eat more meals at a casual dining restaurant (19 per year), as compared to the Nutrition Concerned and the Traditional Couponers, who visit this type of restaurant 13 or 14 times a year.

Of the Fast & Healthy, a quarter report eating at a fast food restaurant once a week or more. In contrast, only 9 percent of the Nutrition Concerned do so. While about half of the Fast & Healthy and Traditional Couponers have visited a fast food restaurant at least once a month or more, slightly more than one-third of the Nutrition Concerned have.

A coffee shop attracts 16 percent of people over 50 once a week or more, but 19 percent of the Fast & Healthy and 14 percent of the Traditional Couponers. While almost half of these two segments eat at a coffee shop at least once a month or more, only about one-third of the Nutrition Concerned do (see Figure 21-1).

PROFILING FREQUENT VISITORS

Forty-four percent of the population over 50 eat out once a week or more, our definition of a frequent restaurant visitor. Only two demographics, marriage and income, substantially differentiate frequent visitors from those who visit less often.

More likely to be married

Compared to the over-50 population (58 percent), more of the frequent restaurant visitors are married (66 percent). Examined by segment, nearly three-quarters of the Nutrition Concerned and Traditional Couponers who are frequent visitors are married. In contrast, only 56 percent of the Fast & Healthy who are frequent visitors are married.

Frequent visitors more affluent

A dramatic contrast in household income exists between those people in the U.S. population over 50 and those within that population who eat at a restaurant at least once a week or more. People over 50 have a median household income before taxes of $25,670. However, those who eat out at least once a week or more have a median

Dining Out

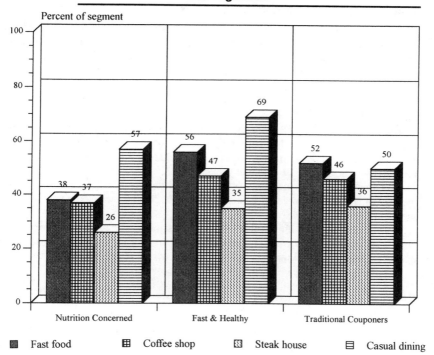

Figure 21-1 Eating Out, 50+ Food Segments

On a multiple-response question on the frequency of eating at five types of restaurants, more of the Fast & Healthy report eating out once a month or more in three of four restaurant categories.

household income of $40,302. By segment, incomes range from $43,129 for the Nutrition Concerned and $43,389 for the Traditional Couponers, as compared to $37,023 for the Fast & Healthy.

Frequent patrons working more

More of those who eat out frequently also work full time as compared to the population over 50. Although our study shows that 40 percent of the U.S. population over 50 works full time, almost half of those who are frequent restaurant patrons do so. More of the Traditional Couponers who are working full time eat out frequently (57 percent) than do any of the other Food segments.

DINING OUT OCCASIONS AND EXPENDITURES

Traditional Couponers spending more

Regardless of occasion, the Traditional Couponers spend slightly more than the other two segments on their own meals. The higher costs of their meals may be related to eating more, including dessert, for example. Since they eat meat frequently, they may be having steak, whereas the Nutrition Concerned may opt for a less-expensive salad.

Spend more with others

Across the board, all of the Food segments spend less on their own meal when they eat alone and more when they eat with someone else, even if that person is a child. When eating out while shopping with a child, people over 50 spend a median of $5.71 on their own meal; the Fast & Healthy spend $6.33.

When eating with a significant other or friend, the over-50 population spends a median of $11.23. However, for this occasion the Traditional Couponers spend a median of $13.31 on their own meal (see Figure 21–2).

Costs increase when dining out with family to a median of $12.61 for the over-50 population and $13.45 for the Traditional Couponers. When dining out for a special occasion, people over 50 spend a median of $17.33, but both the Nutrition Concerned and Traditional Couponers spend 30 to 50 cents more.

Frequent restaurant patrons spend more

Our 50+ data shows that frequent visitors who eat at any type of restaurant once a week or more spend more on their own meals than do those who are infrequent visitors. A couple of examples illustrate this point. When dining out with a significant other or friend, the over-50 population spends a median of $11.23 on their own meal. However, frequent diners spend a median of $11.81. When examined by segment, we see that the Traditional Couponers spend the most, $13.69.

When dining out with family, persons over 50 spend a median of $12.61 for their own meal. In contrast, frequent diners spend $13.74. The Fast & Healthy spend slightly more: a median of $13.94 (see Figure 21–3).

As with the population in general, the amount of money spent by frequent visitors on a restaurant meal is highly dependent on the occasion. While shopping or eating alone even frequent diners spend $5 or less on their own meal. However, when eating with another person the amount spent increases.

POPULARITY OF ETHNIC FOOD

Ethnic restaurants, dishes proliferate

According to the American Restaurant Association, the last five years have seen an explosion of interest in ethnic foods. Two developments support this view. Restaurants in this category increased by 9 percent a year from 1985 to 1990. This

Dining Out

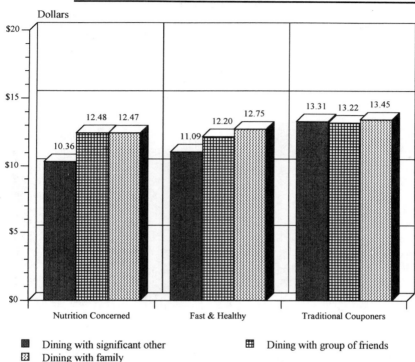

Figure 21-2 Median Expenditures on Own Meal, 50+ Food Segments

■ Dining with significant other   ⊞ Dining with group of friends
▨ Dining with family

The Traditional Couponers spend more on their own meals than the other segments when dining with a significant other and while dining with family.

rate of increase is four times faster than that of the total number of restaurants. In addition, ethnic dishes have proliferated on nonethnic menus. From 1986 to 1991 the number of ethnic dishes appearing on such menus increased by 49 percent. Today 30 percent of all restaurant entrees are ethnic.

Ethnic preferences of the 50+ segments

Of 10 ethnic food types included in the 50+ survey, the top four favorites for the total sample are American, Italian, Chinese, and Mexican. American foods appeal to the vast majority of all segments (91 percent) who say they like it "a lot."

The selection of Italian, Chinese, and Mexican ethnic foods by the 50+ respondents parallels research by the National Restaurant Association which found that these three types of foods accounted for 85 percent of all ethnic entrees. According to the Association, 75 percent of American adults eat these three cuisines.

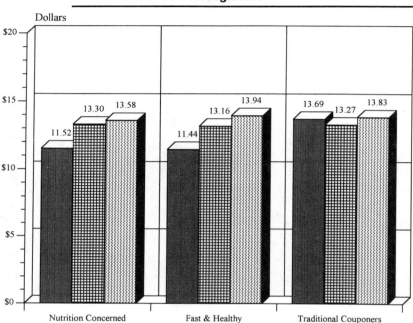

Figure 21–3 Median Expenditures on Own Meal, Frequent Restaurant Visitors, 50+ Food Segments

Frequent restaurant visitors in the Traditional Couponers segment spend more on their own meals when dining with a significant other and with family.

Again we see that some of our 50+ segments clearly prefer certain of the three most popular types of ethnic foods. The Fast & Healthy and Nutrition Concerned are equally open to these foods. Almost a third of each of these segments, for example, says that it likes Mexican food "a lot" (see Figure 21–4) and another third likes Mexican food "somewhat." But 44 percent of the Traditional Couponers don't like Mexican food at all.

All segments like Chinese

Another mainstream ethnic food, Chinese, is liked equally well by all three segments. Forty-four percent of people over 50 like Chinese food "a lot." The same level of acceptance across all segments is also seen with Italian foods.

A second tier of less-accepted ethnic foods is represented by French and German cuisines. German foods are liked "a lot" by almost a quarter of the over-50 population, but by more of the Nutrition Concerned and the Fast & Healthy. French food is preferred by the Nutrition Concerned, 26 percent of whom like it "a lot" compared to only 21 percent of the Fast & Healthy (see Figure 21-4).

Nutrition Concerned open to exotic cuisines

In the realm of more exotic ethnic foods, such as Indian and Vietnamese, the Nutrition Concerned are by far the most adventurous segment. Whereas 4 percent of people over 50 like Vietnamese foods "a lot," 6 percent of the Nutrition Concerned do. An almost identical pattern is evident when it comes to Indian foods. Of the three segments, far fewer of the Traditional Couponers are open to these ethnic foods.

This interest that the Nutrition Concerned have in exotic cuisines parallels that measured within the American adult population by the National Restaurant Association. Between 1986 and 1991, from 5 to 8 percent of Americans had tried these more

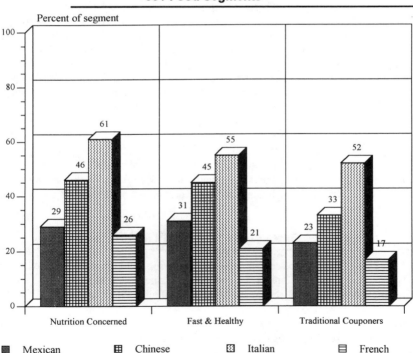

Figure 21-4 Ethnic Food Preferences, 50+ Food Segments

On a multiple-response question on the degree of liking of 10 ethnic foods, fewer Traditional Couponers like Mexican food "a lot." More Nutrition Concerned like Italian and French foods "a lot" as compared to the other two Food segments.

exotic cuisines. In comparison, the Nutrition Concerned have not only tried these foods, but 5 or 6 percent report liking them very much.

The popularity that these exotic foods have for the Nutrition Concerned ties in very well with their commitment to healthful eating. Both cuisines offer many vegetarian dishes, incorporate quantities of fresh vegetables, and use small amounts of meat. In the case of Indian foods, dried beans are also used extensively.

STRESSING VEGETABLES AND SALAD BARS

As we have seen, the Nutrition Concerned are adventurous when it comes to food. A restaurant offering a stir-fry entree would encompass a number of the Nutrition Concerned's needs, from low calorie to exotic, fresh vegetables to low-in-fat.

Vegetarian meals appealing

A recent Gallup poll for the National Restaurant Association reports that one out of five restaurant patrons select the restaurant because it serves vegetarian meals. Health reasons prompted 88 percent of those surveyed to select a vegetarian item at a restaurant. The Association study discovered that vegetarian items are ordered by at least a third of all customers. Of these restaurant visitors only 5 percent are vegetarians.

According to the Association's study, the vegetarian items that they order most frequently include fruit salads, a main-entree salad served with vegetables and grains, and stir-fry vegetables (41 percent). Offering more vegetarian items is something the Association is currently advising its members to do. And doing so will certainly appeal to the Nutrition Concerned segment in the mature population.

Salad bars attract two segments

Both the Nutrition Concerned and the Fast & Healthy would be good targets for salad bars. Unfortunately, in the face of a greater interest in healthful foods, the number of restaurants offering salad bars has actually been declining over the past few years according to the National Restaurant Association. Today only about one in three restaurants has a salad bar.

SUCCEEDING WITH THE 50+ FOOD SEGMENTS

If expenditures on food consumed away from home continue to decline, restaurants should consider mature adults an even more viable target, especially the Fast & Healthy who eat out with the greatest frequency of all the Food segments.

Meeting customers' expectations

Profiting from this attractive market can be achieved by giving them what they want, and this will differ by segment. A survey by the National Restaurant Association reported by *Restaurant Business* shows that customers are willing to spend up to 30 percent more on their meals if their expectations of value are met.

Value differs by segment

As we have seen, the perception of value differs significantly for each 50+ Food segment. The Traditional Couponer sees value as coupons and discounts for seniors. Offering low fat or reduced salt foods will have little value to most of the Traditional Couponers. For many of the Traditional Couponers, ethnic foods have little appeal. As we have seen, even Mexican foods are disliked by almost half of the Traditional Couponers.

Succeeding with the Traditional Couponers means offering traditional American foods such as steak, roasted chicken, and meat loaf. Since Traditional Couponers have a sweet tooth and aren't as concerned about cutting back on fats and sugar as are the other segments, an appetizing dessert menu should be available.

To attract primarily the Nutrition Concerned, menu offerings should be low in fat, calories, salt, sugar, and cholesterol. They should also be high in fiber. Offering discounts and coupons, along with good-for-you foods, means value to this segment.

Restaurant food not healthy

The Nutrition Concerned, the segment that eats out with the least frequency, may do so because they don't view restaurant food as good for them. For example, the Nutrition Concerned, more than the other segments, acknowledge that it is difficult for them to eat at a restaurant and stay on their diet.

Although they don't seek discounts and coupons, the Fast & Healthy do spend less than the other two Food segments on their meals. On the other hand, they eat out more often than the other two segments. Value for this group, which prefers to meet a friend at a restaurant rather than cooking at home, may revolve around a restaurant's ability to foster socializing. For this segment, a restaurant could promote itself as "a great place to get together" by offering reductions on the second meal on the ticket, the one eaten by a friend.

Mature servers don't add value

Although one trend in retailing is hiring and training those over 50 to wait on mature customers, it's an idea that wouldn't add value to a restaurant meal for any Food segment. In fact, compared to the other two segments, the Fast & Healthy are actually negative to being served by someone their own age.

This attitudinal position parallels the Fast & Healthy's belief in aging as a negative experience. Feeling that they don't look as attractive as they used to, more of the Fast & Healthy are interested in cosmetics that will make them look younger. Being served in a restaurant by someone their own age may very well remind them that they are no longer youthful.

CONCLUSION

Savvy restaurant owners will realize that seniors represent an important and viable market. Attitudinal segments within this market have differing needs focused on socializing, nutrition, portion size, and cost. Not every dish or promotion will appeal to every 50+ segment. Restaurant owners should recognize their own strengths and use them to position themselves clearly and strongly to senior targets.

REFERENCES

Arthur, Caroline, "Meatless Menus Rising at Restaurants," *American Demographics,* February 1992, p. 18. Source of Gallup survey on popularity of vegetarian items at restaurants.

Barkema, Alan, Mark Drabenstott, and Kelly Welch, "The Quiet Revolution in the U.S. Food Market," *Economic Review,* May/June 1991, pp. 25-41.

"Customers Not as Price Sensitive as You Think," *Restaurant Business,* April 10, 1992, p. 2. Source of National Restaurant Association survey on customers' willingness to pay more for value.

"Home Cooking Heats Up," *Restaurant Business,* February 10, 1992, p. 2.

"Mature America in the 1990s," special report from *Modern Maturity* magazine and The Roper Organization, 1992, p. 22.

"Salad and Condiment Bars Wilt," *Restaurant Business,* May 20, 1992, p. 2.

Sugarman, Carol, "How Are We Doing?," *Supermarket Business,* September 1991, pp. 30-36.

TWENTY-TWO

REACHING THE FOOD SEGMENTS

The 54-year-old woman sat in front of her television set, switching from channel to channel, looking for images of people like her in a commercial, any commercial. No commercials showed gray-haired women pushing food carts down grocery store aisles. Commercials for coffee-shop type restaurants did not show mature men seated at tables enjoying a slice of pie along with their BLT. A week later, her experiment over, she sighed upon finding that her content analysis was devoid of content. The woman could only conclude that she and her cohorts did not exist.

Their lack of presence, their grayish ghostliness on mass mediums such as network television, stems, in great part, from two misconceptions. Either she and her ilk are so brand loyal they resist buying new products or brands, or they will die before they have the opportunity to switch.

This view toward brand loyalty—or lack of it—among those over 50 opens up possibilities for food and restaurant marketers. Strategies for reaching the Food segments are explored in this chapter. We also examine in detail the media preferences of the segments.

This chapter goes on to contrast senior publications to general interest magazines. How each Food segment reacts to direct marketing overtures is examined and their specific levels of expenditures are detailed. Activities most popular with each segment are outlined as useful in creating targeted advertising.

SEGMENTS DIFFER ON BRAND LOYALTY

We disagree with those who pay for and create advertising and dismiss 26 percent of the population largely because of unfounded conclusions. As we saw in Chapter 13 on promotions for reaching the Self segments, each attitudinal segment's brand loyalty varies according to the product itself. These segments are not equally loyal to all categories of products, or to specific products within these categories.

Fast & Healthy most brand loyal

For every food product, more of the Fast & Healthy, 38 percent of the over-50 population, show themselves to be brand loyal when contrasted with the other two

Food segments. Their high level of brand loyalty ties in quite logically with their lack of interest in newspaper advertising. Being highly loyal to specific food brands, the Fast & Healthy have no need to pay attention to food advertising. It's irrelevant to them.

In contrast, the Nutrition Concerned, 46 percent of the mature population, are average or slightly above average in their loyalty to specific brands. The Traditional Couponers, 16 percent of the over 50, reveal themselves as very low in brand loyalty.

Do brands deliver value?

The question of brand loyalty among the mature population, then, doesn't rest on demographic characteristics, but on attitudes toward the values and benefits that a brand represents. Each of our segments who are category users values brands to a greater or lesser degree, and there are additional variations in loyalty to specific products.

For example, in the case of cereals, more of the Fast & Healthy who are category users buy only one brand (23 percent) than do all people over 50 (19 percent). Only 13 percent of the Traditional Couponers can be defined as loyal to one cereal brand. Of the Nutrition Concerned, 17 percent, buy only one brand of cereal.

In the category of frozen dinners, more of the Fast & Healthy (28 percent) who buy this product are loyal to one brand as compared to those over 50 (24 percent). The Traditional Couponers (24 percent) are average in brand loyalty to the frozen dinner category, while the Nutrition Concerned are below average (17 percent) (see Figure 22–1).

Loyalty depends on product

The extent of brand loyalty varies by product. For example, while we saw that 17 percent of the Nutrition Concerned are loyal to one brand of frozen dinners, 30 percent of them are loyal to one brand of cake mix from a box. These are products that the Food segments have used for years (see Figure 22–1).

Creating new loyalties

Beyond existing brand loyalties, however, the larger question is whether manufacturers of frozen dinners or cake mixes can create new products that large numbers of Nutrition Concerned will try and buy again.

Innovative new products will create entirely new loyalties. ConAgra's Healthy Choice line probably appeals strongly to the Fast & Healthy and less so to the Nutrition Concerned. However, its emerging Life Choice line of frozen dinners should be far more interesting to the Nutrition Concerned.

Nutrition Concerned offer potential

Because they are not as staunchly brand loyal as are the Fast & Healthy and are swayed by advertising, the Nutrition Concerned offer the best potential for restaurant chains and food manufacturers offering new products. Motivated by their commitment

Reaching the Food Segments

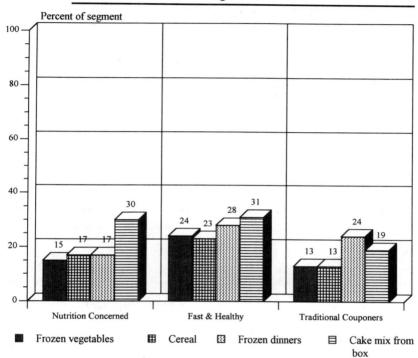

Figure 22–1 Brand Loyalty, 50+ Food Segments

On a multiple-response question on the number of brands bought in eight different product categories, more category users in the Fast & Healthy segment buy only one brand of the four products illustrated. Of the three Food segments, the Traditional Couponers are the least brand loyal.

to healthful eating, the Nutrition Concerned are searching out new, good-for-you foods, including those using artificial sweeteners and fat substitutes.

Marketers who accept the fact that certain attitudinal segments within the over-50 market will try new brands will advertise to them. Selecting the promotional tools that will best reach the target segment is next.

SELECTING FAVORED MEDIUM

Using attitudinal segments

As discussed in Chapter 13 on reaching the Self segments, specific attitudinal segments can be reached most efficiently using their preferred media.

Our 50+ study shows that the Nutrition Concerned are consumers of magazines, radio programming, and books at a higher than average rate. In contrast, the Fast & Healthy strongly favor television and newspapers. The Traditional Couponers segment exhibits an average or below average interest in all forms of media.

We believe that the process of structuring all promotional efforts becomes far more efficient and integrated by identifying the segment(s) that would buy a product or service. The mix of promotional vehicles chosen should reflect the target segment's preferences. It also makes great sense to use meaningful images that tie in to that segment's activities and concerns.

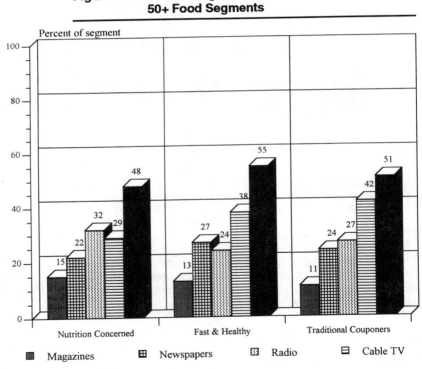

Figure 22-2 Media Usage, 50+ Food Segments

On a multiple-response question on the frequency of use of six types of media, more Fast & Healthy report spending 11 or more hours a week watching network television. More Traditional Couponers view cable television, while more Nutrition Concerned listen to the radio for that amount of time each week.

TELEVISION VIEWING

Although several types of programming may deliver a mature audience, some will be better at reaching a specific segment than others. Denny's, a restaurant chain, has advertised to reach the one-third of its customers who are over 55. In using television, the medium preferred by the Fast & Healthy, Denny's has succeeded in reaching this segment. In the 50+ study, 14 percent of the Fast & Healthy eat at Denny's at least once a month or more, compared to 11 percent of those over 50.

Fast & Healthy prefer television

Of all the Food segments, the Fast & Healthy watch the most hours of television per week. More of this segment (55 percent) watches 11 or more hours of network television per week as compared to those over 50 (52 percent). The Traditional Couponers are heavy consumers of cable television. More of them watch cable 11 or more hours per week (42 percent) as compared to the population over 50 (39 percent).

The Nutrition Concerned consume far less of both cable and network television as compared to the Fast & Healthy. Although this segment says it is influenced by advertising in the food products it purchases, television advertising will be less effective in reaching them than print (see Figure 22–2).

What segments watch

Of the 13 programming options presented in the 50+ questionnaire, the three types of television shows most watched by those over 50 either daily or several times a week are news and current affairs (86 percent), game shows (47 percent), and talk shows (38 percent). But marketers who wish to reach a specific segment will select programming that reflects that segment's interests.

The Fast & Healthy are clearly big fans of game shows (see Figure 22–3). Restaurant or food advertising targeted to the Fast & Healthy, for example, should be broadcast during "Jeopardy" and "Wheel of Fortune." More Fast & Healthy watch such programs daily (28 percent) as compared to the Traditional Couponers (24 percent). The Fast & Healthy's interest in game shows parallels the fact that entering sweepstakes is one of their primary activities.

Although network television is not a preferred medium, more Traditional Couponers watch sports programs (42 percent) several times per week as compared to all seniors (36 percent). In almost all types of television programming, the Nutrition Concerned are less frequent viewers. However, the Nutrition Concerned spend more time than average in viewing movies as compared to those over 50.

NEWSPAPER READERSHIP

Fast & Healthy read most

Besides watching television more than the other segments, the Fast & Healthy also spend more time each week reading newspapers. Compared to all of those over 50 (22 percent), more of the Fast & Healthy (27 percent) spend 11 hours or more a week

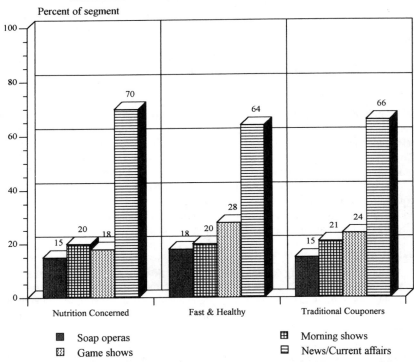

Figure 22–3 Television Programming Watched, 50+ Food Segments

On a multiple-response question on the frequency of watching 13 types of television programs daily, more of the Nutrition Concerned report watching news/current affairs programming, while more of the Fast & Healthy view game shows.

reading newspapers (see Figure 22–2). In contrast, the Nutrition Concerned are average in the hours they spend reading a newspaper.

Segments attracted to different sections

Specific sections of a newspaper attract certain segments. A food manufacturer or restaurant that places an advertisement on the food pages of a newspaper would like knowing that more of the Nutrition Concerned (58 percent) will be reached as compared to only half of those over 50.

Nutrition Concerned read advertising

It is important to note that behaviors quite often parallel attitudes. Advertising sections in the newspaper are read by 37 percent of the Nutrition Concerned, the segment most influenced by advertising. But such sections are regularly read by only

Reaching the Food Segments

30 percent of the Fast & Healthy, the segment that denies that it is influenced by advertising (see Figure 22–4).

Besides the food, entertainment, and advertising sections, more Nutrition Concerned are also reading the travel, home, international news, and opinion and commentary pages. Half of this segment read the business pages as compared to 43 percent of the over-50 population.

Local angle interests Fast & Healthy

More than the other segments, the Fast & Healthy favor local news pages, cartoons, and advice columns in a daily newspaper. As the only segment that favors local news, 93 percent of them read it regularly as compared to 87 percent of people over 50.

Car section read by Traditional Couponers

At a higher rate, the Traditional Couponers are reading advertising, obituaries, sports, cartoons, and advice columns. More Traditional Couponers (21 percent) read newspaper sections on cars and motoring as compared to those over 50 (15 percent). However, fewer Traditional Couponers than the over-50 population read the entertainment and food pages where food-related advertisements typically appear. While half of the population over 50 read the food section, 47 percent of the Traditional Couponers do so (see Figure 22–4).

SELECTING MAGAZINES BY SEGMENT

Measuring magazine readership

Unlike other measures of readership, our 50+ study defined readership more narrowly. The term *reader*, as we have used it here, refers to either the purchaser or subscriber of a magazine, or a respondent who lives in a household where the magazine is regularly available through subscription or purchase.

Nutrition Concerned heavy readers

Marginal television viewers, the Nutrition Concerned are heavy readers of magazines as compared to the other two segments. For example, 15 percent of the Nutrition Concerned report reading magazines from 11 hours or more a week. This is in contrast to the other two segments that are average in the number of hours they spend reading magazines. Advertisers wanting to reach this segment, which says it is influenced by advertising, should focus on print (see Figure 22–2).

More of the Nutrition Concerned also fall into the Magazine Publishers of America's (MPA) definition of a magazine reader: a person who regularly reads five magazines. While 34 percent of the Nutrition Concerned can be classified into this definition, fewer of the Fast & Healthy (30 percent) and Traditional Couponers (28 percent) can.

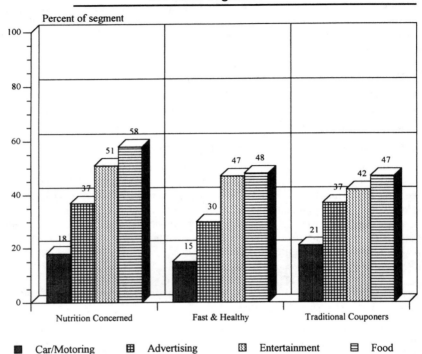

Figure 22–4 Newspaper Sections Read, 50+ Food Segments

On a multiple-response question listing 20 sections of the newspaper, more Nutrition Concerned report reading the food and entertainment sections, while the car/motoring section appeals to greater numbers of Traditional Couponers.

Reaching multiple magazine readers

Among those in the 50+ Food segments who fit the MPA's definition of a magazine reader, five publications are most popular: *Reader's Digest* (69 percent), *Better Homes & Gardens* (57 percent), *Good Housekeeping* (53 percent), *Family Circle* (50 percent), and *Ladies' Home Journal* (37 percent). Even with these readers of multiple magazines, specific publications appeal to each segment. Apparently *Prevention* attracts more of the female Nutrition Concerned (25 percent) as compared to the Fast & Healthy (16 percent). In contrast, *McCall's* is read by more female Fast & Healthy (40 percent) than Nutrition Concerned (31 percent).

Select magazines by segment

Although many magazines offer similar demographics, our 50+ study shows that some magazines are better than others in reaching specific 50+ segments (see Figure 22–5). A food company that sells branded fresh vegetables such as Green Giant or Dole and wants to target the Nutrition Concerned should go beyond demographics and select media that clearly offer more direct access to the segment committed to cooking healthful foods from scratch.

Good bets in planning media campaigns to reach more of the female Nutrition Concerned who are multiple magazine readers are *Better Homes & Gardens* (73 percent) and *Family Circle* (70 percent). However, a manufacturer of expensive kitchen equipment who advertises in *Gourmet* rather than in *Family Circle* will pick up both more Nutrition Concerned and a higher percentage of affluent members of this segment.

More of the Nutrition Concerned who are multiple magazine readers and who read *Gourmet* have annual pre-tax household incomes over $50,000 (74 percent), in contrast to those who read *Family Circle* (33 percent). Our approach to media selection, then, combines the right demographics and the segment's willingness to buy.

National Enquirer popular

While using a magazine to reach the Fast & Healthy is more difficult than targeting the Nutrition Concerned, there are some magazines that the Fast & Healthy prefer. At a rate higher than would be expected, nearly half of the female Fast & Healthy regularly read *Reader's Digest*. Another magazine favored by one out of every ten in this segment is the *National Enquirer*. The Fast & Healthy read this publication at a rate twice that of all of those over 50.

Reaching the Traditional Couponers

Reading a number of magazines at an average rate, the Traditional Couponers are also more difficult to target through specific magazines. And yet associations promoting the consumption of butter, eggs, and red meat have to make sure that they are in those publications heavily read by the Traditional Couponers, reinforcing this segment's overconsumption of these products.

Female Traditional Couponers can be reached through *TV Guide*. About one-third of them read this magazine compared to one-fifth of all those over 50. Male Traditional Couponers favor *National Geographic*, with one out of four reading it regularly compared to one out of five of the total sample.

SENIOR OR MAINSTREAM PUBLICATIONS?

A question facing those companies that want to advertise to the mature market is whether to approach it in general, mainstream publications, such as *Reader's Digest*, or in magazines such as *Modern Maturity* and *New Choices for Retirement Living*, which target the mature market specifically.

Figure 22–5 Magazines Read, 50+ Food Segments

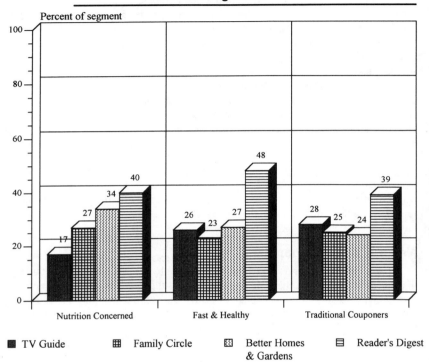

On a multiple-response question listing 39 magazines, more Fast & Healthy report reading *Reader's Digest*. More Nutrition Concerned regularly read *Family Circle* and *Better Homes & Gardens*.

The case for *Modern Maturity*

Of those over 50, 58 percent are members of the American Association of Retired Persons (AARP) and receive its publication, *Modern Maturity*. An impressive 52 percent of our sample reports reading *Modern Maturity* regularly. As we discussed in Chapter 13, the circulation of *Modern Maturity* is so immense it dwarfs that of other senior publications.

Nutrition Concerned favor *Modern Maturity*

From our 50+ study, however, we can show that some 50+ Food segments believe *Modern Maturity* is doing a better job than others. To begin with, far more of the Nutrition Concerned (60 percent) read the publication regularly as compared to 46 percent of the Traditional Couponers.

The study's data on *Modern Maturity* shows that two-thirds of its readers believe that the publication delivers useful health information. However, 69 percent of the Nutrition Concerned share this view. On two other measures, more of the Nutrition Concerned agree that *Modern Maturity* delivers useful financial and lifestyle information. On every measure, the Traditional Couponers are the Food segment least satisfied with *Modern Maturity*.

Other options reach seniors

There is no simple answer as to where to advertise to the mature market. While it is true that *Reader's Digest* is read by 46 percent of persons over 50, only 40 percent of the Nutrition Concerned read it. In contrast, 60 percent of the Nutrition Concerned read *Modern Maturity*. In considering cost per thousand, *Modern Maturity* delivers to more of the Nutrition Concerned, important news for producers of healthful foods like Chiquita and Riceland. By advertising in *Reader's Digest,* however, about a quarter of those Americans under 45 are reached as well.

In advertising to the mature market, another option is to select a publication targeted to a particular product or service, for example, food, health, or travel, which has a base of older subscribers. Some publications in this area are *Gourmet, Prevention,* and *Travel & Leisure*. For example, the median age of subscribers to *Travel & Leisure* is 50 with a median individual income of $43,100 and median household income of $95,200. Almost half of *Prevention*'s subscriber base is over 50, and the magazine publishes a separate edition specifically directed at its subscribers in that age category.

TUNING IN TO RADIO

Nutrition Concerned listen most

Besides magazines, the Nutrition Concerned also favor radio. While the Traditional Couponers are average in the amount of time per week they devote to listening to radio, the Fast & Healthy are actually below average. In contrast to these two segments, almost a third (32 percent) of the Nutrition Concerned spend 11 hours or more a week listening to radio (see Figure 22–2). At a significant level this segment finds a number of types of radio programming appealing, including easy listening (57 percent), news (54 percent), talk shows (31 percent), and classical music (27 percent) (see Figure 22–6).

Segments have favorites

Those of the Fast & Healthy who do listen to radio have one significant favorite: country music (see Figure 22–6). The Traditional Couponers have only one outstanding preference: sports programming on radio.

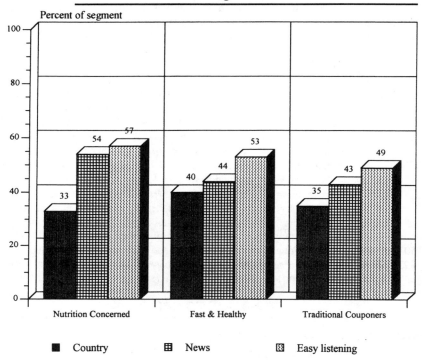

Figure 22–6 Radio Programming Listened to, 50+ Food Segments

On a multiple-response question listing 11 types of radio programming, more Nutrition Concerned favor easy listening. Country music is listened to by more Fast & Healthy.

BOOK READING AND THE 50+ SEGMENTS

Books reach Nutrition Concerned

Although books do not typically carry advertising, they could be offered in a promotional effort focusing on the Nutrition Concerned who are overwhelmingly book readers. Far more of the Nutrition Concerned (27 percent) read books 11 hours or more a week, as compared to the Fast & Healthy (16 percent).

Targets for cookbooks

Because they read, are committed to cooking from scratch, and list gourmet cooking as a favorite activity, the Nutrition Concerned would be a ready readership for a cookbook promoting a new line of good-for-you tomato sauces. However, such a promotion would have little or no appeal to the Fast & Healthy, who cook few meals from basic ingredients. Clearly, not everyone over 50 is interested in knowing more about healthful eating.

Knowing the size of the one segment that would find the cookbook offer very appealing would lead to setting realistic goals on the impact of the promotion and making wise production decisions. In addition, the contents of the book could be truly shaped for the specific interests of the Nutrition Concerned.

IMAGES FOCUS ON ACTIVITIES

Advertising targeted at a specific 50+ Food segment should show them engaged in some of their favorite activities. For the Nutrition Concerned these activities include gardening (45 percent), book reading (39 percent), physical fitness/exercise (33 percent), cultural and art events (17 percent), gourmet cooking and fine foods (17 percent), and health foods (14 percent).

More of the Fast & Healthy than those over 50 and the other two segments list watching cable television (50 percent), grandchildren (44 percent), and entering sweepstakes (21 percent) as their top activities.

As compared to the over-50 population and to the other two Food segments, more of the Traditional Couponers regularly watch sports on television (49 percent); engage in do-it-yourself, home workshop activities (30 percent); fish (16 percent); play golf (14 percent); and hunt and shoot (8 percent).

Tying activities into ads and promos

The second largest percentage of Fast & Healthy listed grandchildren as the activity in which they regularly participate. As more of the Fast & Healthy have the largest number of grandchildren, it's interesting to note which food-related gifts they give them. Two food options, giving either cookies and candy or restaurant meals, were included in a list of 25 gifts given to grandchildren. More of the Fast & Healthy (90 percent) take their grandchildren out for a restaurant meal compared to the national sample (87 percent). Only three-quarters of the Nutrition Concerned and Traditional Couponers do so. In fact, 39 percent of the Fast & Healthy take a grandchild out for such a meal five or more times a year.

Another food-related gift that the Fast & Healthy give their grandchildren is cookies and candy. Over a year's time, 77 percent of them give such a gift as compared to 67 percent of the Traditional Couponers and 69 percent of the Nutrition Concerned.

Sharing a meal

It would be natural and make good marketing sense for a fast food restaurant chain such as McDonald's to create a television advertisement showing a grandmother in the Fast & Healthy segment and her grandchild sharing a meal. As we've seen, the Fast & Healthy is the Food segment that eats in restaurants most frequently and favors fast food.

Another approach would be to encourage cross-generational meals at McDonald's by offering a special frequent-buyer promotion that rewards heavy users.

Encourage new consumers

A chain of candy stores such as Fannie May could promote its candy by giving coupons to grandmothers. The coupon would be redeemed by both grandmother and grandchild upon their next visit. In this way a candy store would reinforce its relationship with a good customer, the grandmother, and introduce the next generation to its products.

If the candy store chain wishes to start a database, information could be requested on the back of the coupon that would classify the grandmother into a segment. In this way the candy store could eventually mail coupons, including cross-generational promotions, to its best customers.

THE DIRECT MAIL OPTION

Correlating attitudes in list purchases

By focusing on behaviors and demographics, the primary parameters in today's databases, direct marketers are limited in their list selection and product development to what is known today, rather than what could be. Targeting a group of people that has never before existed demands the casting of new and wider nets.

Databases for direct marketing can be structured using attitudes that correlate with behaviors and demographics. While we use sophisticated, cutting-edge technologies to find relationships between these various facets, this chapter will indicate how attitudes can be incorporated into the process of selecting and merging and purging a list.

The first question to be answered, however, is how do each of the Food segments react to direct marketing? How good a market are they for such offers?

Men willing to open direct mail

Interestingly, in each Food segment males more than females are willing to open direct mail. For example, while 71 percent of the male Traditional Couponers open direct mail and skim it, only 64 percent of the female Traditional Couponers will do the same.

Reading contents carefully

But more of the Nutrition Concerned will open and read the contents of direct mail carefully (12 percent) and would also put direct mail in a pile to be read later (9 percent). Both of these percentages are slightly higher than those of the population over 50.

Into circular file

Throwing direct mail away unopened is behavior exhibited by more of the Traditional Couponers (20 percent), as compared to the Nutrition Concerned (14 percent) or all of those over 50 (17 percent). In addition, more of those Traditional Couponers over 65 throw away direct mail unopened as compared to those who are 65 and younger.

Turning prospects into customers

Once they have opened a direct mail offer and read its contents carefully, do the Nutrition Concerned actually order? The number of Nutrition Concerned (41 percent) who had made a direct mail purchase over the past three months is virtually the same rate as for all of those over 50 (40 percent).

Age affects response rate

For every segment, the percentage who have made a direct mail purchase over the past three months decreases after 65. For example, while 42 percent of those 65 and younger in the Fast & Healthy segment had made such a purchase, only 35 percent over 65 had done so.

Gender another influence

Although fewer females were willing to open direct mail, in every Food segment more females had made a purchase. While 37 percent of the male Nutrition Concerned had ordered something through direct mail, 45 percent of the females in this segment had done so.

Spending by segment

While more of the Nutrition Concerned have made a purchase through direct mail over the past three months, what they spent is below average. Those over 50 had spent a median of $60.03, but the Nutrition Concerned had spent a median of only $56.70 (see Figure 22-7). Female Nutrition Concerned and those over 65 pull down the average. For example, males in the Nutrition Concerned segment made median purchases of $74.77.

In contrast to the Nutrition Concerned, the Traditional Couponers had spent a median of $111.73 in direct mail purchases over the past three months (see Figure 22-7). However, in this case it's the female Traditional Couponers who are overspenders. Their median purchases come to $204.03.

Mature buyers prefer mail

In the realm of direct marketing, direct mail is more popular with mature consumers than are purchases made via phone solicitations or using the shopping channel on television. Of the population over 50, 40 percent had bought something through direct mail over the past three months compared to only 16 percent of those who had purchased by phone and 10 percent via a television shopping channel.

CREATING TRULY TARGETED MAILING LISTS

Companies such as Figi's of Wisconsin or the Chef's Catalog, selling food or cooking equipment through direct marketing, can target a specific Food segment.

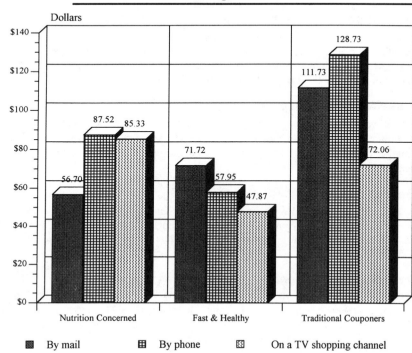

Figure 22–7 Median Expenditures on Direct Marketing, 50+ Food Segments

The Traditional Couponers have spent the most on direct marketing purchases in the last three months, whether by mail, over the phone, or through a television shopping channel.

Using media, car ownership lists

If a direct marketer wanted to target the Traditional Couponers, one in five of those over 50, he or she could begin with mailing lists from magazines read by this segment. These publications would include *TV Guide* and *National Geographic*.

These lists could be selected by income, sex, and census region, probably the Northeast where 43 percent of the Traditional Couponers live. The list could then be merged and purged using the information below to significantly increase the likelihood of locating Traditional Couponers.

In looking at car ownership, information that is available from many states, we know that more of the Traditional Couponers (83 percent) drive American-made cars than the other two Food segments do. In fact, 26 percent of this segment drives either a Plymouth, Dodge, or Cadillac. More of the cars driven by the Traditional Couponers (28 percent) are also 1990 to 1992 models.

Modeling facilitates targeting

In working with marketers, whether they use direct mail or mass media, we use a variety of statistical techniques and other technologies to locate our attitudinally based segments.

Figure 22-8 shows a "tree diagram" produced by KnowledgeSeeker from FirstMark Technologies in Ottawa. This analysis examines a large battery of demographics and behavioral measures, such as magazine readership in this case, and divides the respondents to focus on meaningful characteristics.

This program examines all of the data and splits the sample to magnify differences between subgroups that are made up of either a high or low percentage of the target segment. Each split is then subdivided, again choosing variables to highlight significant differences expressed in percentages of segment members.

By examining the variables that form the tree, marketers are able to better choose the media with which to communicate to the target segments. For example, while 16 percent of the population over 50 is made up of Traditional Couponers, by limiting the selection to males between the ages of 50 and 55 who have incomes of less than $50,000, we increase our probability of reaching the Traditional Couponers to 38.5 percent or by 22.5 percent (see Figure 22-8).

Activities can be mapped

Another type of analysis that would help marketers in list and media selection is seen in Figure 22-8. The map in this figure was produced using a perceptual mapping program called MAPWISE from MarketACTION Research in Peoria, Illinois. This program takes data from a crosstabulation and presents the relationships in graphic form.

Figure 22-9 shows activities in which the segments regularly participate. The activities shown are those that best correlate with each segment. In interpreting the map, note that when a segment is shown to be close to an activity, it means that this segment participates in that activity at a high level. For example, the Nutrition Concerned are highly involved in stock and bond investments.

Conversely, when a segment and an activity are shown at a distance from one another, the segment participates in that activity at a low level. Thus we see that the Traditional Couponers are not involved in community or civic events (see Figure 22-9).

CONCLUSION

Those over 50, if considered by individuals and not households, actually consume as much or more than younger persons. They are not more or less brand loyal. Our 50+ study shows that brand loyalty is not uniform across all the 50+ Food segments. By linking a specific attitudinal segment's media and brand preferences, marketers can increase the efficiency of their promotions. The same approach can be applied to the selection of catalogue products and list selection.

Figure 22–8 KnowledgeSeeker "Tree Diagram," Traditional Couponers

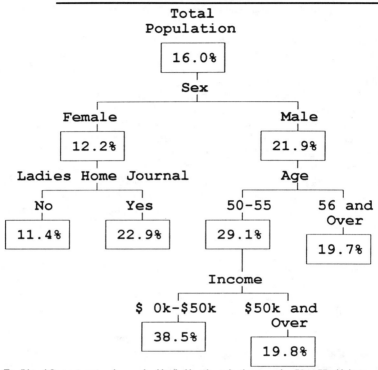

More Traditional Couponers can be reached by limiting the selection to males 50 to 55 with incomes less than $50,000.

Figure 22–9 Perceptual Map by MAPWISE, 50+ Food Segments

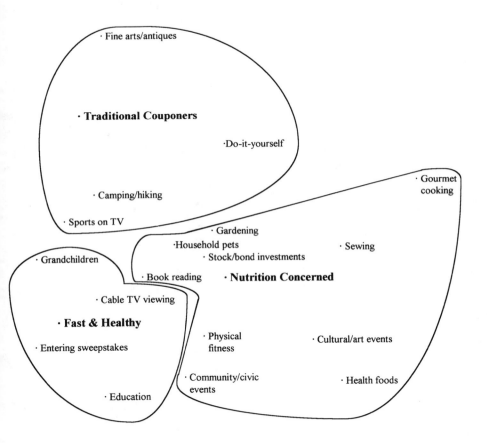

The Nutrition Concerned are more highly involved in stock/bond investments and gourmet cooking compared to the other segments. More Fast & Healthy are frequent viewers of cable television, while Traditional Couponers in greater numbers complete do-it-yourself projects.

INDEX

A

Active Lifestyles, 252
Activities of Seniors, 144-145, 156, 172, 185, 192-193, 245, 261- 263, 266-267, 298, 301, 331, 339-340, 343
   Mapping, 192-193, 266-267, 343
   Those participated in, 185, 261-263, 339
   Tying into ads and promotions, 185, 263, 339-340
Additives in food, 308
Advertising, 173, 179, 183, 185, 256-257, 261, 268, 276, 281, 286, 327-339
   see also Marketing
Advertising Age, 292
Aerobic dance, 245-246, 248
Aerobic exercise, 245-248, 295
Affluent seniors, 21, 34, 50, 88-89, 94-95, 141, 205, 211, 276- 277, 287, 318-319
Age of segments
   Disillusioned, 215
   Faithful Patients, 205
   Fast & Healthy, 281
   Financial Positives, 48-49
   Insecure, 38
   Nutrition Concerned, 276
   Optimists, 210
   Proactives, 200
   Threatened Actives, 45
   Traditional Couponers, 286
   Upbeat Enjoyers, 33
Aging, 111, 119-125, 235, 295
Air travel, 38, 48, 60-62
   Senior airline services, 60-61
Alaska, 63
Alcoholic beverages, 245, 300-302
   Beer, 245, 289, 300, 302, 315
   Hard liquor, 245, 300-302
   Wine, 4, 245, 300-302
Allstate Insurance (auto club), 75-76
American Association of Retired Persons (AARP), 23-24, 115, 134- 138, 182, 336

American Automobile Association (AAA), 75-78
American Cancer Society, 245
American Demographics, 3, 13, 22
American Express Company, 106-109, 192
American Greetings Corp., 152
American Pet Products Manufacturers Association (APPMA), 163, 168
Amoco (auto club), 75-76
Antacids, 232, 251
Apartments, 116, 142
   see also Housing
Appearance
   see Beauty; Cosmetics;
Aroma in food, 305, 310, 315
Arthritis, 220, 227-229, 237, 252, 313
   Drugs for, 227-229, 252, 313
Artificial sweeteners
   see Sugar
Asia, 54, 64, 72
Asp, Elaine, 311, 314
Assets of segments
   Disillusioned, 215
   Faithful Patients, 205
   Fast & Healthy, 282
   Financial Positives, 50
   Insecure, 41
   Nutrition Concerned, 277
   Optimists, 211
   Proactives, 201
   Threatened Actives, 45
   Traditional Couponers, 287
   Upbeat Enjoyers, 34
ATM
   see Automatic teller machines
Attitudes of seniors, 3-9
   Aging, 119-120
   Alcohol, 245, 300-302
   Attitudinal segments, 29-52, 197-217, 271-288
   Automobile clubs, 75-78
   Automobiles, 67-75

Beauty, 119-121
Book reading, 154, 261, 338-339
Brand loyalty, 173-176, 251-254, 327-329
Change, 111-125
Children, 37-38, 48, 117, 137
Computers, 114
Cosmetics, 31, 43, 47, 119-121, 143-144, 174-175, 253-254
Credit cards, 106-109
Crime, 32, 38, 43, 48
Death, 121-124, 236-237
Direct mail, 186-191, 263-264, 341
Disease and illness, 220-222
Drugs, 199, 204, 209, 214, 226-233, 248
Exercise, 198, 203, 208-209, 245-249, 295
Financial services, 97-109
Food and nutrition, 239-244, 271-345
Food segments, 271-288
Gardening, 144, 185, 261, 267, 339
Gift giving, 145-157
Grandchildren, 145-157
Health, 197-268
Healthcare, 136-137, 222-233, 235-236
Health segments, 197-217
Hospitals, 226
Housing, 115-118
Insurance, 100, 199-200, 204, 209, 214, 222-226
Investments, 84-95, 98-106
Living wills, 121
Magazines, 179-182, 257-259, 333-335
Media usage, 176-194, 254-267, 329-345
Microwave ovens, 274, 278-279, 285, 299-301
Moving, 117-118
Newspapers, 178-179, 257, 331-333
Packaging, 172, 175, 271, 274, 278-279, 284-285, 313-314
Pets, 159-172
Photography, 185
Politics and government, 127-138
Radio, 184-185, 261, 337-338
Recreational vehicles, 23, 31, 43, 60
Restaurants, 275, 280, 285, 317-326
Retail stores, 141-157
Retirement, 32-48, 79-96
Retirement communities, 43, 48, 117
Self segments, 29-52

Senior clubs, 102-106
Sewing, 185, 262
Shopping, 275, 280-281, 285, 289-290, 305-315
Smoking, 245-246, 249
Sports, 247-248, 262
Suicide, 124, 236-237
Technology, 111-115
Telephone and telephone services, 111-113
Television, 177-178, 254-256, 331
Tobacco, 245-246, 249
Travel, 53-65
Vitamins, 311-312
Volunteers, 36, 41, 46, 51, 132-133, 203, 208, 213, 217, 277, 282, 287
Attitudinal segments, 8, 176-177, 254, 329-330
  see also Attitudes
Automatic teller machines (ATM) cards, 99, 113
Automobile clubs, 67, 75-78
  Membership in, 75-76
  Services provided by, 76-78
Automobiles, 4, 6, 24, 38, 42-43, 60, 67-78, 178-179, 192, 333
  see also Travel
  Age of, 69-71
  Attitudes toward, 67-68
  Auto clubs, 67, 75-78
  Brand choice, 71-75
  Drivers licenses, 31, 38, 42, 47
  Foreign-made, 72, 75
  Loans, 99
  Luxury cars, 72
  Mailing lists, 192, 264, 342
  Miles driven, 69
  Newspaper ads, 178-179
  Ownership, 68, 192, 264, 342
  Sales, 72-75, 67-68
  Vacations, 60-61
Automobiles, types of, 71-75
  Buick, 71, 74-75
  Cadillac, 72, 75, 264, 342
  Chevrolet, 71, 74
  Dodge, 71, 342
  Ford, 71, 73-74, 78
  Honda, 72, 192

Index

Lincoln, 264
Oldsmobile, 71, 72, 74
Plymouth, 342
Toyota, 72, 75
Volvo, 72
Automotive industry, 4, 6, 67-75

B

Baby boomers, 22-23, 25, 290, 311, 317
Bakery goods, 239
Bank of America, 235
Banks and banking, 88, 98, 100-106, 108
  see also Financial institutions, services provided by
Bartos, Rena, 13-14
Beauty, 31, 38, 43, 47, 119-121, 245
  see also Cosmetics
Beef, 291, 310
  see also Meat
Beer, 245, 289, 300, 302, 315
Better Homes and Gardens, 181-182, 259, 334-335
Beverages, 245, 297, 300-302, 309
  see also Alcohol
  Fruit juice, 297
  Milk, 291, 309
Birds, 161, 163
  see also Pets
Bone, Paula Fitzgerald, 12
Books and bookreading, 150, 154, 185, 261, 338-339
  Cookbooks, 338-339
Bowling, 245
Brand loyalty, 24, 173-176, 251-254, 276, 327-329
Bristol Myers Squibb Inc., 252
Bufferin, 252
Buick, 71, 74-75
Bush, George, 129
Business Week, 182
Butter, 239, 292, 297, 311, 335

C

Cable television
  see Television
Cadillac, 72, 75, 342
Cafeterias, 318
Cake mixes, 174-175, 315, 328
Campbell's Low Sodium Soups, 315

Canada, 63
Cancer, 221, 238, 244-245, 295
  Lung, 245
  Skin, 244-245
Candy, 150, 153-154, 239, 297, 339-340
Candy stores, 340
Careers
  see Employment
Caribbean, 63
Carnation's Perform, 168
Cars
  see Automobiles
Cash as gifts, 150, 155-156
Casual dining, 318
Cats, 159, 161, 163-164, 168
  see also Pets
Cauliflower, 309
Census region of segments
  Disillusioned, 215
  Faithful Patients, 205
  Fast & Healthy, 282
  Financial Positives, 51
  Insecure, 41
  Nutrition Concerned, 277
  Optimists, 212
  Proactives, 201
  Threatened Actives, 45
  Traditional Couponers, 287
  Upbeat Enjoyers, 34
Cereals, 174-175, 239, 297, 305, 312, 328
Certificates of deposit (CDs), 99
Change, 37, 42, 47, 111-125
Cheese, 239, 291-292
Chef's Catalog, 341
Chevrolet, 71, 74
Chicken, 295, 325
Children, 37-38, 48, 117, 137, 185
  see also Grandchildren
  Living with, 48, 117
  Seniors activities with, 185
Chinese food, 321-323
Chips, 241, 297
Cholesterol, 242-244, 296-297, 306
Clinton, Bill, 129-130
Clothing, 31, 43, 141-142, 145-147, 150, 152-153, 185, 191
  As gifts to grandchildren, 146, 152-153
  Personal expenditures on, 142

Clubs
    see Automobile clubs
    see Senior banking clubs
Coffee shops, 318
Cohen, Bernard, 115
Cold medicines, 232
Comfortably Yours Inc., 264
Computers, 114, 193
ConAgra Inc., 310, 328
Condominiums and townhouses, 32, 43, 115-117, 142
    see also Housing
Constipation, 221, 295
Consumer Expenditure Survey, 53
Consumers, seniors as, 23
Containers
    see Packaging
Contributions, political, 133-135
Convenience foods, 274, 278, 285, 297-300, 310, 312-313, 315
Cookbooks, 338-339
Cookies, 150, 153-154, 239, 310, 339
Cooking at home, 297-300, 338-339
    Gourmet, 338-339
Cosmair's Lancôme, 144
Cosmetics, 24, 31, 43, 47, 119-120, 143, 174-175, 245, 253-254, 325
    see also Beauty
    Brand loyalty, 24, 174-176, 253-254
    Face cream, 7, 24, 120, 175, 253-254
    Hair coloring, 24, 121, 174-175
    Moisturizers, 120, 253
    Skin products, 24, 38, 120, 175, 253
    Sunscreens, 245
Coupons, 75-76, 179, 252, 257, 281, 284-286, 325, 340
    see also Discounts to seniors
Credit cards, 99-100, 104-109, 143, 192
    American Express, 106-109, 192
    Department store, 107, 143, 192
    Diners Club, 107, 109
    Discover Card, 109
    Master Card, 107, 109
    Multiple cards, 107
    Payment methods, 109
    Usage, 107-109
    Visa, 104, 107-109, 192
Crime, 32, 38, 43, 48, 115, 144

Crisco, 175
Cruises, 23, 31, 58-59, 298
    see also Travel

D

Dabels, John, 75
Dahlberg Inc., 221
Dairy products, 239, 242, 291-292, 297, 309, 311, 335
Darvocet (drug), 228
Davis, Brian, 17
Day, Ellen, 17
Day's Inn, 23
Death and dying, 111, 121-125, 236-238, 299
    see also Suicide
    Living wills, 121, 224
    Right to die, 121-123
Debit cards, 100
Debt of seniors, 97-98, 109
Democratic Party, 129-132, 137
    see also Politics and government
Demographic segmentation
    see Segmentation
Demographic characteristics of seniors, 5, 12-14
    Disillusioned segment, 214-217
    Faithful Patient segment, 205-208
    Fast & Healthy segment, 281-283
    Financial Positive segment, 48-52
    Insecure segment, 38-42
    Nutrition Concerned segment, 276-277
    Optimists segment, 209-213
    Proactive segment, 200-203
    Threatened Active segment, 45-46
    Traditional Couponer segment, 286-288
    Upbeat Enjoyer segment, 32-37
Denny's Restaurants, 331
Department stores, 142-143, 145, 156
    see also Retail stores
Diet
    see Food
Diners Club, 107, 109
Direct mail, 186-191, 263-264, 340-341
    Expenditures for, 188-190, 263-264, 341
    Mailing lists, 191-193, 264-265, 341-343
    Responses to, 187, 263, 341
Disabled
    see Handicapped
Discounts to seniors, 53, 76, 325, 340

Index

see also Coupons
Attitudes toward, 32, 38, 44
For travel, 76
Discover Card, 109
Discretionary income, 21, 29, 127, 141
  see also Income
Diseases and illness, 219-233
  Angina, 144
  Arthritis, 220, 227-229, 237, 252, 313
  Cancer, 221, 238, 244-245, 295
  Constipation, 221, 295
  Eyes, 209, 221
  Heart disease, 220, 238
  Hemorrhoids, 221, 237, 295
  High blood pressure, 144, 219-220, 222, 226-227, 237
  Indigestion, 221
  Stroke, 220
  Terminally ill, 37, 42, 47, 121-123
Disillusioned segment, 213-217
  Demographic characteristics, 214-217
  Disease and illness, 214, 219-249
  Drug use, 214, 226-233
  Eating habits, 239-244
  Exercise, 245-248
  Healthcare, 214, 222-226
  Media use, 254-261
Diversification, 4-5
Divorce, 33, 36, 39, 49, 52, 201, 215, 217, 286-287
Do-it-yourself projects, 145, 185, 262, 339
Doctors
  see Physicians
Dodge (automobile), 71, 342
Dodge, Robert D., 11
Dogs, 159, 161-163, 168
  see also Pets
Dole Inc., 335
Dorothy Lane Markets, 313
Dove, Rhonda, 17
Downhill skiing, 247
Drugs, 199, 204, 209, 214, 220, 226-233, 248, 252, 264
  Antacids, 232, 251
  Brand loyalty, 252
  Cold medicines, 232
  Estrogen replacement, 228-230, 233
  Generic, 204, 252

High blood pressure, 226-227
Laxatives, 232
Over-the-counter drugs, 199, 204, 209, 214, 230-233, 248
Pain killers, 228, 232
Prescription, 23, 144, 204, 220, 226-233, 248, 252
Vitamins, 144, 169, 232, 240-241, 273, 278, 285, 306, 311-312
Drug stores, 143-144
Dyazide (drug), 226

E

Eating
  see Food
Eating out
  see Restaurants
Eddie Bauer Inc., 191
Education of segments
  Disillusioned, 215
  Faithful Patients, 205
  Fast & Healthy, 282
  Financial Positives, 49-50
  Insecure, 39-40
  Nutrition Concerned, 276
  Optimists, 210
  Proactives, 201
  Threatened Actives, 45
  Traditional Couponers, 286
  Upbeat Enjoyers, 34
EFT
  see Electronic funds transfer
Eggs, 239, 335
Electronic funds transfer (EFT), 112-113
Elizabeth Arden Company, 120, 253
Employment of segments
  Before retirement
    Disillusioned, 215-216
    Faithful Patients, 206
    Fast & Healthy, 282
    Financial Positives, 51
    Insecure, 41
    Nutrition Concerned, 277
    Optimists, 212-213
    Proactives, 201
    Threatened Actives, 46
    Traditional Couponers, 287
    Upbeat Enjoyers, 34

Hours worked
  Disillusioned, 215
  Faithful Patients, 206
  Fast & Healthy, 282
  Financial Positives, 51
  Insecure, 41
  Nutrition Concerned, 277
  Optimists, 212
  Proactives, 201
  Threatened Actives, 46
  Traditional Couponers, 287
  Upbeat Enjoyers, 34
Since retirement
  Disillusioned, 216
  Faithful Patients, 207
  Fast & Healthy, 282
  Financial Positives, 51
  Insecure, 41
  Nutrition Concerned, 277
  Optimists, 213
  Proactives, 202
  Threatened Actives, 46
  Traditional Couponers, 287
  Upbeat Enjoyers, 34-35
Employment of seniors, 23-24, 80-84, 325
  Restaurant workers, 325
  Types of jobs, 82-83
Enrenberg, Andrew, 173
Entitlements, government, 128, 137
Estee Lauder Inc., 253
Estrogen replacement, 228-230, 233
Ethnic food, 320-324
  see also Restaurants; Food;
  Chinese food, 321-323
  French food, 322-323
  German food, 322
  Indian food, 323-324
  Italian food, 321-323
  Mexican food, 321-323
  Vietnamese food, 323
Europe, 55, 63-64
Exercise, 179, 198, 203, 208-209, 245-249, 261, 295, 339
  see also Sports
  Aerobic, 245-248, 295
  Dance, aerobic, 245-246, 248
  Machines, 179, 246
  Media targeting, 261
  Weightlifting, 248
Exercise machines, 179, 246
Exotic foods, 323-324
  see also Restaurants; Food
Experience, Inc., 310
Exter, Thomas, 22
Eye care, 209, 221
Eye doctors, 209

F

Face cream, 7, 24, 120, 175, 253-254
Faithful Patient segment, 203-208
  Demographic characteristics, 205-208
  Disease and illness, 203-204, 219-249
  Drug use, 204, 226-233
  Eating habits, 239-244
  Exercise, 203, 245-248
  Healthcare, 204, 222-226
  Media use, 254-261
Family Circle, 181, 191, 259, 334-335
Fannie Mae, 340
Fast & Healthy segment, 278-283
  Alcohol use, 300-302
  Demographic characteristics, 281-283
  Eating habits, 278, 289-326
  Media use, 329-340
  Microwave use, 278-279, 299-301
  Restaurant use, 280-281, 317-326
  Shopping habits, 280, 289-316
Fast foods, 275, 281, 285, 318-319, 339
  see also Food; Restaurants;
Fat and fat substitutes, 198, 203, 208, 235, 239-242, 271, 273, 285, 291-292, 295-297, 306, 309-310, 312-313, 324, 329
Fela, Leonard, 15
Fiber in food, 273, 278, 285, 294, 307, 324
Field & Stream (magazine), 192
Figi's of Wisconsin, 341
Financial advice, sources of, 85-89
  Affluent, 88-89
  Bankers, 87-88
  Financial planners, 86-88
  Insurance agents, 87-88
  Media, 88
  Mutual fund reps, 87, 89
  Reasons for avoidance, 89-92
  Risk involved, 93-95
  Stockbrokers, 89

Index

Financial institutions, services provided by, 97-109
  ATM cards, 99, 113
  Car loans, 99
  Certificates of deposit, 99
  Credit cards, 99-100, 104-109
  Debit cards, 100
  Insurance, life, 100
  Mortgages, 118
  Safe deposit boxes, 100
  Savings instruments, 98-99
  Stocks and bonds, 101
Financial planners, 86-87
Financial planning, 84-95
  see also Retirement
Financial Positives segment, 47-52
  Aging, 47, 119-121
  Cars and auto clubs, 67-78
  Change, 47, 111-125
  Credit cards, 106-109
  Death and dying, 121-124
  Demographic characteristics, 48-52
  Finance and investments, 47, 85-92
  Financial services, 47, 85-92
  Gifts to grandchildren, 145-156
  Media use, 176-186
  Pets, 159-172
  Political activities, 127-138
  Residence, 48, 115-118
  Retail expenditures, 48, 142-145
  Retirement, 48, 79-96
  Senior banking clubs, 102-104
  Technology, 111-115
  Travel plans, 48, 53-65
FirstMark Technologies, 192, 265, 343
Fish, 161, 163-164, 300
  see also Pets
Fishing, 339
Flesch, Ed, 16
Food, 239-244, 271-345
  Additives, 308
  Aroma, 305, 310, 315
  Bakery products, 239
  Butter, 239, 292, 297, 311, 335
  Cake mixes, 174-175, 315, 328
  Candy, 150, 153-154, 239, 297, 339-340
  Cauliflower, 309
  Cereal, 174-175, 239, 297, 305, 312, 328
  Cheese, 239, 291-292
  Chicken, 295, 325
  Chips, 241, 297
  Cholesterol, 242-244, 296-297, 306
  Convenience foods, 274, 278, 285, 297-300, 310, 312-313, 315
  Cookies, 150, 153-154, 239, 310, 339
  Cooking at home, 297-300, 338-339
  Eating habits, 239-244, 271-326
  Eggs, 239, 335
  Ethnic foods, 320-324
  Exotic foods, 323-324
  Expenditures, 290, 320
  Fast foods, 275, 281, 285, 318-319, 339
  Fat, 198, 203, 208, 235, 239-242, 271, 273, 285, 291-292, 295-297, 306, 309-310, 312-313, 324, 329
  Fiber, 273, 278, 285, 294, 307, 324
  Fish, 300
  Frozen, 299, 328-329
  Fruit, 239, 273, 278, 285, 291-294, 297, 310
  Groceries, 290
  Health foods, 257, 261, 291-296, 305-312, 324
  Home cooking, 290, 325
  Home delivery, 314
  Ice cream, 242
  Importance of attributes, 306-309
  Labels, 272, 278, 291, 313
  Meat, red, 239, 273, 278, 285, 291-292, 295, 297, 308, 310, 335
  Milk, 291, 309
  New products, 305-315
  Nutrition, 239-240, 271-316
  Packaging, 172, 175, 271, 274, 278-279, 284-285, 313-314
  Pesticides in food, 241, 305, 307-308
  Pet food, 167-172
  Pizza, 239
  Popcorn, 241, 297, 311
  Preservatives, 240, 307
  Restaurants, 9, 150, 154, 271, 275, 280-281, 284-285, 298, 317-326, 328, 331-332, 339
  Salad bars, 324
  Salty foods, 271, 273, 278, 285, 289, 297, 315

Seafood, 295
Shelf-stable, 279, 285, 312
Shopping habits, 275, 280-281, 285, 289-290, 305-315
Snacks, 292, 297, 311, 315
Sugar and artificial sweeteners, 242, 273, 278, 285, 289, 291, 297, 325, 329
Supermarkets, 144, 167-169, 171, 271, 280, 284, 289, 292, 310-315
Vegetables, 239, 273, 278, 285, 291-294, 297, 299, 312, 324, 335
Vitamins, 144, 169, 232, 240-241, 273, 278, 285, 306, 311-312
Food Marketing Institute, 292
Food segments, 271-288
Fast & Healthy segment, 278-283
Nutrition Concerned segment, 271-277
Traditional Couponer segment, 284-288
Food Trends and the Changing Consumer, 311, 314
Ford (automobile), 71, 73-74, 78
Francese, Peter, 3
French, Warren, 17
French food, 322-323
Freshtables (Pillsbury product), 312
Frito-Lay, 311
Frozen foods, 299, 328-329
Fruits, 239, 273, 278, 285, 291-294, 297, 310

G

Gallup Poll, 324
Games, 154-155
Gardening and garden supplies, 144, 185, 261, 267, 339
Garry, Philip, 295
Gasoline, 176
Gelb, Betsy, 127, 134, 136
Gender of segments
Disillusioned, 214
Faithful Patients, 205
Fast & Healthy, 281
Financial Positives, 48
Insecure, 38
Nutrition Concerned, 276
Optimists, 209
Proactives, 200
Threatened Actives, 45
Traditional Couponers, 286

Upbeat Enjoyers, 32
General Mills Inc., 294, 312, 315
Generic products, 204, 252, 264
German food, 322
Gerontographics, 16-17
Gifts by seniors to grandchildren, 145-157, 339-340
Albums/tapes/CDs, 150
Books, 150, 154
Cash, 150, 155-156
Clothing, 145-147, 150, 152-153
Coins, 156
Cookies and candy, 150, 153, 339-340
Games, 154-155
Greeting cards, 150-152
Most frequent, 150-156
Movie tickets, 150-151
Restaurant meals, 154
Stocks and bonds, 156
Toys, 154
Travel, 156
Golden Palace (TV show), 256
Golf, 15, 185, 247-248, 262, 295, 339
Good Housekeeping, 181, 191, 259, 334
Gourmet, 258, 335
Gourmet cooking, 301, 338-339
Government
see Politics and government
Grandchildren, 145-147, 185, 263, 339-340
see also Children
Ages, 149
Gifts to, 145-157, 339-340
Numbers of, 147-149
Seniors activities with, 154, 156, 339
Grandtravel, 156
Gray Panthers, 127, 134
Green Giant Foods, 335
Greeting cards, 150-152
Groceries
see Foods
Grocery stores
see Supermarkets

H

Hair coloring, 24, 121, 174-175
Hallmark Cards, 152
Handicapped, 313
Hardware stores, 144
see also Retail stores

Index

Hawaii, 63
Health, 197-268
  Current state of, 236
  Habits, 244-245
  Health segments, 197-217
  Market, 219-234
  Mental health, 236
  Motivation, 236-238
Health foods, 257, 261, 291-296, 305-312, 324
  see also Food
Health insurance
  see Medical insurance
Health Insurance Association of America (HIAA), 225
Health Magazine, 258
Health maintenance organizations (HMOs), 199, 219, 223-224
Health segments, 197-217
  Disillusioned segment, 213-217
  Faithful Patient segment, 203-208
  Optimist segment, 208-213
  Proactive segment, 197-203
Healthcare, 136-37, 219, 222-233, 235-236, 248-249
  Costs, 235-236
  Market, 219-220
Healthy Choice (ConAgra dinners), 310, 328
Health Valley Natural Foods, 310
Hearing and hearing aids, 199, 203, 209, 214, 221, 237, 254
Heart disease, 220, 238
Hemorrhoids, 221, 237, 295
High blood pressure, 144, 219-220, 222, 226-227, 237
  Drugs for, 226-227
  Prevalence of, 220
Hill's Pet Products Inc., 168
Home cooking, 290, 325
Home delivery of food, 314
Home furnishings, 142
Home improvement
  see Do-it-yourself projects
Homes
  see Housing
Honda, 72, 192
Honeywell's Total Home, 115
Hormel Foods' Top Shelf, 312, 314-315

Hospitals, 226, 257
  Selection factors, 226
  Costs, 226
Hotels and motels, 38, 48, 53, 59-60, 76, 104, 175
  see also Travel
Housing, 115-118
  Apartments, 116, 142
  Children, living with, 48, 117
  Condominiums, 32, 43, 115-117, 142
  Mobile homes, 117
  Mortgage, reverse, 118
  Moving, 117-118
  Ownership, 115-116
  Renting, 115-116
  Retirement communities, 43, 48, 117
  Two residences, 117
Humana Seniors Association, 23
Hunting/shooting, 185, 193

I

Ice cream, 242
Illness
  see Diseases and illness
Income of segments
  Disillusioned, 215
  Faithful Patients, 205
  Fast & Healthy, 282
  Financial Positives, 50
  Insecure, 40
  Nutrition Concerned, 276-277
  Optimists, 211
  Proactives, 201
  Threatened Actives, 45
  Traditional Couponers, 287
  Upbeat Enjoyers, 34
Indian foods, 323-324
Indigestion, 221
Insecure segment, 37-42
  Aging, 37-38, 119-121
  Cars and auto clubs, 38, 67-78
  Change, 37, 111-125
  Credit cards, 106-109
  Death and dying, 121-124
  Demographic characteristics, 38-42
  Finance and investments, 37, 85-92
  Financial services, 85-92
  Gifts to grandchildren, 145-156
  Media use, 176-186

Pets, 159-172
Political activities, 127-138
Residence, 37, 115-118
Retail expenditures, 38, 142-145
Retirement, 37, 79-96
  Senior banking clubs, 102-104
  Technology, 111-115
  Travel plans, 38, 53-65
Insurance
  see Life insurance; Medical insurance; Long-term care insurance
Insurance agents, 87-88
International Mass Retail Association, 141-142
Investments, 84-95, 98-106
  Bonds, 88, 93, 106, 155-156, 185, 343
  Certificates of deposit, 99
  Mutual funds, 88, 93, 106, 175
  Risk toleration, 94-95
  Stocks, 88, 101, 155-156, 185, 343
Italian food, 321-323

J

Japan, 54, 64
Jeopardy (TV show), 331
Jogging, 295
Journal of Retailing, 11

K

Kellogg Company, 294
Kinsey, Jean, 311, 314
Kiplinger's Personal Finance Magazine, 183
KnowledgeSeeker®, 192-193, 265-266, 343-344

L

Labels on food packages, 272, 278, 291, 313
Labor source, seniors as, 23-24
  see also Employment
Ladies Home Journal, 181, 334
Land's End Inc., 191
Lansbury, Angela, 252
Lasix, 226
Laxatives, 232
Lear's, 182
Levitt, Theodore, 4
Life, 259
Life Choice (ConAgra dinners), 310, 328

Life insurance, 100
Lifestyle segmentation
  see Segmentation
Lincoln (automobile), 264
Liquor
  see Alcoholic beverages
Living wills, 121, 224
Long-term care insurance, 22, 224-225
Longevity, 24
Loomis, Lynette M., 14
L'Oréal, 143
Lumpkin, James R., 14

M

Magazine Publishers of America (MPA), 180, 259, 333
Magazine readership, 179-184, 257-261, 333-337
  Advertising, 257-261
  Frequent readers, 180, 333
  Measuring readership, 179, 257-258, 333
  Multiple readers, 180-181, 259, 334
  Selecting by segment, 179-182, 257-259, 333-335
  Senior publications, 182-184, 259-261, 335-337
Mail order
  see Direct mail; Mailing lists
Mailing lists, 191-193, 264-265, 341-343
  see also Direct mail
MAPWISE, 192, 194, 266-267, 343, 345
Market Research Corporation of America (MRCA), 173
A Marketer's Guide to Discretionary Income, 21
MarketACTION Research, 192, 266, 343
Marketing, 3-9, 11-26
  see also Advertising
  Mass marketing, 3-4, 25
  Niche marketing, 5
  Target marketing, 25-26, 143, 149
Markle Foundation, 13
Marital status of segments
  Disillusioned, 215
  Faithful Patients, 205
  Fast & Healthy, 281
  Financial Positives, 49
  Insecure, 39
  Nutrition Concerned, 276

Index

Optimists, 210
Proactives, 201
Threatened Actives, 45
Traditional Couponers, 286
Upbeat Enjoyers, 33
Marriott Corp., 23
Martin, Claude R., Jr., 15
Martineau, Pierre, 7
Mass marketing, 3-4, 25
MasterCard, 107, 109
Mature Market Report, 23
McCall's, 334
McDonald's Corp., 23, 128, 310, 339
Mead Johnson and Company, 228
Meat, 239, 273, 278, 285, 291-293, 295, 297, 300, 308, 310, 320, 324, 335
  Beef, 291, 310
  Chicken, 295, 325
  Lunch meat, 239
  Meat loaf, 325
Media usage by seniors, 17-18, 176-194, 254-267, 329-345
  Books, 261, 338-339
  Direct mail, 186-193, 263-267, 340-343
  Magazines, 179-184, 257-261, 333-337
  Newspapers, 178-179, 257, 331-333
  Radio, 184-185, 261, 337-338
  Television, 177-178, 254-256, 331
Medicaid, 22, 224
Medical insurance, 22, 197, 199-200, 204, 209, 213-214, 219, 222- 226
  Coverage, 222
  HMOs, 199, 219, 223-224
  Long-term care, 224-225
  Medicaid, 22, 224
  Medicare, 22, 136, 222, 224, 235
  Medigap, 219, 224-225
  PPOs, 223-224
Medicare, 22, 136, 222, 224, 235
Medicare Catastrophic Coverage Act, 136
Medigap insurance, 219, 224-225
Mental health, 236
Mexican food, 321-323
Michelob Dry beer, 315
Microwave ovens, 274, 278-279, 285, 299-301, 311, 315
Midwest Living, 182
Milk, 291, 309

Mobile homes, 117
  see also Housing
Modern Maturity, 75, 182-184, 260, 317, 335-337
Moisturizers, 120, 253
Money, 182
Morrisett, Lloyd N., 13
Morrison, Rosanna Mentzer, 309
Mortgages, 98-99, 116, 118, 124
Moschis, George, 16-17
Motels
  see Hotels and motels
Motivational research, 7
Motrin, 227
Moving, 117-118
Mutual fund representatives, 87, 89
Mutual funds, 87-89, 93, 106, 175

N

Naprosyn, 227
National Association of Home Builders, 117
National Association of Realtors, 117
National Automated Clearing House Association, 112
National Cancer Institute, 239
National Enquirer, 335
National Geographic, 182, 259, 335, 342
National Restaurant Association, 320-321, 323-324
Nautilus exercise equipment, 248
Net worth of seniors, median, 21, 29, 127
New Choices for Retirement Living, 182, 335
New Mexico Aging Process Study, 295
New York Times Magazine, 182
Newspaper readership, 178-179, 257, 331-333
  Advertising, 179, 257, 332
  Business, 179, 333
  Cartoons, 257
  Car/Motoring, 178-179, 333
  Coupon, 179, 257
  Editorials, 257
  Entertainment, 257, 333
  Food, 179, 257, 332
  Home, 333
  International news, 257, 333
  Letters to the editor, 257
  Opinion and commentary, 257, 333
  Sports, 257

Travel, 179, 333
Niche marketing, 5
　see also Segmentation
NordicTrack, 179
Nostalgia, 24
Nursing homes, 22, 172, 224-225
Nutrition, 239-240, 242-244, 271-316
　see also Food
Nutrition Concerned segment, 271-277
　Alcohol use, 300-302
　Demographic characteristics, 276-277
　Eating habits, 273-275, 289-326
　Media use, 329-340
　Microwave use, 274, 299-301
　Restaurant use, 275, 317-326
　Shopping habits, 272-276, 289-316

O

Oil of Olay, 120
Older Women's League (OWL), 134
Oldsmobile, 71-72, 74
Optimist segment, 208-213
　Demographic characteristics, 209-213
　Disease and illness, 208, 219-249
　Drug use, 209, 226-233
　Eating habits, 208, 239-244
　Exercise, 208, 245-248
　Healthcare, 209, 222-226
　Media use, 254-261
Organic foods
　see Health foods
Orville Redenbacher popcorn, 311
Orvis Inc., 191
Oshkosh B'Gosh Inc., 146-147
Over-the-counter drugs
　see Drugs

P

Packaging, 172, 175, 271, 274, 278-279, 284-285, 313-314
　Design, 175, 271, 278-279, 313-314
　Difficulty in opening, 279, 285, 313
　Labels on, 272, 278, 291, 313
　Portion size, 314
Pain killers, 228, 232
Pearlin, Leonard, 12
Pensions, 93, 137
　see also Retirement
People, 259

Perot, Ross, 129
Pesticides in food, 241, 305, 307-308
Pet food, 167-172
　Packaging, 172
　Targeting owners, 171
　Types, 170-171
　Where purchased, 167-169
　Why purchased, 169-170
Pet industry, 159-172
　Stores, 167-169
　Veterinarians, 159, 168-169, 171-172
Pets, 159-172
　see also Birds; Cats; Dogs; Fish
　Contribution to quality of life, 164-166
　Expenditures on, 159
　Food, 167-172
　Ownership of, 159-161
Pharmaceuticals
　see Drugs
Pharmacies
　see Drug stores
Phone-based services, 112-114
Photography, 185
Physical fitness
　see Exercise; Sports
Physicians, 112, 121, 197, 199-200, 204, 209, 214, 220, 222-223, 226, 231, 237, 242-244, 248, 296
　Hospitals, 226, 257
　Proactive medicine, 248
　Visits to, 199, 204, 222-223, 248
Pillsbury Company, 312, 315
Pizza, 239
Plymouth (automobile), 342
Politics and government, 22, 127-138
　Activity, 134-136
　Bush, George, 129
　Clinton, Bill, 129-130
　Contributions, campaign, 133-135
　Democratic Party, 129-132, 137
　Economic issues, 137
　Entitlements, 128, 137
　Gender vote, 129-132
　Gray Panthers, 127, 134
　Independent voters, 129-131, 133
　Intergenerational issues, 137
　Lobbying, 134-136
　Men, 128-132

Index

1992 election, 129
Older Women's League (OWL), 134
Party affiliation, 129
Perot, Ross, 129
Political power of seniors, 128
Republican Party, 129-133, 137
Socialist and Libertarian Parties, 129
Volunteers, 132-133, 135
Voting, 128-129
Women, 128-132
Popcorn, 241, 297, 311
Preferred Provider Organizations (PPOs), 223-224
Prescription drugs
    see Drugs
Preservatives in food, 240, 307
Prevention, 258-259, 337
Proactive medicine, 248
Proactive segment, 197-203
    Demographic characteristics, 200-203
    Disease and illness, 199, 219-249
    Drug use, 199, 226-233
    Eating habits, 198, 239-244
    Exercise, 198, 245-248
    Healthcare, 199-200, 222-226
    Media use, 254-261
Psychographic segmentation
    see Segmentation
Psychographics, 12, 15-16, 18
Putnam, Judith Jones, 309

Q

Quaker Oat Bran, 315

R

Radio listening, 184-185, 261, 337-338
    Classical, 185, 261, 337
    Country, 185, 337-338
    Easy-listening, 185, 261, 337-338
    Jazz, 185
    News, 261, 337-338
    Sports, 337
    Talk shows, 261, 337
Ramada (Inn) Best Years Program, 175
Reader's Digest, 180-183, 259, 334-337
Reading
    see Books; Magazines; Newspapers
Records, phonograph, 262
Recreational vehicles, 23, 31, 43, 60

Renting, 115-116
Republican Party, 129-133, 137
    see also Politics and government
Restaurants, 9, 150, 154, 271, 275, 280-281, 284-285, 298, 317-326, 328, 331-332, 339
    American, 321
    Cafeterias, 318
    Casual dining, 318
    Chinese, 321-323
    Coffee shops, 318
    Customer expectations, 324
    Desserts, 320
    Ethnic food, 320-324
    Exotic food, 323-324
    Expenditures, 320
    Fast foods, 275, 281, 285, 318-319, 339
    French, 322-323
    Frequent patrons, 320
    German, 322
    Meals as gifts to grandchildren, 154, 339
    Hiring of seniors, 325
    Indian, 323-324
    Italian, 321-323
    Mexican, 321-323
    Patronage, 317
    Salad bars, 324
    Steak houses, 318
    Vegetarian meals, 324
    Vietnamese, 323
Retail stores and trade, 141-147, 156, 167-169, 171, 271, 280, 284, 289, 292, 310-315
    see also Department stores; Drug stores; Hardware stores; Supermarkets
    Candy, 340
    Clothing, 142
    Department stores, 142-143, 145, 156
    Drug stores, 143-144
    Hardware stores, 144
    Home furnishings, 142
    Pet foods, 167-169
    Purchases, 142
    Supermarkets, 144, 167-169, 171, 271, 280, 284, 289, 292, 310-315
Retirement, 32-48, 79-96
    Attitudes toward, 32, 34-35, 43, 48, 80-84
    Basis for segmentation, 15

Financial advice, 85-88
Financial planning, 85-88
Pensions, 93
Rates, 79-80
Returning to work, 80-84
Saving for, 84
Sources of support, 92-93
Retirement-based segmentation
   see Segmentation
Retirement communities, 43, 48, 117
Retirement plan, 84-85
Retirement status of segments
   Disillusioned, 215-216
   Faithful Patients, 206-208
   Fast & Healthy, 282
   Financial Positives, 51
   Insecure, 41
   Nutrition Concerned, 277
   Optimists, 212-213
   Proactives, 201-203
   Threatened Actives, 46
   Traditional Couponers, 287
   Upbeat Enjoyers, 34-35
Reverse mortgages
   see Mortgages
Risk in investment
   see Investments
Riskey, Dwight, 311
Roper Organization, 299
Ruder Finn Inc., 231
RVs
   see Recreational vehicles
Rx REMEDY, 261

S

Safe deposit boxes, 100
Safety
   see Crime
Salad bars, 284
Salt and sodium, 271, 273, 278, 285, 289, 297, 315
Savings of seniors, 84, 98
Savings bonds, 155-156
Schewe, Charles D., 13
Schwab, Charles and Company, 175
SCORE (Service Corp. of Retired Executives), 24
Seafood, 295

Segmentation, 3, 5-9, 11-18, 176-177, 254, 329-330
   Attitudinal, 8, 176-177, 254, 329-330
   Behavioral, 6
   Demographic, 12-14
   Financial status, 13-14
   Geographic, 5-6
   Lifestyle, 14
   Media usage, 17-18
   Psychographic, 15-17
   Retirement, 15
Self segments, 29-52
   Financial Positive segment, 47-52
   Insecure segment, 37-42
   Threatened Active segment, 42-46
   Upbeat Enjoyer segment, 29-37
Senauer, Ben, 311, 314
Senior airline services, 60-61
Senior banking clubs, 102-106
   Membership in, 103
   Services offered by, 103-104
Senior market, 21-26
   growth of, 22-23
   marketing fallacies, 24-25
   source of spending, labor, 23-24
Senior publications, 24, 182-184, 335-337
Seniors as consumers, 23-24
Sewing, 185, 262
Shearson Lehman Brothers, 90
Shelf-stable foods, 279, 285, 312
   see also Convenience foods
Shopping, 275, 280-281, 285, 289-290, 305-315
   see also Retail stores
Shopping lists, 271
Shuffleboard, 245, 295
Skin products
   see Cosmetics; Face cream; Moisturizers;
Smithsonian, 182-183
Smoking
   see Tobacco
Snacks, 292, 297, 311, 315
Social Security, 81, 92, 136-137
Sorce, Patricia, 14
South America, 63, 124
Southern Living, 182
Soviet Union, 63

Index

Sports, 177, 179, 185, 247-248, 256-257, 262, 331, 333, 337, 339
   see also Exercise
   Aerobic dance, 245-246, 248
   Bowling, 245
   Downhill skiing, 247
   Exercise machines, 179, 246
   Fishing, 339
   Golf, 15, 185, 247-248, 262, 295, 339
   Hunting/shooting, 185, 193
   Jogging, 295
   Shuffleboard, 245, 295
   Tennis, 248
   Walking, 245-247
   Watching on TV, 177, 256, 331
   Weightlifting, 248
SRI International, 16
Stat-One Hydrogen Peroxide Gel, 175
Steak houses, 318
Stockbrokers, 8, 87, 89
Stocks and bonds, 88, 93, 101, 106, 155-156, 185, 343
   see also Investments
Strategic Directions Group, Inc., 7-8
Stroke, 220
Suddenly Salad (General Mills product), 315
Sugar and artificial sweeteners, 242, 273, 278, 285, 289, 291, 297, 325, 329
Sugarman, Carol, 292
Suicide, 124-125, 236, 299
Sullivan, Michael P., 97
Sunchips (Frito-Lay product), 311
Sun damage, 244-245
Sunscreen lotions, 245
Sunset, 182
Supermarket Business, 144, 292
Supermarkets, 144, 167-169, 171, 271, 280, 284, 289, 292, 310-315
   Carts, 280, 284, 289
   Deli sections, 313
   Home delivery, 314
   Parking lots, 314
Sweepstakes, 262, 331, 339

T

Target marketing, 25-26, 143, 149
Tavris, Carol, 11
Technology, 111-115
   see also Automatic teller machines; Computers; Electronic funds transfer; Phone-based services; Television; Video phones
Telephone
   see Phone-based services; Video phones
Television viewing, 177-178, 254-256, 331
   Cable programming, 177, 255, 331
   Educational, 256
   Game shows, 177, 255, 331
   Morning shows, 331
   Movies, 331
   Network programming, 177, 254-255, 331
   News/current affairs, 177, 255, 331
   Situation comedies, 256
   Soap operas, 177
   Sports, 177, 256, 331
   Talk shows, 177, 255, 331
Tennis, 248
Threatened Actives segment, 42-46
   Aging, 43, 119-121
   Cars and auto clubs, 43, 67-78
   Change, 42, 111-125
   Credit cards, 106-109
   Death and dying, 42, 121-124
   Demographic characteristics, 45-46
   Finance and investments, 42, 85-92
   Financial services, 85-92
   Gifts to grandchildren, 145-156
   Media use, 176-186
   Pets, 159-172
   Political activities, 127-138
   Residence, 43, 115-118
   Retail expenditures, 44, 142, 145
   Retirement, 43, 79-96
   Senior banking clubs, 102-104
   Technology, 42, 111-115
   Travel plans, 43, 53-65
Tobacco, 245-246, 249
Top Shelf (Hormel product), 312, 314-315
Tours, 23, 38, 48, 57-58
   see also Travel
Towle, Jeffrey G., 15
Toyota, 72, 75
Toys, 23, 146-147, 150, 154
   expenditures on, 146
   frequency of giving, 154

Traditional Couponer segment, 284-288
  Alcohol use, 300-302
  Demographic characteristics, 286-288
  Eating habits, 284-285, 289-326
  Media use, 329-340
  Microwave use, 285, 299-301
  Restaurant use, 285, 317-326
  Shopping habits, 285-286, 289-316
Travel, 15, 23, 31, 53-65, 76, 103-105, 109, 156, 183, 298, 333
  see also Automobiles; Air travel
  Airlines, 38, 48, 60-62
  Alaska, 63
  Asia, 54, 64
  Canada, 63
  Caribbean, 63
  Cruises, 23, 31, 58-59, 298
  Destinations, 62-64
  Domestic trips, 54
  Eastern United States, 63
  Europe, 55, 63-64
  Expenditures, 53
  Foreign trips, 55, 63-64
  Frequency, 54-55
  Frequent domestic travelers, 54-55, 63
  Frequent foreign travelers, 55-58, 63
  Future, 64
  Gifts to grandchildren, 156
  Hawaii, 63
  Hotel stays, 53, 59-60
  Japan, 54, 64
  Length of trips, 54-55
  Midwestern United States, 62
  Preferences, 57-59
  Segmenting of, 54-55
  Selection factors, 55-57
  Senior services, 60-61
  South America, 63
  Southern United States, 63
  Soviet Union, 63
  Tours, escorted, 57-58
  Transportation, 60
  Western United States, 63
Travel & Leisure, 55, 259, 337
Travelers Insurance Company, 23
TV Guide, 259, 266, 335, 342
Tylenol, 228
Tyler, Philip R., 14

U

Uncles, Mark D., 173
Upbeat Enjoyers segment, 29-37
  Aging, 31, 119-121
  Cars and auto clubs, 31, 67-78
  Change, 30, 111-125
  Credit cards, 106-109
  Death and dying, 121-124
  Demographic characteristics, 32-37
  Finance and investments, 30-31. 85-92
  Financial services, 85-92
  Gifts to grandchildren, 145-156
  Media use, 176-186
  Pets, 159-172
  Political activities, 127-138
  Residence, 32, 115-118
  Retail expenditures, 32, 142-145
  Retirement, 32, 79-96
  Senior banking clubs, 102-104
  Technology, 30, 111-115
  Travel plans, 31, 53-65

V

Vacations
  see Travel
VALS (and VALS 2), 16
Vegetables, 239, 273, 278, 285, 291-294, 297, 299, 312, 324, 335
Vegetarian meals, 324
Veterinarians, 159, 168-169, 171-172
Video phones, 113
Videos, 245, 262
Vietnamese food, 323
Visa, 104, 107-109, 192
Vitamins, 144, 169, 232, 240-241, 273, 278, 285, 306, 311-312
Vogue, 258
Volunteering by seniors, 23-24, 32, 36, 41-42, 46, 51-52, 131-135, 137, 185, 203, 208, 213, 217, 277, 282-283, 287-288, 299
  Political, 131-135
Volunteerism of segments
  Disillusioned, 217
  Faithful Patients, 208
  Fast & Healthy, 282
  Financial Positives, 51
  Insecure, 41
  Nutrition Concerned, 277
  Optimists, 213

Proactives, 203
Threatened Actives, 46
Traditional Couponers, 287
Upbeat Enjoyers, 36
Volvo, 72
Voter Research and Surveys, 129
Voting by seniors, 128-132
    1992 vote, 129

W

Walking, 245-247
Wall Street Journal, 314
Wealth
    see Affluent
Weiffenbach, James, 310
Weightlifting, 248
Wheel of Fortune (TV show), 331
Widows and widowers, 17, 33, 36, 39, 42, 45-46, 49, 165, 201, 203, 205, 208, 215, 226, 276, 281, 283, 286-287, 298
Wine, 4, 245, 300-302
Work
    see Employment
Wyeth-Ayerst Laboratories, 228

Y

Yankelovich Clancy Shulman, 23, 24
Youth, 31, 38, 47, 119-121, 124-125
    see also Aging; Beauty; Cosmetics

About the Publisher

PROBUS PUBLISHING COMPANY

Probus Publishing Company fills the informational needs of today's business professional by publishing authoritative, quality books on timely and relevant topics, including:

- Investing
- Futures/Options Trading
- Banking
- Finance
- Marketing and Sales
- Manufacturing and Project Management
- Personal Finance, Real Estate, Insurance and Estate Planning
- Entrepreneurship
- Management

Probus books are available at quantity discounts when purchased for business, educational or sales promotional use. For more information, please call the Director, Corporate/Institutional Sales at 1-800-998-4644, or write:

Director, Corporate/Institutional Sales
Probus Publishing Company
1925 N. Clybourn Avenue
Chicago, Illinois 60614
FAX (312) 868-6250